[illegible]

March 30, 1992

$42 48

5-90

Recent Developments in Insect Neurohormones

Recent Developments in Insect Neurohormones

Marie Raabe

National Center for Scientific Research
Pierre and Marie Curie University
Paris, France

Illustrated by Colette Montmory

Plenum Press • New York and London

Library of Congress Cataloging in Publication Data

Raabe, Marie, 1921–
Recent developments in insect neurohormones.

Bibliography: p.
Includes index.
1. Insect hormones. 2. Neurohormones. I. Title.
QL495.R29 1989 595.7′01927 89-15998
ISBN 0-306-43175-0

A Division of Plenum Publishing Corporation
233 Spring Street, New York, N.Y. 10013

Printed in the United States of America

Preface

The most striking fact revealed by investigations of insect neurohormones is that insects are as well supplied with neurohormones as mammals, since neurohormones regulate not only the functioning of the endocrine glands, prothoracic gland, and corpora allata, but also most physiological processes.

Our knowledge of neurohormones developed originally from anatomocytological investigations and experimental studies. Today, accurate bioassays have been devised for studying both *in vivo* and *in vitro* physiological processes, and RIA determination has yielded knowledge of titer modifications of humoral factors. Much is also known about neurohormone purification, and several neurohormones have even been identified in different species.

Immunocytochemistry has made it possible to demonstrate in their origin and release sites the presence of insect neurohormones whose structure has been elucidated. Moreover, the presence of vertebrate and invertebrate neuropeptides has been demonstrated in insects.

As regards biogenic amines, methods of detection have been greatly refined and it is now possible to identify the cell bodies and axons of the main biogenic amines.

Other new methods, such as cobalt chloride impregnation or Lucifer yellow staining, have revealed the axonal pathways and the location of particular neurons.

The mechanisms of action of neurohormones have been investigated in several cases and the results of these investigations will be related in the chapters which follow.

Neurohormones are produced in neurons as neurotransmitters and neuromodulators; they are by definition transported by the hemolymph, but besides their neurohormonal role, some of them exert local neurotransmitter or neuromodulator action. In this volume, the term *neurohormone* will be employed in a broad sense that includes biogenic amines and also neurohormonal effects that take place at the level of the target tissues.

As my first book on the subject (*Insect Neurohormones*, Plenum Press) was published in 1982, the present volume will mainly focus on the most recent research. Investigations made before 1980 will be dealt with more briefly, except in cases where the evolution of the subject or its future prospects are of particular interest. The literature covered includes 1987 references; 1988 data are included in the Addendum.

The present volume begins with a short review that mostly concerns vertebrates and that aims to present the evolution of concepts in the field of neuropeptides during the last 50 years. Subsequent chapters, with the exception of Chapter 2, which summarizes immunocytochemical data, are devoted to the regulatory mechanisms of the varied insect physiological functions. The volume concludes with a comparison of the regulatory processes existing in insects and other invertebrates, crustaceans in particular.

The aim of this volume is to provide an easy to use, up-to-date and complete survey of an important, growing field: that of insect neurohormones.

Marie Raabe

Paris

Acknowledgments

I am most indebted to all of the authors who contributed illustrations to this book by providing me with prints or allowing me to reproduce their published and unpublished micrographs: Drs. N. Baudry-Partiaoglou, P. Cassier, A. Panov, A. Provansal-Baudez, Ch. Rémy, and J. Veenstra.

I wish to thank those co-workers, colleagues, friends, and relatives from whom I received support during the writing of this book: I am particularly grateful to Jean-Pierre Grillot for his help in preparing the reference section; Florence Jacquemin for typing the manuscript; Colette Montmory for preparing most of the illustrations; M. Dreyfus and N. Marshall for revision and translation; and Susan Woolford for the beautiful editing work.

This work was supported by a grant from the Centre National de la Recherche Scientifique.

M.R.

Abbreviations

20-HE	20-Hydroxyecdysone
α-MDH	(DL)-α-Methyl-α-hydrazino-β/3,4-dihydroxyphenyl propionic monohydrate (a dopa decarboxylase blocker)
ACTH	Adrenocorticotropic hormone
AKH	Adipokinetic hormone
AKHI	Adipokinetic hormone inhibitor
ANO	Abdominal ns organs
APUD	Amine precursor uptake decarboxylation
ARF	Anterior retraction factor
AVP	Arginine-vasopressin
CA	Corpora allata
CC	Corpora cardiaca
CCSF	CC stimulatory factor
CCK	Cholecystokinin
CI	Common inhibitory motoneurons
CLIP	Corticotropin-like intermediate peptide
CRF	Corticotropin-releasing factor
DALA	D-Ala_2-met-enkephalinamide
DDC	Dopa decarboxylase
DH	Diapause hormone
DNES	Diffuse neuroendocrine system
DS	Coxal slow depressor
DUM	Dorsal unpaired motoneurons
DUMET	Dorsal unpaired motoneurons to the extensor tibiae
DUMOV	Dorsal unpaired median oviductal neurons
E	Ecdysone
EDNH	Egg development neurosecretory hormone
EH	Eclosion hormone

ETi	Locust extensor tibiae
ETSH	Chloride transport-stimulating hormone
FABS	Fast-atom bombardment mass spectrometry
FET	Fast excitatory motoneuron
FSH	Follicle-stimulating hormone
G	Gastrin
GH	Growth hormone
GHRH	Growth hormone-releasing hormone
GIP	Gastric inhibitory peptide
GRP	Growth-releasing peptide
HCG	Human chorionic gonadotropin
HCS	Head capsule slippage
HGH	Hyperglycemic hormone
HPLC	High-performance liquid chromatography
HSF	Hemolymph stimulatory factor
HSH	Hindgut-stimulating hormone
HTF	Hypertrehalosemic factor
HTH	Hypertrehalosemic hormone
IBMX	3-Isobutyl-1-methylxanthine (a phosphodiesterase inhibitor)
JH	Juvenile hormone
JHA	Juvenile hormone analogs
JHE	Juvenile hormone esterases
LH	Luteinizing hormone
LHRH	Luteinizing hormone-releasing hormone
LMS	Leucomyosuppressin
LPH	Lipotropin
LPK	Leucopyrokinin
LSK	Leucosulfalkinin
LW	Lateral white neurons
MAGF	Male accessory gland factor
MCH	Melanocyte-concentrating hormone
MDH	Melanin-dispersing hormone
MIH	Molt-inhibiting hormone
MSH	Melanocyte-stimulating hormone
NCA	Nervi corporis allati
NCC	Nervi corporis cardiaci
nh	Neurohemal
ns	Neurosecretory
OSH	Oviposition-stimulating hormone
PaSH	Parturition-stimulating hormone
PG	Prothoracic gland

PI	Pars intercerebralis
PIF	Prolactin release-inhibiting hormone
PIF	Pupariation-immobilizing factor
PO	Perisympathetic organs
PP	Pancreatic polypeptide
PRL	Prolactin
PSF	Puparium stimulatory factor
PTF	Puparium-tanning factor
PTH	Parathormone
PTTH	Prothoracicotropic hormone
RER	Rough endoplasmic reticulum
RIA	Radioimmunoassay
RPCH	Red pigment concentrating hormone
RPHPLC	Reverse phase HPLC
SER	Smooth endoplasmic reticulum
SETi	Slow extensor tibiae motoneurons
SRIF	Somatotropin release-inhibiting hormone
TRH	Thyrotropin-releasing hormone
TSH	Thyroid-stimulating hormone
VIP	Vasopressin intestinal peptide
VP	Vasopressin

Contents

1

Synthesis and Release Sites of Neurohormones

Investigations concerning neurohormones and neurosciences have been numerous and have gathered increasing momentum over the past 40 years. The first concepts such as the existence of neurosecretory (ns) processes, i.e., the existence of glandular activity inside certain neurons, have now been generally accepted, although with difficulty.

Thanks to the continuous appearance of new techniques, a large body of data has been acquired for both vertebrates and invertebrates. Some of these data were controversial, mainly because of the diversity of the anatomical systems studied (e.g., in vertebrates the differences between mammals and fishes). However, this diversification often helped the evolution of the concepts and they became less rigid. Today many conventional ideas have been swept away: thus, ns products are considered to include peptidergic as well as aminergic molecules, which are both produced not only in the ns areas but also in the entire nervous system and even outside it. Lastly, it has been proved that the same neuron can synthesize several different types of molecules, i.e., two or three neuropeptides, or neuropeptides and monoamines, or again, neuropeptides and conventional neurotransmitters such as acetylcholine, glutamate, or GABA.

The clear distinction initially established between neurohormones and neurotransmitters has lost much of its significance, and the concept of neuromodulation has been introduced. At the same time it was demonstrated that the same substance can intervene in any of three ways, i.e., as a neurohormone, neurotransmitter, or neuromodulator, depending on its site of release.

Having said this, it is clear that the terminology used still has to be defined. The choice made here was to use the terms *neurohormone, neurotransmitter,* and *neuromodulator* in a functional sense, since, as indicated above, neither the chemical structure of the substances considered nor their sites of origin allows a real distinction to be drawn between them; thus, a neurohormone is a factor which mostly acts on distant effectors and is transported by the bloodstream or eventually diffused in a connective stroma, whereas the neurotransmitter, on the contrary, acts only at the synaptic cleft. The neuromodulator modifies the conditions of the emission and reception of nerve impulses. The terms *ns products* and *ns cells* refer to cytological and ultrastructural data demonstrating the existence of a secretory activity inside the neurons. These terms do not concern a particular category of molecules, e.g., ns cells may be peptidergic but are not necessarily so, and ns products may act not only as neurohormones but as neurotransmitters or neuromodulators as well. In many cases, however, the term *ns cells* refers to well-known secretory neurons in which pathways were shown to terminate in neurohemal organs.

The fundamental progress achieved brought radical conceptual changes, and it is worth including in the first chapter of this work a short historical survey of the evolution of the data and concepts. This survey mostly concerns studies performed in vertebrates; studies of insects will be described subsequently in much greater detail.

1.1. FROM THE ORIGINAL CONCEPT OF NEUROSECRETION TO CONTEMPORARY VIEWS (TABLE I)

1.1.1. The Concept of Neurosecretion

Before considering the historical evolution of concepts concerning neurosecretion and neurohormones, a few words will be said in order to situate the early endocrinological discoveries.

At the end of the eighteenth and beginning of the nineteenth centuries, fundamental discoveries were made demonstrating the existence in certain organs and tissues of factors later called hormones and neurotransmitters, which modified the functioning of other organs or tissues. The presence of these factors was demonstrated in particular in the pancreatic islets, the small intestine, and the hypophysis, whose extracts act on the blood sugar level, the release of pancreatic juice, and blood pressure plus uterine contractions, respectively. In 1921 the first known neurotransmitter was discovered.

Although the observation of the appearance of secretory activity inside certain neurons in both vertebrates and invertebrates had already been described in 1919 by Speidel and in 1928 by E. Scharrer, these unusual developments were only understood in 1937, when E. and B. Scharrer clearly expressed the idea that certain neurons display glandular activity.

A few years later, it was shown that the material produced by the glandular neurons moves along their axons and is stored in the glandular organs where these axons terminate. Thus, the existence of neuroglandular complexes, includ-

TABLE I. Historical Survey of Concepts from Neurosecretion to Peptidergic Neuron[a]

Time	New techniques	New concepts	Some identified neuropeptides
1937		Neurosecretion	
1944		Neuroendocrine complexes (hypothalamus-hypophysis; pi-CC)	
1949	Chrome hematoxylin phloxine		Insulin
1953		Neurohemal organs	Oxytocin
1954	Electron microscopy Autoradiography	Hypothalamus–median eminence–adenohypophysis complex	Vasopressin
1957		Elementary granules	
1959	RIA	Synaptic vesicles	Glucagon
1960		Hypothalamohypophyseal releasing factors	
1962	Histochemical amine detection	Monoamine regulation in hypothalamus and median eminence	
1963		Electrical activity of ns cells	
1966		Neurosecretomotor junctions	
1968			
1969	HPLC	Neurosecretory product release by exocytosis	TRH
1973		Neuropeptide ubiquity	LHRH
1974	PAP immunocytochemistry		
1975			Enkephalins
1975			Proctolin
1976			AKH
1977		Neuropeptide colocalization	
1977		Neuromodulation	

[a]AKH, adipokinetic hormone; HPLC, high-performance liquid chromatography; LHRH, luteinizing hormone-releasing hormone; pi-CC, pars intercerebralis–corpora cardiaca; RIA, radioimmunoassay; TRH, thyrotropin-releasing hormone.

ing neurons located in the central nervous system, nerves, and terminal glandular storage organs, was demonstrated (Scharrer and Scharrer, 1944). The hypothalamohypophyseal complex of vertebrates (Table II), the pars intercerebralis (PI) of the brain—corpora cardiaca (cc)–corpora allata (ca) of insects, and, later, the X organ–sinus gland of crustaceans (Hanström, 1953) were all shown to belong to the same category of organ complexes, i.e., the neuroendocrine complexes.

Thus the concept of neurosecretion was born. It was to be the basis of numerous studies and lead to the discovery of the physiological role and biochemical structure of the various important neurohormones produced in the ns cells.

TABLE II. Vertebrate Terminology

Hypothalamus
Supraoptic nucleus
Paraventrical nucleus
Hypophysiotropic area
Median eminence
Hypophysis or pituitary
Neurohypophysis or posterior lobe or pars nervosa or neural lobe
Adenohypophysis
Anterior lobe
Pars distalis
Pars tuberalis
Intermediate lobe or pars intermedia

The study of ns cells and ns complexes has been greatly facilitated since 1949 by new histological methods which very clearly demonstrated the presence of ns products in both the cell body and the axon of the ns cells. Gomori's chrome hematoxylin phloxine staining was adapted to neurosecretion with much success by Bargmann (1949), so that the black staining obtained with this method was often considered as a criterion for demonstrating the presence of ns products which were sometimes termed Gomori-positive products. Several other techniques appeared later including paraldehyde fuchsin and fluorescent pseudoisocyanin (see Section 1.2.2).

The concept of neurosecretion was established mostly on histological grounds; it was supported by experimental findings showing that the biologically active factors were not only present in the ns cell bodies but also in their terminals and that sectioning of the pituitary stalk in vertebrates, insect nervi corporis cardiaci (NCC)(B. Scharrer, 1952), and crustacean x organ–sinus gland tract (Enami, 1954) caused important modifications. After such sectioning, the terminal organs were progressively deprived of their secretory products while a large accumulation of stainable material was observed at the cut end of the ns pathways, confirming the idea that the ns products migrate along the axons.

1.1.2. The Blood Pathway—Neurohemal Organs

Once the transport of the ns material to terminal organs had been demonstrated, the next step was to show that the ns products pass into the blood vessels, a passage which for a number of years had been suspected to occur in the median eminence and neurohypophysis (see Scharrer and Scharrer, 1954).

The term "neurohemal" (nh) organ was proposed by Carlisle and Knowles

(1953) "to describe any system of fibre terminations in crustaceans, which bears close spatial relationships to the blood vascular system and does not supply any muscle." This term has been used ever since for the terminal part of any ns complex in vertebrates and invertebrates, and the ns cells were not considered to form synapses with other neurons or effector organs, as their axons ended in blood spaces into which they released the ns material.

While the relationships between the hypothalamus and the neurohypophysis or neural lobe were in the process of being understood, the functional relationships between the hypothalamus, median eminence, and adenohypophysis were also being grasped, as was the significance of the hypophyseal portal vessels (Table II; Figure 2). Numerous ns fibers were seen to terminate in the median eminence close to the hypophyseal portal vessels, whose destruction was first shown to prevent the release of gonadotropic (Benoit and Assenmacher, 1953) and corticotropic factors from the adenohypophysis, thus indicating the control exerted by the hypothalamus on the adenohypophysis through the median eminence. Since that time, experimental studies combined with biochemical investigations gradually led to the identification of the numerous hypothalamic releasing factors of the median eminence that control the activity of the adenohypophysis.

1.1.3. Developmental Phase of the Investigations

Since the end of the 1950s, investigations devoted to neurosecretion have developed considerably thanks to new techniques such as electron microscopy, autoradiography, and biogenic amine detection. Such techniques allowed fine analysis of the mechanisms governing the synthesis, transport, and release of ns products.

Ultrastructural investigations concerning both vertebrates and invertebrates demonstrated that the ns cells and fibers contained numerous electron-dense, membrane-bound granules, 100–300 nm in diameter. These elementary granules (Palay, 1957) were revealed to contain both oxytocin and vasopressin activity (Heller and Lederis, 1962), and therefore to be the cell organelles involved in the storage and transport of neurohormones.

Besides neurohormones, the elementary granules were shown to contain in vertebrates a large carrier protein responsible for the staining properties of the ns products. Both neurohormones and carrier protein are included in a large precursor molecule which is synthesized in the cell body of the ns neuron. During the transport of the elementary granules along the axons, a process of maturation takes place which consists of the progressive separation of the neurohormones and carrier protein by enzymatic cleavage. At the release site, the neurohormones and carrier protein are released simultaneously (see Pickering, 1980).

The first steps in the processes of elementary granule synthesis were demonstrated to take place in the endoplasmic reticulum and then in the cisternae of the Golgi complex, so that the ns cells were seen to behave like other glandular cells.

Once formed, the elementary granule is transported along the axon. This was observed in vertebrates and invertebrates both *in vivo* and *in vitro,* and also after injection of radiolabeled cystine. In the rat, the labeled material appeared within 5 min of injection in the hypothalamus and within 30 min in the neurohypophysis where the release process takes place. Using colchicine or vinblastine, which both inhibit microtubule functioning, it was shown that microtubules were involved in elementary granule transport.

Numerous studies were devoted to the mechanisms regulating the release of ns products. First of all, the ns cells were shown, like other neurons, to be electrically active, although their action potential appeared to be of especially long duration (Yagi and Bern, 1963). It was established that the arrival of the action potential in the ns axon terminal initiated the release of the ns products by increasing membrane permeability and inward calcium movement (Douglas and Poisner, 1964). The electrical stimulation of nerves was shown to induce the release of ns products from nh storage organs, an effect also exerted by high K^+ concentrations, which open the Ca^{2+} channels.

The occurrence in the ns axon terminals of clusters of electron-lucent small vesicles has been observed since the first ultrastructural investigations (Palay, 1957). The close morphological resemblance of these vesicles to synaptic vesicles led to the introduction of the terms *synaptoid vesicle* and *synaptoid contact* or site. Some authors considered these small vesicles as true acetylcholine synaptic vesicles which might regulate the discharge of ns material at the ns axon terminals (De Robertis, 1965) in causing permeability changes. Mostly, however, it was thought that these vesicles were byproducts of ns granule release and that their presence indicated the location of release sites. They were observed to be transient structures whose number increased after electrical stimulation.

The origin of the synaptoid vesicles was under debate for a long time. Certain authors considered them as residues of the fragmentation of the elementary granules while others thought they constituted retrieval material from the granule membrane occurring after exocytosis (Fig. 1).

The exocytosis process itself was not easily accepted. On the one hand, dedensification and budding of the elementary granule was observed, suggesting that synaptoid vesicles originate in such processes and that the ns material diffuses from the granule to the membrane through the cytoplasm. On the other hand, elementary granules adhering to the plasmatic membrane as well as omega profiles exhibited by the membrane were also observed; they were assumed to correspond to the release sites of the elementary granules. Finally, the exocytosis concept was accepted for the axon terminals of the ns cells and it was proposed that the term synaptoid vesicles be replaced by that of microvesicles. It was

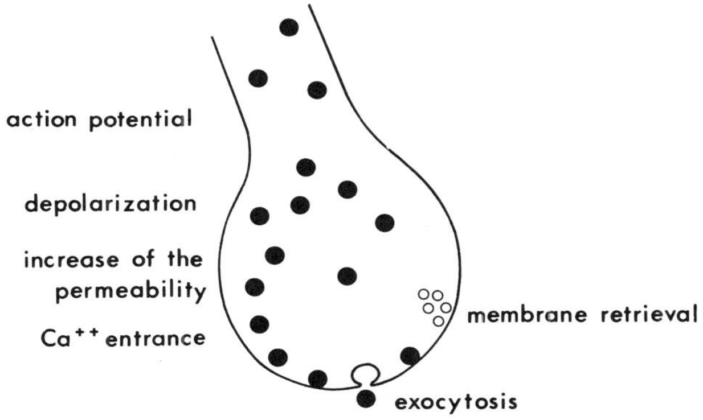

FIGURE 1. Schematic representation of the processes involved in the release of neurosecretory products at the axon terminal. The arrival of the nerve impulse causes depolarization followed by Ca^{2+} ion penetration. The elementary granule moves toward the axon membrane with which it fuses. Its content is then released outside the ending and the retrieval of the membrane in excess takes place.

therefore clear that ns cells do exhibit the same release mechanisms as glandular cells.

1.1.4. New Problems

The late 1960s brought a new awareness of the complexity of the data as shown by the reviews published at the time (Bargmann, 1966; Knowles and Bern, 1966; B. Scharrer, 1968).

1.1.5. The Neuroeffector Junctions

One of the main problems of this period concerned the significance of certain ns terminals. Instead of being located close to the blood vessels as in nh organs like the neurohypophysis and median eminence, they were observed, despite previous views, in glandular parts of the hypophysis near the secretory cells or in close contact with them, thus forming the so-called secretomotor junctions observed in mammals, teleosts, and insects (E. Scharrer, 1965; B. Scharrer, 1965a).

The existence of these junctions, already observed in histological investigations, was confirmed in electron microscopy studies, in which it was observed that the ns axons may terminate in close contact with glandular cells, nerve cells, and even muscle cells, as well as with the connective tissue. The latter was suggested to act as a network of channels allowing the ns products released to reach a number of glandular cells. The best known examples of secretomotor junctions were those in the intermediary and distal lobes of the teleost hypophysis (Knowles and Vollrath, 1966a,b), where two types of ns products (A and B) were found to be present in nerve terminals. It was later proved that the A and B ns products respectively corresponded to peptidergic and aminergic fibers which were shown to innervate the hypophysis.

The demonstration of the existence of the secretomotor junctions, also termed *neuroeffector junctions,* made it necessary to modify in part the concept of neurosecretion, and it was suggested that "the discharge of hormone into the blood stream could no longer be considered as the sole function of ns neurons" (Knowles and Bern, 1966); accordingly, it was proposed that "the common final pathway from the central nervous system to an endocrine gland was achieved by the ns peptidergic neuron in two different ways, either by ending in a nh organ or on the gland cell itself" (E. Scharrer, 1965).

1.1.6. Nonconventional Neurosecretory Cells

While the neuroeffector junctions were puzzling investigators, another point became the focus of numerous discussions. It concerned the ns cells behaving nonconventionally as reflected by their staining and ultrastructure properties.

In these cells, the ns products did not stain with chrome hematoxylin but did stain with acidic dyes (see Gabe, 1954). Such ns products were found in the hypothalamus, the median eminence, the pars intermedia and pars distalis, the urophysis, and also the central nervous system of various invertebrates. They were revealed by ultrastructural studies which demonstrated the presence of granules or vesicles different from the conventional elementary granules (see Bargmann, 1966), and some authors considered that these cells were not "true" ns cells. In fact, a new kind of ns product, the biogenic amines, was discovered.

1.1.7. Aminergic Neurosecretory Cells

In the early 1960s, a highly specific and sensitive fluorescence histochemical technique was devised by Falck and co-workers (1962) which made it possible to demonstrate the presence of monoamines, dopamine, norepinephrine, and serotonin in the cell bodies and fibers of the nervous system.

This demonstration was the origin of a revolutionary change in the ideas prevailing at the time because it revealed the extensive presence of the monoamines throughout the nervous system as well as in areas like the hypothalamus, which contain the cell bodies of the ns cells, and in nh areas such as the median eminence, which contain their terminals. The biogenic amines thus emerged as very important substances controlling the functioning of the ns cells and regulating, together with the peptides, the secretory activities of the glandular cells of the adenohypophysis. This regulation was shown to be done through the vascular route in the higher vertebrates and through neuroeffector junctions in teleosts.

In fact, biogenic amines had already been suggested to regulate pars distalis activity in experimental studies of the relationships between the median eminence and the gonadotropic function; electron microscopic investigations had demonstrated their presence in the median eminence and intermediary and anterior lobes of the hypophysis of nonconventional nerve terminals loaded with vesicles larger than the synaptic vesicles, which often contain a dense core surrounded by an electron-lucent zone.

The Falck fluorescence method was followed by other new methods (see 1.6.1) such as radioautography and immunocytochemistry using antibodies raised against the biogenic amines and the enzymes controlling their synthesis, such as tyrosine hydroxylase and dopamine β-hydroxylase.

It was shown that a dense norepinephrine plexus terminates in the supraoptic and paraventricular nuclei of the hypothalamus, where the axons end on the dendrites or the soma of the ns cells (Fig. 2). Furthermore, dopaminergic terminals were shown to be present close to the blood vessels of the median eminence (Carlsson *et al.,* 1962; Fuxe, 1963); the passage of dopamine into the blood vessels was only demonstrated later, when it was discovered that dopamine was in fact the prolactin release-inhibiting hormone (PIF), which behaves like the other peptidergic releasing factors and has its own specific receptors in the anterior pituitary.

In teleosts, which lack a portal blood circulation, the ns fibers were demonstrated to end near or in contact with the glandular cells of the pars intermedia and pars distalis, and aminergic fibers were seen to be involved, as in mammals, but terminating in the adenohypophysis instead of the median eminence.

In addition to the regulatory functions biogenic amines exert on the secretory activity of the adenohypophysis, they exert other roles. Since synaptoid axoaxonal contacts between aminergic preterminals and median eminence endings were observed, it was inferred that the amines might influence the release of certain releasing factors in the median eminence by acting on the peptidergic endings.

The cell bodies of the aminergic neurons were identified either in the hypothalamus, in the case of dopaminergic neurons projecting into the median eminence and intermediary lobe, or outside the hypothalamus, in the case of neurons projecting into the hypothalamus.

Biogenic amines therefore appeared, at least in some cases, to play the same part as the peptidergic ns products, as they were also observed to be released either into blood vessels or at neuroeffectory junctions. Moreover, they were found to play a very important role in regulating the functioning of the peptidergic ns cells of the hypothalamus, with which they form numerous axonic, dendritic, and somatic contacts.

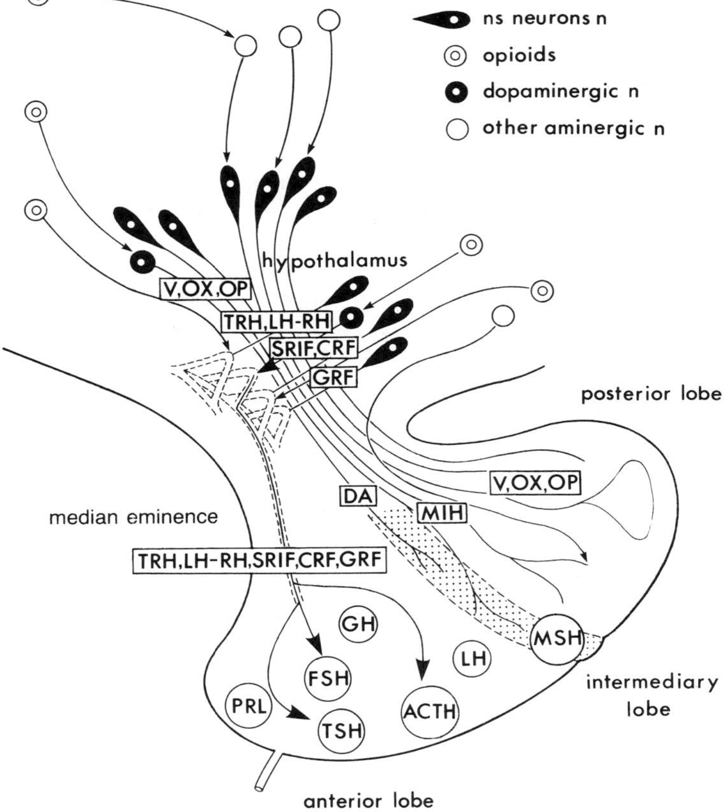

FIGURE 2. Schematic representation of the regulation of neurohormone synthesis and release in the hypothalamohypophyseal system. Aminergic neurons, mostly dopaminergic, are involved in both the cell body of the neurosecretory neurons (ns) and at terminal level in the median eminence.

1.1.8. Endorphins

Although monoamines play an important part in regulating ns cell function, other molecules, such as the recently discovered opioid endorphins (Hughes, 1975), are also important in this respect.

Endorphins are present in the brain and spinal cord, and in the nerves and ganglia of the autonomic nervous system. They mostly consist of β-endorphin, and met- and leu-enkephalins. The enkephalins are small molecules whose precursor has not yet been identified. Their action is fast and they are rapidly degraded. β-Endorphin, on the contrary, is a large molecule produced in the anterior and intermediary pituitary lobes; its precursor is proopiomelanocortin, which is also the precursor of both adrenocorticotropic hormone (ACTH) and melanocyte-stimulating hormone (α-MHS).

The endorphins appear to act by inhibiting excitatory nerve transmission. They function like analgesics and control blood pressure, respiration, and sensory mechanisms. They also regulate the release of the hypothalamic neurohormones into the median eminence and neural lobe, either directly or indirectly, by acting on the hypothalamic monoamine neurons. It has been suggested that leu-enkephalin might be present in the vasopressin and oxytocin terminals.

1.1.9. Characterization of Peptidergic Neurohormones

While the investigations described above were taking place, other biochemical studies were developing whose goal was the characterization of active factors. Oxytocin, involved in parturition, was the first peptidergic factor to be identified and synthesized in 1953 (Du Vigneaud *et al.*, 1953). During the 20 years that followed, the main peptidergic neurohormones were identified. They included vasopressin (VP), thyrotropin-releasing hormone (TRH), luteinizing hormone-releasing hormone (LHRH), somatotropin release-inhibiting factor (SRIF), and other releasing factors from the median eminence (see Schally *et al.*, 1973). This represented a new fundamental step in the field of neuropeptides. Several anterior lobe hormones, including thyroid-stimulating hormone (TSH), prolactin (PRL), and ACTH, were also identified during the same period (Fig. 2).

1.1.10. Extensive Localization of Neuropeptides

The availability of synthetic neurohormones made it possible to produce specific antibodies allowing the use of two important techniques: radioimmunoassay and immunocytochemistry. Radioimmunoassay permitted measurement of hormonal levels in both blood and tissues, while immunocytochemistry allowed detection of the sites of neuropeptide synthesis, transport, and release by investigations performed at the cellular and ultrastructural levels. Surprising discoveries were made thanks to these and other related methods.

Following the introduction of immunocytochemistry it gradually became clear that, contrary to expectations, the hypothalamic neuropeptides vasopressin, oxytocin, and releasing factors, as well as the adenohypophyseal peptides and other biologically active peptides, were all present outside their known places of origin; thus, they were found in large areas of the nervous system, including the brain, spinal cord, peripheral nervous system, and the nerves and ganglia of the autonomic nervous system (Table III). It was also discovered that they were

TABLE III. Distribution of the Main Neuropeptides[a]

Hypothalamus–neurohypophysis–median eminence	Adenohypophysis	Gastroenteropancreatic system	Central and peripheral nervous system
Vasopressin			Vasopressin
Oxytocin			Oxytocin
Enkephalins		Enkephalins	Enkephalins
Somatostatin	Somatostatin	Somatostatin	Somatostatin
LHRH			LHRH
TRH		TRH	TRH
CRH		CRH	CRH
GRF		GRF	GRF
	TSH		TSH
	PRL	PRL	PRL
	GH	GH	GH
	ACTH	ACTH	ACTH
	β-Endorphin	β-Endorphin	β-Endorphin
	MSH		MSH
Neurotensin		Neurotensin	Neurotensin
Substance P		Substance P	Substance P
VIP		VIP	VIP
CCK		Gastrin-CCK	Gastrin-CCK
		Secretin	
		Motilin	
		Bombesin	Bombesin
			Angiotensin
		Insulin	
		Glucagon	
		Pancreatic polypeptide	

[a]ACTH, adrenocorticotropic hormone; CCK, cholecystokinin; CRH, corticotropin-releasing hormone; GH, growth hormone; GRF, growth hormone-releasing factor; LHRH; luteinizing hormone-releasing hormone; MSH, melanocyte-stimulating hormone; PRL, prolactin; TRH, thyrotropin-releasing hormone; TSH, thyroid-stimulating hormone; VIP, vasoactive intestinal peptide.

present outside the nervous system, namely, in the gastroenteropancreatic tract, and that digestive tract peptides were present in the nervous system, especially in the brain (see Emson, 1979).

These discoveries raised numerous questions, the more so as it was discovered at the same time that the same neuron might contain either several distinct neuropeptides, or a neuropeptide and a biogenic amine, or again a neuropeptide and other substances such as "classical neurotransmitters" (see Chan-Palay *et al.*, 1978; Hökfelt *et al.*, 1977, 1987).

1.1.11. Colocalization

The first demonstration of colocalization was made in peripheral norepinephrine neurons, which were shown to react to the antibodies raised against the neuropeptide somatostatin, and in the spinal cord, where serotonergic neurons responded to the antibodies raised against substance P. Later on, it was demonstrated that a number of central and peripheral neuropeptidergic neurons contained not only monoamines but also acetylcholine, GABA, and various neuropeptides (see Table IV).

The coexistence of neuropeptides and monoamines is particularly evident in the endocrine cells of the gastrointestinal tract, and this gave rise to the concept of the existence of an amine precursor uptake and decarboxylation (APUD) system (Pearse, 1966), which implied that cells which are able to synthesize both peptides and amines and are located either in the gastroenteropancreatic system or in the nervous system have a common embryological origin. This concept was extended to other cells to constitute the diffuse neuroendocrine system (DNES) (Pearse and Takor-Takor, 1979). APUD or DNES cells which synthesize both amines and neuropeptides are not all neurons, but they often display sensory properties and seem able to make contacts with other cells; they are sometimes referred to as paraneurons.

Colocalization of several neuropeptides was not only observed inside the neuron but also in the elementary granule. The neuropeptides colocalized included met-enkephalin, which was found in vasopressin, oxytocin, and somatostatin granules; neuropeptides were also found to be colocalized with amines (Hökfelt *et al.*, 1987).

In certain cases the peptides which are colocated and which exhibit common sequences belong to the same biochemical family and are derived from each other by enzymatic cleavage. This especially applies to ACTH, MSH, and β-endorphin, which are cleavage products of the large proopiomelanocortin molecule. Adrenocorticotropic hormone and β-lipotropin in turn give rise to α-MSH and CLIP, respectively, and to β-MSH and endorphins. Enkephalins, however, are derived from a different precursor molecule, although they share amino acid sequences with the β-endorphin.

The colocalization of different molecules in the same neuron raises questions concerning its functioning, including the synthesis, transport, and recovery of the different secretory products at the terminals. Whereas peptides are produced in the cell body, this is not so for amines and other neuromediators which are synthesized at the endings or incorporated into the granule during its transport from the cell body to the terminals. In the axon terminals, both amines and "classical" neurotransmitters are recovered, but apparently neuropeptides are not.

TABLE IV. Colocalization of Certain Neuropeptides with Other Neuropeptides, Amines, or Neurotransmitters[a]

First neuropeptide	Second neuropeptide	Epinephrine	Norepinephrine	Dopamine	Serotonin	Acetylcholine	GABA
Hypothalamus–neurohypophysis–median eminence							
Vasopressin	Substance P						
	Dynorphin						
	CRH		+	+			
Oxytocin	CCK		+	+			
	Enkephalins					+	
ACTH	β-endorphin						
TRH	Substance P				+		
Somatostatin	Enkephalins						+
	CCK						+
Pituitary anterior lobe							
FSH, LH				+			
TSH				+			
GH, PRL			+				
MSH					+		
Pituitary intermediary lobe							
ACTH			+	+	+		
MSH			+	+	+		
β-Endorphin			+	+	+		
Calcitonin			+	+	+		

			Central and autonomous nervous system				
Enkephalins			+				
MSH					+		
Substance P	Enkephalins					+	
	TRH				+		
	TRH				+		+
Somatostatin			+				
							+
Neurotensin				+			
CCK				+			
VIP						+	
			Adrenal medulla				
Enkephalins		+	+				
Somatostatin			+				
Neurotensin			+				
			Endocrine pancreas				
Insulin					+		
Glucagon					+		
Somatostatin				+			
PP				+			
Enkephalins				+			
			Gastrointestinal tract				
Glucagon					+		
Substance P					+		

[a]ACTH, adrenocorticotropic hormone; CCK, cholecystokinin; CRH, corticotropin-releasing hormone; FSH, follicle-stimulating hormone; GABA, γ-aminobutyric acid; GH, growth hormone; LH, luteinizing hormone; MSH, melanocyte-stimulating hormone; PP, pancreatic polypeptide; PRL, prolactin; TRH, thyrotropin-releasing hormone; TSH, thyroid-stimulating hormone; VIP, vasoactive intestinal peptide.

Whether or not present in the same neuron, the neuropeptides are spread over large areas of the nervous system and one may wonder about their physiological significance in these areas. Behavioral effects of neuropeptides are known as regards learning and memory (vasopressin, ACTH, endorphins), sexual behavior (LHRH), and satiety (CCK). However, the mechanisms intervening in these processes have not yet been identified.

When the sites of neuropeptide release are located close to the circulatory system in nh areas such as the median eminence or neurohypophysis, there is reason to think that they function like neurohormones, but it is also possible that they regulate the release of other peptides located in neighboring axons. Thus, in the neurohypophysis, the release of vasopressin from the axon terminals might be inhibited by leu-enkephalin and facilitated by substance P and neurotensin.

In addition, there is at the present time much evidence indicating that in the nervous system, several hypothalamic and gut-brain peptides intervene in neurotransmission in the same way as nonpeptidergic neurotransmitters. Oxytocin and vasopressin fibers increase the rate of firing of certain interneurons and form synapses in various parts of the brain, while met-enkephalin fibers were observed to form synapses on ns terminals.

1.1.12. Neuromodulation

Besides acting as neurohormones and neurotransmitters, neuropeptides were shown to modify nerve transmission in a different way from that usually described, which led to the formulation of the new concept of neuromodulation.

The neuromodulators do not induce the transsynaptic passage of the nerve impulse but modify its characteristics, facilitating or inhibiting synaptic transmission. They act in a more diffuse and remote way than the neurotransmitters and their effects last longer. The sites of neuromodulator release are synaptic or extrasynaptic, and neuromodulators intervene on both pre- and postsynaptic specific receptors located on terminals, axons, dendrites, or soma. The neuromodulators do not act through Na^+ ions but, like neurohormones, act through Ca^{2+} ions. Numerous neuropeptides were shown to be neuromodulators, but monoamines and conventional neurotransmitters such as acetylcholine, glutamate, or GABA may also function as neuromodulators.

In the axon terminals where different molecules are colocated, release processes may take place separately or simultaneously. In the first case it seems that the regulation is done by high- or low-frequency nerve impulses. When the different molecules are released together, each of them has a distinct role; while one acts as a neurotransmitter, the other or others act as neuromodulators, inhibiting, for example, the feedback effects of the neurotransmitter (Fig. 3). The occurrence of presynaptic receptors was demonstrated for several neuromodulators, including the opiate receptors of aminergic neurons.

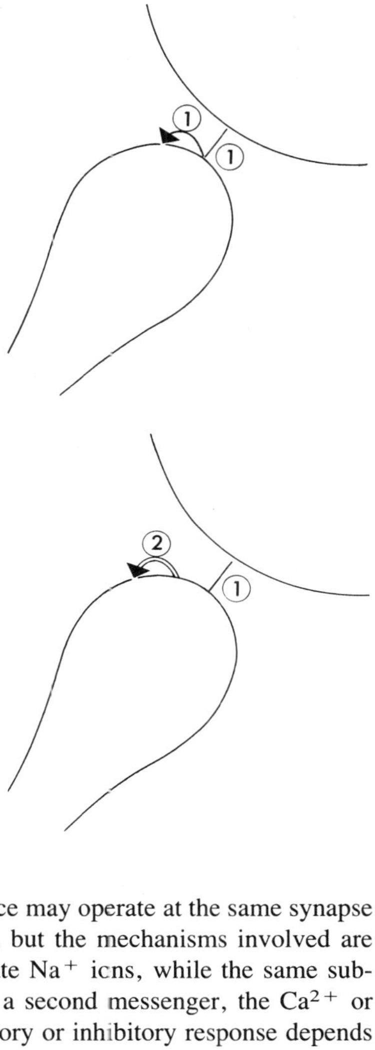

FIGURE 3. Presynaptic feedback exerted by the neurotransmitter itself (1) or by the neurotransmitter (1) and a neuromodulator (2).

As already mentioned, the same substance may operate at the same synapse as a neurotransmitter and a neuromodulator, but the mechanisms involved are different; thus glutamate and GABA implicate Na^+ icns, while the same substances acting as neuromodulators implicate a second messenger, the Ca^{2+} or cyclic nucleotides. In both cases the stimulatory or inhibitory response depends on the nature and number of receptors.

1.1.13. Regulation of Peptidergic Neuron Activity

It has already been indicated that biogenic monoamines are involved in regulating the synthesis and release processes of ns cells, as shown by their presence in the hypothalamus on the one hand, and in the neurohypophysis on the other. Although many points still require elucidation, it seems to be well established that the main forms of control are exerted by dopamine and norepinephrine, with serotonin and epinephrine also intervening in some controls.

The opioids, on the other hand, regulate the same processes in two ways: they regulate the functioning of the dopamine and norepinephrine neurons involved in controlling the hypothalamic ns neurons, and they act directly on the ns terminals in the median eminence and neurohypophysis.

1.1.14. Conclusion

Progress from the beginning of the century to now reveals how, starting from the idea that neurons may display secretory activity and synthesize ns products released into the circulatory system, it gradually became possible to show that, in addition to ns cells, very large numbers of neurons were synthesizing neuropeptides which acted both as neurohormones and neurotransmitters or neuromodulators. More than 100 neuropeptides were indexed whose chemical structure is known and whose receptors have been identified, at least in a number of cases. Among these peptides, the opioid family, like the biogenic amines, has been shown to play an important part in controlling secretion of hypothalamic neurohormones.

The neuropeptides are synthesized in large areas of the central and peripheral nervous system. They are often produced in neurons which also produce other neuropeptides or monoamines or conventional neurotransmitters.

An important idea which has been generally accepted as a result of the most recent investigations is that there are no clear-cut differences between neurohormones, neuromodulators, and neurotransmitters, since each of these functions can be carried out by the same molecule.

The mechanisms implicated in nerve transmission are therefore no longer considered in the same way as formerly; the importance of modulation is known, and it is also known that neuropeptides may act as neurotransmitters or neuromodulators.

1.2. CLASSICAL INSECT NEUROSECRETORY CELLS

The processes studied in this section concern the classical ns cells detected by histology, histochemistry, and electron microscopy, their pathways and release sites, the regulatory mechanisms controlling the ns processes and aminergic neurons. Data concerning the numerous recent investigations using vertebrate antisera will be discussed in Chapter 2, since not all the reactive neurons appear to be identical to previously identified ns cells.

The abundant literature concerning ns cells makes it necessary to operate a selection. As a rule, the authors quoted will firstly be those who initially showed the existence of the processes described and secondly authors of publications which have appeared since 1980. Other references may be found in earliers reviews (Raabe, 1982; Orchard and Loughton, 1985).

The morphoanatomical data and distinctions between the different ns cells have reached a fairly advanced stage of development. Given the strides presently being made in immunocytochemical research, knowledge of the anatomy of the ns system and the diversity of its component parts has acquired renewed importance, as it constitutes the basis of this research (see Chapter 2).

1.2.1. General Features

1.2.1.1. Morphology

The morphology of ns cells does not differ very much from that of other neurons: the cell body is not necessarily large, and the nucleocytoplasmic ratio is variable. The dendrites originate in the proximal part of the axon and are difficult to distinguish from the collaterals. They penetrate the neuropil, where they make synaptic contact with interneurons running from the sensory or association centers. Synaptic contact may also occur elsewhere along the axon pathway or within the nh organs. Recently, the pathways of the ns cells have been extensively studied thanks to the new methods of cobalt chloride filling (Pitman *et al.*, 1973) and intracellular iontophoretical microinjection of Procion yellow (Stretton and Kravitz, 1968), and Lucifer yellow (Stewart, 1981), which revealed their abundant arborizations.

The ns cells' axons run long distances before entering nh organs, where they branch repeatedly. At each distal end they form a preterminal swelling which precedes the ending proper. The latter is in contact with a thin connective tissue sheath which separates the axon terminals from the extracellular fluid into which the ns products are released. In other places, a glial investment forming a number of folds surrounds the ns axons.

Most of the ns cells are located inside the central nervous system but some have been found among the peripheral neurons. They are multipolar (Finlayson and Osborne, 1968) and sometimes multinucleate (Reinecke *et al.*, 1978).

1.2.1.2. Production of Neurosecretory Material

Synthesis of ns material is initiated in the rough endoplasmic reticulum, and continues in the Golgi apparatus. The material is gradually condensed, and electron-dense, membrane-bound granules bud off from the extremities of the Golgi saccules. These processes are linked to modification of the nucleus, the rough endoplasmic reticulum, cell body size, and the nucleus/cytoplasm ratio (Thomsen, 1965; Bloch *et al.*, 1966; McCaffery and Highnam, 1975a,b; Morris and Steel, 1975).

Because the ns products of certain ns cells contain cystine or cysteine, [^{35}S]cysteine can be used in injections to investigate their functioning. The rate of its incorporation into the perikaryon reflects the level of synthetic activity, while the speed of transit toward the CC supplies information about release mechanisms. Such investigations have been carried out on different species (see Raabe, 1982).

The presence in the elementary granules of carrier proteins associated with the neurohormones has been established in vertebrates, as shown above. In insects, putative neurophysins were isolated from the brain and CC of locusts (Friedel *et al.*, 1980a, 1981; Orchard *et al.*, 1981; Girardie *et al.*, 1985, 1987a).

1.2.1.3. Colocalization

The coexistence in the same neuron of peptides and amines has been demonstrated in insects; in the locust, for instance, the A-type ns cells of the PI were demonstrated to contain biogenic amines, probably dopamine (Vieillemaringe *et al.*, 1982, 1984; Takeda *et al.*, 1986), and certain neurons in the ventral cord were suggested to contain both biogenic amines and proctolin (O'Shea and Adams, 1981).

1.2.1.4. Electrical Activity

Like other neurons, ns cells display electrical activity and produce action potentials (for details, see Orchard and Loughton, 1985). This was demonstrated by both extracellular and intracellular recordings performed in the central nervous system, including the brain, of several insects such as *Periplaneta* (Cook and Milligan, 1972; Krauthamer, 1985a), *Manduca* (Carrow *et al.*, 1984), and *Bombyx* (Miyazaki, 1980), and in ventral nerve cord ganglia. Recordings were also made from the NCC in *Rhodnius* (Orchard and Steel, 1980) and CC in flies (Normann, 1973) and *Locusta* (Bloemen and De Vlieger, 1985) and from nh areas in *Carausius* (Finlayson and Osborne, 1968) and peripheral ns cells, also in *Carausius* (Orchard and Finlayson, 1976).

The ns cells appear to be spontaneously active, with a resting potential of 20–50 mV and an action potential of about 60 mV, similar to that of non-ns neurons but much slower (Cook and Milligan, 1972 in *Periplaneta*; Finlayson and Osborne, 1970 in *Carausius*; Miyazaki, 1980 in *Bombyx*; Normann, 1973 in *Calliphora*). Their axon terminals seem able to generate action potentials, at least in *Carausius* and *Rhodnius* (Finlayson and Osborne, 1968).

When the nerve impulses reach the nerve terminals, they trigger the release of neurohormones, which might also be obtained by electrical stimulation of the brain. Electrical stimulation of the NCC or CC was also performed in cockroaches and locusts. The effects of this stimulation on neurohormonal release were observed histologically and ultrastructurally (Scharrer and Kater, 1969), and in several bioassays demonstrating that it triggered the release of cardioaccelerators and ecdysiotropic factors in cockroaches (Kater, 1968; Gersch *et al.*, 1970) and of diuretic (Maddrell and Gee, 1974), gonadotropic, and metabolic factors in locusts (Girardie and Girardie, 1977; Orchard and Loughton, 1981).

The way in which depolarization acts on neurohormonal release was demonstrated in vertebrates, in which it involves the penetration of Ca^{2+} ions into the axon terminal. This has been confirmed in insects, where the presence of Ca^{2+} is essential for hormone release by electrical or potassium stimulation in cockroaches (Kater, 1968; Gersch *et al.*, 1970), locusts (Friedel *et al.*, 1981), *Calliphora* (Normann and Duve, 1969; Normann, 1974), *Glossina* (Maddrell and Gee, 1974), and *Rhodnius* (Berlind, 1981). The Na^{+} ions are also very important for the functioning of the soma and terminals of the ns cells, as observed in *Rhodnius* (Orchard and Steel, 1980), *Bombyx* (Miyazaki, 1980), and *Periplaneta* (Krauthamer, 1985a), whose hemolymph Na level is very high. However, this does not appear to be the case in *Manduca* (Carrow *et al.*, 1984).

1.2.1.5. Transport of Neurosecretory Products

Neurosecretory products are transported along the axons to the nh organs such as the CC, the perisympathetic organ (PO) or nh areas, where the release processes take place by exocytosis. This was demonstrated by ligating or sectioning the nerves of the CC which causes the accumulation of ns material on the proximal side of the damaged nerve (Scharrer, 1952).

The speed of transport of ns material has been estimated in *Schistocerca* using labeled cysteine; it oscillates between 1 and 3 mm/hr in adult females, depending on their reproductive stage (Highnam, 1976). In *Rhodnius,* electron microscopic and electrophysiological investigations demonstrated that the speed of axonal transport of the elementary granules increases during the period of increased bursting activity of the ns cells (Orchard and Steel, 1980), with the pattern and intensity of electrical activity determining the timing of the release. In starved *Rhodnius,* low continuous activity was recorded by suction electrodes from the CC nerves, but within minutes of the blood meal spikes representing high-frequency bursts were recorded (Orchard and Steel, 1980).

1.2.1.6. Release of Neurosecretory Products

Neurohormone release is elicted by the arrival of a nerve impulse which depolarizes the axon terminal. Depletion of the ns products is then observed as well as an increase in omega profiles and synaptoid vesicles (Normann and Duve, 1969; Scharrer and Kater, 1969).

As in vertebrates, the mechanisms triggering the release of ns products are considered to involve exocytosis and retrieval of membranes giving rise to synaptoid vesicles and synaptoid sites (vesicle clusters) that correspond to the release sites which appear and disappear with the release of neurohormones (Smith and Smith, 1966; Scharrer, 1968; Normann, 1969, 1970; and other authors, see Raabe, 1982).

1.2.1.7. Resorption of Neurosecretory Products

All the ns material synthesized in the perikaryon is not transported to the axon terminals: thus, any excess material synthesized is destroyed in the cell body by lysosomes or multivesicular bodies. These organelles are generally considered to be involved in the resorption processes but they have also been reported to play a part in the process of maturation of granules after their release from the Golgi apparatus into the cytoplasm. Neurosecretory material is also destroyed at the axon endings in the preterminal swellings. These swellings may, however, intervene in a different process—that of granule storage; the granules stored are thus available if the fast release of a large amount of material is required.

1.2.2. Neurosecretory Cell Diversity

The best known insect ns cells are those located in the PI, whose existence was first demonstrated by Weyer in the bee (1935) and which have turned out to be a mosaic of neurons with very diverse staining and immunocytochemical affinities.

The techniques used to study ns cells were the same as in vertebrates (see Table V). In all these techniques except azan, the ns products which are basic are submitted to an oxidation procedure before staining. While some remain basic, the others become acidic and are subsequently stained by basic stains like chrome hematoxylin; the ns products which remain basic are stained by the acidic counterstains such as phloxine, light green, or picroindigo carmine. It is thus possible to distinguish two kinds of ns products—basophilic and acidophilic, often termed A- and B-type ns products.

These techniques are not histochemical techniques, but staining by basic dyes is often due to the presence of SS or SH groups in the ns products. Further, aldehyde and dicarboxylic groups stain intensely with paraldehyde fuchsin, the results of which are not identical to those of chrome hematoxylin; paraldehyde fuchsin stains certain ns cells that are not stained by chrome hematoxylin phloxine.

TABLE V. Main Techniques Used to Demonstrate the Presence of Neurosecretory Cells

Chrome hematoxylin phloxine	Gomori (1941)
Paraldehyde fuchsin	Gomori (1950)
	Halmi (1952)
	Dawson (1953)
	Gabe (1953)
	Ewen (1962)
Paraldehyde thionine-phloxine	Panov (1964)
Paraldehyde thionine-paraldehyde	Panov (1976)
fuchsin	Sterba (1961)
Pseudoisocyanin	Adams and Sloper (1956)
Alcian blue	Ravetto (1964)
Alcian blue–Alcian yellow	Delphin (1965)
Alcian blue–phloxine	Heidenhain (1916)
Azan	Girardie and Rossi (1978)
Victoria blue–paraldehyde fuchsin	
In toto	
Performic acid–Victoria blue (PAVB)	Dogra and Tandan (1964)
Performic acid–resorcin fuchsin (PARF)	Ittycheriah and Marks (1971)
Vital staining with acridine orange	Beattie (1971)

The significance of A- and B-type ns cells has often been debated. Some authors have stated that B cells are inactive A cells, while others consider that two different kinds of cell are involved. In fact, the PI contains not only A and B cells, but also other ns cell types and immunoreactive neurons (see Chapter 2) (Figs. 4, and 5).

1.2.2.1. Type A Neurosecretory Cells

Among the PI-type A ns cells, two categories of similar ns cells were described and termed A and A′ or A_1 and A_2. These cells are located close to one another and when stained with chrome hematoxylin phloxine, they can usually be distinguished because A cells are heavily loaded whereas A′ cells appear to contain little ns material (Baehr, 1968, in *Rhodnius*). Further observation of both types during an intermolt period in bugs showed that their activity cycles are asynchronous (Furtado, 1979; Robert, 1979).

A clear distinction between the two categories was made possible by using the paraldehyde thionine phloxine technique or paraldehyde thionine–paraldehyde fuchsin method (Panov and Davydova, 1976 in mecopterans and neuropterans; Panov, 1976 in the dipterans *Calliphora* and *Lucilia*; Raabe *et al.*, 1979 in various species). The A_1 ns cells reveal a strong affinity for paraldehyde thionine because they contain strong acids, while the A_1 cells are mostly stained

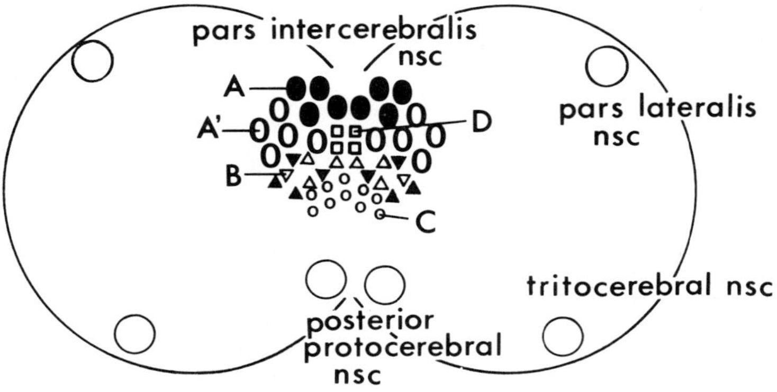

FIGURE 4. Main locations and diversity of brain neurosecretory cells. nsc: neurosecretory cells.

by the paraldehyde fuchsin because they contain weak acids. The investigations conducted with paraldehyde fuchsin and Victoria blue also revealed that while all the type A cells were stained by paraldehyde fuchsin, only some of them appeared with Victoria blue (Ganagarajah and Saleuddin, 1970; Girardie and Rossi, 1978). Both types stained by this method appeared significantly different by electron microscopy. The occurrence in the PI of two distinct type A ns cells (A_1 and A_2) was confirmed in a number of insects, namely, dermapterans (Panov and Krjutchkova, 1983), coleopterans (Panov and Melnikova, 1986), and lepidopterans (Panov, 1983).

Two other categories of type A ns cells were described in the PI, i.e., the C-type cells described in *Oncopeltus* (Johansson, 1958) and *Locusta* (Girardie and Girardie, 1966), whose distribution is mostly peripheral, and the D type ns cells, four in number which always occupy the same anatomical position in the medioventral part of the PI.

1.2.2.2. Type B and Cr Neurosecretory Cells

The B-type acidophilic ns cells stain with phloxine or other counterstains. When several counterstains are used simultaneously, different colors may be observed among the B-type ns cells stained. Certain ns neurons can mostly be visualized by azocarmine staining, following fixation in Helly's fluid (Raabe, 1965). This method revealed bright red "Gomori-negative" Cr ns cells in the PI, the tritocerebrum, and all the ganglia of the ventral nerve cord (Raabe, 1963, 1965).

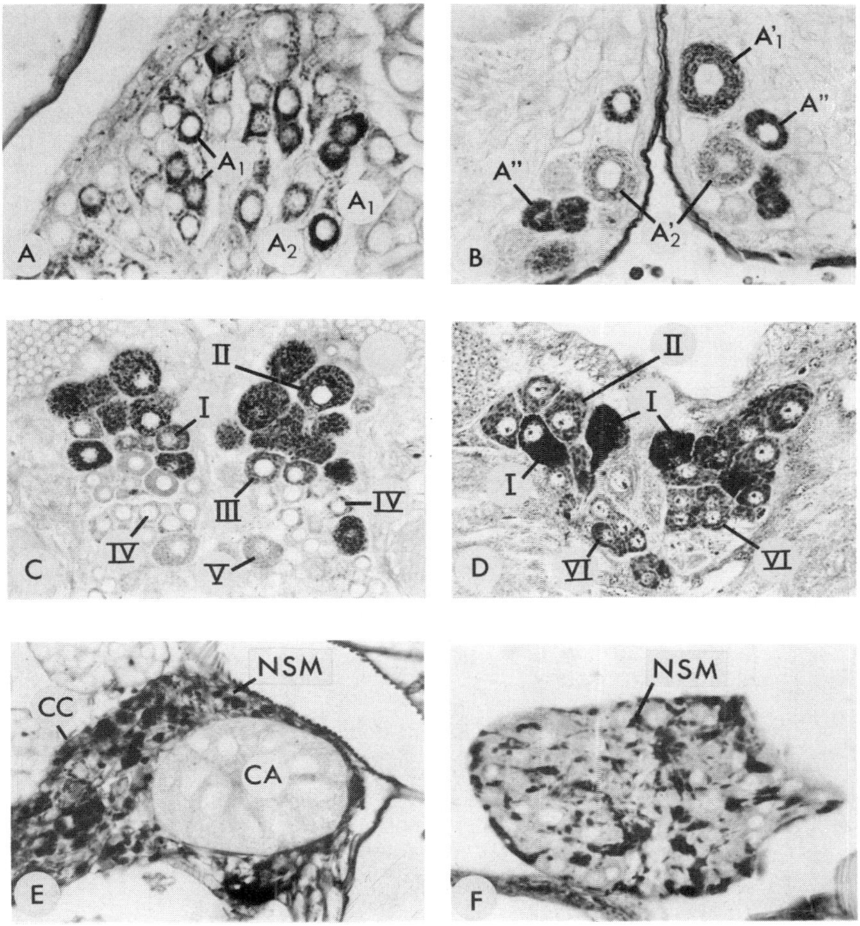

FIGURE 5. Diversity of neurosecretory cell types in the pars intercerebralis of insects. (A) *Schistocerca gregaria* adult male (Alcian blue–paraldehyde fuchsin); (B) *Malacosma neustria* larva (paraldehyde thionine–paraldehyde fuchsin); (C) *Pterostichus niger* adult (Alcian blue–paraldehyde fuchsin); (D) *Broscus cephalotes* adult (paraldehyde thionine–phloxine); (E,F) neurohemal area surrounding and penetrating the corpora allata in the coleopterans *Ontophagus uniformis* and *Margarinotus cadaverinus.* (Courtesy of Panov.)

1.2.2.3. Histochemistry

A number of histochemical investigations have been performed on species belonging to the main insect orders (see Arvy and Gabe, 1962; Raabe, 1982). The brain ns cells were shown to contain few carbohydrates and lipids and large amounts of proteins. The latter often contain sulfhydryl and disulfur groups, mostly in the type A ns cells of the PI. A recent study in *Schizodactylus monstrosus* showed that both type A_1 and type A_2 ns cells responded positively to ferric-ferricyanide staining, which reveals the presence of sulfydryl groups, but that the techniques used to evidence the disulfur groups gave much stronger reactions in the A_1 than the A_2 cells (Choudhuri *et al.*, 1984).

It is interesting to point out that the lateral type A ns cells in the ventral nerve cord display inconstant reactions to staining methods designed to demonstrate the presence of SS and SH groups. On the other hand, two medioventral subesophageal ns cells definitely contain these groups.

The B-type ns cells of the PI also react inconstantly as regards their sulfydryl or disulfur groups. The Cr cells are clearly negative in both brain and ventral nerve cord.

The phenol groups appear to be present in fairly large amounts in the type A ns cells and sometimes also the B ns cells of the PI. Type B cells have been shown to be rich in basic amino acids, i.e., histidine and arginine (Rehm, 1955). The data concerning the other types of ns cells in the nervous system as a whole are not sufficient to draw any conclusions in this respect.

1.2.2.4. Electron Microscopy

In insects as in vertebrates, the ns cell bodies, as well as their axons and terminals, have been shown to contain elementary granules ranging from 100 to 300 nm in diameter. These elementary granules vary not only in size but also in shape, and in the density of their content; most elementary granules are electron-dense but some are electron-lucent (Scharrer, 1963, 1968; Normann, 1965; Smith and Smith, 1966; Cassier and Fain-Maurel, 1970a; Cazal *et al.*, 1971). In addition to electron-dense and electron-lucent granules or vesicles, pale granules have also been described, which might correspond to a particular phase in the synthetic activity cycle or in the release of ns material (Schooneveld, 1974a).

Several investigations were performed on various insects, and the results mostly confirm the histological data since they showed the presence of a number of ns cell types in the PI of *Locusta, Acheta, Leptinotarsa, Musca,* and *Rhodnius* (see Raabe, 1982). In *Ostrinia,* several types were distinguished using a histological double-staining procedure (Victoria blue–paraldehyde fuchsin and electron microscopy) (Lavenseau *et al.*, 1985). In *Locusta,* types A_1 and A_2 ns cells were clearly distinguished ultrastructurally by both the diameter (140–200 and

250–300 nm, respectively) and the shape of their granules (Girardie and Rossi, 1978).

1.2.3. Anatomy of the Neurosecretory System

The main ns complex discovered in insects consists of the PI, the CC, and the NCCI. Although the ns center of the PI is quite large, it does not contain all the ns cells in the insect brain; there are also two paired lateral protocerebral centers and two paired tritocerebral centers. The lateral protocerebral and tritocerebral ns cells are respectively connected to the CC by the NCCII and NCCIII. Neurosecretory cells are also found in the anteromedial part of the protocerebrum, in the optic lobes, all along the ventral nerve cord, and in the anterior sympathetic nervous system of some species; lastly, some ns cells are located outside the central nervous system among the peripheral ns cells.

1.2.3.1. Brain

The PI has been investigated by many authors in many insect species. The major works dealing with the most common laboratory insects concerned flies (Thomsen, 1965; Bloch *et al.*, 1966; Panov, 1976, 1979a), bugs (Johansson, 1958; Baehr, 1968; Morris and Steel, 1975; Panov, 1979b), locusts (Girardie and Girardie, 1967), and beetles (Schooneveld, 1974a; Panov and Melnikova, 1987). It contains several type of ns cells as shown in Section 1.2.2.1. (Figs. 4 and 5). Large variations in the number of ns cells have been observed among different species (see Rowell, 1976). In dictyopterans, phasmopterans, and orthopterans, the type A ns cells in the PI are small and very numerous (1000–2000) while in lepidopterans, dipterans, heteropterans, there are, on the contrary, only a few large ns cells (Panov and Kind, 1963). The PI is not always homologous from one species to another or from one instar to another. In a number of insects, the ns cells of the PI, which form a single group in the adult, are divided into distinct groups in the larva (see Section 1.2.3.6).

Some medial posterior ns cells, which constitute one or two groups, are known to be present in several insect species, i.e., odonata, cockroaches, orthopterans, and lepidopterans. They were first identified by McLeod and Beck (1963) and Schaller and Meunier (1968) and recently studied in lepidopterans (Awasthi and Singh, 1981; Panov, 1983). According to several authors, they are incorporated in the PI ns area during metamorphosis. Their axonal pathways may terminate in a nh area of the brain (Geldiay and Edwards, 1973).

Ns cells are present in the lateral parts of the protocerebrum, near the calices of the pedunculate bodies. Few in number, they often form two distinct groups on each side at the origin of the NCCII. They are often paraldehyde fuchsin-positive.

Neurosecretory cells have not often been described in the optic lobes but their presence has been reported in some species (Barde, 1981; Prasad, 1981; Panov, 1985a,b). They are paraldehyde fuchsin-negative and azocarmine-positive. The destination of the axonal tract of optic lobe ns cells is not known. It seems to run toward the brain, where ns axons seem to form synapses with micrgranular axons (Beattie, 1971).

Deutocerebral ns cells have been described in a few cases and correspond to a distinct pair of CC nerves, the NCCIV (Brousse-Gaury, 1967) at least in *Periplaneta*. A few ns cells are present in the tritocerebrum (Dupont-Raabe, 1957; Raabe, 1963a; Girardie, 1975; Awasthi and Singh, 1981). They most often belong to the Cr type. The axons of the tritocerebral ns cells enter the NCCIII (Raabe, 1963b) and also the NCCI as observed by cobalt chloride filling in *Schistocerca* (Mason, 1973) and *Manduca* (Koontz and Edwards, 1980).

1.2.3.2. Ventral Nerve Cord

Neurosecretory cells are present in all the ganglia of the ventral nerve cord (Fig. 6). They are fewer in number and more widely dispersed than those of the brain, but they belong to several types and form rich, diversified systems, namely, in the subesophageal ganglion (see Raabe, 1982). Four major categories of ns cells were found, which are approximatively organized in the same pattern in most of the insect species studied.

The first category consists of a single pair of A-type ns cells, located medioventrally in the subesophageal ganglion. The axons of these cells, which were recently shown to be immunoreactive to several vertebrate antisera (see Chapter 2), follow a particular path as seen in *Locusta* (see Section 1.3.3).

Another type A ns cell category is exemplified by large, paired thoracoabdominal ns cells first described in the very early studies of Geldiay (1959, *Blaberus*) and Füller (1960, *Blatta, Periplaneta, Chaoborus crystallinus*). In certain insects, these cells are confined to the thoracic region, but in others they may extend to the subesophageal ganglion and abdominal ganglia, where they are present at all levels or only in the first four ganglia, as in cockroaches and stick insects.

The third kind of ns cell is confined to the abdominal part of the central nervous system. It consists of rather small lateral cells which form small paired groups.

The fourth kind of ns cell is represented by the azocarminophilic Cr ns cells. The ventral nerve cord contains a large number of these cells, which are present in all the insect investigated. They form large ventromedian areas in the subesophageal ganglion and abdominal ganglia, and small lateral groups in the thoracic ganglia (Raabe, 1965, stick insects; Chalaye, 1967, locusts; De Bessé,

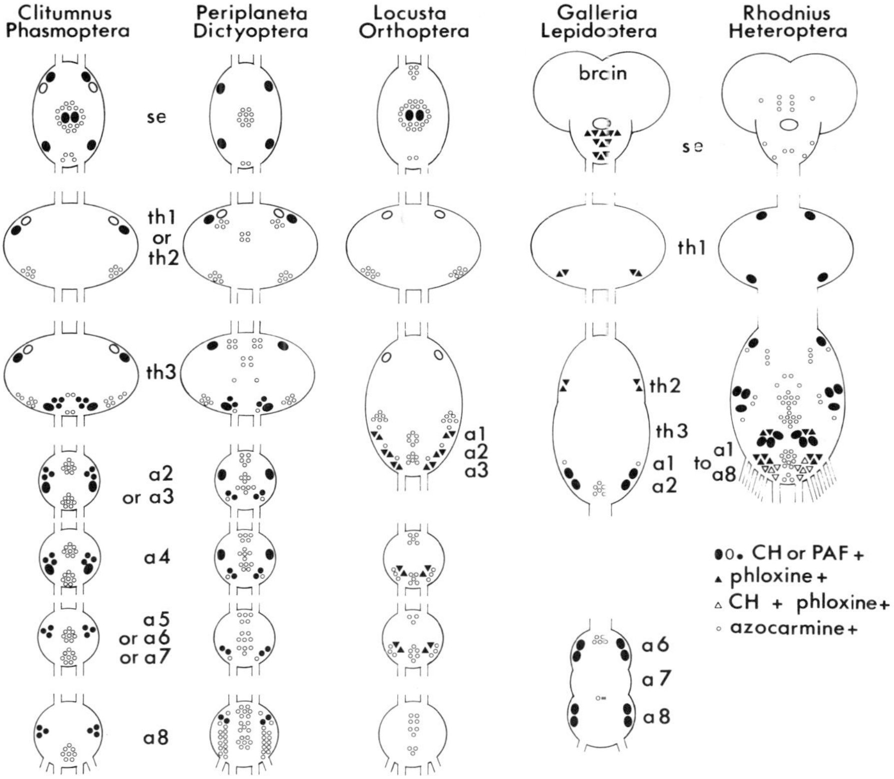

FIGURE 6. Neurosecretory cells in the ventral nerve cord of certain insects. a, Abdominal ganglion; se, subesophageal ganglion; th, thoracic ganglion; CH, chrome hematoxyln; PAF, paraldehyde fuchsin. (From the data of Raabe, 1965; De Bessé, 1967; Fréon, 1964; Chalaye, 1967; Delépine, 1965; Baehr, 1968; Baudry, 1968.)

1967, cockroaches; Baehr, 1968, Baudry, 1968, bugs). Some of these Cr ns cells might be immunoreactive to certain vertebrate antisera.

1.2.3.3. Anterior Sympathetic Nervous System

Neurosecretory cells are not often found in the sympathetic nervous system. However, certain histological studies and most electron microscopic investigations reported the presence of either peptidergic or aminergic ns cells in the sympathetic ganglia (see Section 1.6). Neurosecretory cells have been shown to occur in the frontal and hypocerebral ganglia in cockroaches and in the frontal ganglion of moth larvae. They have also been reported in the frontal ganglion of the bug *Lygaeus* (Awasthi, 1980). Electron microscopic studies of the hypocerebral ganglion of *Periplaneta* showed that several types of ns products were present including an aminergic product. All the ns products present in the sympathetic nervous system do not originate in the ns cells of the ganglia; some come from the brain as suggested by Cazal *et al.* (1971) for *Locusta,* Dorn (1976) for *Carausius,* Ude *et al.* (1978) for *Periplaneta,* and Kirby *et al.* (1984) for *Acheta.*

The final destination or the ns products of the sympathetic ganglia is not known precisely. Those of the frontal ganglion run either toward the brain or toward the recurrent nerve from where they probably reach esophageal targets. Neurosecretory granules have been reported in the esophageal nerves of *Melanoplus* and *Calliphora.*

1.2.3.4. Peripheral Neurosecretory Cells

Peripheral ns cells were first identified in the vicinity of viceral organs such as the heart (Bowers and Johnson, 1966; Johnson, 1966; Miller and Thomson, 1968), the alary muscles, the digestive tract, and the abdominal ns organs (Kuster and Davey, 1981).

They were also discovered along various abdominal or thoracic nerves, such as the link nerves which connect the transverse and segmental nerves in cockroaches (Chervin and Baudry-Partiaoglou, 1984; Raabe, 1986a) stick insects (Finlayson and Osborne, 1968) (Fig. 7), lepidopterans (Griffiths and Finlayson, 1982; Wasserman, 1985), and dipterans (Robert *et al.,* 1984).

The peripheral ns cells are multipolar neurons. Their secretory products invade the different branches, whose terminal portion is often devoid of glial sheath, and are released along narrow and diffuse nh areas, in the PO, and at the level of visceral muscles (Anwyl and Finlayson, 1973; Wasserthal and Wasserthal, 1980).

Peripheral ns cells differ from non-ns peripheral cells in their electrical activity, which resembles that of ganglionic ns cells (Orchard and Finlayson, 1977), but they may fulfill a double sensory and secretory role. In *Periplaneta, Carausius,* and *Locusta,* the link nerve peripheral ns cells have been revealed to be immunoreactive to mammalian glucagon (see Chapter 2).

1.2.3.5. Identified Neurons

The identification of neurons displaying well-defined characteristics as regards their general morphology, i.e., branching of the collaterals, axon pathways, as well as their pattern of electrical activity is important because these neurons may be further recognized and characterized for their physiological

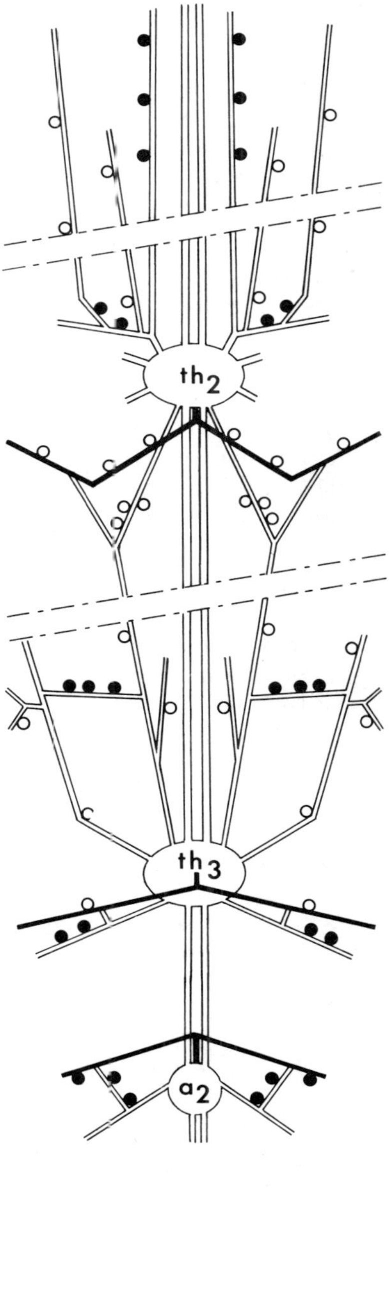

FIGURE 7. Schematic representation of peripheral neurons (○) and peripheral neurosecretory cells (●) in thoracal and abdominal segments of *Carausius*. (Modified from Finlayson and Osborne, 1968; Orchard and Finlayson, 1977.)

activity. Several specific methods are used for this purpose: recording of electrical activity, demonstration of the entire morphology of the neuron by $CoCl_2$ backfilling or intracellular injection of horseradish peroxidase or Lucifer yellow, immunocytochemistry and bioassay of the content of isolated neurons.

1.2.3.5a. *Dorsal Unpaired Motoneurons.* The first discoveries were those concerning the dorsal unpaired motoneurons (DUM) located in the dorsal midline of the thoracic and abdominal ganglia which were found to be octopaminergic (Hoyle and Barker, 1975; Evans and O'Shea, 1978) and to intervene in the regulation of neuromuscular transmission in the extensor tibiae muscle in the hind leg of the locust. In the last abdominal ganglion of the same species two octopaminergic DUM neurons were shown to innervate the oviduct muscles (Lange and Orchard, 1984; Orchard and Lange, 1986) (see Chapters 6 and 7).

1.2.3.5b. *Bursicon.* The neurons involved in the production of bursicon were identified in *Manduca* (Taghert and Truman, 1982b) (see Chapter 8). They are three to four lateral paired monopolar neurons which branch in the median ventral neuropil and send their axons through the segmental nerves into the transverse nerve PO (Fig. 8).

1.2.3.5c. *Diuretic Hormone Neurons.* In *Rhodnius,* a histological study of the thoracoabdominal nerve mass revealed the occurrence of four types of ns cells, one of which displays significant changes in its load during diuresis (Baudry, 1969). The diuretic activity of isolated neurons showed that these ns cells do exhibit large modifications in their diuretic hormone content following a blood meal (Berlind and Maddrell, 1979).

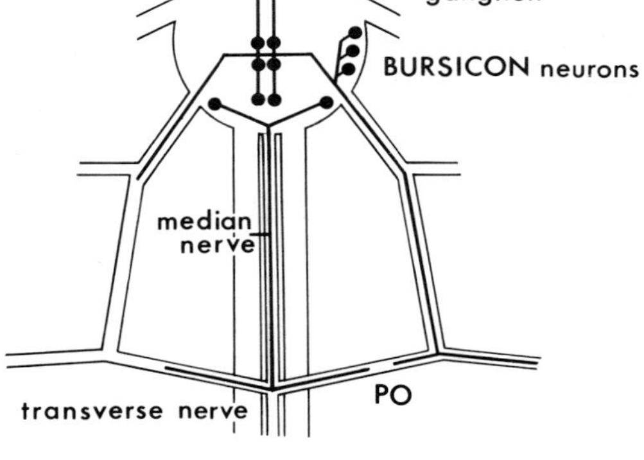

FIGURE 8. Identification of bursicon and CAP neurons in *Manduca.* The axons run through the segmental nerves into the transverse nerve perisympathetic organs. Two other paired neurons reach the perisympathetic organs via the median nerve pathway. (Modified from Taghert and Truman, 1982b.)

In *Locusta,* the paired medioventral ns cell in the subesophageal ganglion (see 1.2.3.2) was shown to contain a vasopressin-like factor (Rémy and Girardie, 1980) also present in the hemolymph, where its amount varies during diuresis, suggesting that it has a physiological role in this process (Proux and Rougon-Rapuzzi, 1980).

1.2.3.5d. *Eclosion Hormone.* The brain ns cells involved in the synthesis of eclosion hormone were determined by bioassay of eclosion hormone activity. Immunocytochemical staining with an antisera raised against purified eclosion hormone showed that the ns cell involved belonged to group Ia of lateral ns cells (Copenhaver and Truman, 1986a). Their electrostimulation causes the release of the hormone from the CC.

1.2.3.5e. *Proctolin Neurons.* Immunocytochemical investigations demonstrated the presence of several proctolin-like neurons in cockroaches (Eckert *et al.*, 1981; Bishop and O'Shea, 1982; Agricola *et al.*, 1985). Using intracellular recording and dye filling, two large, lateral, bluish white (LW) abdominal neurons of *Periplaneta* were observed to project into the cardiac nerve. They exhibited anatomical and electrophysiological features enabling them to be characterized as identified neurons and were demonstrated to contain large amounts of proctolin (O'Shea and Adams, 1981). Another large proctolin-containing motoneuron, the Ds neuron, which innervates a skeletal muscle of the cockroach leg, was identified by the same methods (O'Shea and Bishop, 1982).

1.2.3.5f. *Cardioaccelerators (CAP 1 and 2).* Certain neurons were identified in *Manduca* by cobalt chloride impregnation from the transverse nerve PO. They are located in the medioventral part of the abdominal ganglia and display cardioacceleratory activity which is released following intracellular stimulation of their soma (Tublitz and Truman, 1985a).

1.2.3.5g. *Paired Medioventral ns Cells of the Subesophageal Ganglion.* These ns cells have been suggested to be involved in various physiological processes such as diuresis (Proux and Rougon-Rapuzzi, 1980), embryonic diapause (Fukuda and Takeuchi, 1967), and reproduction (Fréon, 1964; Huignard, 1964).

1.2.3.6. Metamorphic Changes in the Anatomy of the Neurosecretory System

As already mentioned, the PI of the adult holometabolous insects results from the fusion of distinct ns areas of the larvae. This was observed in several

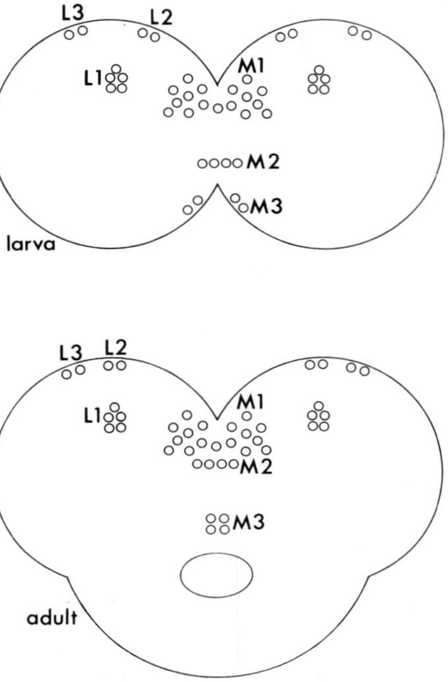

FIGURE 9. Neurosecretory cell migration during development in a moth *Spilosoma menthastri*. The median M2 and M3 clusters are far from the M1 clusters in the larvae but come closer together during growth. The lateral neurosecretory clusters (L1, L2, and L3) also change their position during development. (Modified from Panov and Kind, 1963.)

orders such as the lepidopterans (Panov and Kind, 1963) (Fig. 9), trichopterans, coleopterans (Melnikova and Panov, 1981), and dipterans (Panov and Kind, 1963; Dogra and Tandan, 1965; see also Khurad and Thakare, 1983).

It was noticed that in several lepidopterans the number of brain ns cells increases during development (see Awasthi and Singh, 1981). It is only in the second instar that type B ns cells differentiate and type A ns cells divide into several distinct types. The repositioning of the cell groups and their dendritic field at metamorphosis was also studied in *Manduca* (Copenhaver and Truman, 1986b) (Fig. 10).

In the ventral nerve cord the same process of progressive differentiation of the ns cells was observed to take place in an orthopteran (Singh and Arif, 1981) as in a lepidopteran (Srivastava and Prasad, 1985).

According to Panov, who studied a large number of species of the main insect orders, not only does the number of ns cells increase during evolution but this increase is accompanied by the diversification of cells types.

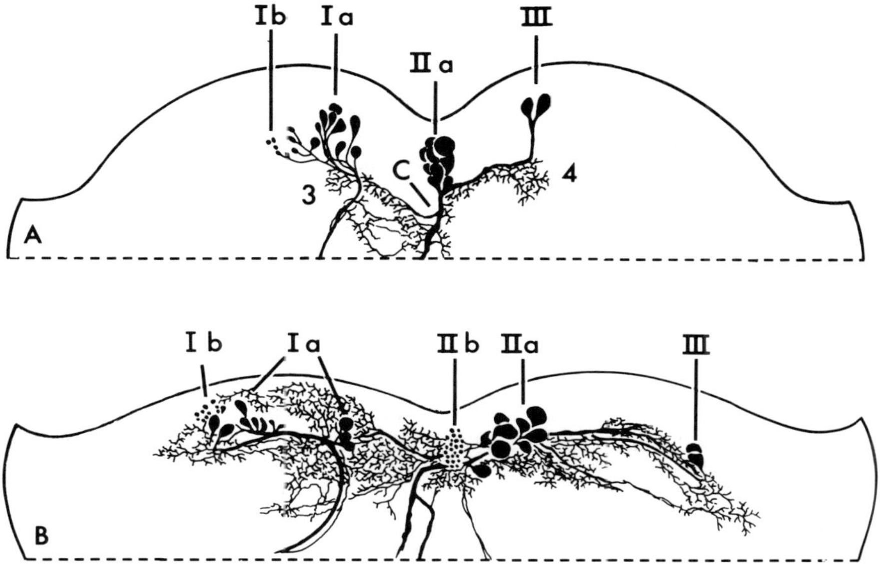

FIGURE 10. Repositioning of the neurosecretory cell groups (Ia, Ib, IIa, IIb, and III) at metamorphosis in *Manduca*. (A) Pupal brain; (B) adult brain. (Modified from Copenhaver and Truman, 1986b.)

1.3. NEUROSECRETORY PATHWAYS

The main pathways of the ns cells have been known for a long time to be localized in the NCC, but a number of new pathways were revealed recently by cobalt chloride filling, Lucifer yellow intracellular injection, and immunocytochemistry which resulted in the staining of the whole neuron, including cell body and collaterals. This showed that the pathways of the ns cells of the PI were not as simple as was initially believed.

1.3.1. Neuropil Arborization of Brain Neurosecretory Cells

In *Schistocera* the axons of the PI ns cells were observed to form wide loops entering the neuropil before resuming the ipsilateral and contralateral axon pathways (Highnam and West, 1971). On the other hand, it was also reported that the collaterals of the ns cells enter the protocerebral neuropil, where they branch repeatedly, thus creating extensive dendritic fields; this was observed in several species (see Raabe, 1982 and, for recent investigations, Buys and Gibbs, 1981; Zaretsky and Loher, 1983; Carrow *et al.*, 1984; Krauthamer, 1985b).

It is interesting to note that the branches of the PI ns cells come close to those of the protocerebral lateral ns cells (Fraser and Pipa, 1977; Rademakers, 1977b; Koontz and Edwards, 1980) and of the tritocerebral ns cells (Zaretsky and Loher, 1983). We shall see below that immunocytochemistry also reveals this type of connection between ns cell groups.

It might therefore be that a transfer of information takes place between the different groups of ns cells. Another possibility is that the release of certain ns products may occur in the neuropil itself and that these products might modulate nerve activity in certain parts of the brain.

Some ns cells in the PI of a cricket were shown to have no axons or collaterals extending outside the boundaries of the brain. They branch extensively in the protocerebrum and have terminal arborizations in a posterior ventromedial part of the brain suggested to be the brain's neurohemal area (Zaretsky and Loher, 1983) (see Section 1.4.3.5).

1.3.2. Pathways to the Corpora Cardiaca and Corpora Allata

The brain ns cell pathways most often studied are those linking these cells to the CC by three pairs of nerves originating in the PI (NCCI), in the lateral protocerebral ns cells (NCCII), and in the tritocerebral ns cells (NCCIII). It is also known that the CA are innervated by axons coming from the CC (the nervi corporate allati 1, NCA1) and by neurons of the subesophageal ganglion whose axons constitute the NCA2. Simple morphological data mask a fairly complex reality (Table VI). Electron microscopy studies and cobalt chloride filling reveal that the CA are innervated by ns cells both from the PI and from the lateral parts of the protocerebrum.

As for the glandular cells of the CC, they are in contact with ns fibers whose ultrastructure has an aminergic appearance (Scharrer, 1963, 1968 in cockroaches; Cassier and Fain-Maurel, 1970; Cazal *et al.*, 1971; Rademakers and Beenakkers, 1977 in locusts). In *Locusta,* these fibers were shown to originate from the lateral protocerebral ns cells and to travel along the NCCII (Rademakers, 1977b).

Lastly, ns fibers from the brain migrate to the nh zone of the CC, where aminergic axons were also observed (Lafon-Cazal and Arluison, 1976; Lafon-Cazal, 1976; Klemm and Falck, 1978). Ns axons from the CC are known to exit in sympathetic nerve in cockroaches (Gundel and Penzlin, 1980), locusts (Strong, 1966), and crickets (Raziuddin *et al.*, 1979; Zaretsky and Loher, 1983; Kirby *et al.*, 1984). In some species they proceed toward the aorta walls or into the cardiac nerves.

TABLE VI. Identified Axon Pathways to Corpora Allata and Corpora Cardiaca[a]

Insect	Reference	Median nsc	Lateral nsc	Trito nsc	So nsc	Other neurons	Putative aminergic neurons
Periplaneta	Fraser and Pipa, 1977	CA	CA				
	Pipa and Novak, 1979				CA		
	Gundel and Penzlin, 1980	CA	CA				
Locusta	Cassier and Fain-Maurel, 1970b	CA	CA				CC gdc
	Rademakers, 1977b	CA	CA			CC gdc	CC gdc
	Chalaye, 1967				CA		
Schistocerca	Mason, 1973		CA	CA	CA		
Manduca	Nijhout, 1975[b]	CA	CA				
	Carrow *et al.*, 1984[b]	CA	CA			CA, CC	
	Copenhaver and Truman, 1986c	CA	CA				
Calliphora	Normann, 1965						CC gdc

[a]CA, corpora allata; CC, corpora cardiaca; gdc, glandular cells; nsc, neurosecretory cells; so, subesophageal ganglion; trito, tritocerebrum.
[b]In lepidopterans it is difficult to assess whether the ns axons entering the corpora allata are innervating the glandular cells of these organs or only spread over them in a neurohemal area.

1.3.3. Pathways to the Prothoracic Gland

For obvious reasons the pathways of the ns and non-ns neurons innervating the prothoracic gland (PG) have been intensively studied. Since the position of this gland varies in insects, its innervation is not always the same. The PG can be innervated by the subesophageal ganglion only, by both the subesophageal and prothoracic ganglia, or by several thoracic ganglia.

An interesting feature is the presence in the PG of ns terminals which are often of both the peptidergic and aminergic types. Moreover, in orthopterans, dermapterans, and lepidopterans, the PG is in close contact with the PO, which suggests a functional interrelationship between both organs (see Section 1.4.2).

1.3.4. Pathways in the Central Nervous System

Neurosecretory pathways were also discovered in the circumesophageal connectives and the connectives of the ventral nerve cord in *Acheta* (Geldiay, 1959), mosquitoes (Füller, 1960; Burgess, 1973; Panov, 1979c), aphids (Bowers and Johnson, 1966; Steel, 1977), and bugs (Seshan, 1968; Kudryavtseva, 1986). The cells from which these long axons originate were found in the brain and it

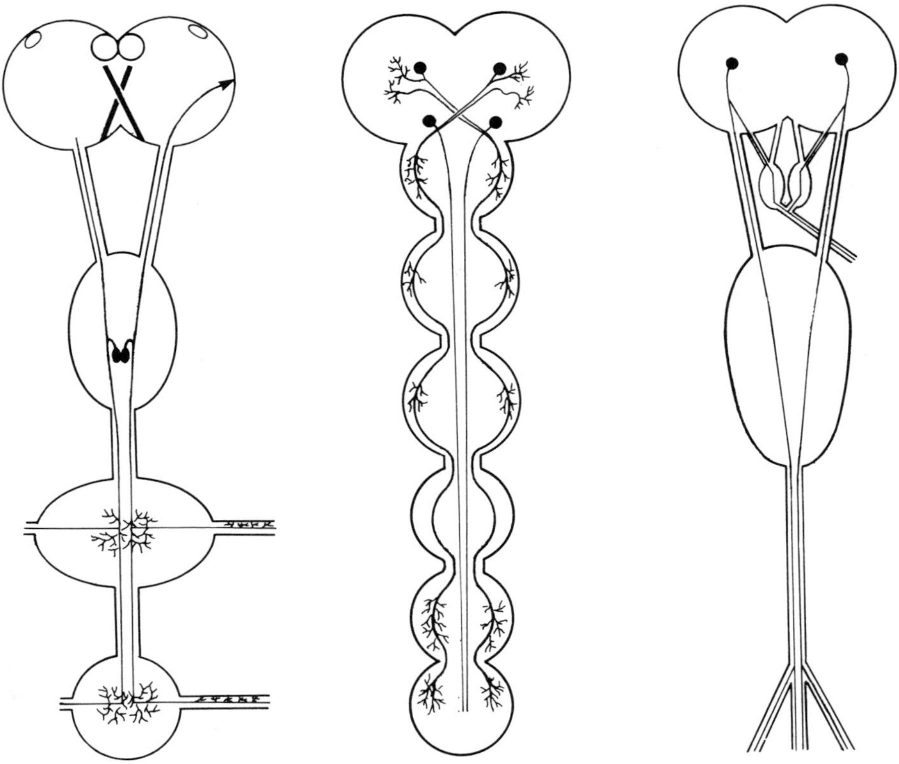

FIGURE 11. Pathway of certain cerebral neurosecretory axons sending collaterals to ventral nerve cord ganglia. From left to right: in *Locusta, Chironomus,* and *Myzus.* (From the data of Rémy and Girardie, 1980 for *Locusta;* Panov, 1979c for *Chironomus;* and Johnson, 1963 for *Myzus.*)

was observed that the same ns cell may simultaneously send out one process to the ventral nerve cord, another into the neuropil, and a third into the CC. A different pathway was demonstrated in stick insects and locusts for the paired medioventral ns cells of the subesophageal ganglion (Fig. 11) (see Section 1.4.3.7). These were shown to send processes to the brain and ventral nerve cord (Rémy and Girardie, 1980).

1.4. RELEASE SITES

The neurohormone release sites in insects are now known to be many and varied. Instead of a single place of release in the CC, there appear to be several distinct release systems. They may be centrally located in the CC and PO or distributed more diffusely in nh areas and inside target organs.

1.4.1. The Corpora Cardiaca

The CC are mostly located in the head behind the brain, but are sometimes found in the neck or anterior part of the prothoracic segment. They are paired but may fuse into a single unpaired organ. They often merge with the lateroventral walls of the aorta (Fig. 12), but sometimes they are only partly fused with it and sometimes they are not connected to it at all, as in certain lepidopterans, in which case their position becomes lateral instead of medial. In apterygote insects they constitute swellings on the aortic nerves (Cassagneau and Juberthie, 1983; Bitsch, 1986). The CC are associated to the sympathetic hypocerebral ganglion which innervates the gut. In certain insects this ganglion is fused with the CC whereas in others it is located close to these organs, with which it exchanges nerve fibers. In some cases there are no links between the CC and the hypocerebral ganglion and the fibers pass from the CC to the recurrent nerve, which contains axons from the frontal ganglion and also from the brain. Like the CC, the CA fuse in a number of species. In the larvae of higher dipterans, the CC, CA, and PG are incorporated into a single organ, the ring gland or Weismann's ring. Except for ephemeropterans, whose CA are innervated by the subesophageal ganglion only, constant connections exist between the CA and CC. The CC are connected to ns centers of the brain by the NCC, which contain ns aminergic and cholinergic axons. There are several types of ns axons, as revealed by electron microscopy, in cockroaches (Scharrer, 1963, 1968; Scharrer and Kater, 1969), the blowfly (Normann, 1965), stick insects (Smith and Smith, 1966), locusts (Girardie and Girardie, 1967; Cassier and Fain-Maurel, 1970b; Cazal *et al.*, 1971; Lafon-Cazal, 1976), and gryllids (Awasthi, 1977). The nerve fibers entering the CC form a large bundle in the center of the organ, which sends out a large number of branches which reach the connective membrane or its invaginations, where the ns products are released as stated above. The glial cells are well developed.

The CC contain a few neurons and glandular cells (also termed intrinsic or parenchymatous cells) with short processes which come into contact with the connective sheath or its invaginations. They are often paraldehyde fuchsin-positive but not always stained by chrome hematoxylin–phloxine. Electron microscopy clearly reveals the glandular cells of the CC, which contain well-developed synthetic organelles and electron-dense granules which may be distinctly larger than the ns granules originating in the brain in cockroaches (Scharrer, 1968), locusts (Cassier and Fain-Maurel, 1970b; Cazal *et al.*, 1971), and *Calliphora* (Normann, 1965). Gap junctions have been demonstrated to occur between them (Orchard and Shivers, 1986).

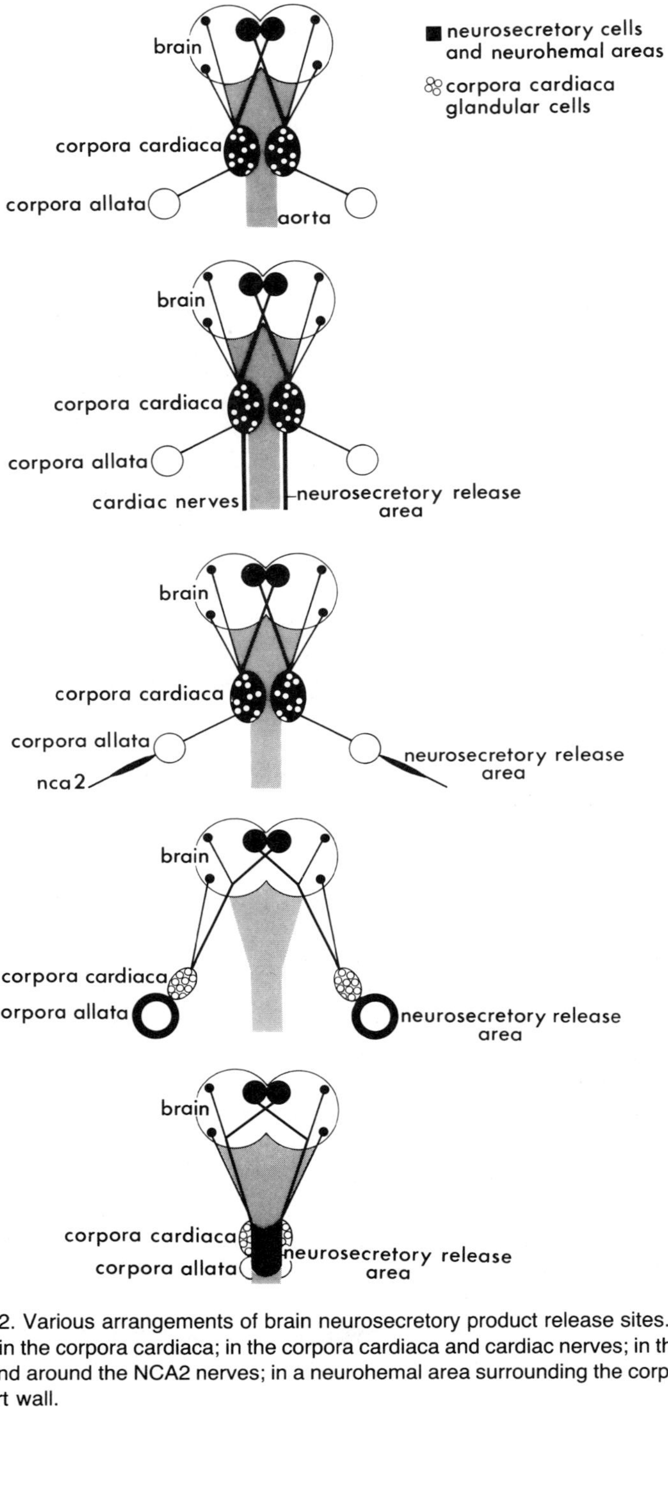

FIGURE 12. Various arrangements of brain neurosecretory product release sites. From top to bottom: in the corpora cardiaca; in the corpora cardiaca and cardiac nerves; in the corpora cardiaca and around the NCA2 nerves; in a neurohemal area surrounding the corpora allata; in the heart wall.

The glandular cells of the CC are often intermingled with the ns axons coming from the brain. In some insects, however, they are completely separated from them, forming a glandular lobe apposed to the nh storage lobe, which is the case in locusts (Highnam, 1961a). In other insects such as heteropterans, dermapterans, and certain dipterans, the CC only contain glandular cells and the ns endings are located in the aorta walls.

Although ns product release is elicted by the arrival at nerve terminals of bursts of nerve impulses transported by the ns axon itself, it is probably also influenced by other axons making axoaxonal contact. Axoaxonal or axosomatic contacts between ns fibers and the processes or somata of glandular cells were indeed observed in several insects, i.e., cockroaches (Scharrer, 1963, 1968), locust (Cassier and Fain-Cazal, 1970a; Cazal *et al.*, 1971), *Calliphora* (Normann, 1965), and the tsetse fly (Awiti *et al.*, 1985).

We saw above that the NCC and the nh and glandular parts of the CC contain aminergic fibers. Recent research has shed light on the nature of these amines, which are essentially octopamine, serotonin, and dopamine (see Section 1.6.6).

Acetylcholine was also shown to be present in the CC and to play a part in the release of certain neurohormones from this organ. In *Periplaneta* the release of neurohormone D (Gersch, 1972) was elicted by electrical stimulation of the NCCI, which involves cholinergic components. It is worth noting that in the ventral nerve cord, cholinergic compounds stimulate the release of heart accelerators.

1.4.2. Perisympathetic Organs

The PO are nh organs associated with ventral nerve cord which play the same role as the CC (Raabe, 1965; Raabe *et al.*, 1974; Provansal-Baudez, 1983; Grillot *et al.*, 1987). Their main feature is their metameric distribution. They owe their name to their location on the nerves of the unpaired sympathetic system which probably helps to regulate nh release; the presence of sympathetic nerve fibers is also known in the CC. The PO contain ns axons which repeatedly branch and form numerous terminals. Numerous glial cells ensheath these axons up to their endings. The neural lamella is thin and may form invaginations, creating sinuses where the ns material is released through the neural lamella. The ns products contained in the PO originate in the posterior nerve ganglion, as shown in experiments in which the median nerve was cut, thus inducing an accumulation of ns material at the sectioned end and depletion of the PO (Raabe, 1965); moreover, certain axons terminating in the PO were shown in *Manduca* to originate in the anterior ganglion, and to reach the PO by the posterior median nerve or by both the segmental nerve and link nerves (Taghert and Truman, 1982b; Tublitz and Truman, 1985a) (Fig. 7). The occurrence of neurons or glandular cells in the PO is not a general feature but both were reported in several species (Grillot, 1976a,b; Baudry-Partiaoglou, 1978, 1983b).

The PO display great morphological variation, depending on developmental stage and anatomical level (see Grillot, 1983) (Fig. 13). In primitive insects, these organs consist of swellings located on the median nerve, the transverse nerves, or both. In evolved insects such as hymenopterans and coleopterans the

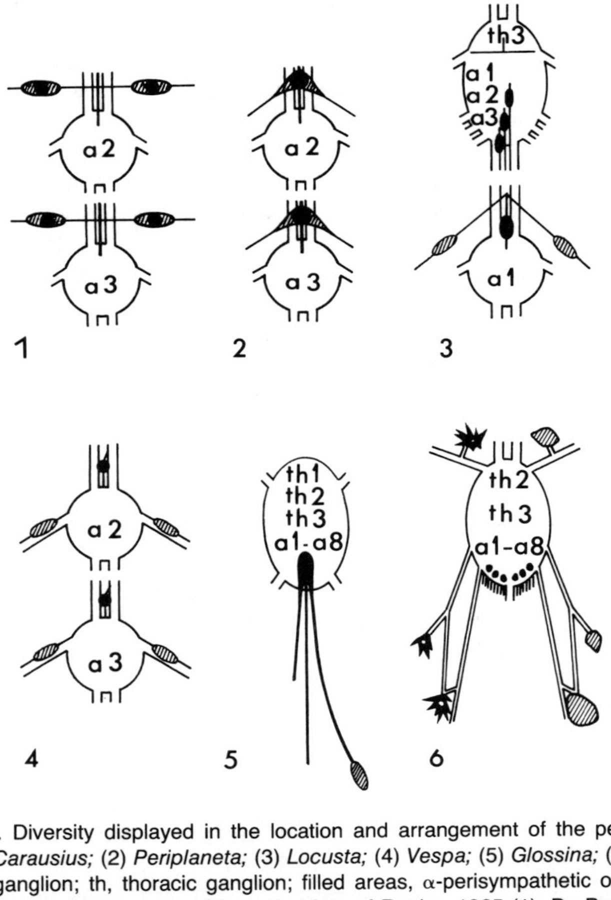

FIGURE 13. Diversity displayed in the location and arrangement of the perisympathetic organs. (1) *Carausius;* (2) *Periplaneta;* (3) *Locusta;* (4) *Vespa;* (5) *Glossina;* (6) *Roscius.* a, Abdominal ganglion; th, thoracic ganglion; filled areas, α-perisympathetic organs; striped areas, β-perisympathetic organs. [From the data of Raabe, 1965 (1); De Bessé, 1967 (2); Chalaye, 1967 (3); Provansal, 1968 (4); Grillot and Raabe, 1973 (5); Baudry-Partiaoglou, 1978 (6); and Baudry-Partiaoglou, 1983b, 1986; Provansal-Baudez and Baudry-Partiaoglou, 1983 for α- and β-structure demonstration.]

PO migrate inside the segmental nerves or inside the ganglia, where they constitute metameric nh areas, as in heteropterans, or a single nh area, as in the tsetse fly.

PO have been classified into three types, according to their anatomical position: median, transverse, and mediotransverse, each of which has been divided into subtypes, according to the degree of PO fusion with the segmental nerve or nerve ganglia (Grillot, 1976a,b, 1983).

Ultrastructural studies were performed in a number of species first in stick insects (Raabe and Ramade, 1967; Brady and Maddrell, 1967) and in cockroaches (Scharrer, 1968), then a number of other species (see Baudry-Partiaoglou, 1983a, 1986; Provansal-Baudez and Baudry-Partiaoglou, 1983). These investigations revealed that the PO comprise two clearly distinct ultrastructural types termed α- and β-PO, respectively, which were demonstrated to coexist in the same metamere in all the insect species so far investigated, including *Vespa* (Provansal *et al.*, 1970; Raabe *et al.*, 1970), coleopterans (Provansal-Baudez and Baudry-Partiaoglou, 1983), and the tsetse fly (Baudry-Partiaoglou and Grillot, 1975) (Fig. 14). This was especially clearly shown in heteropterans (Baudry-Partiaoglou, 1983b), cockroaches, stick insects, and locusts (Baudry-Partiaoglou, 1986).

The β-PO are characterized by bare axon terminals, a loose structure, and numerous sinuses. The ns products present in the β-PO belong to three or four distinct types. In the β-type PO, the release processes occur in places where the ns terminals are in direct contact with the basement membrane or its invaginations. The α-type PO that contain a single type of elementary granule display a compact structure. The ns terminals are invested by glial infoldings and the discharge mode is different from that of β-PO; indeed, the ns products are discharged into the intercellular spaces at a certain distance from the neural lamella. In most insect species, α and β structures are located in distinct median and transverse PO. In some species, however, both structures are colocated in the same PO as in *Periplaneta* and *Carausius* (Baudry-Partiaoglou, 1986), while in the locust, distinct median and transverse PO correspond to α and β structures, respectively.

The proximity of the PG and of the anterior well-developed PO corresponding to the subesophageal ganglion has been observed in both dermapterans (Raabe, 1972) and orthopterans (Thomas and Raabe, 1974). Recently, investigations were devoted to the relationships of both organs in several lepidopteran larvae where the PO are particularly well developed (H organs) (Abou-Halawa and Slama, 1986) and disappear within a few days during pharate adult development.

As in the CC, several biogenic amines, i.e., dopamine, octopamine, and serotonin, were demonstrated to occur in the PO (see Section 1.6.7).

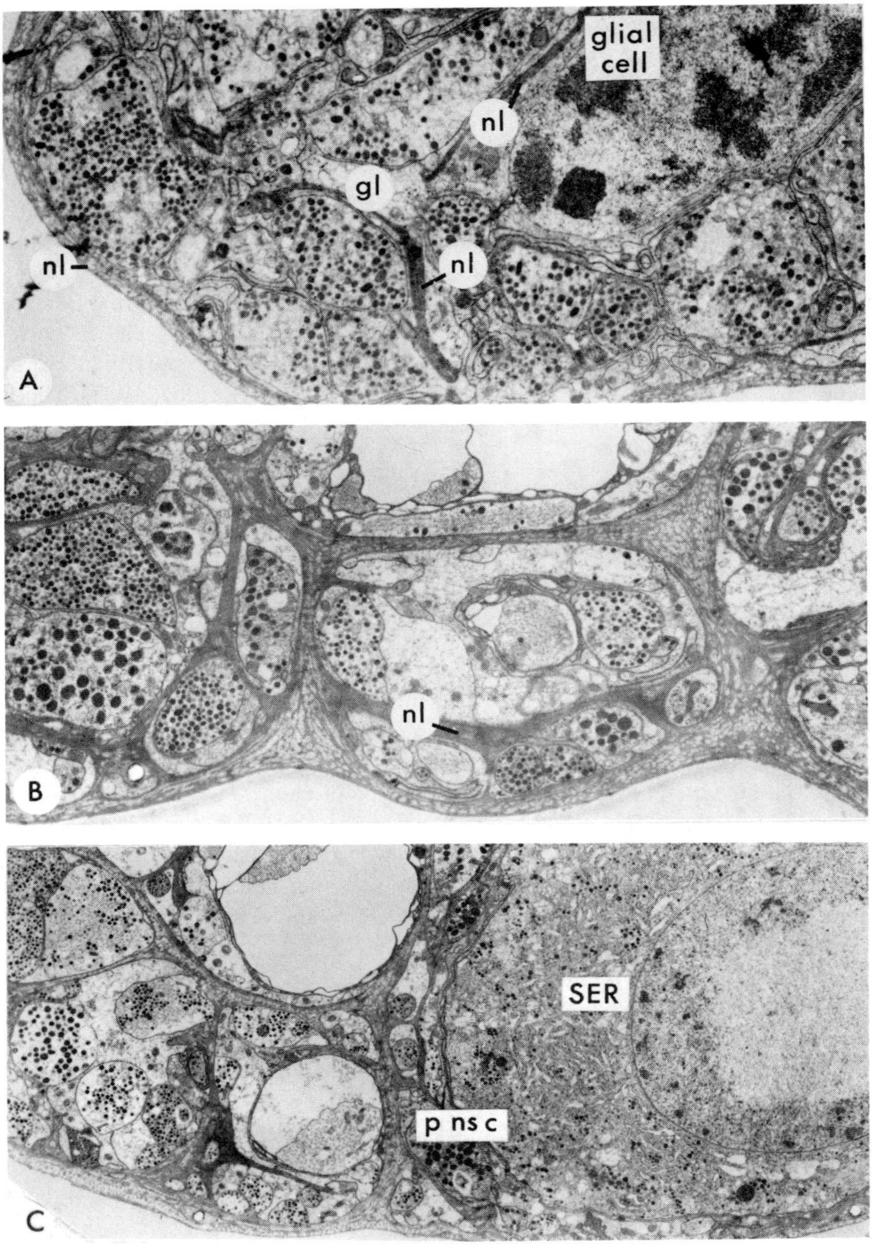

FIGURE 14. Structural diversity of α- and β-perisympathetic organs in *Locusta.* (A) Median α-perisympathetic organ containing a single type of ns terminals surrounded by glial processes (gl); the neural lamella (nl) invaginates deeply. (B) Lateral β-perisympathetic organ containing bare ns terminals of different types. (C) A neurohemal area containing a pe-

1.4.3. Neurohemal Areas

The CC and PO are not the only sites of ns product release. Other fairly diffuse nh areas are also observed along various nerves, often together with peripheral ns cells, where bare axons are facing the neural lamellar. More than one type of terminal may be distinguished ultrastructurally.

1.4.3.1. Aorta Walls—Antennal Ampulla

In some species, the CC mainly contain the glandular intrinsic secretory cells, the ns axons from the brain, terminating in the wall of the aorta. This arrangement was first described in heteropterans (Johansson, 1958; Singh and Narain, 1980a), then in mosquitoes, flies (Normann, 1965; Johnson, 1966; Awiti *et al.,* 1985), and dermapterans (Awasthi, 1975; Singh and Narain, 1980b; Panov and Krjutchkova, 1983) and has also been reported in a coleopteran (Ray *et al.,* 1980; Melnikova, 1985). Synaptic contacts have also been observed between ns terminals and muscle of the aortic wall in *Periplaneta* (Johnson, 1966) and the tsetse fly (Awiti *et al.,* 1985).

In the cockroach *Periplaneta,* a nh structure has been described in the wall of the antennal pulsatile ampulla.

1.4.3.2. NCA2 and Corpora Allata

The NCA2 joining the CA to the subesophageal ganglion form small nh organs in *Periplaneta* (Pipa and Novak, 1979) and rather well-developed ones in crickets (Thomsen, 1943).

A surprising feature is that in some species, the ns endings are located in a peripheral area surrounding the CA (Nijhout, 1975 in *Manduca*). It occurs in lepidopterans and coleopterans where CC only contain glandular cells (Tombes and Smith, 1970; Srivastava *et al.,* 1975; Singh and Awasthi, 1981; Sedlak, 1981). This morphological arrangement explains why in lepidopterans the PG is stimulated by CA implants. The origin of prothoracicotropic hormone (PTTH) in the nh area surrounding the CA was demonstrated by Agui and co-workers (1979) in *Manduca* (see Chapter 3).

1.4.3.3. Various Nerves and Plexuses

In other insects, a few ns axons terminate in various regions outside the CC, although these organs play a normal nh role. In several species, ns endings were

ripheral neurosecretory cell (pnsc) and numerous neurosecretory endings of various types. Note the neurosecretory granules and well-developed SER in the peripheral neurosecretory cell. A, ×5000; B, ×9000; C, ×7000. (Courtesy of Baudry-Partiaoglou.)

found against the sheath of the nerves leading from the brain to the CC, e.g., in *Leptinotarsa* (Schooneveld, 1974b) and *Calliphora* (Kaiser, 1979). The esophageal or cardiac nerves issuing from the CC are nh in *Periplaneta* (Johnson, 1966), *Blaberus* (Gersch *et al.*, 1974; Novak and Pipa, 1986), *Myzus* (Bowers and Johnson, 1966), *Calliphora* (Thomsen, 1969), and the primitive dipteran *Chironomus* in which the source cells of these nh areas were identified in the PI (Panov, 1979c).

In *Periplaneta* a fairly well-developed nh plexus occurs that include nerves reaching and leaving the CC-CA complex (Adiyodi, 1974; Novak and Pipa, 1986). It somewhat resembles the nh NCA2 described in crickets. In the dipteran *Aedes,* a similar extensive nh network spread over NCC, NCA, and esophageal nerves has been observed for brain ns products while the secretory cells of the CC appear to release their products in the CA nerves (Clements *et al.,* 1985).

At the level of the last abdominal ganglion of *Periplaneta* and *Teleogryllus* a commissural ring nerve connecting both cercal nerves was described in females only. It contains axons with granular and agranular vesicles which stem from median bifurcating neurons. The latter may be homologous to octopaminergic DUM neurons (Pipa, 1988).

In eight dipterans from four different families, discrete nh structures were recently revealed in the walls of the pericardial system, also termed dorsal diaphragm. Nh branches are found throughout the metathorax up to the fifth abdominal segment. They originate from segmental nerves and send small branches to the heart that form neuroeffector junctions (Meola and Cook, 1986).

1.4.3.4. ANO

In *Rhodnius,* four pairs of small ganglia were seen to be segmentally arranged in abdominal segments 1–5 and innervated by a branch of the abdominal nerves (Davey and Kuster, 1981). They consisted of a pair of ns neurons surrounded by a few glial cells and connective tissue. They branched out in several directions and nh sites were seen containing isolated axons surrounded by a thin layer of connective tissue only (Kuster and Davey, 1981) (Fig. 8). These ganglia appear to be involved in stretch perception and the release of antigonadotropins (see Chapter 6). They resemble small ganglia located in the vicinity of the uterus in the tsetse fly (Robert *et al.,* 1984).

1.4.3.5. Brain Neurohemal Area

Finally, in the cricket *Acheta,* the ventral superficial part of the brain was found to have a nh structure (Geldiay and Edwards, 1973; Geldiay and Karaçali, 1980). In this insect, the ns axons penetrate between the perilemma cells up to the basement membrane, and synaptoid sites can be observed. The source cells of this nh cerebral region are supposed to be located in the brain and in the subesophageal ganglion. The axonal pathway of the medioventral ns cells of the subesophageal ganglion was traced to the brain through the circumesophageal connectives, both by *in toto* staining in *Teleogryllus* (Weinbörmair *et al.*, 1975) and by immunochemistry in *Acheta* (Strambi *et al.*, 1979) and *Locusta* (Rémy and Girardie, 1980). Moreover, sectioning of circumesophageal connectives induced noticeable changes in the nh area (Breitfuss *et al.*, 1980; Rémy and Girardie, 1980). The occurrence of such nh structures has seldom been reported, perhaps because of the lack of appropriate studies. A possibly similar structure was described in another orthopteran, *Poekilocerus pictus* (Prasad and Srivastava, 1984).

1.4.3.6. Link Nerve Neurohemal Areas

A widespread system of thin nh nerves was described in the stick insect (Finlayson and Osborne, 1968; Fifield and Finlayson, 1978) in the vicinity of certain peripheral ns cells. It is associated with somatic and sympathetic nerves, sometimes located near PO, often close to visceral organs like the heart or gut.

Besides ns products from the peripheral ns cells, the link nerve nh areas appear to contain ns products from the ventral nerve cord, as demonstrated in stick insects (Fifield and Finlayson, 1978). In these insects as well as in cockroaches and locusts (Baudry-Partiaoglou, 1987), four types of elementary granules were observed, one of which was also present in the peripheral ns cells. Using immunocytochemistry it was shown that the link nerve nh areas of *Carausius, Periplaneta,* and *Locusta* do contain ns products reactive to adipokinetic hormone (AKH) antiserum not found in the peripheral ns cells and glucagon-like substances produced in these cells (Raabe, 1986a).

1.4.3.7. The Neurohemal System of the Paired Medioventral Subesophageal Neurosecretory Cells

Very particular features are exhibited by the pathways of the medioventral ns cells of the subesophageal ganglion. Their axons run all along the ventral nerve cord, sending out lateral processes at the level of each nerve ganglion. These processes terminate at the proximal surface of the segmental nerves, thus determining metameric nh areas (Rémy and Girardie, 1980) which strongly recall structures described in the lobster (Beltz and Kravitz, 1986).

1.4.3.8. Serotonergic Networks

Recently, the utilization of antiserotonin antiserum revealed the existence of a large serotonergic plexus located at the periphery of the buccal and sympathetic nerves in several insects (see Section 1.6.8). A morphologically similar system was described in *Locusta,* formed by thin nerves originating in the subesophageal ganglion, which follow the mouth part nerves and branch extensively at their outer periphery (Bräunig, 1987).

1.4.4. Neuroeffector Junctions

A number of observations indicate that the ns products are released not only in nh organs and areas but also in the target tissues themselves, where ns terminals are located. These terminals are called neuroeffector junctions. Those in the CA and PG have inspired numerous studies, including early pioneering investigations (Arvy and Gabe, 1953; Normann, 1965; Scharrer, 1964a,b; Thomsen and Thomsen, 1969) and more recent research (Panov, 1986; Panov and Melnikova, 1986) (see Raabe, 1982).

In the CC, synaptoid and synaptic contacts were reported between brain ns fibers and glandular cells or their processes in *Calliphora* (Normann, 1965), *Periplaneta* (Scharrer, 1968), and locusts (Cassier and Fain-Maurel, 1970b; Cazal *et al.,* 1971). Neurosecretory terminals were also observed in many visceral muscles, where they probably intervene in the release of myotropic factors (see Chapter 7). They were reported to occur in the heart and associated muscles such as the alary muscles, the hyperneural muscle, and the diaphragm muscles. Neurosecretory terminals were also found in the foregut, midgut, and hindgut. Neuroeffector junctions were described in the salivary glands, Malpighian tubules, oviducts (Kiss *et al.,* 1984; Robert *et al.,* 1984; Orchard and Lange, 1985), spermatheca, light organs, and termite sternal glands. Even somatic and skeletal muscles such as the ventral intersegmental muscles of *Schistocera, Carausius,* and *Phormia* (Osborne *et al.,* 1971) contain neuroeffector junctions.

Histology and electron microscopy revealed that in a number of cases two or more distinct ns products are involved in neuroeffector junctions. In electron microscopy, one of them displays the appearance of peptidergic substances, while the other resembles the small dense-cored aminergic vesicles. As regards the peptides, immunocytochemical investigations demonstrated that at least some of the axon terminals in the hindgut of *Periplaneta* are proctolin-like (see Chapter 2).

The question of whether the neuroeffector terminals make synaptic contact with the efferent cells has not really been answered. Peptidergic synapses have been described in heart muscle of *Periplaneta* (Johnson, 1966), rectal pads (Gupta and Berridge, 1966), uterine muscles (Robert *et al.,* 1984), and in some other organs, but ns terminals were also observed to be in contact with the connective sheath of certain organs.

The significance of the presence of neuroeffector junctions in addition to nh organs and areas is difficult to understand. When ns products are released in contact with connective tissue, it may be supposed that the active substances move around in the lacunae of this connective tissue and come within reach of a number of cells. Such a humoral release in the vicinity of target cells would allow reduction of the amount of substance necessary to transmit a message, since no dilution takes place in the hemolymph, and would speed up the transmission of information. Another explanation that might also be proposed is that the substances released at neuroeffector junctions act as neuromodulators.

As was just mentioned, the ns terminals in various organs might have a modulatory role. This suggestion is also valid for a number of swollen terminals loaded with ns products which have been described in the neuropil of the brain and the ventral nerve cord ganglia in numerous species. These terminals do contain electron-dense elementary granules, dense cored vesicles, and immunoreactive peptides (see Chapter 2). Among the collaterals described, it was shown that met-enkephalin collaterals occur in *Locusta* at the periphery of the tractus of the NCCI and NCCII (Rémy and Dubois, 1981) and that proctolin-like terminals are abundant in the neuropil of the terminal ganglion of *Periplaneta* (Eckert *et al.*, 1981). In the brain of *Leptinotarsa,* the relationships of the ns processes with other neurons appear to be of two kinds: first, true synapses observed between the ns processes and presynaptic neurons, the ns branches receiving presynaptic inputs, and second, synaptoid sites formed by the ns processes; the formation of the latter sites is suggestive of release processes and also suggests that ns products exert some kind of action, probably modulatory, in the central nervous system (Schooneveld, 1974b).

1.4.5. Embryonic Formation of Perisympathetic Neurohemal Areas

The mechanisms involved in the embryonic development of a nh structure were thoroughly investigated in *Manduca* (Taghert *et al.*, 1986) (Fig. 15) using various monoclonal antibodies; single cells were also dye-filled with microelectrodes containing Lucifer yellow and processed with an anti-Lucifer yellow antiserum.

It appears that the nh transverse nerves of the larva, which includes the arborization of both ns cells of the anterior ganglion and peripheral ns cells, are formed by the outgrowth of neuronal cones guided by nonneural elements, i.e., glial and tracheal precursors which constitute a continuous pathway prefiguring the future transverse nerve. Not only do the future glial elements orientate the direction of the growing axons, but they also appear to substitute for targets. It is possible that they provide signals to the axons to stop growing and to produce numerous branches.

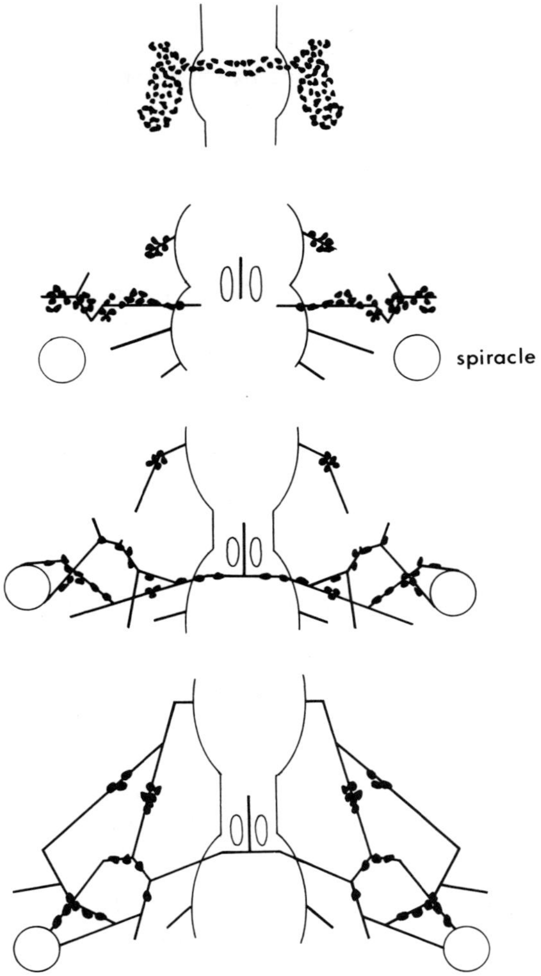

FIGURE 15. Schematic representation of the developmental pattern of a neurohemal structure in *Manduca.* Development stage from top to bottom: 32%; 43%; 57%; 61%; 90%. (Modified from Taghert *et al.,* 1986.)

1.4.6. Discussion

The organization of the nh system of insects, which is diversified and widespread, may result from an evolution linked to the rudimentary circulatory system and to the dispersion of the ns cells.

It is possible that a single neurohormone is released at several distinct sites, as proposed for crustaceans, mollusks, and vertebrates. In insects, the axons from certain ns cells of the PI were demonstrated to send out a process toward the CC and another toward the ventral nerve cord. Moreover, the axons of the median ns cells of the subesophageal ganglion follow the ventral nerve cord and branch in every segmental nerve; further processes from the brain ns cells were observed to leave the CC in sympathetic or aortic nerves, although certain of their branches probably terminate in the nh part of the CC.

The PO display marked ultrastructural duality which suggests different modes of release. The diffusion of ns products in the β-PO follows the usual pattern via the connective tissue, which is in close contact with the hemolymph. In the α-PO, on the other hand, the ns products certainly migrate in narrow spaces delimited by a dense glial network. These opposed modes of release might be linked to the chemical nature of the substances produced and on their lability. Thus, certain neurohormones would be released directly into the general circulation in the β-PO while others would remain for a time in the internal medium of the nervous tissue where they would undergo certain structural modifications and eventually link up with special extracellular carrier proteins. Another possibility is that in the α-PO, presynaptic feedback is of importance, although such feedback might also take place in the neuroeffector junctions but not in the β-PO where neurohormones are dispersed in the circulatory flux.

The multiplication of release sites is certainly an advantage when the circulatory system is not well developed. The closeness of certain release places to certain targets, which is the case for the PG, CA, and the fat body, is also certainly advantageous. In crustaceans too, several release sites are known: they comprise the sinus gland homologous to the CC, the pericardial organs, the postcommissural organs, and the anterior ramifications. In vertebrates, the nh organs display various relationships with the general circulation. The neurohypophysis releases its ns products directly into the general bloodstream whereas the blood pathway of the neurohormones from the median eminence is quite short and in the adrenal glands the blood irrigating the cortex originates directly from the medulla. Lastly, in teleosts, an important nh organ, the urophysis is present in the caudal part of the body. As far as the neuroeffector junctions are concerned, their location in the anterior and intermediate lobe of the hypophysis of nonmammalian vertebrates is well known.

1.5. REGULATION OF NEUROHORMONE PRODUCTION AND RELEASE

The neurohormones regulate both the glandular activity of the PG and CA and a number of physiological processes which must be precisely adjusted to the external conditions and the physiologic state of the insects. In order to fulfill this task, the ns cells display, through their collaterals, a large number of direct or indirect connections with various sensory organs.

1.5.1. Environmental Factors

Temperature, hygrometry, and the availability of food are of importance, but photoperiod and the intensity of the light are the most important factors involved in the regulation of ns activity.

Investigations in cockroaches reveal the existence of connections between the sensory organs and the brain ns centers: ocellar receptors and ns cells of the PI or the pars lateralis, antennal receptors and deutocerebral ns cells, tegumentary photoreceptors, labrum receptors, and deuto- and tritocerebral ns cells. Furthermore, by surgical procedures the action of sensory receptors on the synthesis and the release activity of brain ns cells was confirmed (Brousse-Gaury,

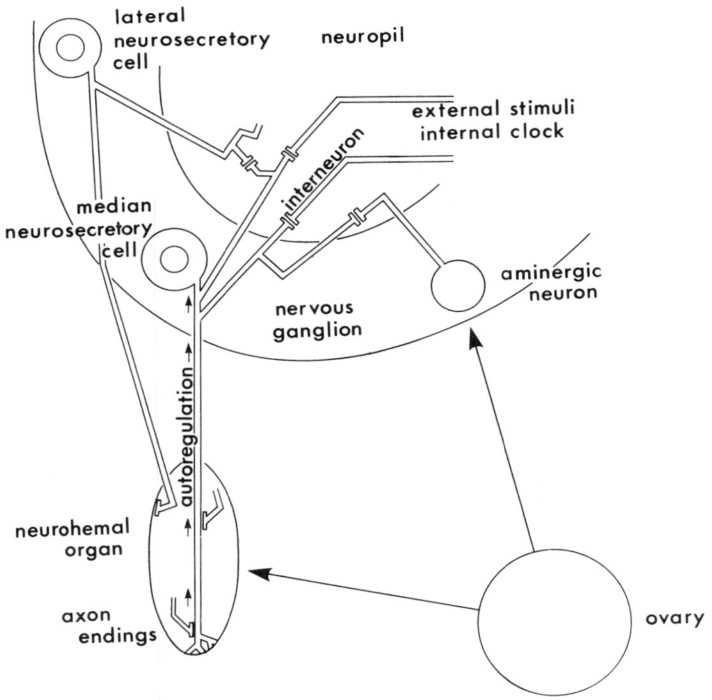

FIGURE 16. Schematic representation of the regulation of the activity of a neurosecretory cell.

1971a,b). In other insects, interneurons from the ocelles and optic lobes have been shown to terminate in the region of arborization of the PI ns cells (Koontz and Edwards, 1980) (Fig. 16).

It was also suggested that the brain itself could act as a photoreceptor. Indeed paracrystalline structures formed by closely packed microtubules were observed in the glial cells of the *Pieris* brain (Kono *et al.*, 1983; L'Helias *et al.*, 1984). They might represent the brain extraocular receptors involved in diapause and morph determination whose existence was suggested in a number of experiments (Williams and Adkisson, 1964; Claret, 1966; Steel and Lees, 1977). Similar rhabdome-bearing structures were described in the CC of a cricket (Geldiay and Karaçali, 1983) where they were found making synaptic contact with ns axons.

1.5.2. Internal Factors

Internal inputs originating from tension receptors are at the inception of many physiological processes regulated by neurohormones, such as diuresis following a blood meal in blood-sucking insects or egg laying after mating. The larva deposition in the tsetse fly when it reaches its final size is dependent on internal receptors, and so is the ability to molt observed in *Manduca* larvae when they reach a certain weight. Proprioceptors too are involved in brain ns cell activation or inhibition during flight, which causes first the lowering of hemolymph trehalose and then the release of AKH from the CC, which in turn induces the release of diglycerides from the fat body.

1.5.3. Circadian Activity—Autoregulation

Neurosecretory cell functioning appears to be governed by a circadian rhythm in some insects. Experiments conducted in locusts further showed that insect ns cells regulated their own activity in accordance with the load carried by their terminals. Thus, they are able to modify their synthetic rhythm, the speed of transport of ns material, and the catabolic processes in response to terminal information (Girardie and Girardie, 1977).

1.5.4. Neurosecretory Cell Interrelationships (Fig. 16)

In some insect species, the arborizations of collateral branches of the axons contained in the NCCI and II bundles have been demonstrated to overlap in certain parts of the brain, which indicates a possible relationship between the two ns regions of the brain. However, the ns cells also synapse with interneurons from their own nerve center or from other central nervous ganglia. Numerous arborizations are observed in the ventral nerve cord ganglia, whose origin is in the ganglia themselves, in other ganglia, or in the brain. Connections between the sympathetic nervous system and the brain were demonstrated by disconnection or removal of the frontal ganglia, which disturbed ns activity.

1.5.5. Substances Involved

As regards the molecules involved in controlling ns cell activity, the existence of cholinergic and, more especially, aminergic contacts was demonstrated. The amount of ns products stored in the ns cells was shown to be influenced by treatments with amine antagonists such as reserpine and disulfiram in *Galleria* (Warton, 1981), and it was experimentally shown in *Rhodnius* that the ovarian ecdysone feedback involves brain aminergic interneurons (Orchard *et al.*, 1983c). Furthermore, the release of the AKH produced in the glandular cells of the CC was demonstrated to be dependent on octopaminergic axons of the NCCII (Orchard and Loughton, 1981).

Besides the cholinergic and aminergic regulation of ns cell activity, it is highly probable that there are also peptidergic forms of regulation, since electron microscopy revealed electron-dense, granule-loaded terminals that synapsed onto ns axons, and immunocytochemistry demonstrated the presence of several peptides in the neuropil of brain and ventral nerve cord ganglia.

1.5.6. Control of the Release of Neurosecretory Products

Neurosecretory product transport and release mechanisms are known to result from the electrical activity of the ns cells and are mimicked by electrical stimulation (see Section 1.2.14). Neurosecretory product release can also be controlled at ns terminals; acetylcholine, biogenic amines, and peptides are involved in this local regulation. Acetylcholine was shown to enhance the action potentials in the PO of *Carausius* and *Rhodnius* (Finlayson *et al.*, 1976) and octopamine was shown to control the release of AKH from the CC (see Chapter 11). In higher vertebrates as well, hypothalamic acetylcholinergic, GABAergic, and dopaminergic axons which terminate in the median eminence make axoaxonic synapses onto the ns fibers from the hypothalamus and might control the release of its ns products.

1.5.7. Hormonal Feedback

The feedback effects of both ecdysone and juvenile hormone (JH) on the functioning of the ns cells were revealed histologically, experimentally, and by electrical recording.

Ecdysone often modifies PTTH release from the cell bodies, for instance, in the cockroach brain *in vitro* (Marks *et al.*, 1972) or in the brain of *Mamestra in vivo* or *in vitro* (Agui and Hiruma, 1977a,b). Extirpation of the PG also causes changes in certain brain ns cells and in type A ns cells of the ventral nerve cord in *Aeshna* (Charlet *et al.*, 1974). In addition, ecdysone acts on brain ns cells by triggering the release of both the eclosion hormone in *Manduca* (Truman, 1981) and the ovulation hormone in *Rhodnius* (Ruegg *et al.*, 1981, 1982).

As regards JH, injection of this hormone or of its analogs was followed in a great many species by histological modifications in the ns cells of the PI. Juvenile hormone stimulated the activity of PI ns cells, an effect substantiated by the nuclear enlargement observed in *Calliphora* (Thomsen and Lea, 1969) and by load changes of type A ns cells in adult locusts (McCaffery and Highnam, 1975a,b). Conversely, allatectomy led to a decrease in the amount of ns products stored in the ns cells in both species (see Chapters 4 and 6).

An interesting point which deserves stress is that the effects of JH differ according to the developmental stage of the insect treated. This was observed, for example, in *Calliphora* (Scheller and Bodenstein, 1981) where the effects of both JH and ecdysone on RNA and protein synthesis in the ns cells depend on the age of the larva.

Both JH and ecdysone therefore exert negative or positive feedback on the ns cells. Their mechanisms of action are still not really clear. It is often considered that they modulate the electrical activity at the terminals. However, other possible sites of action should also be considered, including the cell body receptors of the ns cell themselves or those of the aminergic cells regulating their activity (Orchard *et al.*, 1983c).

The demonstration that feedback mechanisms exist in insects is not surprising, considering the situation in high vertebrates in which, in addition to the regulation exerted by internal and environmental inputs, several feedback loops are known to occur involving the ns cells of the hypothalamus, the glandular pituitary cells, and the target glands or tissues.

1.6. AMINERGIC NEURONS

Very thorough investigations were conducted for the detection of biogenic amines in the nerve centers and nh organs of insects. Since these molecules appear to act more frequently as neurotransmitters or neuromodulators than neurohormones, the studies devoted to them will be briefly outlined, except for the recent ones. Additional data may be found in a number of reviews (Klemm, 1976; Evans, 1985; Pitman, 1985).

1.6.1. Methods (Table VII)

The methods used to investigate biogenic amines have progressed rapidly over the last two decades. The first modern technique to be extensively used was the Flack and Hillarp fluorescence method, which allows catecholamines to be

TABLE VII. Methods Used in Amine Identification

	Catecholamines			Phenolamine	Indolamine
Methods	Epinephrine	Norepinephrine	Dopamine	Octopamine	Serotonin
Falck and Hillarp	+	+	+	0	+
Glyoxidic acid	+	+	+	0	+
Immunofluorescence					+
Neutral red		+	+	+	+

distinguished from indolamines since the latter exhibit light yellow fluorescence while catecholamines display bright green fluorescence. The histochemical glyoxylic acid fluorescence method applied subsequently revealed both catecholamines and indolamines, the latter more distinctly than with the Falck and Hillarp method. Autoradiography and radioenzymology, which are more sensitive, are also used for amine detection. To some extent electron microscopy permits terminals containing amines to be detected from the appearance of the vesicles (often dense-cored). One technique which is fairly simple but requires confirmation by other methods is *in toto* detection with neutral red. Most sensitive are the immunocytochemical techniques. At the present time only antiserotonin antisera have been prepared (Steinbusch *et al.*, 1978), and their sensitivity is 1000 times greater than that of the Falck and Hillarp method.

To conclude, it must be mentioned that treatments of insects with reserpine cause disappearance of the biogenic amines and thus allow verification of the aminergic nature of the substance studied.

1.6.2. Occurrence in the Central Nervous System

The early studies using the Falck and Hillarp fluorescence technique revealed the presence of catecholamines and indolamines in the central nervous system, chiefly in the neuropilar structures of the brain (Plotnikova and Govyrin, 1966; Frontali, 1968; Mancini and Frontali, 1970; Klemm, 1968; Klemm and Axelsson, 1973; Klemm and Falck, 1978). These results were confirmed by biochemical studies demonstrating the occurrence of the main biogenic amines in the central nervous system (Gersch *et al.*, 1961; Colhoun, 1963; and many others) (Fig. 17). At the same time, electron microscopy showed the presence of microgranular neurons in the PI of several insect species, i.e., flies (Bloch *et al.*, 1966; Ramade and L'Hermite, 1971), locusts (Girardie and Girardie, 1966), crickets (Geldiay and Edwards, 1973), and the coleopteran *Leptinotarsa* (Schooneveld, 1974a).

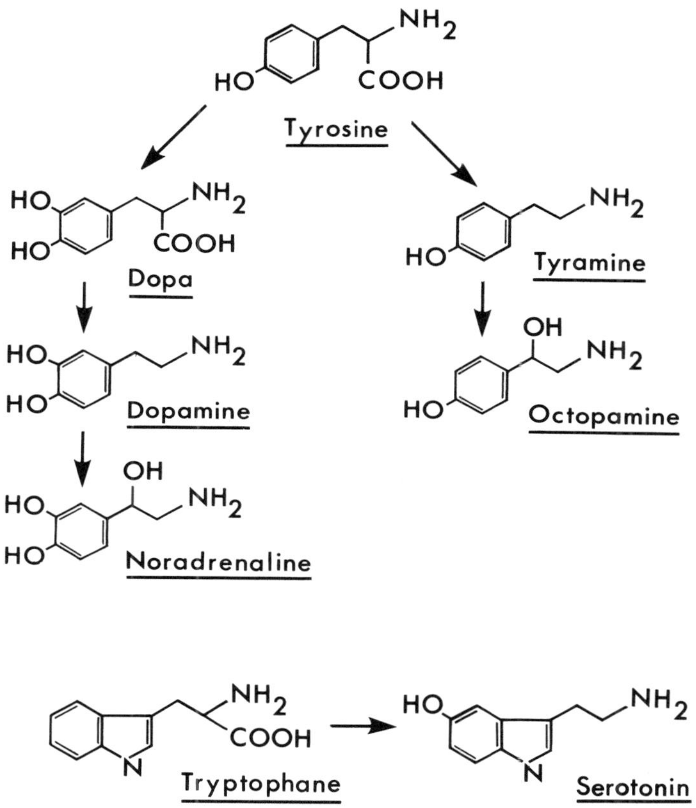

FIGURE 17. Main biogenic amines and their synthetic pathways.

1.6.3. Quantitative Determinations

Quantitative determination of the different biogenic amines in the main parts of the central nervous system (Table VIII) allows the following conclusions to be drawn: norepinephrine, dopamine, octopamine, and serotonin are present in the entire central nervous system but the amounts of norepinephrine are small, as observed in the early investigations (Frontali and Häggendal, 1969; Hiripi and S-Rozsa, 1973). Relatively large amounts of dopamine and octopamine were revealed throughout the central nervous system and particularly in the brain, but dopamine was only determined in three insect species, i.e., *Periplaneta, Apis,* and *Calliphora,* whereas octopamine was investigated in seven species.

The high dopamine and octopamine content of the brain is probably due to

TABLE VIII. Distribution of Biogenic Amines in the Central Nervous System and Neurohemal Organs of Several Insect Species[a,b]

Insect	Norepinephrine	Dopamine	Octopamine	Serotonin	Reference
Periplaneta					Evans, 1978; Dymond and Evans, 1979
brain	5.53	26.81	14.68		
op	0.36		7.24		
sgg	0.33	3.29	5.60		
th2	0.23	2.25	4.33		
a_1	0.23	1.12	1.56		
CC	0.11	0.19	1.20		
PO	0.14	0.33	0.07		
Schistocera					
brain			11.58		
op			8.58		
sgg			4.7		
th2			6.23		
a_1			1.04		
CC			0.47		
thPO			0.25–0.15		
aPO			0.05		
thconn			0.70		
aconn			0.36		
Locusta					David and Lafon-Cazal, 1979
brain			40.41		
op			14.53		
sgg			6.59		
th3			6.05		
hypogg			2.38		
CC nh			0.47		
CC gd			0.95		

Apis							Mercer *et al.*, 1983
brain	1.12	24.96		7.62		21.37	
op	0.48	7.56		2.49		9.27	
antl		0.75		0.48		3.03	
sgg		5.28		0.16		5.56	
Mamestra			Larva	Pupa	Adult		Bodnaryk, 1980
brain				0.45	12.6		
op					6.6		
Manduca							Davenport and Wright, 1986
brain			7.56				
protoc				9.01	11.34		
op				2.94	8.53		
antl				2.19	2.55		
sgg			2.09	2.40	3.67		
th3			3.43	6.05	9.22		
a_6			3.06	5.42	5.79		
Spodoptera							Davenport and Wright, 1986
brain			1.51				
sgg			0.97				
th3			0.99				
a_6			1.04				
Calliphora							Nässel and Laxmyr, 1983
brain		2.4		4.0		1.8	
op		0.92		1.6		0.78	
vgg		1.4		1.3		1.1	

[a]The data are expressed in pmol/tissue.

[b]a, abdominal ganglion; aconn, abdominal connectives; antl, antennal lobes; aPO, abdominal perisympathetic organs; CC, corpora cardiaca; CCgd, glandular part of the corpora cardiaca; CCnh, neurohemal lobe of the corpora cardiaca; hypogg, hypocerebral ganglion; op, optic lobes; PO, perisympathetic organs; protoc, protocerebrum; sgg, subesophageal ganglion; th, thoracic ganglion; thconn, thoracic connectives; thPO, thoracic perisympathetic organs; vgg, ventral ganglia.

the fact that it is the major coordination center in insects, and also that it is larger than the other nerve centers.

The octopamine concentration in the optic lobes is high and clearly exceeds the concentration in the antennal lobes. The optic lobes also contain a high level of serotonin (Maxwell *et al.*, 1978 in *Manduca*). The connectives in the ventral nerve cord contain a nonnegligible amount of octopamine.

Norepinephrine, dopamine, octopamine, and serotonin were demonstrated in the CC and PO. In the CC octopamine predominates, while in the PO dopamine appears to be most abundant. In the locust CC, the glandular lobe is richer in these substances than the nh lobe but both lobes contain 10 times less of them than the brain. The glandular lobe contents of dopamine, octopamine, and serotonin for males and females are 26–86, 46–74, and 217–173 pmol tissue, respectively, so that indolamine thus appears to have important functions in this organ (Orchard *et al.*, 1986). The hypocerebral sympathetic ganglion also contains large amounts of serotonin in *Locusta*.

The comparative study of pupae and freshly eclosed adults of *Mamestra* and *Manduca* shows that, contrarily to the other parts of the nervous system, the octopamine content of the optic lobes increases markedly after eclosion.

The data just examined are those expressed in pmol/tissue; other results, expressed in pg/mg, gave values of the same order of magnitude for the octopamine content of the central nervous system of *Periplaneta* (1300 pg/mg) (Dymond and Evans, 1979), *Schistocerca* (2430) (Robertson and Juorio, 1976), *Mamestra* (2800) (Bodnaryk, 1980), *Spodoptera* (2311) and *Manduca* (2596) (Davenport and Wright, 1986).

1.6.4. Distribution in Central Nervous System

Numerous studies devoted to the distribution of biogenic amines in the neuropils and neurons of the nerve centers were carried out by the Falck and Hillarp method. The glyoxylic acid method was also used and the presence of serotonin was investigated by immunocytochemistry.

1.6.4.1. Serotonin

The distribution of the serotonergic neurons reveals certain characteristics (Table IX) shared by phylogenetically distant species (dragonflies, cockroaches, flies, and bugs), such as a large number of serotonergic neurons in the optic lobes, and the presence in all thoracic and abdominal ganglia of two pairs of large lateral neurons (Fig. 18); more pairs occur in the subesophageal and last abdominal ganglia, which both result from the fusion of several ganglia. These neurons, which are located in the same position as the proctolinergic neurons, are, however, distinct from the latter (Bishop and O'Shea, 1983). Median clusters of serotonergic neurons are occasionally observed in the subesophageal

TABLE IX. Distribution of Serotonin and Catecholamine Neurons and Fibers: A Few Recent Studies[a]

Insect	Brain	Sgg	Ventral nerve cord					Miscellaneous	Methods	Reference
			th1	th2	th3 a_1	a	last a			
					Serotonin					
Dragon flies		8ln	4ln	4ln	8ln	4ln, mcl	12ln			Longley and Longley, 1986
Periplaneta	opnpil								immuno, FH	Nässel and Klemm, 1983
	opn, opnpil, prn, prnpil, dn dnpil	8ln, mcl	4ln	4ln	10ln	4ln	12ln	conn	immuno	Bishop and O'Shea, 1983
Locusts and crickets	opn, opnpil						mn++	80frggn, npil	immuno, EM	Takeda *et al.*, 1985; Klemm *et al.*, 1986
										Hustert and Topel, 1986
										Elekes *et al.*, 1987
Acheta							4ln, 4ln+		immuno	
							12ln, mn		immuno	
Schistocerca	opnpil						6–8mn++		immuno, FH	Nässel and Klemm, 1983
Apis	2dn	12ln								Rehder *et al.*, 1987
Calliphora	26mn	22ln	6ln	4ln	4ln ←	4ln × 7	→		immuno	Nässel and Cantera, 1985
Rhodnius		6ln, 4mn	4ln ←		8ln		→		immuno, FH	Flanagan, 1986
					Catecholamines					
Cockroaches	opn, opnpil, prnpil, dn, dnpil							frggn, npil	FH	Klemm, 1983
Acheta							6ln		immuno	Elekes *et al.*, 1987
Locusts	sev 100 n								FH	Klemm and Falck, 1978
Locusta	opn, nsc, prn, prnpil								immuno	Vieillemaringe *et al.*, 1984
Bombyx	pin, pln	n								Takeda *et al.*, 1986
Apis	prn, dn, tn, npil	n							FH	Mercer *et al.*, 1983
Rhodnius	75opn, 44pn, 48dtn	3mn, 4ln	3mn ←		1mn, 12ln				gly ac	Flanagan, 1984

[a]a, Abdominal ganglion; an, abdominal neurons; ane, abdominal nerves; conn, connectives; dn, dnpil, deutocerebral neurons, neuropil; E, electron microscopy; FH, Falck and Hillarp method; frggn, npil, frontal ganglion neurons, neuropil; gly ac, glyoxilic acid method; hyggn npil, hypocerebral ganglion neurons, neuropil; last a, last abdominal ganglion; ln, lateral neurons; medne, median nerve; man, median neurons; mcl, median cluster; mn, median neurons; n, neuron; ne, nerves; npil, neuropil; opn, opnpil, optic lobe neurons, neuropil; pin, pars intercerebralis neurons; pln, pars lateralis neurons; prn, prnpil, protocerebral neurons, neuropil; sev 100, several hundreds; Sgg, subesophageal ganglion, th1, 2, 3, prothoracic, mesothoracic, metathoracic ganglion; tn, tnpil, tritocerebral neurons, neuropil; +neurons innervating the genital organs; ++neurons innervating the hindgut; °dopaminergic neurons.

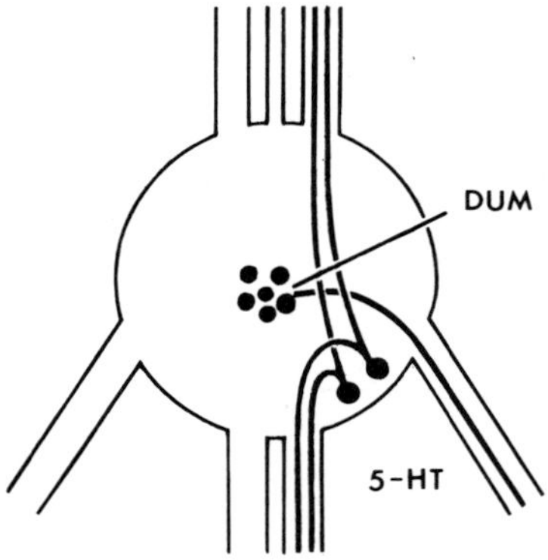

FIGURE 18. Schematic representation of the distribution of serotonin and octopamine neurons in an abdominal ganglion. (From the data of Longley and Longley, 1986 in dragonflies; Smalley, 1970; Bishop and O'Shea, 1983 in *Periplaneta;* Hustert and Topel, 1986 in *Acheta.)*

ganglion or in the last abdominal ganglion. In the brain, numerous serotonergic neurons were described in *Periplaneta* and *Calliphora,* but the main staining was observed in the neuropil structures such as the calyces of the mushroom bodies. It shall be noted that in the honeybee a large deutocerebral serotonergic neuron runs into the subesophageal ganglion and further in the ventral nerve cord (Rehder *et al.*, 1987).

Except for two pairs of lateral motoneurons innervating the genital organs and for six to eight median motoneurons innervating the hindgut, the serotonergic neurons of the ventral nerve cord are interneurons which branch profusely in the ganglionic neuropil (Elekes *et al.*, 1987) and travel along the connectives. They are not observed in the peripheral nerves, except the buccal nerves, however, which contain serotonergic fibers which branch off and create extensive peripheral networks (see Section 1.6.8).

A detailed study of the postembryonic development of serotonin-immunoreactive neurons in both optic lobes and ventral nerve cord have been performed in *Calliphora* (Ohlsson and Nässel, 1987; Cantera and Nässel, 1987). It shows that all the adult serotonin-like neurons derive from larval neurons but that the thoracic larval processes regress at metamorphosis while new, more extensive processes develop.

1.6.4.2. Octopamine

The localization of octopamine neurons is mostly based on quantitative determination because there are no other methods for their detection, although it is possible to estimate the number of octopamine neurons by substracting the number of neurons stained by histochemical fluorescence from the number stained by neutral red. The latter method revealed the presence in the ventral nerve cord of locusts, cockroaches (Evans and O'Shea, 1978), and the lepidopterans *Spodoptera* and *Manduca* (Davenport and Wright, 1986) of clusters of dorsomedian unpaired neurons, the DUM. The octopaminergic neurons are also found among the globuli cells of the brain, which are important interneurons whose branches constitute the calyces of the mushroom bodies (Dymond and Evans, 1979 in cockroaches; Mercer *et al.*, 1983 in bees). These data agree with the quantitative data which showed the presence of octopamine all along the ventral nerve cord. Contrarily to serotonin, the octopaminergic neurons of the ventral nerve cord innervate peripheral target organs.

1.6.4.3. Catecholamines

According to Klemm and Falck (1978), several hundreds of catecholamine neurons are located in the PI of locusts, a result confirmed by an immunocytochemical study (Vieillemaringue *et al.*, 1984) which shows that the axons of these catecholaminergic neurons are connected with the CC. A number of other catecholaminergic neurons were also described, for instance, in the cluster of median subocellar ns cells, around the mushroom bodies, in the optic lobes, and in the deutocerebrum. Intense fluorescence of the neuropils was frequently reported in the protocerebrum, optic lobes, and deuto- and tritocerebrum (Mercer *et al.*, 1983; Vieillemaringe *et al.*, 1984).

Although catecholaminergic neurons are probably mostly interneurons, it has been observed that they send axons through the NCCI (to the CC nh lobe). They have also been observed to innervate both the subesophageal ganglion and the frontal ganglion (Klemm, 1983 in cockroach).

Double staining of the brain of *Locusta* with paraldehyde fuchsin, histochemical fluorescence, and immunocytochemistry using an antidopamine antiserum revealed that dopamine was colocated in type A ns neurons of the PI (Vieillemaringe *et al.*, 1982, 1984). In *Bombyx* larvae and pupae dopaminergic immunoreactive neurons were found in the PI, pars lateralis, and subesophageal ganglion. The dopamine-immunoreactive neurons of the subesophageal ganglion were shown to be immunoreactive to α-endorphin (Takeda *et al.*, 1986).

In the abdominal ganglia of the ventral nerve cord of *Periplaneta*, dopaminergic neurons were found in three median clusters composed of 24 neurons whose processes terminated in the PO (Smalley, 1970). In *Rhodnius* (Flanagan, 1986) incubation of the ventral nerve cord with dopamine made it possible to recognize a number of median dopaminergic neurons in the prothoracic ganglion and thoracoabdominal nerve mass. Branches of these neurons are present in the neuropil, interganglionic connectives, and thoracic and abdominal nerves.

1.6.5. Sympathetic Nervous System

The presence of amines has been shown in the sympathetic nervous system in locusts, trichopterans, cockroaches, and several other species. These amines are assumed to be catecholamines (Chanussot and Pentreath, 1973) and serotonin (Klemm, 1983).

The aminergic terminals found in the digestive tract probably originate in aminergic neurons of the sympathetic ganglia; they consist of catecholamines in *Schistocerca* (Klemm, 1972) and serotonin in locusts and crickets (Klemm *et al.*, 1986), *Blaberus* (Andriès and Tramu, 1985b), and *Chironomus* (Johansson *et al.*, 1986).

It was recently demonstrated that the sympathetic nerves constitute a nh plexus which extends along the digestive tract in *Chironomus* (Johansson *et al.*, 1986), locusts, and cockroaches (Klemm *et al.*, 1986). In locusts and crickets, serotonergic fibers from the frontal ganglion were also observed to innervate the Malpighian tubules, salivary glands, and CC. Similar large serotonergic plexuses were observed in the periphery of the mouth part nerves (see 1.6.8). Their physiological significance is not yet understood.

1.6.6. Corpora Cardiaca

The occurrence of biogenic amines in the CC was described some years ago. At the time, these amines were said to be serotonin and dopamine (Klemm, 1971; Lafon-Cazal *et al.*, 1973; Lafon-Cazal and Arluison, 1976; Lafon-Cazal, 1976). An exhaustive study of both the glandular and nh lobes of the CC of locusts allowed a parallel to be established between the fiber types distinguished by electron microscopy and the amine locations revealed by autoradiography. The nh lobe axons, which contain dense granules 100 nm in diameter, appear to incorporate large amounts of tritiated serotonin (Lafon-Cazal and Arluison, 1976) and smaller amounts of dopamine. They also contain some octopamine, but the glandular lobe contains much more (David and Lafon-Cazal, 1979; Goosey and Candy, 1982; Orchard *et al.*, 1986). In this lobe, octopamine plays a major role in the release of AKH produced in the glandular cells (see Chapter 11). Quantitative determination by a radioenzymatic method confirms the presence of octopamine and slight traces of dopamine in the CC of the locust (Goosey and Candy, 1982). Moreover, high-performance liquid chromatography (HPLC) determination of the dopamine, octopamine, and serotonin content of the glandular lobe of the CC and of the NCCII showed that serotonin and octopamine were both present in the NCCII and glandular lobe, which also contained dopamine.

1.6.7. Perisympathetic Organs

In the PO, as in the CC, aminergic axons have been revealed. The PO, the median nerve, and their cells of origin were shown to incorporate tritiated dopamine (Smalley, 1970). It was demonstrated then by a radioenzymatic assay that the PO of *Periplaneta* and *Schistocerca* contain a certain amount of octopamine whose concentration was higher in the mediothoracic PO than in the abdominal PO (0.25 versus 0.05 pmol/organ); the amine content of the CC was 0.47 pmol/organ and that of the brain 11.6 pmol/organ. Dopamine and norepinephrine were also detected; their mean values in the cockroach were respectively estimated at 0.33 and 0.14 pmol in one pair of abdominal PO (Evans, 1978; Dymond and Evans, 1979).

Immunocytological studies in *Rhodnius* (Flanagan, 1984) showed serotonin-containing fibers projecting into the PO. The latter revealed a high-affinity serotonin uptake and responded to depolarization by releasing a labeled serotonin.

1.6.8. Serotonergic Networks

A serotonergic network of fibers was recently discovered by immunocytochemistry on the surface of nerves in *Periplaneta* (Davis, 1985, 1987) and *Calliphora* (Nässel and Elekes, 1985). In *Periplaneta,* it was identified at the periphery of the mouth part nerves. In *Calliphora* it extends to the labrofrontal nerves, the cervical connectives, cervical nerves, and prothoracic nerves. It is also present in the neural sheath of the thoracoabdominal ganglion.

As regards the cell bodies from which these extended plexuses originate, it appears from the study in *Calliphora* that they are restricted to only four somata located in the subesophageal ganglion. These somata also send processes into the neuropil.

The pathway of the serotonergic fibers is a very complicated one, as demonstrated in *Periplaneta* (Fig. 19). The axons which originate from neurons of the subesophageal ganglion extend through the mandibular and link nerves to the maxillary nerve, through which they return to the ganglion and then extend up through the circumesophageal connectives to the tritocerebrum and out again to the nh system of the labral nerve. All along their pathway the serotonergic axons give rise to secondary branches which form the superficial nh meshwork. In *Rhodnius,* too, serotonergic fibers travel along the mandibular nerves (Flanagan 1986).

1.6.9. Mode of Action

The implication of cAMP as a second messenger of biogenic amine action and the role of specific adenylate cyclases was investigated in *Locusta* and *Periplaneta*. The role of cAMP was established in both species, and the amount in the CC was observed to increase after electrical stimulation or incubation with octopamine (Orchard *et al.*, 1983a,b), dopamine, and serotonin (Downer *et al.*, 1984). Further, an increase in adenylate cyclase was detected in the nervous

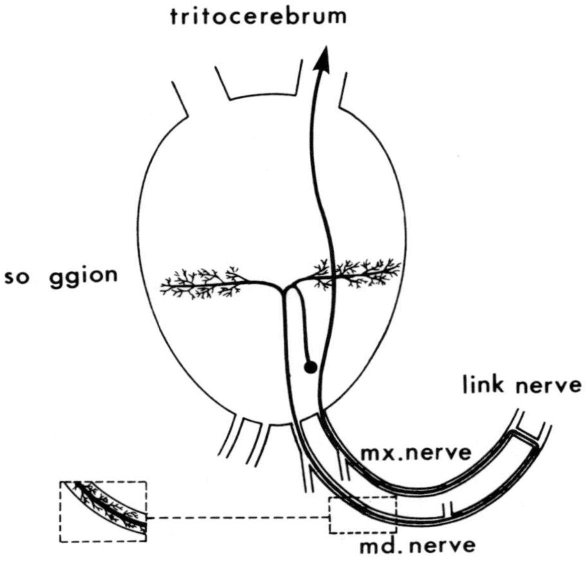

FIGURE 19. The serotonergic network demonstrated in *Periplaneta.* (From the data of Davis, 1985, 1987.)

tissue exposed to biogenic amines, e.g., in the thoracic ganglia of *Periplaneta* (Nathanson and Greengard, 1973), the brain and the subesophageal ganglion of *Mamestra* (Bodnaryk, 1979, 1980), and the ventral nerve cord of *Manduca* (Taylor *et al.,* 1976). Specific adenylate cyclases and receptors were identified for the main amines (Pannabecker and Orchard, 1986b). Octopamine receptors in particular were thoroughly investigated (see Bodnaryk, 1982; Evans, 1985).

1.6.10. Conclusion

Aminergic neurons are widespread in the central and sympathetic nervous systems of insects and it has been demonstrated that dopamine is present in PI ns cells. Most catecholamine- and serotonin-secreting neurons are interneurons involved in neurotransmission or neuromodulation; octopaminergic neurons mostly intervene in peripheral regulation.

The occurrence of biogenic amines in the nh organs PO and CC raises the question of their function. They may simply be released as neurohormones but they may also regulate the release of neurohormones from the ns terminals. In *Rhodnius,* the kinetics of the release of serotonin was shown to be characteristic of nh release, thus supporting the second possibility (Flanagan and Berlind, 1984).

The data concerning invertebrates other than insects also suggest that the biogenic amines dopamines, norepinephrine, octopamine, and serotonin occur ubiquitously in the nervous system and nh organs (see David and Coulon, 1985; Walker, 1984). As far as their physiological significance is concerned, both octopamine and serotonin were demonstrated to exert both central and neurohormonal effects in crustaceans (see Beltz and Kravitz, 1986). Octopamine is involved in muscle contraction, especially of the heart, in mollusks, crustaceans, and other invertebrates.

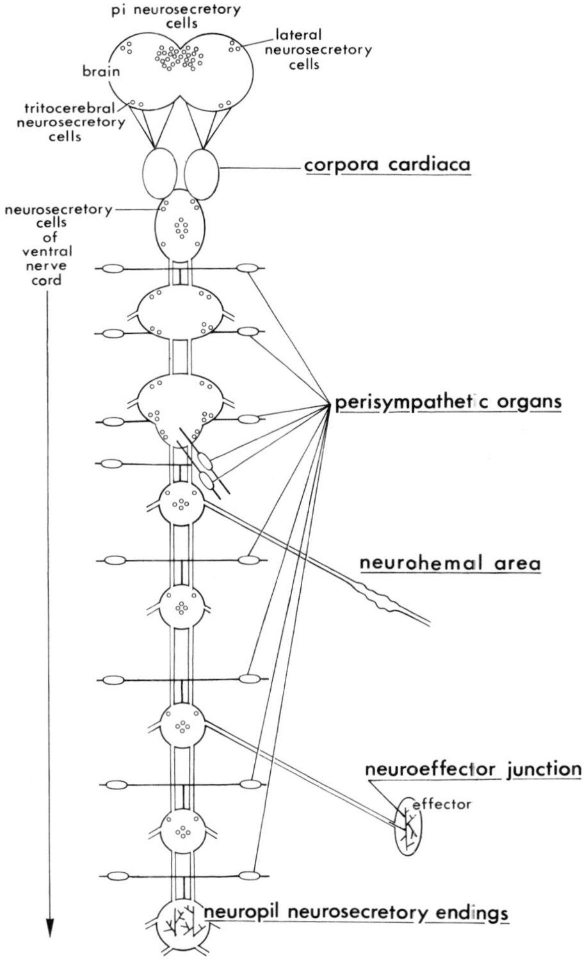

FIGURE 20. Release sites of neurohormones in insects, a recapitulation. pi: Pars intercerebralis.

1.7. CONCLUDING REMARKS

Insect ns cells have been the focus of a large number of investigations devoted to their morphology, location, and functioning using mostly histochemistry, electron microscopy, and histology. Numerous ns cells were discovered in different parts of the brain, the ventral nerve cord, and the sympathetic nervous system, and the existence of peripheral ns cells was revealed. New techniques that allow staining of the whole neuron have demonstrated the morphological complexity of ns cells and showed that they ramify extensively in the neuropil and are in contact with each other.

Several types of ns cell were distinguished in insects, mostly by histological methods. Electron microscopy and histochemistry confirm the diversity of the ns products as does immunocytochemistry (see Chapter 2).

The sites of neurohormonal release are numerous and diversified (Fig. 20). Besides the CC and the PO, there are nh areas along different nerves and neuroeffector junctions inside the target organs. Among recent results, the following findings stand out: the nh part of the CC is sometimes far from the glandular part of these organs, since it may be located in the aortic walls or around the CA. The PO located in a single metamere exhibit an ultrastructural duality, which implies the existence of two opposite modes of release. Certain axons travel long distances and ramify a number of times, thus creating diverse nh areas. Serotonergic networks originate in the subesophageal ganglion and extend to the periphery in mouth part nerves.

The functioning of the ns cells is extremely dependent on internal and environmental inputs mediated by cholinergic, aminergic, and probably peptidergic neurons. Juvenile hormone and ecdysone also modify the functioning of the ns cells.

Biogenic amines are involved in interneuron communication and neurohormone release. They are also discharged into the bloodstream, where they act as neurohormones.

2

Vertebrate and Invertebrate Neuropeptides in Insects

A new way of understanding the molecular structure of neurohormones was recently developed in insects and other groups using radioimmunoassay (RIA) and chiefly immunocytochemistry with antisera raised against vertebrate peptides and certain invertebrate neurohormones.

Numerous positive results were obtained whose physiological significance is still mostly unknown. Positive immunocytological reactions do not in fact mean that the reactive material is identical to the immunogenic substance but indicate instead the occurrence of common antigenic sequences. The investigations concerned will be reported here in detail because they constitute valuable starting points for research on the identification of neurohormones, thus opening up new lines of investigation. Most of these investigations concern about 10 species belonging to the main insect orders including, in particular, the cockroaches *Leucophaea* and *Periplaneta,* the locusts, the coleopteran *Leptinotarsa,* the lepidopterans *Thaumetopoea, Bombyx,* and *Manduca,* and the dipterans *Eristalis* and *Calliphora.* Antisera raised against the 31 following peptides were used in these studies: adrenocorticoptropic hormone (ACTH), bombesin or gastrin-releasing peptide (GRF), calcitonin, corticotropin-like intermediate peptide (CLIP), corticotropin-releasing factor (CRF), follicle-stimulating hormone (FSH), gastric inhibitory peptide (GIP), gastrin/cholecystokinin (G/CCK), growth hormone (GH), growth hormone-releasing hormone (GHRH), glucagon, human chorionic gonadotropin (HCG), insulin, luteinizing hormone (LH), luteinizing hormone-releasing hormone (LHRH), melanocyte-stimulating hormone (MSH), motilin, neurophysin I and II, neurotensin, various opioids, oxytocin, pancreatic polypeptide (PP), prolactin (PRL), parathormone (PTH), secretin, somatotropin release-inhibiting factor or somatostatin (SRIF), substance P, thyrotropin-releasing hormone (TRH), thyroid-stimulating hormone (TSH), vasopressin, vasotocin, and vasoactive intestinal peptide (VIP). The mollusk FMRFamide, insect adipokinetic hormone (AKH), proctolin, and eclosion hormone antisera were also used (see Figs. 21 and 22).

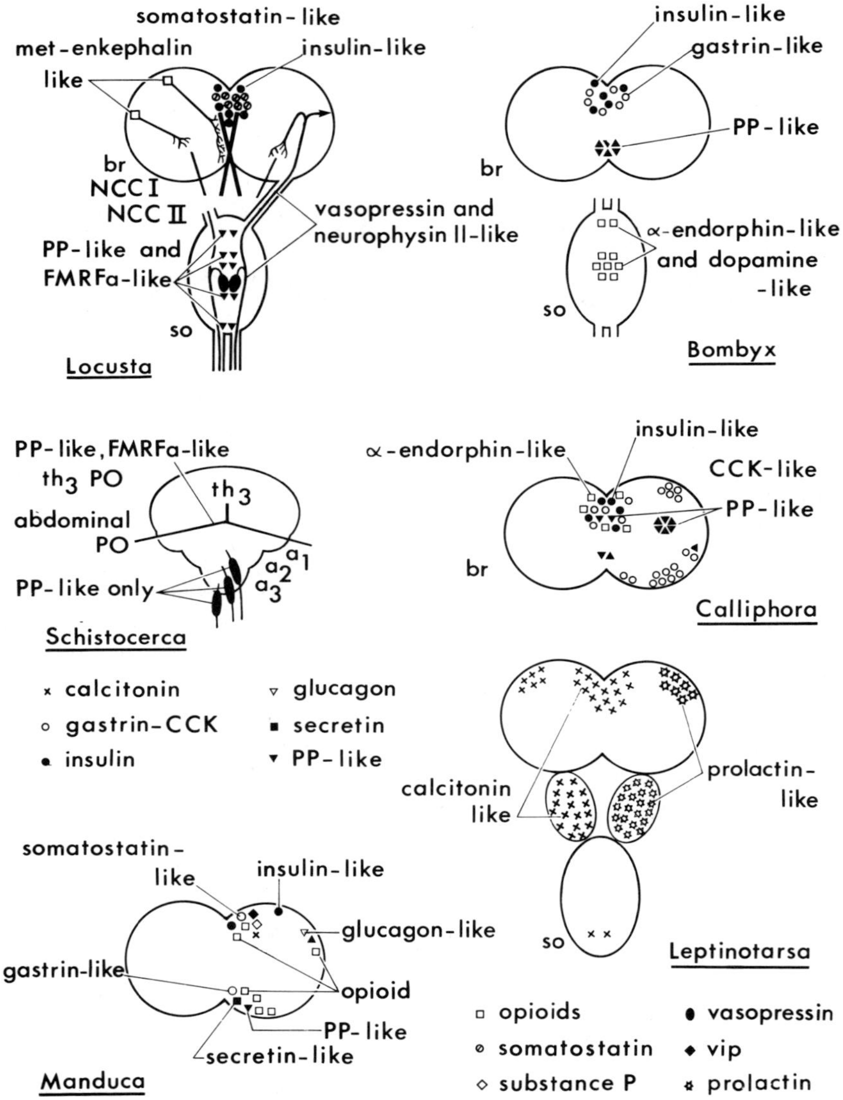

FIGURE 21. Schematic representation of immunoreactive neuron distribution in the brain, subesophageal ganglion, corpora cardiaca, and perisympathetic organs of *Locusta, Schistocerca, Leptinotarsa, Manduca, Bombyx,* and *Calliphora vomitoria* and *erythrocephala.* (From the data of Doerr-Schott *et al.,* 1978; Duve and Thorpe, 1979, 1980, 1983, 1984; El-Salhy *et al.,* 1983; Myers and Evans, 1985; Orchard and Loughton, 1980; Rémy *et al.,* 1979; Rémy and Dubois, 1981; Takeda *et al.,* 1986; Veenstra *et al.,* 1985a; Yui *et al.,* 1980.)

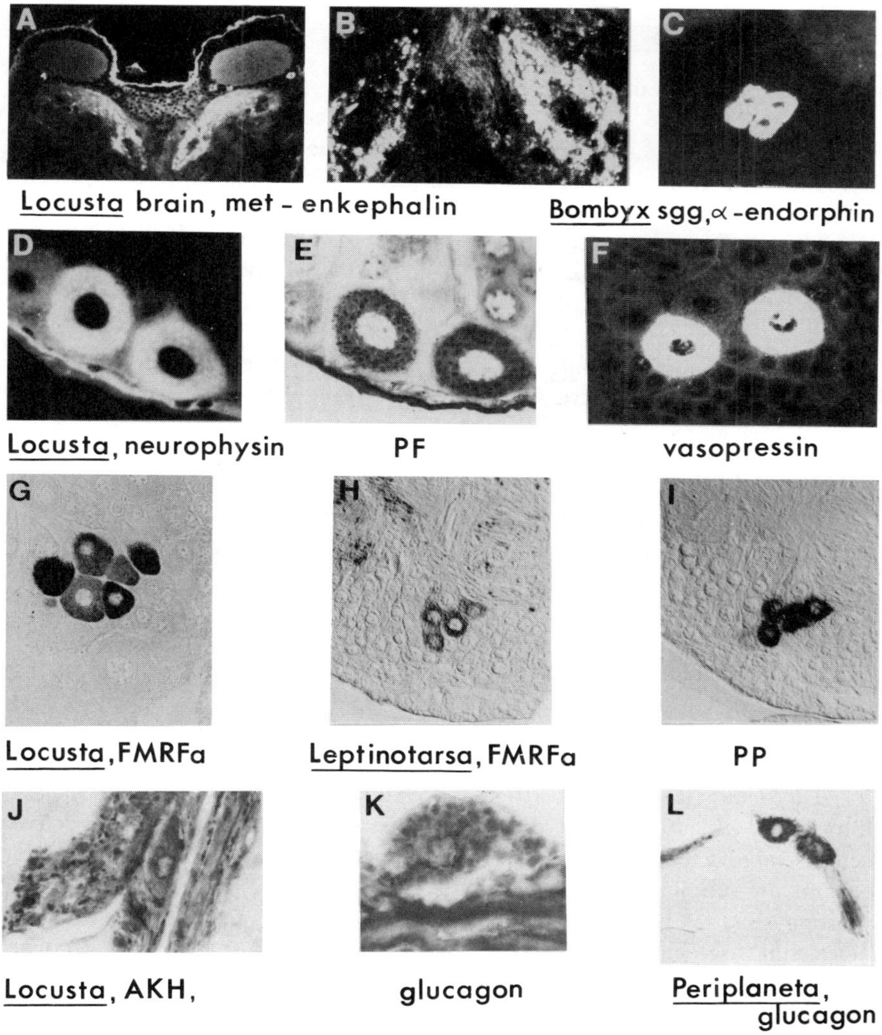

FIGURE 22. Immunocytochemical responses of certain neurons and axons to several antisera. (A,B) Met-enkephalin-like reactivity in a protocerebral lateral neuron and its arborizations located in the vicinity of the NCCI tractus in the brain of *Locusta.* (C) α-endorphin-like neurons in the subesophageal ganglion of *Bombyx.* (D,E,F) Paired medioventral neurosecretory cells in the subesophageal ganglion of *Locusta* treated with antisera raised against neurophysin (D) and vasopressin (F); (E) an adjacent section stained with paraldehyde fuchsin. (G,H) Ventral cluster of FMRFamide-like neurons in the subesophageal ganglion of *Locusta* and *Leptinotarsa.* (I) The same neurons reacting with PP antisera. (J,K,L) Peripheral neurosecretory cells and neurohemal areas in *Locusta* (J, K) and *Periplaneta* (L). Note the immunoreactivity to glucagon antisera of the neurosecretory cells and neurohemal areas in both species. Adipokinetic hormone antiserum gives positive results in the neurohemal area but not in the neurosecretory cells. PF, Paraldehyde fuchsin; sgg, subesophageal ganglion. (A–F Courtesy of Remy. G–I Courtesy of Veenstra and Schooneveld. J–L Courtesy of Raabe.)

The results obtained concern the brain and other nerve ganglia, the nh organs, and also the gut. They will be reported according to peptide groups related by their origin or molecular structure. A review will follow regarding the immunocytological reactions obtained in insects with certain invertebrate peptides and with the insects' own. Results will be grouped together for the localization of immunoreactive elements and for the relations between these elements and the ns cells. Lastly, the problem of colocation will be considered.

2.1. INSECT NEURONS IMMUNOREACTIVE TO VERTEBRATE PEPTIDE ANTISERA (FIGS. 2.1 AND 2.2)

2.1.1. Neurophysins—Vasopressin—Oxytocin (Table X)

Positive reactions to neurophysin II were observed in the paired medioventral ns cells of the subesophageal ganglion in *Locusta* and the stick insect *Clitumnus* (Rémy *et al.,* 1978, 1979), but no immunoreactive neurons were revealed in the brain. In the cockroach *Periplaneta,* however, several neurophysin I- and neurophysin II-like neurons were observed in the brain, subesophageal ganglion, and CC (Verhaert *et al.,* 1984b). On the other hand, in the Colorado beetle *Leptinotarsa* and in the lepidopterans *Thaumetopoea* and *Bombyx* no neurophysin-like neurons were discovered in either the brain or the subesophageal ganglion (Veenstra *et al.,* 1984; Rémy *et al.,* 1977, 1979).

Clitumnus (Rémy *et al.,* 1977), *Locusta* (Rémy *et al.,* 1979), and *Acheta* (Strambi *et al.,* 1979) were all tested with vasopressin antisera. No reactive neurons were detected in the brains of these species, but the paired medioventral cells of the subesophageal ganglion (see Section 1.2.3.2), which reacted with neurophysin antisera, also reacted strongly with vasopressin antisera and sometimes with vasotocin antisera as well. Double staining in *Clitumnus* revealed that these two neurons were the two medioventral ns cells which were previously described in the same insect species (Raabe, 1965) and are present in many others (see Chapter 1). Since these neurons are stained by antibodies to both neurophysin II and vasopressin, it might be that neurophysin II plays the part of a carrier protein as in vertebrates. These ns neurons have very distinct pathways toward the brain and along the ventral nerve cord up to the segmental nerves that are beautifully revealed by immunohistochemistry. Among the other species investigated, no positive reaction to vasopressin or oxytocin was observed in *Thaumetopoea* (Rémy *et al.,* 1978), but some brain neurons and two subesophageal neurons were stained in *Periplaneta* (Verhaert *et al.,* 1984b). In *Leptinotarsa* (Veenstra *et al.,* 1984), two median ns cells, probably homologous to those described in *Clitumnus* and *Locusta,* and a number of other median neurons, possibly Cr-type ns cells, were reactive with vasopressin, vasotocin, and oxytocin antisera, which is not surprising since these molecules exhibit marked similarities. In fact, only certain vasopressin antisera revealed the Cr-type neurons, which are also PP- and FMRFamide-like (Veenstra and Schooneveld, 1984; Veenstra *et al.*, 1984).

As regards the CC, the nh part of these organs was reported to be immunoreactive to vasopressin in *Leucophaea* (Hansen *et al.*, 1982), and to vasopressin and oxytocin in *Periplaneta* (Verhaert *et al.*, 1984b) and the Colorado beetle (Veenstra *et al.*, 1984). Moreover, axons immunoreactive to vasopressin and oxytocin were observed in the CA of *Periplaneta* (Verhaert *et al.*, 1984) and *Leptinotarsa* (Veenstra *et al.*, 1984).

It was interesting to observe that vasopressin was detectable in the central nervous system of locust embryos as early as 3 days before hatching. Both the cell bodies and axon pathways in the ventral nerve cord as well as the terminal arborizations were stained at this early stage (Romeuf *et al.*, 1986).

The presence of a vasopressin-like molecule was detected in the hemolymph of *Locusta* (Proux *et al.*, 1980) and it was shown to exert diuretic effects (Proux *et al.*, 1982) (see Chapter 10).

The occurrence of a vasopressin-like peptide in the subesophageal ganglion, the entire central nervous system, and the hemolymph was confirmed by RIA in *Locusta* (Camier *et al.*, 1980; Proux *et al.*, 1980) and *Acheta* (Strambi *et al.*, 1978, 1979). Using HPLC it was possible to reveal the presence of two separate vasopressin-like peptides, only one of which was active (Proux *et al.*, 1987).

2.1.2. Opioids and Related Molecules (Table XI)

The use of antisera raised against several opioids such as met-enkephalin, leu-enkephalin, and α- and β-endorphins as well as antisera raised against α- and β-MSH and ACTH gave inconstant results for the brain, CC, and subesophageal ganglion.

The clearest reactions were obtained with endorphins. The reactive cells were those of the PI, mostly ns neurons in *Leptinotarsa* (Veenstra *et al.*, 1985a), *Bombyx* (Takeda *et al.*, 1986), *Eristalis* (El-Salhy *et al.*, 1983), and *Calliphora* (Duve and Thorpe, 1983). In the latter species, endorphin- and enkephalin-like neurons (Duve and Thorpe, 1988) were identified by double staining to type A ns cells. The same method made it possible to observe that these cells are partly distinct from those that reacted with antisera to insulin, PP, and G/CCK.

The presence of endorphin-like material was also observed in median cells of the subesophageal ganglion in the lepidopterans *Bombyx* and *Thaumetopoea,* in which the reactions to antibodies to α-endorphin were relatively specific, since these cells did not react to antisera to β-endorphin or to α- and β-MSH (Rémy *et al.*, 1978, 1979).

In *Bombyx,* the α-endorphin-like cells of the subesophageal ganglion also

TABLE X. Immunoreactive Neurons to Neurophysin I, Neurophysin II, Vasopressin, Vasotocin, and Oxytocin Antisera[a]

Insect	Brain	Sgg	VNC	CC	CA	Reference
			Neurophysin I			
Periplaneta	10MNC, 30ln, 21dn, tn, npil	n		+	+	Verhaert *et al.*, 1984b
Clitumnus	0	0				Rémy *et al.*, 1977
Locusta	0	0				Rémy *et al.*, 1979
Leptinotarsa	0	0		0	0	Veenstra *et al.*, 1984
Thaumetopoea	0	0				Rémy *et al.*, 1978
Bombyx		0				Rémy *et al.*, 1979
			Neurophysin II			
Periplaneta	20MNC, 10ln, npil	n		+	0	Verhaert *et al.*, 1984b
Clitumnus	0	2NSC				Rémy *et al.*, 1977
Locusta	0	2NSC				Rémy *et al.*, 1979
Leptinotarsa	0	0		0	0	Veenstra *et al.*, 1984
Thaumetopoea	0	0				Rémy *et al.*, 1978
Bombyx		0				Rémy *et al.*, 1979
			Vasopressin			
Leucophaea				+	0	Hansen *et al.*, 1982
Periplaneta	10MNC, ln, npil	2n		+	+	Verhaert *et al.*, 1984b
Clitumnus	0	2NSC				Rémy *et al.*, 1977

Locusta	npil	2NSC			Rémy *et al.*, 1979
Acheta	npil	2NSC	0		Strambi *et al.*, 1979
Leptinotarsa	12opn	2NSC, 14NSC	+	0	Veenstra *et al.*, 1984
Thaumetopoea	0	0			Rémy *et al.*, 1978
Bombyx		0			Rémy *et al.*, 1979
		Vasotocin			
Clitumnus	0	2NSC			Rémy *et al.*, 1977
Locusta	0	0			Rémy *et al.*, 1979
Leptinotarsa	0	2NSC, 14NSC	0	0	Veenstra *et al.*, 1984
Thaumetopoea	0				Rémy *et al.*, 1978
Bombyx		0			Rémy *et al.*, 1979
		Oxytocin			
Leucophaea			+	0	Hansen *et al.*, 1982
Periplaneta	10MNC, 6ln, npil	2n	+	+	Verhaert *et al.*, 1984b
Leptinotarsa	LNC	2NSC, 14NSC	0	+	Veenstra *et al.*, 1984

[a]an, abdominal neurons; ane, abdominal nerve; aone, aortic nerve; CA, neuroseretory terminals in vicinity of the glandular cells of the corpora allata only; the neurohemal tissue surrounding the corpora allata in certain species is not mentioned; CC, neurohemal part of the corpora cardiaca; CCgd, glandular part of the corpora cardiaca; conn, connectives; dn, deutocerebral neurons; frggn, neurons of the frontal ganglion; hyggn, neurons of the hypocerebral ganglion; ln, lateral neurons; LNC, lateral neurosecretory cells; medne, median nerve; mn, median neurons; miscell, miscellaneous; MNC, median neurosecretory cells; NCC, nervi corporis cardiaci; ne, nerves; NEJ, neuroeffector junctions; NHA, neurohemal areas; npil, neuropil; NSC, neurosecretory cells, opn, optic lobe neurons; pnsc, peripheral neurosecretory cells; PO, perisympathetic organs; Sgg, subesophageal ganglion; spne, sympathetic nerves; thn, thoracic neurons; tn, tritocerebral neurons; VNC, ventral nerve cord; ( ), uncertain data; /, or.

TABLE XI. Insect Immunoreactive Neurons to Vertebrate Opioid Peptide Antisera[a]

Insect	Brain	Sgg	VNC	CC	CA	Reference
			Met-enkephalin			
Periplaneta	26MNC/mn, 4tn, npil	0		+	+	Verhaert and De Loof, 1985a
Leucophaea				+	0	Hansen *et al.*, 1982
Locusta	2ln, 2opn, npil					Rémy and Dubois, 1981
Leptinotarsa	0	0		0	0	Veenstra *et al.*, 1985a
Drosophila	(n)	n	n			Pagés *et al.*, 1983
Chironomus		Rare axons				Johansson *et al.*, 1986
			Leu-enkephalin			
Leptinotarsa	0	0		0	0	Veenstra *et al.*, 1985a
Drosophila	mn	n	n			Pagés *et al.*, 1983
			Enkephalins			
Manduca	4mn, 34ln			+	+	El-Salhy *et al.*, 1983
Eristalis	4mn, 6ln	0	14n, npil	0	0	El-Salhy *et al.*, 1980
Calliphora	2MNC, 2LNC, 16opn	2mn	8n	0	0	Duve and Thorpe, 1988
			α-Endorphin			
Locusta	0	0				Rémy *et al.*, 1979
Leptinotarsa	MNC	0		+	0	Veenstra *et al.*, 1985a
Thaumetopoea	0	12n (type C)		+		Rémy *et al.*, 1978
Bombyx		7n (type C)				Rémy *et al.*, 1979
	8MNC, 8ln	7n				Takeda *et al.*, 1986

Manduca	6ln			+	0	El-Salhy *et al.*, 1983
Calliphora	12MNC			+	0	Duve and Thorpe, 1983
			β-Endorphin			
Periplaneta	3-5opn					Takeda *et al.*, 1986
Leucophaea				+	+	Hansen *et al.*, 1982
Locusta	0	0				Rémy *et al.*, 1979
Leptinotarsa	MNC, ln	0		+		Veenstra *et al.*, 1985a
Manduca	8mn, 10ln			+	0	El-Salhy *et al.*, 1983
			Endorphins			
Eristalis	10mn	0	0	0	0	El-Salhy *et al.*, 1980
			α- and β-Melanocyte-stimulating hormone (MSH)			
Locusta	0	2n	n	0	0	Veenstra *et al.*, 1984
α-MSH	4opncl, npil	2n, npil		0	0	Schoofs *et al.*, 1987
β-MSH	npil					
Leptinotarsa	0	2n		0	0	Veenstra *et al.*, 1984
Thaumetopoea	0	0				Rémy *et al.*, 1978
Bombyx		0				Rémy *et al.*, 1979
Sarcophaga α-MSH	20ln	20n				Schoofs *et al.*, 1987
β-MSH	10mn, 8ln, 2opn	n				
			Adrenocorticotropic hormone (ACTH)			
Periplaneta	20MNC, 6dn			+	+	Verhaert *et al.*, 1984a
Leptinotarsa	6ln, npil	0		+	0	Veenstra *et al.*, 1985a

[a]For abbreviations, see Table X.

reacted positively to an antidopamine serum, but in this species the dopaminergic cells of the brain are not α-endorphin-like (Takeda *et al.*, 1986).

As regards enkephalins, reactions have been reported in the PI of *Periplaneta* (Verhaert and De Loof, 1985a), *Manduca* (El-Salhy *et al.*, 1983), *Eristalis* (El-Salhy *et al.*, 1980), *Drosophila* (Pagès *et al.*, 1983), and *Calliphora* (Duve and Thorpe, 1988), but not in other species in which enkephalin-like neurons are located in the lateral regions of the protocerebrum.

It is worth noting that in most cases the CC were stained by antisera raised against opioids and that the latter sometimes also stained nerve terminals in the CA and neuropil. The observations made by Rémy and Dubois (1981) in the neuropil of *Locusta* are interesting because the antisera to met-enkephalin revealed two protocerebral lateral neurons forming large arborizations near the internal NCCI bundle and two other optic lobe neurons which branch in the vicinity of the NCCII emergence point. This suggests that these met-enkephalin-like neurons may intervene in the release of ns products.

In *Schistocerca* (Davenport and Evans, 1986) the met-enkephalin content was determined in the brain, subesophageal, thoracic, and abdominal ganglia of adult males and females. It appears that female tissues contain more than twice as much met-enkephalin-like immunoreactive material as male tissues and that the largest concentrations are present in the subesophageal and thoracic ganglia, the amount in the brain being much smaller; no immunoreactivity to met-enkephalin was found in the abdominal ganglia, Malpighian tubules, gut, or hemolymph.

The distribution of enkephalin- and endorphin-like material was also investigated by RIA in several regions of the brain of *Locusta* (Gros *et al.*, 1978). No β-endorphin was found but leu- and met-enkephalin were both present, the latter in larger amounts than the former. The concentrations found were as follows: optic lobes < CC < brain < subesophageal ganglion. In *Locusta* embryos both met-enkephalin and somatostatin were demonstrated in the brain and CC (Romeuf and Rémy, 1984).

From the physiological point of view, it may well be that opioids regulate the release of certain neurohormones, as suggested by the presence of met-enkephalin-like material in the nerve terminals located near NCCI and NCCII intracerebral pathways; it has also been proposed that opioids regulate locomotor activity in the cockroach *Leucophaea* (Ford *et al.*, 1986), where the presence of high-affinity binding sites was revealed in the brain with D-ala$_2$-met-enkephalin amide (DALA), an enkephalin analog (Stefano and Scharrer, 1981).

Staining with ACTH and α- or β-MSH was done in a few species. It sometimes revealed a distribution similar to that of opioids, which may be explained by the fact that these peptides share common sequences. It is interesting to note that in locusts, ACTH mimics the activity of a diuretic factor from the CC (Rafaeli *et al.*, 1987) and was able to reestablish imaginal development in

Barathra brassicae pupae deprived of their brain–CC–CA complex (Chernysh, 1983).

As regards the molecular structure of insect opioids, investigations have only concerned enkephalins. In *Locusta,* the molecular weight of the immunoreactive molecule was found to be higher than that of standard enkephalin (Gros *et al.*, 1978), but in *Leucophaea,* the insect met-enkephalin-like substance was reported to have the same chromatographic properties as the vertebrate opioid. In *Drosophila,* the presence of a complex pattern of immunoreactive compounds was demonstrated, one of which displayed chromatographic properties similar to but distinct from those of met-enkephalin (Pagés *et al.*, 1983).

2.1.3. Gastroenteropancreatic Peptides (Table XII)

Many positive reactions were observed in insects using antisera to vertebrate gastroenteropancreatic peptides.

2.1.3.1. Gastrin/Cholecystokinin

Gastrin and cholecystokinin share a 5-amino-acid sequence and an amidated carboxy terminal; their antisera therefore react with the same antigenic material.

Substances immunoreactive to the antisera of G/CCK were found in the central nervous system of six insect species. They are present in the nh part of the CC, the PO of the thoracoabdominal nerve mass, and the glandular cells of the CC in *Calliphora* (Duve and Thorpe, 1984).

In *Manduca* and *Calliphora,* the activity of the antisera was shown to be confined to the portion directed against G/CCK C terminal and not the G/CCK mid-portion or the N terminal (Duve and Thorpe, 1981; El-Salhy *et al.*, 1983).

As regards the molecular structure of the peptide whose presence was demonstrated by G/CCK antibodies, extraction, separation, and RIA were performed in *Manduca* (Kramer *et al.*, 1977b) and *Calliphora* (Dockray *et al.*, 1981a). The presence of a gastrin-like material in the CC–CA complex of *Manduca* was confirmed. In *Calliphora* the brain immunoreactive material to G/CCK was shown to elute in two peaks specific for the C terminal of gastrin and cholecystokinin. This terminal resembles the corresponding mammalian C-terminal peptide but its chromatographic properties do not correspond exactly to any previously characterized forms of mammalian gastrin or cholecystokinin. Thus the immunoreactive material might not be identical in *Manduca* and *Calliphora.*

Double staining demonstrated that the neurons reactive to G/CCK antisera are not identical to A-type ns cells in *Calliphora* (Duve and Thorpe, 1981) and that some of the neurons that reacted with G/CCK antisera were also reactive with PP antisera. In addition, the large cells of the CC were revealed to be

TABLE XII. Insect Immunoreactive Neurons to Vertebrate Gastroenteropancreatic Peptides[a,b]

Insect	Brain	Sgg	VNC	CC	CA	Miscellaneous	Reference
			Gastrin/cholecystokinin (g/CCK)				
Leucophaea				+	+		Hansen *et al.*, 1987
Periplaneta	3-5opn						Takeda *et al.*, 1985
Periplaneta	n						Fujita *et al.*, 1981
Leptinotarsa	4ln, npil						Veenstra *et al.*, 1985a
Bombyx	10mn	0	2thn	0	0		Yui *et al.*, 1980
Manduca	8ln			+	0		El-Salhy *et al.*, 1983
Chironomus	0	0	0				Johansson *et al.*, 1986
Eristalis	mn, ln	0	6n	0	0		El-Salhy *et al.*, 1980
Calliphora er	8mn, 14ln, 8dn, npil			+			Duve and Thorpe, 1980, 1983
Calliphora vo			18thn, 45an, npil			CCgd, PO, ane, aone	Duve and Thorpe, 1984, 1988
			Pancreatic polypeptide (PP)				
Periplaneta	15opn						Takeda *et al.*, 1985
	MNC, LNC, dn, tn	+	n	+	+		Fujita *et al.*, 1981
	mn, ln, opn, dn, npil	8n, npil	n	+	+	conn, frggn	Endo *et al.*, 1982
	57mn, 114ln, opn, dn, tn, npil	6n, npil		+	+		Verhaert *et al.*, 1985
Locusta		16n					Veenstra *et al.*, 1984
Schistocreca		12n, npil	thn, an			medne, PO	Myers and Evans, 1985
Leptinotarsa	2mn, 2ln, 10opn, npil	14n, npil		+			Veenstra, 1984; Veenstra and Schooneveld, 1984
Bombyx	12mn			0	0		Yui *et al.*, 1980
Manduca	14ln			+	(+)		El-Salhy *et al.*, 1983
Eristalis	2mn, 2ln	0	n	0	0		El-Salhy *et al.*, 1980
Calliphora er	mn, ln, opn	12n					Duve and Thorpe, 1980
Calliphora vo			6thn, 8an			PO, conn, ne, hyggn	Duve and Thorpe, 1982
			Insulin				
Periplaneta				+	0		Raabe, 1986a
Carausius				+	0		Raabe, 1986a
Locusta	MNC			+			Orchard and Loughton, 1980

Locusta				+	0		Raabe, 1986a
Leptinotarsa	MNC						Veenstra *et al.*, 1985a
Bombyx	8MNC	0	0	+	+		Yui *et al.*, 1980
Manduca	18MNC, 8LNC			+	+	frggn	El-Salhy *et al.*, 1983, 1984
Ostrinia	8MNC						Lavenseau *et al.*, 1984
Eristalis	10MNC	0	0	+	0		El-Salhy *et al.*, 1980
Calliphora vo	8MNC						Duve and Thorpe, 1979
			Glucagon				
Periplaneta				+	0	pnsc, NHA	Raabe, 1986
Clitumnus				+	0	pnsc, NHA	Raabe, 1986
Locusta				+	+	pnsc, NHA	Raabe, 1986
Leptinotarsa	0	0	0	0	0		Veenstra *et al.*, 1985a
Bombyx	0		0	0			Yui *et al.*, 1980
Manduca	12ln			+	0		El-Salhy *et al.*, 1983
Eristalis	4ln	0	0	0	0		El-Salhy *et al.*, 1980
			Gastric inhibitory peptide (GIP)				
Leptinotarsa	0	0	0				Veenstra *et al.*, 1985a
Manduca	4mn, 6ln			(+)	0		El-Salhy *et al.*, 1983
Eristalis	0	0	0	0	0		El-Salhy *et al.*, 1980
			Secretin				
Leptinotarsa	0	0		0	0		Veenstra *et al.*, 1985a
Manduca	6ln			+	0		El-Salhy *et al.*, 1983
Eristalis	ln	0	6n	0	0		El-Salhy *et al.*, 1980
			Vasoactive intestinal peptide (VIP)				
Aeshna	n						Andriès *et al.*, 1984
Periplaneta	ln, tn, npil	+	+	0	+		Fujita *et al.*, 1981
Leptinotarsa	0	0		0	0		Veenstra *et al.*, 1985a
Manduca	8mn			+	+		El-Salhy *et al.*, 1983
Chironomus	0	0	0				Johansson *et al.*, 1986
Eristalis	0	0	0	0	0		El-Salhy *et al.*, 1980

[a] *Calliphora er, vo, Calliphora erythrocephala, vomitoria.*
[b] For abbreviations, see Table X.

G/CCK-like and secretin-like, but neither PP-like nor glucagon-like (Duve and Thorpe, 1984). In *Leptinotarsa,* G/CCK neurons reacted with other antisera, namely, those directed against FMRFamide (Veenstra *et al.,* 1985a) (see Section 2.2).

It should be noted that no gastrin-like peptide was found either in the hemolymph or in the gut of *Manduca* (Kramer *et al.,* 1977b, 1980).

An interesting similarity of molecular structure was observed between G/CCK and certain cockroach hindgut myotropic factors (see Section 2.7 and Chapter 7).

2.1.3.2. Pancreatic Polypeptide

A number of studies were made using bovine PP antisera. They concerned the brain, CC, and CA, and also the ventral nerve cord. Cockroaches, locusts, lepidopterans, and flies were studied. It appears that in these species PP-like neurons are present in different regions of the brain, including the optic lobes. They were also found in the subesophageal ganglion, the thoracic and abdominal ganglia, and the hypocerebral or frontal ganglia of the sympathetic nervous system. In addition, axons stained by PP antisera were observed in the neuropil, CC, and PO of *Calliphora* (Duve and Thorpe, 1982) and *Schistocerca* (Myers and Evans, 1985). Investigations of the pathways by cobalt backfilling from the CC demonstrated that some brain immunoreactive products migrate along the NCC up to the CC, whereas others do not leave the brain and end in the central neuropil (Duve and Thorpe, 1983).

Certain studies made with an antiserum raised against the molluscan cardioexcitatory peptide FMRFamide and against PP allowed precise comparison of the immunoreactive neurons with both antisera. From the results obtained in several insect species, including *Periplaneta, Locusta, Schistocerca, Leptinotarsa, Manduca,* and *Pieris* (Boer *et al.,* 1980; Veenstra, 1984; Verhaert *et al.,* 1985; Myers and Evans, 1985; Carroll *et al.,* 1986), it appears that a number of neurons as well as the nh organs, CC, and PO are stained by both these antisera, but that about half the neurons react only with PP antisera. Cross-reactivity of PP-like neurons with FMRFamide antisera has also been shown in vertebrates (Kubben *et al.,* 1986). Besides this cross-reactivity, it was also observed in *Calliphora* (Duve and Thorpe, 1982, 1984) and *Leptinotarsa* (Veenstra *et al.,* 1985a) that some of the thoracoabdominal PP-like neurons were G/CCK like.

The occurrence of two distinct substances reacting with PP antisera is thus probable, but it is not certain that one of them is FMRFamide. The cross-reactions might be explained by the fact that PP and FMRFamide share an arginine and an amidated aromatic amino acid in common positions. Certain selected antisera raised against FMRFamide did not in fact stain PP-like neurons at all (see Verhaert *et al.*, 1985 in *Periplaneta*; Carroll *et al.*, 1986 in *Manduca*).

To sum up, when considering the molecular structures of recently identified peptides, it appears that the immunocytochemical responses of insect neurons and nh tissues to PP, FMRFamide, and even G/CCK may be due to the presence of the myotropic peptides, leucomyosuppressin or leucosulfakinin I or II whose Arg-Phe-NH_2 C-terminal sequences are the same as that of FMRFamide (see Section 2.2).

Immunoreactivity to PP antisera is not limited to neurons and nh organs. Neuroendocrine cells reactive to PP were demonstrated in the midgut of *Periplaneta* (Iwanaga *et al.*, 1981; Endo *et al.*, 1982), *Calliphora* (Duve and Thorpe, 1982), and *Aedes* (Brown *et al.*, 1986). In the latter species they also reacted positively to FMRFamide antisera (Brown *et al.*, 1986).

Neither PP nor FMRFamide has been demonstrated to be present in the hemolymph although in *Manduca* the release of a FMRFamide-like substance from the CC–CA complex was evoked by high potassium depolarization *in vitro* (Carroll *et al.*, 1986). In addition, recent studies showed the physiological effects of FMRFamide as a skeletal muscle neuromodulator in *Schistocerca* (see Section 2.2 and Chapter 7).

In *Calliphora,* purification of the PP-like substance and the study of its physicochemical properties, as revealed by Sephadex gel filtration and polyacrylamide gel electrophoresis, showed that they are almost identical to those of bovine PP. Determination of the amino acid composition of this substance showed a close resemblance between insect and vertebrate PP (Duve *et al.*, 1982).

2.1.3.3. Insulin

Very interesting findings are reported in immunocytochemical studies using insulin antisera. The great similarity of results obtained in various species pointed to the existence of certain insulin-like median type A ns cells common to them all. The species concerned included *Locusta, Leptinotarsa, Bombyx, Manduca, Ostrinia, Calliphora,* and *Eristalis.*

Immunocytochemistry as well as RIA revealed positive reactions to insulin antisera in the CC or CC–CA complexes of *Periplaneta* and *Manduca* (Tager *et al.*, 1976). In addition, immunoreactive nerve fibers were observed in the CA, and anti-insulin B chain-like neurons were identified in the frontal ganglion of *Manduca* (El-Salhy *et al.*, 1984), which did not react with 19 other antisera. However, neither the peripheral ns cells nor their associated nh areas were found to contain insulin-like products (Raabe, 1986a).

The presence of insulin was also demonstrated by RIA in whole head of

Tenebrio (Teller *et al.*, 1983), *Plodia* and *Manduca* (Kramer *et al.*, 1980), in the body of *Drosophila* (Kramer *et al.*, 1980), in the midgut of *Periplaneta* (Moreau *et al.*, 1982), in several hymenopterans (Ishay *et al.*, 1976; Moreau *et al.*, 1981), and in *Tenebrio* (Teller *et al.*, 1983). In the honey bee, insulin-like peptides were found in the head but not in the body (O'Connor and Baxter, 1985).

Of special interest is the fact that an insulin-like molecule was identified in the hemolymph of several insect species including a number of hymenopterans (Ishay *et al.*, 1976; Moreau *et al.*, 1981), *Manduca* (Kramer *et al.*, 1980, 1982), and *Calliphora* (Duve and Thorpe, 1982), which suggests that insulin-like material might play a physiological role in insects (Table XIII).

A few experiments were undertaken to investigate this question, using both vertebrate insulin and the insulin-like fractions isolated from the CC–CA complexes or from the gut of insects. The results were as follows: vertebrate insulin lowered glucosemia or trehalosemia in *Locusta, Vespa,* and *Apis* (Moreau *et al.*, 1982; Ishay *et al.*, 1976; Bounias and Pacheco, 1979) but had no effect on *Tenebrio* or *Manduca* (Bahjou *et al.*, 1986; Tager *et al.*, 1976); an insulin-like

TABLE XIII. Demonstration by RIA of the Presence of Insulin-like and Glucagon-like Peptides in Insect Tissues

Insect	Head	Whole body	Hemolymph	Midgut	CC-CA	Reference
			Insulin			
Periplaneta					+	Tager *et al.*, 1976
Locusta				+		Moreau *et al.*, 1982
Apis, Vespa, other hymenopterans			+	+		Ishay *et al.*, 1976
Apis			+	+		Moreau *et al.*, 1981
	+					O'Connor and Baxter, 1985
Tenebrio	+			+		Teller *et al.*, 1983
Plodia	+					Tager *et al.*, 1976
Manduca					+	Tager *et al.*, 1976, 1982
	+		+			Kramer *et al.*, 1980
Drosophila		+				LeRoith *et al.*, 1981
Calliphora			+			Duve and Thorpe, 1982
			Glucagon			
Periplaneta					+	Tager *et al.*, 1976
Apis, Vespa, other hymenopterans			+	0		Ishay *et al.*, 1976
Plodia	+					Tager *et al.*, 1976
Manduca					+	Tager *et al.*, 1976
				+		Tager and Kramer, 1980
			+			Kramer *et al.*, 1980

fraction isolated from the CC–CA complex of *Manduca* (Tager *et al.*, 1976) and the head of *Calliphora* (Duve and Thorpe, 1979) also reduced glucosemia or trehalosemia, as did an intestinal insulin-like peptide from *Locusta* (Moreau *et al.*, 1982) and *Apis* (Bounias *et al.*, 1986). In *Tenebrio*, the intestinal insulin-like peptide reduced the glucogenesis which involves, in this insect, certain amino acids (see Chapter 11).

The insulin-like material in *Calliphora* was shown to act on the fat cells of the rat (Meneses and Ortiz, 1975). Besides these metabolic effects, it was suggested that insulin acts as a differentiation hormone in *Drosophila* (Seecof and Dewhurst, 1974), and it was discovered that both the A and B chains of human insulin have certain sequences in common with the 4K PTTH from *Bombyx* (see Chapter 3). Indeed, insulin injections cause diapause break in *Pieris* diapausing pupae (Arpagaus, 1987).

Comparison of tissue immunoreactivity to the A and B chains of insulin showed that in *Manduca* only the antisera to the B chain were positive (El-Salhy *et al.*, 1983, 1984). Purifications followed by RIA were performed, starting with material of different origins, i.e., the heads of *Plodia* (Tager *et al.*, 1976), *Calliphora* (Duve and Thorpe, 1979), *Apis* (O'Connor and Baxter, 1985), the head and body of *Drosophila* (LeRoith *et al.*, 1981), the head and midgut of *Tenebrio* (Teller *et al.*, 1983; Teller and Pilc, 1986), the CC–CA complex of *Manduca* and *Periplaneta* (Tager *et al.*, 1976), the gut of *Apis* (Moreau *et al.*, 1981), the hemolymph of *Manduca* (Kramer *et al.*, 1980, 1982), and the royal jelly of *Apis* (Kramer *et al.*, 1977a).

In most species, one or two immunoreactive fractions of insulin-like substances were revealed whose migration patterns were about the same as that of vertebrate insulin. The molecular weight of these insulin-like substances was in the range of that of vertebrate insulin (Kramer *et al.*, 1977a) except for the hemolymph and gut factors, whose respective weights in *Manduca* and *Apis* were about 5000 and 18,000 (Kramer *et al.*, 1980; Moreau *et al.*, 1981). In extracts of *Tenebrio* heads and midguts, both a 6000- and a 18,000-Da fraction were separated (Teller *et al.*, 1983; Teller and Pilc, 1986).

Analysis of the amino acid composition of the insulin-like material of the hemolymph of *Manduca* showed great similarities with porcine insulin, except for the serine, leucine, and phenylalanine residues. In *Calliphora*, the immunoreactive material was shown to displace ^{125}I-labeled porcine insulin bound to liver/plasma membrane–insulin receptors (Duve and Thorpe, 1979).

2.1.3.4. Glucagon

Fewer investigations have been devoted to glucagon than to insulin and the results were less homogeneous. In *Manduca* and *Eristalis*, lateral glucagon-like neurons were revealed in the brain; the CC were shown to be positive in this respect in *Manduca* but not in *Eristalis,* although these organs were demonstrated by RIA to contain a glucagon-like substance in *Periplaneta* and *Manduca,* as well as the head of *Plodia* (Tager *et al.*, 1976). In *Periplaneta* (Table IX), *Carausius,* and *Locusta,* the peripheral ns cells located around the abdominal link nerves were shown to contain a glucagon-like substance which was also found in the nh areas associated with them; however, the PO did not react with glucagon antisera. The glucagon-like peripheral ns cells and nh areas did not react either with insulin or AKH antisera (Raabe, 1986a).

By RIA, the whole midgut showed immunoreactivity to glucagon antisera in *Manduca* (Tager and Kramer, 1980) but not in various hymenopterans (Ishay *et al.*, 1976). Detailed immunocytochemical investigations were performed in *Periplaneta* (Iwanaga *et al.*, 1981); they showed that, in the hindgut, neither the glandular cells nor the nerve fibers responded to glucagon C terminal or glucagon mid-portion antisera but did respond to glucagon N-terminal and glicentin 80–100 antisera.

The hemolymph of various hymenopterans and of *Manduca* was seen to contain a glucagon-like substance, suggesting that it may have a physiological function (Ishay *et al.*, 1976; Kramer *et al.*, 1980).

Extraction, purification, and RIA application revealed several glucagon-like peaks in the CC–CA complexes of *Manduca,* in the CC–CA of *Periplaneta,* and in the head of *Plodia* and the hemolymph of *Manduca* (Tager *et al.*, 1976; Kramer *et al.*, 1980) (Table XIII).

It is interesting to note that two metabolic hormones—the adipokinetic hormone and the hyperglycemic factor MII—exhibit a short-sequence homology with glucagon (see Chapter 11).

Glucagon 12	His-Ser-Glu-Gly-Thr-Phe-Thr-Ser-Asp-Tyr-Ser-Lys
MII	pGlu-Leu-Thr-Phe-Thr-Pro-Asn-Trp-NH_2

In the CC–CA complexes of *Periplaneta* and in the head of *Plodia,* the smallest factor has a molecular weight similar to that of bovine glucagon (Tager *et al.*, 1976). In the hemolymph of *Manduca* (Kramer *et al.*, 1980), two large peptides, possibly precursor molecules, are present with molecular weights of 3000 and 5000, respectively. A factor with a much higher weight was identified in the midgut of *Manduca* (Tager and Kramer, 1980), which recalls the situation encountered in vertebrates.

Glucagon was reported to have some physiological effects in insects. It increases the trehalose concentration in *Apis* (Bounias and Pacheco, 1979) and *Locusta* (Moreau *et al.*, 1982), but does not do so in *Manduca* (Tager *et al.*, 1976). On the other hand, a glucagon-like peptide from the CC–CA complexes of *Manduca* was shown to cause a decrease in the glycogen content of the fat body, which can be considered as a hyperglycemic effect. An *in vitro* fat body bioassay devised for the demonstration of hyperglycemic activity, however, failed to disclose that glucagon induced such activity (Hayes *et al.*, 1986).

2.1.3.5. Gastric Inhibitory Peptide, Secretin, Vasointestinal Peptide

Gastric inhibitory peptide (GIP), secretin, and the vasointestinal peptide (VIP), which exhibit great similarities to glucagon and may thus easily produce cross-reactions, were tested in particular on *Leptinotarsa, Manduca,* and *Eristalis*. The results were inconstant except for *Manduca,* in which certain brain neurons and the CC reacted with the three antisera. In *Aeshna* and *Periplaneta,* the presence of VIP-like neurons was demonstrated. The glandular cells of the CC were shown to contain secretin-like substances in *Calliphora* (Duve and Thorpe, 1984).

2.1.4. Miscellaneous

A number of other antisera raised against vertebrate peptides were also assayed in a few insect species. The results, reported in Table XIV, show that for certain peptide antisera no reactions were observed. This was the case for bombesin, corticotropin-like intermediate peptide (CLIP), epithelial and nerve growth factors, HCG, neurotensin, polypeptide Y, PTH, TRH, and TSH. In other cases, clear positive results were obtained in one or a few insect species.

The distribution of certain immunoreactive neurons may be similar in different species as is the case for neurons containing somatostatin, which are mostly located in the median part of the protocerebrum. For other peptides the distribution is less regular. Among the reactions observed, it is of interest to note that in *Leptinotarsa,* both antennal nerves and deutocerebral neuropils were stained by the antiserum raised against corticotropin-releasing factor (CRF), which might indicate the existence of CRF peripheral cells.

In some cases, the reactive neurons were believed to be median or lateral ns cells but this was not established in all species investigated. The nh part of the CC and ns axon fibers in the CA were reported to react positively to several antisera. Conversely, some of the neurons reacting with motilin, prolactin, and substance P antisera were reported to be present in the brain, although the nh part of the CC was devoid of the corresponding substances. Further, the glandular cells of the CC were observed to contain, at least in *Leptinotarsa,* substances reactive to calcitonin, growth hormone, motilin, and prolactin antisera (Veenstra *et al.*, 1985a). All these results may seem confusing but they are nevertheless of interest because they reveal new peptidergic neurons and lead to further experimental and biochemical investigations.

It should not, however, be forgotten that a peptidergic-like substance is not necessarily the peptide itself, since numerous cross-reactions are known to occur

TABLE XIV. Insect Neurons Immunoreactive to Miscellaneous Vertebrate Peptide Antisera[a]

Species	Brain	Sgg	VNC	CC	CA	Miscellaneous	Reference
			Bombesin/gastrin-releasing peptide (GRP)				
Leucophaea				+	0	CCgd	Hansen *et al.*, 1982
Periplaneta	4mn, tn	npil				Conn	Veenstra and Yanaihara, 1984
Locusta	4mn, 500opn, 40tn, npil	npil					Veenstra and Yanaihara, 1984
Extatosoma	4mn, 4MNC, 4ln, npil	npil				Conn	Veenstra and Yanaihara, 1984
Apis	0	0					Veenstra and Yanaihara, 1984
Leptinotarsa	4MNC	0		0	0		Veenstra and Yanaihara, 1984
Galleria	0	0					Veenstra and Yanaihara, 1984
Pieris	0	0					Veenstra and Yanaihara, 1984
Manduca	0			0	0		El-Salhy *et al.*, 1983
Eristalis	0	0	0	0	0		El-Salhy *et al.*, 1980
Musca	4MNC						Veenstra and Yanaihara, 1984
Dysdercus	4MNC, 2mn, 2ln,12tn, npil	2n, npil					Veenstra and Yanaihara, 1984
			Calcitonin				
Leucophaea				(+)	0		Hansen *et al.*, 1982
Leptinotarsa	MNC, LNC, ln	NSC		+	+	CCgd	Veenstra *et al.*, 1985a
Manduca	6ln			+	0		El-Salhy *et al.*, 1983
			Corticotropin-like intermediate peptide (CLIP)				
Leptinotarsa	0	0	0	0	0		Veenstra *et al.*, 1985a
			Corticotropin-releasing factor (CRF)				
Periplaneta	30MNC, mn, 10LNC, 20ln, dn, tn, npil			+	+		Verhaert *et al.*, 1984a
Leptinotarsa	MNC (B type), opn	6n		+			Veenstra et al., 1985a
			Epithelial and nerve growth factors				
Manduca	0			0	0		El-Salhy *et al.*, 1983

		Follicle-stimulating hormone (FSH)					
Periplaneta	56mn, 6ln, opn, dn, tn	n, npil		+	0		Verhaert and De Loof, 1986
		Growth hormone (GH)					
Leptinotarsa	MNC, LNC, opn	n		+		CCgd	Veenstra *et al.*, 1985a
		Growth hormone-releasing hormone (GHRH)					
Aeshna	12mn, 10ln, 100opn, 8tn	n, npil		+			Andries *et al.*, 1984
		Human chorionic gonadotropin (HCG)					
Leptinotarsa	0	0		0	0		Veenstra *et al.*, 1985a
		Luteinizing hormone (LH)					
Periplaneta	60mn			+	0		Verhaert and De Loof, 1986
		Luteinizing hormone-releasing hormone (LHRH)					
Leucophaea				(+)	+		Hansen *et al.*, 1982
Periplaneta	70mn, 10ln, 40pn, 7dn, 12tn, npil	10n, npil		+	0	Conn	Verhaert and De Loof, 1986
		Motilin					
Leptinotarsa	8ln, 12opn	0		0	0	CCgd	Veenstra *et al.*, 1985a
Manduca	0			0	0		El-Salhy *et al.*, 1983
		Neurotensin					
Leucophaea				+	0		Hansen *et al.*, 1982
Leptinotarsa	0	0		0	0		Veenstra *et al.*, 1985a
Manduca	0			0	0		El-Salhy *et al.*, 1983
Chironomus	0	0	0				Johansson *et al.*, 1986
Eristalis	0	0	0	0	0		El-Salhy *et al.*, 1980
		Prolactin (PRL)					
Periplaneta	0			0	0		Veenstra *et al.*, 1985a
Locusta	0			0	0		Veenstra *et al.*, 1985a

[a]For abbreviations, see Table X.

(continued)

TABLE XIV. (*Continued*)

Species	Brain	Sgg	VNC	CC	CA	Miscellaneous	Reference
Leptinotarsa	LNC			0	+	CCgd	Veenstra *et al.*, 1985a
	opn			0	0		Veenstra *et al.*, 1985a
Tenebrio	0			0	0		Veenstra *et al.*, 1985a
Musca	0			0	0		Veenstra *et al.*, 1985a
Polypeptide Y							
Manduca	0			0	0		El-Salhy *et al.*, 1983
Parathyroid hormone (PTH)							
Leptinotarsa	0	0		0	0		Veenstra *et al.*, 1985a
Somatostatin							
Periplaneta	MNC, LNC, tn	n	n	+	+		Fujita *et al.*, 1981
Leucophaea				0	0	CCgd	Hansen *et al.*, 1982
Locusta	MNC			+			Doerr-Schott *et al.*, 1978
Thaumetopoea	0	0					Rémy *et al.*, 1978
Leptinotarsa	ln, npil			0	0		Veenstra *et al.*, 1985a
Manduca	8mn			+	+		El-Salhy *et al.*, 1983
Chironomus	0	0	0				Johanssen *et al.*, 1986
Eristalis	8mn, ln, npil	0	n	0	0		El-Salhy *et al.*, 1980
Substance P							
Leucophaea				0	+	CCgd	Hansen *et al.*, 1982
Periplaneta	5mn, 12ln, 20pn, 5dn, 28tn, npil	4n		0	0	Conn, frgg	Verhaert and De Loof, 1985b
Locusta	mn, npil						Benedeczky *et al.*, 1982
Leptinotarsa	0	0		0	0		Veenstra *et al.*, 1985a
Manduca	8mn			+	(+)		El-Salhy *et al.*, 1983
Chironomus	0	0	0				Johansson *et al.*, 1986
Eristalis	4ln	0	0	0	0		El-Salhy *et al.*, 1980
Thyrotropin-releasing hormone and thyroid-stimulating hormone (TRH and TSH)							
Periplaneta	0	0		0	0		Verhaert *et al.*, 1986
Leptinotarsa	0	0		0	0		Veenstra *et al.*, 1985a
Chironomus	0	0	0				Johansson *et al.*, 1986

and it is well established that several peptides, including bombesin, growth hormone-releasing hormone (GRP), substance P, ACTH, PTH, urotensin, and somatostatin, share short common sequences.

Experimental investigations would provide further information, for example, as regards somatostatin. In the honey bee, this peptide was shown to inhibit the effects of the insulin-like midgut peptide on the glucose and fatty acid concentrations of the hemolymph (Bounias *et al.*, 1985); it was also seen to antagonize the effect of glucagon and to exert hypoglycemic effects (Bounias, 1986).

2.2. INSECT NEURONS IMMUNOREACTIVE TO INVERTEBRATE NEUROHORMONES

Few invertebrate peptides have yet been isolated and consequently only a few antisera raised against invertebrate peptides have been assayed on insect tissues, including the antisera raised against the hyperglycemic hormone of crustaceans and the molluscan cardioaccelerator peptide FMRFamide.

Products immunoreactive to crustacean hyperglycemic hormone were identified in the PI and the nh part of the CC in *Carausius* (Jaros and Gäde, 1982).

A number of reports concern the molluscan tetrapeptide FMRFamide, which was sequenced by Price and Greenberg in 1977. FMRFamide exerts a cardioexcitatory role in *Aplysia* and inhibitory or excitatory effects on certain central neurons. As shown above, FMRFamide antisera when assayed in insects prove to stain ns cells in the brain, the subesophageal ganglion, and nh organs such as the CC and PO in seven insect species. In *Drosophila,* the nerve fibers innervating the ring gland were observed to be FMRFamide-like and it was shown that the immunoreactive neurons remain immunoreactive throughout the larval stages and persist in the adult. In most cases FMRFamide antisera were seen to cross-react at last partly with PP-like neurons (Myers and Evans, 1985, 1987).

In addition to being present in nervous and neuroendocrine tissues, an FMRFamide-like substance was also demonstrated to be a potent modulator of the slow excitatory motoneurons (SETi) of the locust extensor tibiae muscle (Evans and Myers, 1986) (see Chapter 7) and to act as a neurohormone since its release *in vitro* from the CC–CA complex was evoked by high potassium depolarization (Carroll *et al.*, 1986).

The activity of FMRFamide was shown to depend mostly on the C-terminal structure of the peptide, i.e., Arg-Phe-NH2 (Walther *et al.*, 1984; Evans and Myers, 1986). In mollusks detailed investigations concerning structure–activity relations (Kobayashi and Muneoka, 1986; Muneoka and Saitoh, 1986) showed that the C-terminal amide and the Arg and Phe residues were necessary for FMRFamide-like activity on the heart.

It should be noted that a FMRFamide-like peptide was also discovered in the enterogastrointestinal tract and brain of vertebrate species, including fish, frog, chicken, cow, dog, rat, and mouse (Boer *et al.*, 1980; Dockray *et al.*, 1981b). The amino acid sequence of FMRFamide resembles the C-terminal tetrapeptide of G/CCK and is identical to the primary sequence of the C-terminal tetrapeptide of a met-enkephalin-related heptapeptide, as shown below:

FMRFamide	Phe-Met-Arg-Phe-NH_2
Gastrin/cholecystokinin C-terminal tetrapeptide	Trp-Met-Asp-Phe
Met-enkephalin terminal tetrapeptide	Phe-Met-Arg-Phe-OH

Cross-reactivity between FMRFamide-like and PP-like neurons and fibers was also shown in vertebrates (Triepel and Grimmelikhuijzen, 1984), which recalls the results obtained in insects and might be due to the fact that both peptides have in common one arginine and one amidated aromatic amino acid residue.

In insects, myotropic neuropeptides isolated from the head of the cockroach *Leucophaea* have in common the C-terminal sequence of FMRFamide and might correspond to the PP/FMRFamide-like peptides.

FMRFamide	Phe-Met-Arg-Phe-NH_2
LSK I	Glu-Gln-Phe-Glu-Asp-Tyr-(SO^3H) -Gly-His-Met-Arg-Phe-NH_2
LSK II	pGlu-Ser-Asp-Asp-Tyr-(SO^3H) -Gly-His-Met-Arg-Phe-NH_2
LMS	pGlu-Asp-Val-Asp-His-Val-Phe-Leu-Arg-Phe-NH_2

2.3. INSECT NEURONS IMMUNOREACTIVE TO INSECT NEUROHORMONES

A few investigations were made using antisera to the first two neurohormones isolated in insects, and also to brain and CC tissues.

2.3.1. Proctolin

After the identification and synthesis of proctolin, the hindgut neurotransmitter/neuromodulator, specific antisera were raised and the immunocytochemical localization of proctolin-like neurons was investigated.

Several distinct types of proctolin-like neurons were described in the last abdominal ganglion of *Periplaneta,* in both the lateral and median groups of neurons that either project into the proctodeal nerve and hindgut muscles or terminate in the neuropil (Eckert *et al.*, 1981). The presence of about 100 neurons of nine different ultrastructural types was demonstrated (Agricola *et al.*, 1985). In *Schistocerca* about 70 proctolin-like neurons were revealed in the whole central nervous system, including both visceral and skeletal motoneurons and interneurons (Keshishian and O'Shea, 1985).

Other studies using whole-mount immunohistochemistry (Bishop and O'Shea, 1982) pointed to the widespread occurrence of proctolin-like neurons in thoracic and abdominal ganglia of *Periplaneta*. These neurons are located bilaterally, in symetrical pairs, and also medially both dorsally and ventrally. In the brain, immunoreactive neurons are only present in the tritocerebrum. Radioimmunoassay investigations (Bishop *et al.*, 1981) showed that the distribution of proctolin-like products in the central nervous system correlates with the distribution of immunoreactive neurons, since proctolin-like immunoreactivity is low in the brain and very high in the last abdominal ganglion.

In *Leptinotarsa* the distribution of proctolin-like neurons was studied throughout the entire nervous system, and it was shown that in the ventral nerve cord numerous proctolin-like neurons occur in the subesophageal ganglion, and that they are also present in the tritocerebral part of the brain, the CC, and the frontal ganglion. In addition, proctolin-like axon terminals were observed in the muscles of the hindgut and foregut (Veenstra *et al.*, 1985b).

The presence of proctolin-like neurons was also demonstrated in the penultimate abdominal ganglion of *Locusta* (Lange *et al.*, 1986). They innervate the basal part of the lateral oviducts and enhance myogenic activity of these organs (see Chapter 7).

In *Calliphora,* numerous proctolin-like neurons were described in the entire central nervous system and in abdominal nh lateral nerves (Nässel and O'Shea, 1987).

Proctolin-like terminals are widespread in various areas of the central nervous system; in most of the visceral muscles such as the foregut, hindgut, or oviduct; in skeletal muscles; and in the neuropil where proctolin probably acts either as a neurotransmitter or as a neuromodulator. They were also observed in nh areas suggesting neurohormonal function.

A proctolin-like bioactivity was confirmed with the locust bioassay which consists of recording the rhythm of the extensor tibiae muscle contractions. This rhythm is increased by proctolin. The extracts of two distinct pairs of immunoreactive neurons from the cockroach *Periplaneta* were studied in this way: the LW neurons, which are abdominal neurons innervaling the heart (O'Shea and Adams, 1981), and the DS neurons, which are the slow coxal depressor motoneurons of the proximal segment of the cockroach leg (O'Shea and Bishop, 1982). Both extracts, as well as their HPLC-purified fractions, induced proctolin-like effects in the extensor tibiae preparation and the active elution fraction compared with the elution fractions of [^{3}H]proctolin revealed to be identical.

2.3.2. Adipokinetic Hormone

An antiserum raised against a synthetic derivative of adipokinetic hormone, [Tyr^1]AKH I, was assayed on the CC and the central nervous system of *Locusta* and several other insect species (Schooneveld *et al.*, 1983, 1985). In the CC, the glandular cells were stained, but not always with the same intensity. Strong reactions were observed in *Locusta* and *Schistocerca* but weaker or irregular reactions were obtained in *Acheta, Carausius,* and *Periplaneta*; in other species, including hymenopterans, coleopterans, lepidopterans, and dipterans, the CC did not stain at all. In fact, as shown below, antisera raised against N-terminal AKH gave different results and a good staining of the CC glandular cells.

Immunoreactive neurons were observed with the C-terminal antiserum in various parts of the brain, including the PI, subesophageal ganglion, and ventral nerve cord ganglia, and positive axons and terminals were revealed both in the main protocerebral associative centers (central body and protocerebral bridge) and in the optic lobes. In addition, immunoreactive pathways were detected in the circumesophageal connectives, the ventral connectives, and the anterior sympathetic nervous system, and also in the CA of some species (*Acheta, Blattella*). In the nh part of the CC, some axons were stained but their relationships with the aorta remained questionable. Furthermore, immunoreactive terminals were observed in abdominal nh areas associated with peripheral ns cells (Raabe, 1986a). Since the peripheral neurosecretory cells did not reveal immunoreactivity against AKH antiserum, it is probable that these AKH-like products originate in the ventral nerve cord.

Further comparative investigations were carried out in *Periplaneta* and *Locusta* using antisera raised either against [Tyr^1]AKH C-terminal (Tyr-Leu-Asn-Phe-Thr-Pro-Asn-Trp-Gly-Thr-NH_2) or against N-terminal AKH (1–4) (pGlu-Leu-Asn-Phe-$(CH_2)_6$-NH^2-HCl) (Schooneveld *et al.*, 1986). They indicate that most of the glandular cells of the CC of *Locusta* were immunoreactive to both the C- and N-terminal antisera but that the neurons were stained only by the C-terminal antiserum. In *Periplaneta,* all the glandular cells of the CC were only revealed by the N-terminal antiserum. It therefore appears that at least two substances are present in the glandular cells of the CC of these insects and that the substance contained in the immunoreactive neurons of the brain is different from the material revealed in the glandular cells of the CC (Schooneveld *et al.*, 1987a).

These results suggest that the specific antigenes might be identical to AKHI and AKHII in *Locusta,* and possibly to MI or MII in *Periplaneta.* However, preabsorption experiments revealed that MII was not involved in the staining of both *Locusta* and *Periplaneta* cells.

Locusta AKHI	pGlu-Leu-Asn-Phe-Thr-Pro-Asn-Trp-Gly-Thr-NH_2
Locusta AKHII	pGlu-Leu-Asn-Phe-Ser-Thr-Gly-Thr-NH_2
Periplaneta MI	pGlu-Val-Asn-Phe-Ser-Pro-Asn-Trp-NH_2
Periplaneta MII	pGlu-Leu-Thr-Phe-Thr-Pro-Asn-Trp-NH_2

An electron microscopic study using the protein A-gold method demonstrated that AKHI and AKHII were colocated in the same glandular cells and even in the same granules (Diederen *et al.*, 1987).

2.3.3. Eclosion Hormone

An antiserum raised against a purified extract of eclosion hormone from *Manduca* was applied to sections of the brain of this insect. It stained five large cell bodies located in the lateral parts of the protocerebrum which were identifiable as belonging to group Ia ns cell. Their terminals were found in both CC and the nh part of the CA (see Chapter 8).

2.3.4. Neuroparsins A and B

Antisera were raised against the peptides neuroparsins A and B isolated from the PI of *Locusta*. They only stain the type A ns cells of the PI. Both antisera recognize the two peptides, which suggests that they are chemically related (Bourême *et al.*, 1987).

2.4. DISTRIBUTION OF IMMUNOREACTIVE PRODUCTS

2.4.1. Nervous and Neuroendocrine System

To sum up, the presence of immunoreactive cells was demonstrated in the PI, the pars lateralis, and several other regions of the brain, including the optic lobes. They were also found in the subesophageal ganglion, in thoracic and abdominal ganglia, and in the anterior sympathetic nervous system. Certain peripheral ns cells and the glandular cells of the CC were immunoreactive to the antisera of certain peptides.

The immunoreactive material was revealed in the NCC and in the nh part of the CC, in the median nerve and PO (G/CCK, PP), and in nh areas of nerves (glucagon). This indicates that the substances stained are probably neurohormones which are released in the hemolymph.

In some cases, however, there are apparently no immunoreactive axons exiting from the nerve ganglia, and substances revealed probably act as neuromodulators or neurotransmitters. Such observations were made for the motilin-like neurons of *Leptinotarsa* (Veenstra *et al.*, 1985a) and the bombesin-like neurons of seven species belonging to the main insect orders (Veenstra and Yanaihara, 1984). The same may also apply to the somatostatin-like neurons of *Leptinotarsa* (Veenstra *et al.*, 1985a) and *Eristalis* (El-Salhy *et al.*, 1980).

Since immunocytochemistry clearly shows the positive elements and those elements only, it appears to be a particularly suitable means of revealing the processes of immunoreactive neurons which display extensive branching patterns in the neuropil and different types of pathways covering long distances beyond the CC, in the CA or sympathetic ganglia, and along the ventral nerve cord.

The immunoreactive neurons that send their axons into the CC often also display an immunoreactive branching pattern inside the brain, e.g., for G/CCK in *Calliphora* (Duve and Thorpe, 1983), and for CRF and LHRH in *Periplaneta* (Verhaert *et al.*, 1984a; Verhaert and De Loof, 1986). Such a pattern is also observed around the NCCI bundles, i.e., met-enkephalin-like axons in *Locusta* (Rémy and Dubois, 1981) or G/CCK axons in *Calliphora* (Duve and Thorpe, 1983). Moreover, immunoreactive nerve fibers originating in the brain were seen in nerves issuing from the CC and running along pathways leading to the anterior sympathetic nervous system and CA.

Some vasopressin-like products migrate in a retrograde way from the subesophageal ganglion to the brain through the periesophageal connectives and branch repeatedly in the deutocerebrum and optic lobe neuropils. Long pathways along the ventral nerve cord were shown to be followed by a vasopressin-like substance in *Locusta* (Rémy and Girardie, 1980), a PP-like substance (Endo *et al.*, 1982 in *Periplaneta*; Myers and Evans, 1985 in *Schistocerca*), and an AKH-like molecule (Schooneveld *et al.*, 1983, 1985 in *Leptinotarsa*).

It is probable that the substances present in these pathways act as neuromodulators and neurotransmitters. In *Periplaneta,* for example, PP-like pathways leading from the brain to the CC, ventral nerve cord, and sympathetic nervous system were described (Endo *et al.*, 1982).

On the whole, the immunoreactive products appear to be produced in various regions of the nervous system and released at different nh sites and also within the neuropil.

2.4.2. Midgut (Table XV)

Immunocytochemical reactions have been observed in the wall of the gut in several insect species. They concern two distinct elements: endocrine cells and nerve fibers. The endocrine cells reach the lumen of the digestive tract where

TABLE XV. Occurrence of Immunoreactive Cells and Axons in the Midgut of Some Insects[a]

	1	2	3	4	5	6	7	8	9	10	11	12
ACTH		++										
Bombesin								+				
Calcitonin	00											
CCK	+0	+										
CRF	00	++										
β-Endorphin		++										
Leu-enkephalin	+0	++										
Met-enkephalin	+0	++						+				
Gastrin		00					$\underline{0}$			0	0	
Glicentin		+0										
Glucagon		+0							00	00		
GRF	+0	+										
Insulin		00	$\underline{+}$		$\underline{+}$	$\underline{+}$			00	00		
LHRH	+0	+										
Motilin		00										
α-MSH		++										
Neurotensin	00	00										+
PP	+0	++		+					+	00		
Secretin		00								0		
Somatostatin	+0	++			+				0	0		
Substance P	+0	+								+		
VIP	++											

[a]Underlined, RIA demonstration; left column, endocrine cells; right column, axons, 1, *Aeshna* (Andriès *et al.*, 1984; Andriès and Tramu, 1985a); 2, *Periplaneta* (Iwanaga *et al.*, 1981; Andriès and Tramu, 1985b; Schols *et al.*, 1987); 3, *Locusta* (Moreau *et al.*, 1982); 4, *Gryllus* (Iwanaga *et al.*, 1986); 5, *Apis* (Ishay *et al.*, 1976; Moreau *et al.*, 1981; Bounias and Dubois, 1982); 6, *Tenebrio* (Teller *et al.*, 1983); 7, *Manduca* (Kramer *et al.*, 1977b); 8, *Chironomus* (Johansson *et al.*, 1986); 9, *Aedes* (Brown *et al.*, 1986); 10, *Eristalis* (El-Salhy *et al.*, 1980); 11, *Musca* (Kramer *et al.*, 1977b); 12, *Calliphora* (Duve and Thorpe, 1982).

they form microvilli and send out processes to the digestive cells. The nerve fibers terminate in the vicinity of the circulatory muscles where they appear to act in regulating muscle activity and at the same time releasing their secretory products into the hemolymph.

Using both immunocytochemistry and RIA, a number of vertebrate brain–gut-like peptides were detected in insect midgut, i.e., CCK, opioid-related peptides, gastrin, insulin, LHRH, PP, SRIF, substance P, urotensin, VIP, and CRF. Antisera to other vertebrate peptides such as glucagon, motilin, neurotensin, and secretin were not effective, and neither were the antisera raised to insect AKH and proctolin (Brown *et al.*, 1986).

PP-like cells of the midgut were demonstrated to be immunoreactive to

FMRFamide, a cross-reactivity also observed in the central nervous system (Brown *et al.*, 1986). These cells were distinct from the urotensin-like cells described in *Gryllus* and *Periplaneta* (Iwanaga *et al.*, 1986).

More peptides were detected in glandular cells than in nerve fibers, which were only shown to contain enkephalin-, PP-, somatostatin, and VIP-like molecules.

2.5. IMMUNOREACTIVE NEURONS AND NEUROSECRETORY CELLS

In several studies it was not possible to say definitely whether or not the immunoreactive neurons discovered had already been identified as ns cells. In some cases, however (namely, for the type A ns cells of the PI), it was possible to demonstrate that some of them were identical to immunoreactive neurons. This was shown in *Calliphora* for the insulin-like and α-endorphin-like neurons which are distinctly A-type ns cells (Duve and Thorpe, 1979, 1983), and in *Ostrinia* for insulin-like neurons (Lavenseau *et al.*, 1984). In addition, the two ventromedian neurons of the subesophageal ganglion which reacted with vasopressin and neurophysin antisera correspond to ns cells, at least, in *Clitumnus, Locusta* (Rémy *et al.*, 1977; 1979), and *Leptinotarsa* (Veenstra *et al.*, 1984; Veenstra and Schooneveld, 1984). Comparison of immunoreactive neurons with B- or Cr-type ns cells, which are dispersed and numerous, is difficult, and at the present time it is not known whether these do indeed correspond to such neurons. The presence of immunoreactive neurons not belonging to A type, however, has often been reported in the PI and it might well be that they are identical to B- or C-type ns cells. In the subesophageal ganglion groups of Cr-type azocarmin-positive ns cells located medioventrally have been described in a number of insect species (see Chapter 1). Their distribution recalls that of α-endorphin-like neurons discovered in the lepidopterans *Thaumetopoea* and *Bombyx* (Rémy *et al.*, 1978, 1979). Moreover, PP/FMRFamide-like neurons appear to be located in the same position in *Locusta, Schistocerca,* and *Leptinotarsa* (Veenstra and Schooneveld, 1984; Veenstra *et al.*, 1984).

The distribution of immunoreactive neurons, particularly in the deutocerebrum and optic lobes, does not recall the distribution of the ns cell observed by histological methods, and it seems thus that immunocytochemistry reveals more peptidergic neurons than histology. This is important because the peptidergic neurons are probably involved in neurohormone production. Immunocytochemistry thus appears to hold out good prospects for the future.

2.6. CROSS-REACTIONS–COLOCATION OF IMMUNOREACTIVE PEPTIDES

Attempts have been made to determine whether the same neurons were involved in some of the immunocytological reactions observed.

Cross-reactions were reported for related substances such as vasopressin, vasotocin, and oxytocin (Rémy *et al.*, 1977 in *Clitumnus*; Veenstra *et al.*, 1984 in *Leptinotarsa*; Verhaert *et al.*, 1984a in *Periplaneta*). They were also observed for antisera raised against peptides like PP and FMRFamide in *Locusta* and *Leptinotarsa* (Veenstra *et al.*, 1984; Veenstra and Schooneveld, 1984). Only half the PP-like elements proved to be FMRFamide in *Schistocerca* (Myers and Evans, 1985) and *Periplaneta* (Verhaert *et al.*, 1985), which indicates that the same neurons produce two different peptides.

Further, certain PP/FMRFamide-like neurons of *Leptinotarsa* and *Locusta* are also α-MSH-like (Veenstra *et al.*, 1984), whereas in *Leptinotarsa* some of the PP/FMRFamide-like neurons are G/CCK-like and proctolin-like (Veenstra *et al.*, 1985a,b). In *Calliphora*, certain thoracoabdominal neurons and the PO, which are immunoreactive to PP antisera, are also immunoreactive to G/CCK and met- and leu-enkephalin antisera (Duve and Thorpe, 1982, 1984, 1988). However, the PP-like neurons of the PI are different from the G/CCK-like neurons. In addition, the glandular cells of the CC are G/CCK-like and secretin-like, but not PP-like. This is surprising, as the G/CCK and secretin sequences are completely different. It suggests that, in this case, two different peptides are colocated in the same cell. Another example of colocation might be provided by the two medioventral cells of the subesophageal ganglion which are vasopressin-like and neurophysin II-like in *Clitumnus* and *Locusta*. However, in *Periplaneta,* similar cells were only positive for vasopressin and oxytocin (Verhaert *et al.*, 1984b).

In the case of the glandular cells of the CC, which are both G/CCK-like and secretin-like, these two peptides might well be colocated. In other cases, however, there may simply be cross-reactions because the antigenic sites may be fairly short and consequently form part of very different peptides.

It was also shown, using double immunostaining, that neurons immunoreactive to different antisera are indeed different. This, for example, was the case for the median ns cells of *Calliphora* which are α-endorphin-, G/CCK-, insulin-, and PP-like; the medioventral pair of ns cell in the subesophageal ganglion, which are vasopressin-, vasotocin-, and oxytocin-like but do not react with PP, FMRFamide, or α-MHS antisera in *Leptinotarsa* (Veenstra *et al.*, 1984). Moreover, the neurons stained by antisera raised against ACTH, AKH, bombesin, CRF, FMRFamide, FSH, gastrin, LH, LHRH, met-enkephalin, neurophysins I and II, oxytocin, PP, substance P, and vasopressin might all be different in *Periplaneta* (Verhaert and De Loof, 1985b, 1986) and *Leptinotarsa* (Veenstra *et al.*, 1984, 1985b).

2.7. CONCLUDING REMARKS

In insects, immunocytochemical investigations revealed the occurrence in the central nervous system, nh organs, and midgut of many elements that reacted with the antisera raised against vertebrate peptides (Figs. 18 and 19), a situation also encountered in other invertebrates (see Rémy, 1982; Kramer, 1985). Immunoreactivity to a molluscan peptide, FMRFamide, and to the insect peptides AKH and proctolin were also shown.

A certain heterogeneity was observed in the results for different species that might partly be due to the utilization of antisera of different origins, which react with different sequences of the antigenic molecule. It might also be due to interspecific diversity of the molecules produced by the insects. Such diversity is known in vertebrate and has also been shown in insects, in which one or more amino acid residues of the AKH molecule differ depending on the species (see Chapter 11).

Certain immunoreactive neurons appear to be ns cells. Immunochemistry has thus revealed new peptidergic neurons and this is important. Neurons immunoreactive to several vertebrate peptides send out their axons to the nh organs, CC, and PO, which suggests that the immunoreactive peptides have a neurohormonal function. Quite often, however, material immunoreactive to vertebrate, molluscan, or insect peptides was observed in the connectives and in the neuropil of the brain and ventral nerve cord ganglia, where extensive arborizations were revealed. Certain immunoreactive peptides might therefore act as neuromodulators and/or neurotransmitters.

It is well known that positive reactions to a given antiserum do not prove that an identical antigen is involved but only that this antigen possesses one or several identical immunogenic sequences. It is therefore necessary to combine immunocytological studies with other methods such as RIA, chromatography, or HPLC in order to purify the insect immunoreactive peptides and to compare them to vertebrate peptides. Such investigations have been carried out for several peptides. They showed that insect immunoreactive peptides such as gastrin, PP, and vasopressin have a molecular structure close to that of vertebrate peptides even almost identical. For other peptides, such as glucagon, insulin, and met-enkephalin, the similarity is still under debate.

A few insect hormones are presently being purified or have already been characterized (see Chapters 7 and 11). It is interesting to compare their molecular structure to that of the immunoreactive peptides. Sequence homologies were found between the insect metabolic and cardioexcitatory peptides AKHI, MII and glucagon, secretin and VIP, as illustrated below:

AKHI	pGlu-Leu-Asn-Phe-Thr-Pro-Asn-Trp-Gly-Thr-NH_2
MII	pGlu-Leu-Thr-Phe-Thr-Pro-Asn-Trp-NH_2
Glucagon 12	His-Ser-Gln-Gly-Thr-Phe-Thr-Ser-Asp-Tyr-Ser-Lys-
Secretin 12	His-Ser-Asp-Gly-Thr-Phe-Thr-Ser-Glu-Leu-Ser-Arg-
VIP 12	His-Ser-Asp-Ala-Val-Phe-Thr-Asp-Asn-Tyr-Thr-Arg-

In addition, marked homologies were also observed between FMRFamide and three insect myotropic hormones isolated from the head of *Leucophaea*, i.e., leucosulfakinin I and II (LSKI, II) and leucomyosuppressin (LMS). Further LSKI and LSKII share amino acid sequences, a tyrosine SO_3H residue, and an amidated C terminal with the gastrin/cholecystokinin C terminal, which might explain the cross-reactions observed in certain neurons with G/CCK and FMRFamide antisera (see below):

LMS	pGlu-Asp-Val-Asp-His-Val-Phe-Leu-Arg-Phe-NH_2
FMRFamide	Phe-Met-Arg-Phe-NH_2
LSKI	Glu-Gln-Phe-Glu-Asp-Tyr-(SO_3H)-Gly-His-Met-Arg-Phe-NH_2
LSKII	pGlu-Ser-Asp-Asp-Tyr-(SO_3H)-Gly-His-Met-Arg-Phe-NH_2
CCK 8	Asp-Tyr-(SO_3H)-Met-Gly-Trp-Met-Asp-Phe-NH_2
Gastrin 8	Glu-Ala-Tyr-(SO_3H)——————Gly-Tyr-Met-Asp-Phe-NH_2

Therefore it seems that identical or closely related peptides have often been preserved in the animal kingdom throughout evolution, as is the case for biogenic amines and other neurotransmitters.

As shown above, the distribution of immunoreactive nerve fibers in the nerve centers often suggests that a number of peptides might act as neuromodulators or neurotransmitters. Other data such as the presence of certain peptides in the nh organs and hemolymph (e.g., gastrin, glucagon, insulin, PP, and VP) indicate that they may also act as neurohormones. Some experimental research indeed demonstrated that certain vertebrate peptides had a physiological effect in insects. Thus, VP exerts diuretic effects while insulin, glucagon, and SRIF appear to act on trehalosemia, and the molluscan FMRFamide behaves like a neuromodulator in the insect extensor tibiae muscle.

Future research will certainly demonstrate the physiological role of other vertebrate peptides, but it might be that during evolution certain neuropeptides have acquired function, quite different from those they originally possessed.

3

Control of Prothoracic Gland Activity

The existence of a brain factor involved in the control of molting was discovered by Kopec (1922) in the gypsy moth *Lymantria dispar*. In fact, ligation experiments on the caterpillars demonstrated that only the anterior half of the insect was transformed into a pupa, while the posterior half remained a caterpillar. Furthermore, brain extirpation experiments conducted at different times showed that this organ was vitally important in the regulation of growth because its absence suspended development up to a certain critical period.

These results were more or less ignored until about the 1930s. With the blood-sucking bug *Rhodnius,* whose development could be dated accurately, Wigglesworth (1934, 1940) in his famous parabiosis and decapitation experiments demonstrated the existence of a molting hormone originating in the dorsal part of the protocerebrum, where ns cells were identified (Hanström, 1938). The existence of the critical period of brain action was again established.

However, investigations conducted in another lepidopteran, *Bombyx* (ligatures and implantations), identified another tissue site of a molting hormone, the prothoracic gland, also called the molting gland (Fukuda, 1944), which tended to confuse the situation.

The understanding of the mechanisms involved in the regulation of growth and of brain–prothoracic gland interactions was ultimately procured by the studies of Williams (1946, 1947, 1948) on a large saturniid, *Platysamia cecropia,* in which the existence of a long diapause offered considerable advantages. In these investigations, it was demonstrated that the regulation of growth required the presence of two distinct hormones, a brain hormone and a prothoracic gland hormone, whose secretion was regulated by the brain hormone.

The brain molting hormone–prothoracic gland molting hormone scheme was subsequently confirmed in different species and especially *Rhodnius* (Wigglesworth, 1952). Later (Agui, 1975), the stimulatory role of the brain on the prothoracic gland was established *in vitro,* and this facilitated the development of extremely accurate investigations.

In the early studies, the presence of the molting hormone had been demonstrated by the *Calliphora* test, consisting of injections into the posterior part of the body of ligated larvae of extracts designed to induce the formation of the puparium. The identification of the molecular structure of ecdysone, synthesized in the prothoracic gland, enabled the accurate and quantitative determination of its concentration in the tissues and in the hemolymph.

Ecdysone was isolated by Butenandt and Karlson in 1954 from pupae of the silkworm *Bombyx.* Subsequent studies demonstrated the existence of two different hormones: ecdysone (E) and 20-hydroxyecdysone (20-HE) (Fig. 23). The former is synthesized in the PG from cholesterol and is hydroxylated to 20-HE in the fat body and midgut; 20-HE is considered as the main physiological form, at least in the regulation of growth. Several degraded forms are known to occur for ecdysone and 20-HE, whose pathways of biosynthesis are well known today (see Hoffmann, 1986). It has been possible to determine ecdysteroid titer by RIA since 1974 (Borst and O'Connor, 1974).

In certain insects, such as hymenopterans and hemipterans, the prothoracic gland synthesizes makisterone A instead of ecdysone (see Kelly *et al.*, 1981a).

Besides its role in the regulation of growth, ecdysone, which is also synthesized by the follicular cells of the ovary, plays a role in the regulation of reproduction. Its production is controlled by a brain factor, the egg development neurosecretory hormone (EDNH), which could be none other than the stimulatory brain factor of the prothoracic gland.

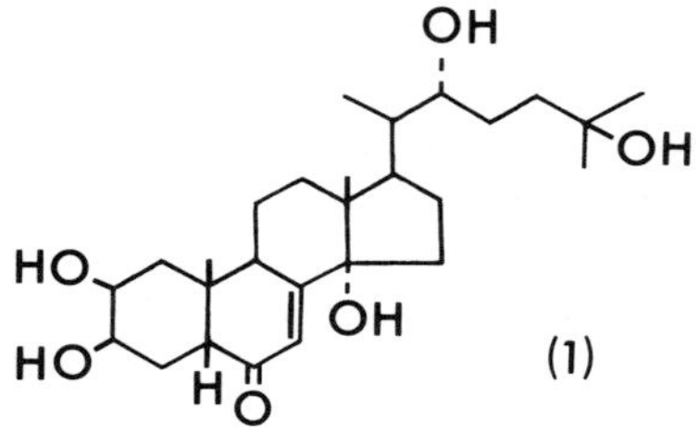

(1)

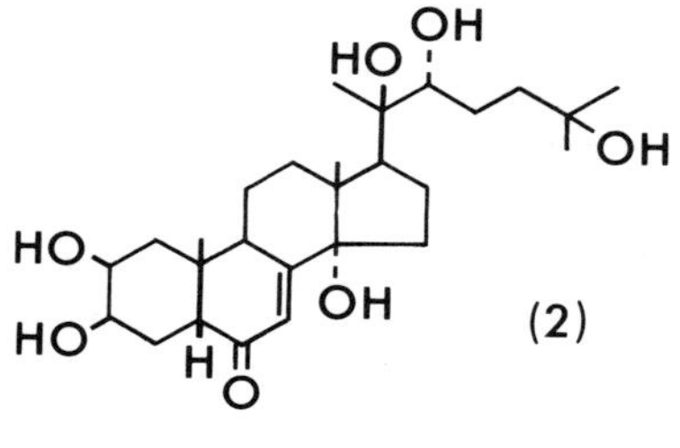

(2)

FIGURE 23. Molecular structure of ecdysone (1) and 20-hydroxyecdysone (2).

From the standpoint of their transport and degradation, it may be recalled that ecdysteroids, which are very soluble due to their very large number of hydroxyl groups, do not appear to have carrier proteins.

The prothoracic gland, which is located at various sites, has been designated under different names according to its location: thoracic gland, prothoracic gland, ventral gland. The brain factor which causes its activation has also been given various names, i.e., brain hormone, prothoracicotropic hormone (PTTH), thoracotropic hormone, activation hormone, and ecdysiotropin. We shall refer to it as PTTH according to the terminology used in lepidopterans: accordingly, the molting gland shall be termed the prothoracic gland (PG).

3.1. PROTHORACIC GLAND INNERVATION AND ULTRASTRUCTURE

The PG exhibits morphological and cytological characteristics which differ slightly from one species to another. Often formed of many small cells in primitive insects, it is made up of large, polyploidal elements constituting diffuse organs in more evolved insects.

3.1.1. Prothoracic Gland Innervation

The PG is richly innervated by ventral nerve cord ganglia, sometimes by the subesophageal ganglion only, sometimes by the subesophageal and prothoracic ganglia, and sometimes by several thoracic ganglia. Neurosecretory fibers have been described in PGs and their nerves in several insect species, namely, *Leucophaea* (Scharrer, 1964b), several lepidopterans such as *Bombyx* (Beaulaton, 1968a), *Cerura* (Hintze-Podufal, 1970), *Galleria* (Blazsek *et al.*, 1975; Gersch *et al.*, 1975; Benedeczky *et al.*, 1980), and *Calliphora* (Normann, 1965). From the size and electron density of their granules, it appears that mostly peptidergic fibers are involved.

3.1.2. Perisympathetic Organs

The prothoracic PO are located in close contact with the PG in *Gryllus* (Thomas and Raabe, 1974), *Labidura* (Raabe, 1972), and varied lepidopterans (Abou-Halawa and Sláma, 1986). According to the latter authors, these organs, termed H organs, are responsible for the developmental events since the insects (*Galleria* larvae) can undergo normal development after the removal of the PG (Sláma, 1983) but are unable to develop after extirpation of the prothoracic ganglion–H organ. The H organs are observed to degenerate during pharate adult development when PG is degenerating.

3.1.3. Origin Cells

A number of investigations using the cobalt chloride technique were conducted to determine the source cells of the fibers that innervate the PG. In *Galleria* (Granger, 1978; Singh and Sehnal, 1979), the neurons innervating the PG have been identified. They include some ns cells: a median pair located in the subesophageal ganglion, and two lateral pairs located in the prothoracic and mesothoracic ganglia. Hence it is quite probable that the functioning of the PG is regulated, apart from the cerebral PTTH, by nerve impulses and ns products originating in these ventral ganglia. In *Periplaneta* (Birkenbeil and Agricola, 1980), four neurons were identified in the ipsilateral anterior part of the prothoracic ganglion (see Sections 3.8.1 and 3.8.2).

3.1.4. Prothoracic Gland Structure—Variations

Considerable research has been carried out on the structure of the PG and its cyclic variations during developmental cycles.

The histological descriptions are fairly similar from one species to another (see Herman, 1967) and recall those made in *Rhodnius* (Wigglesworth, 1952). Once they are activated, the PG cells experience major transformations. Their cytoplasmic volume and their basophilia increase considerably, while the nuclei take on lobulate forms and undergo a large increase in volume. Many vacuoles also appear in the nucleus and cytoplasm (Herman and Gilbert, 1966).

3.1.5. Ultrastructural Data

One of the most characteristic ultrastructural features of the PG appears to be the presence of many plasmalemmal infoldings and pronounced intercellular lacunae systems which were observed to develop around the critical period in *Leucophaea* (Scharrer, 1964b), *Galleria* (Gersch *et al.*, 1975), and *Manduca* (Sedlak *et al.*, 1983). This results in a large increase in the cell surfaces, probably associated with ecdysteroid discharge processes.

According to Beaulaton (1968b, 1980), the appearance of mitochondria is associated with that of ecdysteroids in the hemolymph in *Antheraea, Bombyx,* and *Rhodnius*. The endoplasmic reticulum initially occurs in the form of ribosomes or polysomes, followed by the appearance of ergastoplasmic saccules which increase in size. In the final phase, the ergastoplasm is replaced by a smooth endoplasmic reticulum (SER), while ribosomes reappear (Fain-Maurel and Cassier, 1968). As for the rough endoplasmic reticulum (RER), it grows at the time of the ecdysteroid peaks, or just afterward. Its involvement has been suggested in the synthesis of carrier protein (McDaniel *et al.*, 1976 in *Hyalophora*; Joly *et al.*, 1977 in *Locusta*).

The ultrastructural aspect of the PG was considered as a function of hemolymphatic ecdysone peaks in a number of insects. In *Manduca* (Sedlak *et al.*, 1983), an increase in the intercellular spaces and the multivesicular sacs is observed at the time of the second ecdysteroid peak of the last larval stage. In *Spodoptera* final larval instar (Zimowska *et al.*, 1985), a positive correlation is observed between hemolymphatic ecdysteroid titer and PG activity, which increases when the ecdysteroid titer reaches its highest levels.

3.1.6. Ultrastructural and Immunocytochemical Data

It is not yet positively known whether the mechanisms entering into play during the activation of the PG are synthesis or release mechanisms. Immunocytochemical studies could clarify this matter and are therefore important. They have been carried out only rarely and the data essentially rely on the work of Birkenbeil (1983) on *Galleria,* using an antiecydsone serum. In the active glands the SER, which is highly developed, sometimes reveals an electron-dense content and sprouts secretory granules that are also electron-dense. Immunostaining of the exocytosis omega profiles is also observed.

3.2. PROTHORACIC GLAND DEGENERATION

The degeneration processes which take place in the adult PG have been observed for many years (Pflugfelder, 1938) and by many authors, including Scharrer (1965b) in *Leucophaea* and Fain-Maurel and Cassier (1969) in *Locusta.*

The Golgi apparatus, which elaborates the lytic structures, plays a vital role in these processes, which are manifested by the appearance of large heterogeneous inclusion bodies. These are autophagic vacuoles whose content is degraded by enzymes and which disappear progressively at the same time as the microtubule areas. The nuclei become picnotic, the cell and nuclear membranes disappear, and many hemocytes penetrate into the PG and phagocytize whatever remains.

The mechanisms which control imaginal degeneration of the PG appear to be complex and are difficult to identify. Degeneration of the PG may result from a JH deficiency, since CA implants and the injection of JH-rich extracts prevent its degeneration in several lepidopterans such as *Hyalophora* (Williams, 1961), *Antheraea* (Gilbert, 1962), and *Samia* (Oberlander *et al.,* 1965), in *Locusta* (Cassier and Fain-Maurel, 1970c), and in *Oncopeltus* (Smith and Nijhout, 1982).

However, the effects of JH appear to differ according to time of application. In *Spodoptera* the PG of insects treated with JH analogs (JHA) before the switchover (see Section 3.8.4.2), i.e., when they inhibit PG activity, may induce its regression (Cymborowski and Zimowska, 1984). The transplantation experiments performed in the dermapteran *Anisolabis maritima* with PG of different ages also indicate a critical period, in the late nymphal instar, from which the glands are found to regress (Ozeki, 1968).

In *Rhodnius,* Wigglesworth (1954) demonstrated that the absence of JH is not sufficient to cause the breakdown of the PG. In order to degenerate, the latter must undergo hormonal stimulation during the imaginal molting period, of which the extracerebral source has not been identified.

In the cockroach *Nauphoeta cinerea,* Lanzrein (1975) confirmed the role of JH, the absence of which programs the degeneration of the PG, but also reported a protective function performed by the CC as indicated previously by Bodenstein (1953a) in *Periplaneta,* and demonstrated the existence of a humoral factor inducing degeneration, which could be ecdysone.

The role of 20-HE in the stimulation of cell death in the PG was demonstrated in *Oncopeltus fasciatus.* The breakdown of the PG was triggered by *in vitro* exposure to this hormone, which could only last 24 hr. Once triggered, the breakdown proceeded in the absence of further stimulation (Smith and Nijhout, 1983).

It remains to be seen whether the regulation of cell death in the PG does not also depend on nerve-transported information. In *Hyalophora,* it was suggested by Herman and Gilbert (1966) that degeneration of the PG could be compared to the degeneration of the intersegmental muscles, and could be due to a decrease of the nerve impulses. On the other hand, it was observed in several lepidopterans that PG degeneration was preceded by the disappearance of the H organ (prothoracic perisympathetic organs) (Abou-Halawa and Slama, 1986).

3.3. PTTH ASSAYS

3.3.1. *In Vivo* Assays

Several bioassays were used to identify the presence of PTTH and to evaluate its amount. These assays dealt with larvae and, above all, pupae in natural or induced diapause, in which the presence of PTTH is marked by a break in diapause. This concerns the pupae of *Hyalophora, Bombyx,* and *Samia* in particular.

In *Hyalophora,* the onset of diapause depends on external factors as shown above. Debrained diapausing pupae were used by Williams (1946) to identify the role of the brain in the break of diapause.

In *Samia,* pupal diapause does not occur in normal development but can be induced by decerebration 20 hr after pupation. This produces a diapausing pupa usable for 1–10 months, in which an injection of brain extract causes complete development of the adult in 20 days (Ishizaki and Ichikawa, 1967); the induction of development can be observed between the fourth and sixth day at the epidermis where the retraction occurs. The *Samia* assay is sensitive, practical, and specific. It allows quantitative evaluations and was widely used by Japanese researchers who purified PTTH. A *Samia* unit was defined corresponding to the minimum amount required to induce the development of a debrained pupa (Suzuki, and Ishizaki, 1986). Each head contains 10 *Samia* units.

In certain races of *Bombyx,* decerebration carried out immediately after pupation causes the arrest of the development favorable to diapause (Kobayashi, 1957; Yamazaki and Kobayashi, 1969). Diapause can be interrupted by extracts containing PTTH.

The *Bombyx* larval *in vivo* assay has been used for many years and the method has been improved several times (Kiguchi and Agui, 1981; Sakurai, 1983; Okuda *et al.,* 1985; Suzuki and Ishizaki, 1986) (Fig. 24). Fourth instar larvae, molted during the same period, are starved and then fed at the same time, neck-ligated between 00:00 and 2:00 on day 2, i.e., just after the activation of the PG, and immediately injected with the test extracts. Thirty to forty-five hours after the injection, the state of advancement of the spiracle apolysis is examined in order to determine the PTTH titer by comparing the responses obtained with a

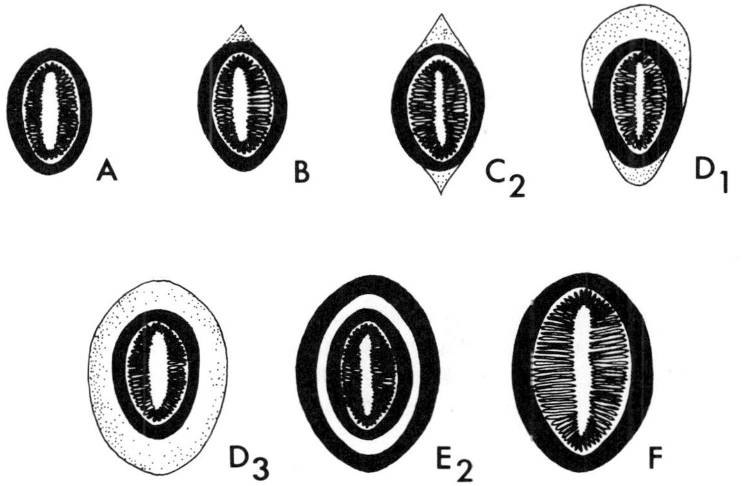

FIGURE 24. Spiracle index as a basis for staging the larval ecdysis in *Bombyx.* (A) Absence of distinctive features around the spiracle. (B) Appearance of a spot on the dorsal side of the spiracle. (C2) Appearance of a new spot on the ventral side. (D1) Fusion of both areas. (D3) Deposition of the new cuticula. (E2) Melanization of the peritreme. (F) Appearance after ecdysis. (From Kiguchi and Agui, 1981.)

standard dose-response curve. A dose inducing a half-maximal response is considered as one *Bombyx* larva unit.

In *Manduca* (Gibbs and Riddiford, 1977), fourth instar, neck-ligated larvae were used just before the head critical period, and the molting process was recorded from the spiracle apolysis. This is highly sensitive and allows a quantitative evaluation of PTTH, which is very important. The percentage of injected larvae exhibiting spiracle apolysis after 48 hr increased linearly with increasing concentrations of injected brain extract, and allowed the determination of PTTH.

In *Galleria,* the test larvae were decapitated and injected with various extracts whose PTTH activity was evaluated from their ability to induce the formation of a new cuticle (Mala *et al.,* 1977).

3.3.2. *In Vitro* Assays

3.3.2.1. Early Investigations

The first PTTH identification tests in culture date back to the 1970s and even earlier because the *in vitro* activation of RNA syntheses in the PG by the brain was demonstrated as early as 1965 by Oberlander, and the activation of the PG of diapausing insects was observed in culture in the presence of active brain (Kambysellis and Williams, 1972).

3.3.2.2. *Chilo* Integument Assay

This assay, designed to measure the amounts of ecdysone present in the incubation medium, uses small fragments of integument of diapausing *Chilo,* which molt as a function of the amounts of ecdysteroids present (Agui *et al.,* 1972).

3.3.2.3. *Manduca* Prothoracic Gland Assay

The *Manduca* PG assay led to advances in our knowledge in various directions. It is based on the RIA determination of synthesized ecdysteroids *in vitro* by activated PG (Bollenbacher *et al.,* 1979, 1983), whose activities compared with that of the contralateral nonactivated PG of the same nonactivated insect (Fig. 25).

The age of the PG to be used was determined by comparing the results obtained with PG of day 3 larvae, day 6–7 larvae, and day 0 pupae. The former are very slightly activated, whereas the latter synthesize a great deal of ecdysone, suggesting that they are already activated. The synthesis kinetics of day 0 pupal PG is intermediate and most convenient. It is incubated for short periods of 2 or 6

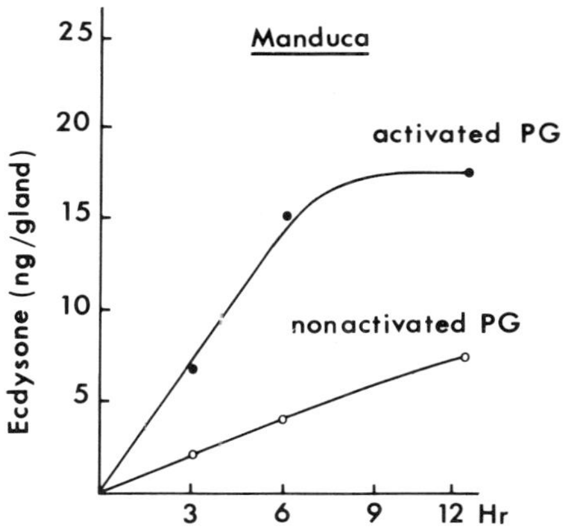

FIGURE 25. Time course of ecdysone synthesis for PTTH activated and nonactivated day 0 pupal prothoracic glands of *Manduca*. (Modified from Bollenbacher *et al.*, 1979.)

hr (Bollenbacher *et al.*, 1983), in order to avoid the depletion of their substratum titer. During this time interval, they synthesize 16–20 ng of ecdysteroids.

The PG can be activated by incubation in the presence of brain, but the effects are irregular. The results are much more constant by using heat-treated brain homogenates, with which activation is dose-dependent and very sensitive. One PG can detect 0.01 brain equivalent. One *Manduca* PTTH unit is the amount of PTTH present in the brain of a 1-day pupa.

The activation of pupal PG by PTTH occurs in a characteristic dose-dependent manner over a very narrow range of PTTH concentration. The maximum activation ratio (A_{max}) varies from 3 to 8, while the PTTH concentration necessary to induce the half-maximal effect on the PG is relatively constant (ED_{50}), so that the calculation can be made from A_{max} and the different samples compared (Fig. 26).

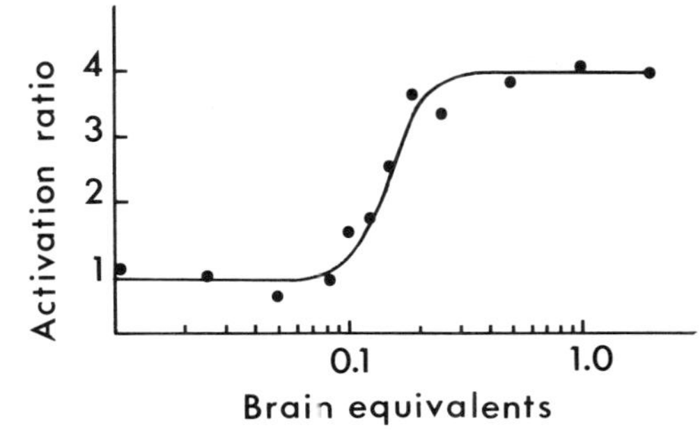

FIGURE 26. Dose-response of PTTH activation of the prothoracic gland in gate II day 3 *Manduca* last instar larvae. (Modified from Bollenbacher *et al.*, 1983.)

3.3.2.4. *Mamestra, Bombyx,* and Other Assays

The effectiveness of *in vitro* assays of the activation of the *Manduca* PG led researchers to apply similar methods to other lepidopterans, i.e., *Mamestra* (Agui *et al.*, 1983), *Bombyx* (Ishizaki *et al.*, 1983a; Sakurai, 1984; Okuda *et al.*, 1985), and *Samia* (Nagasawa *et al.*, 1984a,b). In addition, *in vitro* assays were developed for the *Sarcophaga* ring gland (Roberts *et al.*, 1984) and *Locusta* PG (Reichhart and Charlet, 1986).

3.4. TIMING OF PTTH RELEASE

Adjustment to environmental factors and to certain internal factors is a vital necessity for survival. Hence it is evident that the regulation of growth is dependent on these factors, through the intermediary of the brain, which controls the production of PTTH. Investigations were carried out on several species to determine the external or internal factors involved in PTTH release. These investigations were carried out with cervical ligatures and decerebrations, which helped to determine the brain critical period either by histological and ultrastructural examination of the PG or by determining the titer of hemolymphatic ecdysteroids.

3.4.1. *Hyalophora*

In large saturniids such as *Hyalophora* and *Samia,* the photoperiod combined with temperature plays a very important role in insect development. Its effects are exerted seasonally by inducing diapause phenomena in particular.

In *Hyalophora* and *Samia,* the insects enter into diapause upon their pupation. If they are kept at room temperature, diapause persists for at least 5 months. On the other hand, if the dispausing pupae are placed at 3–5°C for 1½ months and then returned to room temperature, emergence of the ''chilled pupae'' occurs within 1½ months.

The possibility of simultaneously using insects from both batches encouraged experimental investigations that identified the respective roles played by the brain and the PG in insect development. In fact, when parabiosis is carried out on a diapausing pupa and a chilled pupa, the diapausing pupa develops at the same time as the chilled pupa. Further, the implantation of various organs shows that only the brain is capable of inducing the termination of diapause in the given conditions, and that cold causes its reactivation. The PG becomes necessary later.

Stimulation of the ns cells may involve the intervention of cAMP. In *Hyalophora* pupae taken out of the cold and reactivated, Rasenick *et al.* (1976) reported a significant increase in brain cAMP. Similarly, *Antheraea* pupae, transferred from a short photoperiod to a long photoperiod which interrupts diapause, reveal a considerable elevation of cAMP in the PI.

3.4.2. *Manduca*

3.4.2.1. Critical Weight

In *Manduca* the timing of PTTH release is correlated with the weight reached by the insect. This weight is 5 g in the final instar larva (Nijhout and Williams, 1974a,b).

3.4.2.2. Gated Release of PTTH

Once the critical weight is reached, the larva can molt; this occurs at a precise time of day, which differs slightly from one instar to another. Molting is preceded by PTTH release, which also occurs at a specific time (Truman and Riddiford, 1974). Neck ligation experiments, conducted during the light 12/dark 12 cycle, revealed that PTTH release is regulated by a photosensitive circadian clock. Gated PTTH release occurs only during the beginning of the dark period, and the insects that are not sufficiently developed to pass through the first gate wait until the next day and pass through the second. Similar occurrences are observed in *Antheraea* and *Trichoplusia* (Jones *et al.*, 1986) in which hormone releases are also related to the photoperiod.

During the final larval instar of *Manduca*, different morphological and behavioral changes occur which may be correlated accurately with the timing of PTTH release and which allow accurate dating (Nijhout and Williams, 1974a; Truman and Riddiford, 1974) (Fig. 27). During the first part of the instar, feeding takes place until the first PTTH release; then it stops and the insect begins to move continually (wandering stage). Twelve to fifteen hours after PTTH release, the larva makes clearing of the gut or gut purge and heart exposure, a process involving deposition along the dorsum of the larva of a pink ommochrome pigment, and clearing of the tissues around the heart, so that the latter becomes visible along its entire length. After the second PTTH release, which induces the high ecdysone peak, the epidermis retracts and a pupal cuticle is then synthesized.

3.4.2.3. PTTH Titers

Contrary to what may have been expected, in *Manduca* the brain titer of PTTH increases in a stepwise manner following each molt, but no dramatic changes are observed before and after the critical period (Bollenbacher and

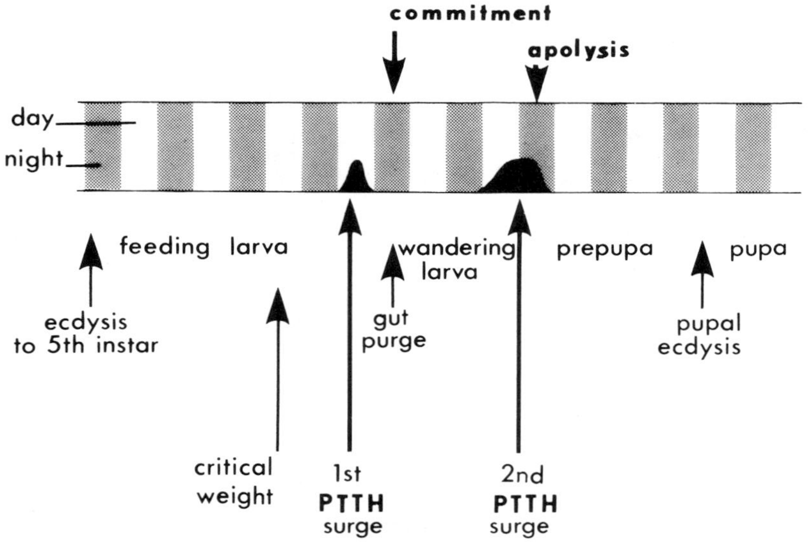

FIGURE 27. Timing of morphological, behavioral, and endocrinological events during the final instar and prepupal stage of *Manduca.* (Modified from Truman and Riddiford, 1974; Nijhout and Williams, 1974a.)

Gilbert, 1981), confirming the results obtained in *Samia* (Ishizaki, 1969). In *Antheraea,* moreover, brain activity was demonstrated as similar in diapausing insects and in insects emerging from diapause (Williams, 1967).

A PTTH peak is observed in the hemolymph of *Manduca* during the antepenultimate instar critical period and final instar second critical period. Its duration is very short and it appears to be due to the presence of big PTTH (see Section 3.6.2). The ecdysteroid titer of the final instar first critical period is quite different. It is characterized by three short releases of small PTTH (Bollenbacher and Gilbert, 1981; Bollenbacher *et al.,* 1981) (see Bollenbacher and Granger, 1985) (Fig. 28).

Thanks to other studies (see Riddiford, 1980), further details were obtained making it possible to introduce JH into the sequence of events. When the final instar critical weight is reached, the JH titer decreases and PTTH release takes place in the next allowable gate, which induces the first ecdysteroid peak, causing commitment. At the end of the instar, the JH increases and the second ecdysone peak occurs, followed by molting. Besides the critical periods of PTTH, a critical period of the JH also exists (Fain and Riddiford, 1976) and is manifested by the juvenile or pupal character of the synthesized cuticle (Fig. 29).

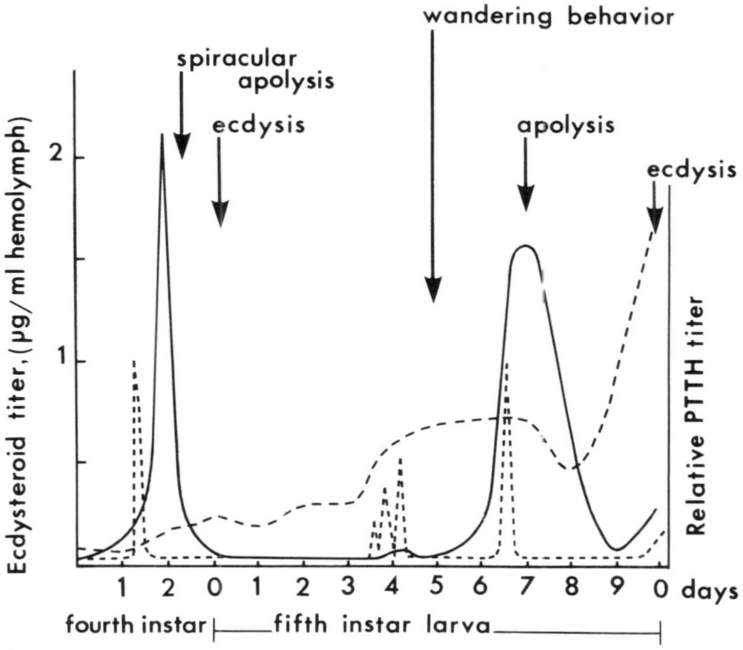

FIGURE 28. Ecdysteroid and PTTH titers during larval and larval-pupal development of *Manduca*. Hemolymph ecdysteroid titer is expressed in 20-HE equivalents (—). Hemolymph (--) and brain (— —) PTTH titers are expressed relative to the highest level of hormone activity found. (Modified from Bollenbacher and Gilbert, 1981.)

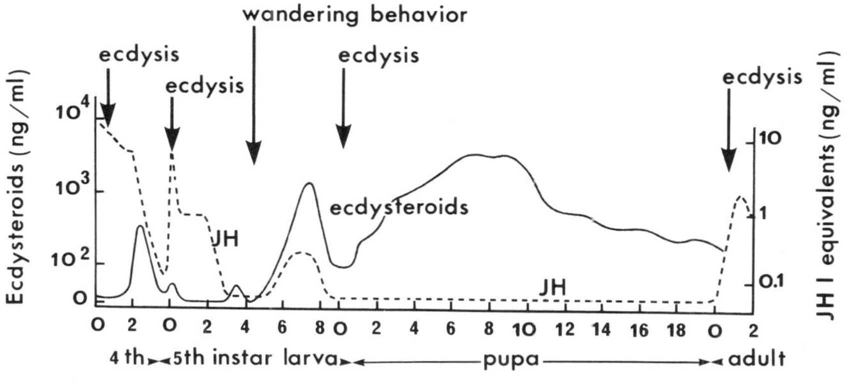

FIGURE 29. Juvenile hormone and ecdysteroid titers during larval growth and metamorphosis of *Manduca*. (Modified from Riddiford, 1980.)

3.4.3. *Bombyx*

The temporal organization of endocrine events was also established in *Bombyx* antepenultimate and penultimate larval instars by head ligation and decerebration at different times, possibly followed by PTTH injection and subsequent determination of ecdysteroid and JH titers. The scoring of development is established by the spiracular index (Sakurai, 1983, 1984).

It appears (Okuda *et al.*, 1985) that the timing of PTTH release is dependent on a circadian rhythm and that the sensitivity of the PG to PTTH varies during the penultimate instar. Two ecdysone peaks occur in the *Bombyx* final instar, as already observed in other lepidopterans (Calvez *et al.*, 1976; Kiguchi and Agui, 1981 in *Bombyx*; Bollenbacher *et al.*, 1975 in *Manduca*).

3.4.4. *Galleria* and *Ephestia*

Proprioceptors appear to be involved in the initiation of the molting cycle in *Galleria,* where external distortions of the larvae cause an inhibition of molting that might be reestablished by injection of ecdysone or 20-HE (Edwards, 1966; Sehnal and Edwards, 1969) (see Section 2.9.1).

In *Ephestia,* the determination of hemolymphatic ecdysteroids in final instar larvae shows that the darkness and isolation factors condition a prolongation of the wandering stage and a delay in pupation related to a low ecdysteroid titer (Giebultowicz *et al.*, 1984).

3.4.5. *Rhodnius*

From the earliest investigations of Wigglesworth, it was shown that in the blood-sucking bug *Rhodnius,* all the hormonal sequences are initiated by a single blood meal taken at each intermolt period. This is perceived by abdominal proprioceptors whose information is known to travel to the brain via the ventral nerve cord because if this is cut, brain stimulation is suspended. Furthermore, a series of small meals proves to be inactive, implying that abdominal distension plays a decisive role.

Decapitation experiments performed on different days after the meal served to determine a critical period during which the brain is necessary to induce growth.

The histological analysis of different tissues of the fourth instar larva found that the activation of the growth of the epidermis, reflected by an increase in the nuclear and nucleolar diameters, an increase in the cytoplasmic volume, and the initiation of the syntheses of nucleic acids and proteins takes place a few hours after the meal (Wigglesworth, 1957). This indicates that PTTH release and PG activation occur nearly immediately after the meal, although the critical period is only 4 days away. Thus the prolonged presence of PTTH appears to be necessary, at least in *Rhodnius*.

Two studies were recently carried out on the bursting activity of the CC and of the hemolymph ecdysteroid titer in the final instar larva of *Rhodnius* (Orchard and Steel, 1980; Steel *et al.*, 1982). The results showed that a correlation exists between the two processes. Moreover, the existence of a circadian rhythm of ecdysis and a daily rhythm in hemolymph ecdysteroid titer was demonstrated in *Rhodnius* raised in alternating light/dark conditions (Ampleford and Steel, 1982). It was also shown that 20-HE injections can elicit an ecdysis rhythm in insects reared in constant light (Ampleford, 1985).

3.4.6. Concluding Remarks

The interaction of external and internal factors and of different endocrine secretions (PTTH, ecdysone, JH) has been intensively investigated in *Rhodnius* and in *Manduca,* in which it has been shown *inter alia* that PTTH release is conditioned by the weight reached by the insect and by a circadian gate.

Recent investigations show that in other lepidopterans molting takes place, as in *Manduca,* in a definite part of the day (Sakurai, 1983 in *Bombyx*; Fujishita and Ishizaki, 1981 in *Samia*).

3.5. PTTH PRODUCTION AND RELEASE SITES

3.5.1. Production Sites

The location of the ns cells producing PTTH appears to vary according to the groups studied. Following the discovery by Hanström (1938) that ns cells occur in the PI of *Rhodnius,* Wigglesworth (1940) showed experimentally that median ns cells were the origin of the brain hormone. This was then confirmed in *Locusta* (Girardie, 1967) and in the dragonfly *Aeshna* (Charlet and Schaller, 1975). In aphids, one or two pairs of median protocerebral ns cells were reported to be involved in molting control (Steel, 1978). In the bug *Panstronglyus,* it appears that brain control of molting is exerted according to two modes by two distinct ns cell types of the PI, an early direct control by the type A ns cells and a delayed control exerted by the type A′ ns cells through the PG (Furtado, 1979). It remains to be seen if such a dual control of molting by the brain exists in other insects (see Chapter 6).

Using monoclonal antibodies against a synthetic N-terminal portion of the A chain of *Bombyx* 4K-PTTH, whose disulfide bonds were reduced, four pairs of median ns cells were revealed in both larvae and pupae, the nh peripheral part of the CA also being stained (Mizoguchi *et al.*, 1987). These ns cells recall the A-type ns cells and also the four pairs of insulin-like neurons demonstrated previously by Yui *et al.* (1980).

According to some experimental data, the lateral ns cells could play a role beside the PI ns cells. In *Hyalophora* pupae, brain action requires the 16 median cells of the PI as well as the 10 lateral cells (Williams, 1951). The function of the lateral ns cells was confirmed in *Locusta* (Girardie, 1974; Girardie and De Reggi, 1978) by the use of electrical stimulations followed by determination of hemolymph ecdysteroid titers, but these cells appear to be of lesser importance in this insect than the ns cells of the PI.

In other insects the lateral ns cells only are involved in PTTH production. This is the case in *Manduca* (Gibbs and Riddiford, 1977). The precise source cells producing the PTTH in this insect were identified by Agui and co-workers (1979), who incubated the PG with different parts of the brain. Ecdysone production was determined, and it was demonstrated that one particular neuron in each lobe of the brain is involved in PTTH production.

The implantation of various brain ns cells in head-ligated final instar *Galleria* larvae showed that only the lateral ns cells were able to induce pupation. The modifications of the load of these cells do not correlate with PTTH releases preceding ecdysteroid peaks. However, the absence of ecdysteroids in the hemolymph, as observed during starvation or in reduced space conditions, was accompanied by a depletion of the ns material in lateral ns cells (Muszynska-Pytel, 1987).

3.5.2. Release Sites

Evidence concerning ns processes implies that PTTH is released from the CC. However, it was observed in some cases that CC implants were less active than PI implants. This may be explained by the fact that isolated CC are formed of separate endings, and can receive release impulses neither from the ns cells themselves nor from any neurons controlling their activity. In addition, we have shown that the release site of brain ns products is not always located in the CC and may be located in other sites, namely, in the aorta wall in heteropterans and around the CA in lepidopterans (see Chapter 1).

The presence of PTTH activity in the CA was observed in *Philosamia* (Ichikawa and Nishiitsutsuji-Uwo, 1959), in *Bombyx* (Kobayashi and Yamashita, 1959), and, more recently, in *Manduca* (Gibbs and Riddiford, 1977), agreeing with the data of Nijhout (1975), demonstrating that the axon pathways of lateral protocerebral ns cells led to the CA. *In vitro* experiments aimed to determine the amount of ecdysone produced by the PG when kept in contact with various organs demonstrated definitely that in *Manduca* the CA were the source of the PTTH (Agui *et al.*, 1980; Carrow *et al.*, 1981).

Evaluation of the titer of PTTH activity in the CA and CC of *Manduca* during the final larval and early pupal instars showed that the changes in CA titer are the same as those of the brain (Agui *et al.*, 1980). Furthermore, the spontaneous release of PTTH was shown to be stimulated by high K^+ incubation medium, while in calcium-free media, spontaneous PTTH release was reduced (Carrow *et al.*, 1981).

The mechanisms regulating the release of PTTH have been studied in some cases. By electrical stimulation of the NCCI or NCCII in *Periplaneta,* it was observed that PTTH release depends exclusively on the NCCII (Gersch *et al.*, 1970). This does not necessarily imply, however, that the synthesis site is located in the lateral parts of the protocerebrum because the stimulated fibers may only regulate the release of ns products. The use of sympathico- and parasympathicomimetics, sympathico- and parasympathicolytics confirms this hypothesis and indicates the existence of aminergic transmission.

The mode of transport of the PTTH to the PG was also investigated. The results obtained suggest that PTTH released into the hemolymph is conveyed by hemocytes (Wigglesworth, 1955). The PTTH might also reach the PG directly via its nerves, for the presence of ns products has often been shown in these nerves as in the PG itself. Moreover, in *Calliphora,* the connections between the brain and Weismann's ring are indispensable to the functioning of the PG (Possompès, 1953), which supports the proposed hypothesis.

3.6. PURIFICATION OF PTTH

The identification of the chemical nature of the PTTH led to many investigations using different donors and assays. Following a period of preliminary research, the investigations focused on *Bombyx* and *Samia* on the one hand, and *Manduca* on the other (see Bollenbacher and Granger, 1985; Suzuki, 1986).

Gersch and his co-workers purified the factors present in the brain, CC, CA, and ventral nerve cord of the cockroach *Periplaneta,* which display several physiological effects. The C and D factors were isolated initially (Gersch and Mothes, 1956), and four different substances were separated from them—factors C1, C2, D1, and D2 (Gersch *et al.*, 1960). The presence of PTTH in the D1 fraction was demonstrated (Gersch, 1962) by injections into ligated abdomens of *Lymantria* and *Calliphora* implanted with Weismann's rings. Two distinct factors inducing molting in decerebrated *Antheraea* pupae—AHI and AHII—were separated from the D factor by Sephadex gel filtration using brain of *Periplaneta* (Gersch and Stürzebecher, 1968). These factors exhibit different properties. The AHI factor, with a molecular weight of about 50,000, stimulates RNA syntheses in the PG of *Periplaneta* but does not exert this action on other tissues; AHII, with a molecular weight of 10,000–20,000, elevates the membrane potential of the PG of *Galleria* and thus induces an increase in permeability (Gersch and Birkenbeil, 1980). Subsequent investigations showed that activation factor I enhances RNA syntheses *in vitro* (Gersch and Bräuer, 1974) and also increases ecdysone synthesis by the PG (Bräuer *et al.*, 1977).

3.6.1. *Bombyx* PTTH

The brains of *Bombyx* pupae were used to isolate 4 mg of active substance, by brainless diapausing *Bombyx* pupal assay, from 8500 and then 220,000 brains (Kobayashi and Kirimura, 1958; Kobayashi *et al.*, 1962). It was also shown (Yamazaki and Kobayashi, 1969) that *Bombyx* PTTH, active at the dose of 20 ng, was heat-stable, inactivated by trypsin, pronase, and nagarse, but not by chymotrypsin and sialidase. Its molecular weight was estimated at about 20,000 and its isoelectric point at 8.35–8.65.

Another research team using the brainless *Samia* pupal assay and aqueous extracts of *Bombyx* brains showed that the active factor was a water-soluble, heat-stable, nondialyzable protein which resists pepsin and trypsin but is inactivated by pronase and nagarse (Ichikawa and Ishizaki, 1961, 1963). In a subsequent work, Ishizaki and Ichikawa (1967) demonstrated the existence of three different proteins with molecular weights of 9000, 12,000, and 31,000.

3.6.1.1. Crude PTTH

In 1969, Ishizaki reported that the head of a *Bombyx* adult contained a high titer of PTTH. Later, Nishiitsutsuji-Uwo (1972) and Suzuki *et al.* (1975) obtained a partially purified PTTH from heads of *Bombyx* adults, with a molecular weight of 5000, that could cause adult development in a brainless *Samia* pupa at a dose of 1 μg. This partially purified PTTH, termed crude PTTH, was prepared from *Bombyx* adult heads kept at −20°C and purified according to a nine-step protocol.

3.6.1.2. Highly Purified PTTH

A 12-step procedure for PTTH purification (Fig. 30) was subsequently devised (Nagasawa *et al.*, 1979). Starting with 96,000 adult *Bombyx* heads, a 28,500-fold purification was performed. Six nanograms of this purified preparation causes adult development in a brainless *Samia*.

Three 4K-PTTHs were identified and complete amino acid sequence of 4K-PTTH II was determined using a 15-step procedure (Fig. 31) (Nagasawa *et al.*, 1984a,b, 1986). It was shown to be composed of two chains: the A chain with 20

96,000 Bombyx heads (fresh wt. 720 g)
↓
Acetone powder
↓ Washed with 80% EtOH; filtered
Residue
↓ Extracted with 2% aqueous NaCl,
Extract
↓ Heated in boiling water for 10 min
Cooled rapidly; filtered
Filtrate
↓ $(NH_4)_2SO_4$ added (80% saturation); filtered
Precipitate
↓ Dissolved in water
Acetone added; filtered
Acetone added to filtrate; filtered
Precipitate
↓ Dissolved in water
Saturated aqueous solution of picric acid added; filtered
Precipitate
↓ Dissolved in 0.1 M Tris buffer (pH 7.8)
Acetone added; filtered
Precipitate ("crude PTTH")
↓ Gel filtration on Sephadex G-50 (fine)
Sephadex G-50 (fine) fraction
↓ Chromatography on DEAE-Sepharose Cl 6B
DEAE-Sepharose fraction
↓ Chromatography on SP-Sephadex C-25
SP-Sephadex fraction
↓ Gel filtration on Sephadex G-50 (superfine)
Highly purified brain hormone

FIGURE 30. Procedure for purification of PTTH from the head of *Bombyx.* (From Nagasawa *et al.,* 1979.)

amino acids, and the B chain, a mixture of four microheterogeneous peptides, of which two have 26 amino acids and two have 28 amino acids:

4K-PTTH II A chain H-Gly-Ile-Val-Asp-Glu-Cys-Cys-Leu-Arg-Pro-Cys-Ser-Val-Asp-Val-Leu-Leu-Ser-Tyr-Cys-OH

4K-PTTH II B chain pGlu-Gln-Pro-Gln-Ala-Val-His-Thr-Tyr-Cys-Gly-Arg-His-Leu-Ala-Arg-Thr-Leu-Ala-Asp-Leu-Cys-Trp-Glu-Ala-Gly-Val-Asp-

Highly purified 4K-PTTH

13 DEAE-Sepharose CL-6B (gradient)

14 Sephadex G-50

15 Reversed phase HPLC

4K-PTTH-I, -II and -III

FIGURE 31. Further purification procedure for 4K-PTTHs. (From Suzuki *et al.,* 1982; Nagasawa *et al.,* 1984a.)

The A chain of the different 4K-PTTH exhibit rather close similarities, especially concerning the presence of disulfide bonds:

	1 5 10 15 19
4K-PTTH I	H-Gly-Val -Val-Asp-Glu-(Cys)-(Cys) -Phe -Arg -Pro-(Cys)-Thr -Leu -Asp-Val-Leu -Leu -Ser -Tyr-
4K-PTTH II	H-Gly-Ile -Val-Asp-Glu- Cys - Cys -Leu -Arg -Pro- Cys -Ser -Val -Asp-Val-Leu -Leu -Ser -Tyr-
4K-PTTH III	H-Gly-Val -Val-Asp-Glu-(Cys)-(Cys) -Leu -Gln -Pro-(Cys)-Thr - ? -Asp-Val-Val -Ala -Thr -Tyr-

Moreover, as shown below, 4K-PTTH II resembles the human insulin A chain and porcine relaxin. As for insulin, it should be noted that immunoreactive neurons to this peptide have been described in the brain and CC of several insect species and that it affects sugar metabolism and growth. It has been observed in

A chains

	1 5 10 15
Human insulin	H-Gly -Ile -Val -Glu -Gln -Cys-Cys-Thr -Ser -Ile -Cys-Ser -Leu -Tyr -Glu -Leu -Glu -Asn-
4K-PTTH II	H-Gly -Ile -Val -Asp-Glu -Cys-Cys-Leu-Arg-Pro -Cys-Ser -Val -Asp-Val -Leu -Leu-Ser -
Porcine relaxin	H-Arg-Met-Thr -Leu -Ser -Glu -Lys -Cys-Cys-Glu -Val -Gly -Cys-Ile -Arg -Lys -Asp-Ile -Ala -Arg -

	20
Human insulin	Tyr -Cys-Asn-OH
4K-PTTH II	Tyr -Cys-OH
Porcine relaxin	Leu -Cys-OH

B chains

	1 5 10 15
Human insulin	H-Phe -Val -Asn-Gln -His -Leu -Cys-Gly-Ser -His -Leu-Val-Glu -Ala -Leu-
4K-PTTH II	pGlu-Gln -Pro -Gln -Ala -Val -His -Thr -Tyr -Cys-Gly-Arg-His -Leu-Ala-Arg -Thr -Leu-
Porcine relaxin	pGlu-Ser -Thr -Asn-Asp-Phe -Ile -Lys -Ala -Cys-Gly-Arg -Glu -Leu-Val-Arg -Leu -Trp-

	20 25
Human insulin	Tyr -Leu -Val -Cys-Gly -Glu -Arg -Gly -Phe -Phe -Tyr-Thr-Pro-Lys-Thr-OH
4K-PTTH II	Ala -Asp-Leu -Cys-Trp -Glu -Ala -Gly -Val -Asp-OH
Porcine relaxin	Val -Glu -Ile -Cys-Gly -Val -Trp -Ser -OH

Pieris that an injection of insulin (A and B chain) was able to break diapause in a dose-dependent manner, inducing ecdysone synthesis even in decapitated insects (Arpagaus, 1987).

Other investigations (Suzuki *et al.*, 1982; Nagasawa *et al.*, 1984a,b) revealed the existence of 22K-PTTH, as well as the 4K-PTTHs of another molecular species.

The purified 4K-PTTHs were subjected to different enzymes and chemical reagents (Ishizaki *et al.*, 1977; Nagasawa *et al.*, 1979, 1984a) and to amino acid analysis by the dansyl method, which showed that the N-terminal amino acid was glycine and that the disulfide bond was indispensable to hormone activity.

As for 22K-PTTH, thorough purification produced 5 mg of purified hormone with a specific activity of 100,000 units/mg. The hormone, labeled with ^{125}I, appeared as a single band with a molecular weight of 18,000 (Matsuo *et al.*, 1985). Further investigation (Kataoka *et al.*, 1987) showed that it is a basic protein containing disulfide bonds whose terminal sequence is

Gly-Asn-Ile-Gln-Val-Glu-Asn-Gln-Ala-Ile-Pro-Asp-Pro-

It has no homology with 4K-PTTH. It is active in brainless *Bombyx* pupae at a dose of 0.11 ng.

3.6.1.3. Significance of PTTH Diversity

Purification of PTTH from *Bombyx* adult heads revealed the existence of four different molecules, 22K-PTTH and 4K-PTTH I, II, and III.

Using different molecules of *Bombyx* on decerebrated pupae of *Bombyx* and *Samia* (Ishizaki *et al.*, 1983a,b; Matsuo *et al.*, 1985), it was found that the active fractions on each of the two insects are not the same because 4K-PTTH acts on *Samia* and 22K-PTTH on *Bombyx* both in the larva and in the pupa (Suzuki and Ishizaki, 1986). In *Samia,* the amount of PTTH required for the development of a pupa is much smaller for 4K-PTTH III (0.1 ng) than for 4K-PTTH II (0.4 ng) (Nagasawa *et al.*, 1984a). Furthermore, the brain PTTH titer in *Bombyx* units (22K-PTTH) and *Samia* units (4K-PTTH) exhibits wide variations during development; the pupa and adult contain many more *Samia* units than *Bombyx* units whereas, by contrast, *Bombyx* units are essentially found in the larva.

3.6.2. *Manduca* PTTH

Research by American scientists aimed at identifying and characterizing the PTTH began with the work of Williams (1967) on the giant silkmoth *Antheraea*. Using tests of enzymatic destruction and spectrophotometric analysis, it was suggested that the hormone could be a mucopolysaccharide. Later, using the

Manduca larval assay, the PTTH of *Manduca* was shown to be heat-stable, water-soluble, and sensitive to pronase (Gibbs and Riddiford, 1977). The development of an *in vitro* assay then led to a rapid growth of knowledge (Bollenbacher and Gilbert, 1981; Bollenbacher *et al.*, 1981). The purification protocol was as follows:

1. 2500 brains from *Manduca* day 1 pupae
2. Extraction in Grace's medium/distilled water (1 : 2)
3. Heating for 2 min at 100°C
4. Centrifugation
5. Ultrafiltration
6. Lyophilization

This achieved a tenfold purification with no loss of activity.

Gel filtration of lyophilized material showed that in *Manduca* the PTTH existed in two forms called the big and small PTTH, their respective molecular weights being 28 and 7 kDa (Bollenbacher *et al.*, 1984) (Fig. 32). They were

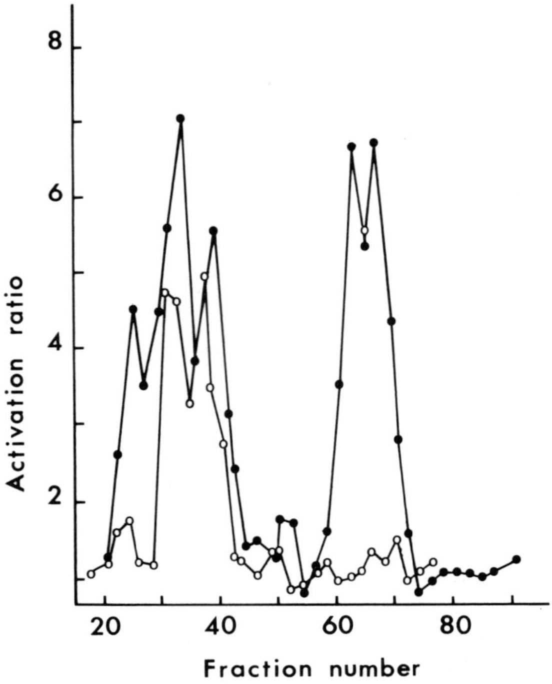

FIGURE 32. PTTH molecular weight determination in *Manduca*. Pupal brains are used and the PTTH activity quantified *in vitro* with prothoracic glands of day 3 final instar larvae (filled circles) and day O pupae (open circles). (Modified from Bollenbacher *et al.*, 1984.)

found to exhibit different activity on the larval and pupal PG, although both were present in the pupa. Thus the big PTTH is active on both larval and pupal PG, but the small PTTH is much more effective on the larva. The ED_{50} of the small PTTH of the pupa is about 20 times greater (3.8 units) than the ED_{50} of the larva (0.16 unit), while in the larva, the ED_{50} of small and big PTTH is the same in both the 3-day larva and the 0-day pupa (Table XVI).

The analysis of the hemolymph titer of small and big PTTH in different physiological stages reveals the presence of the big PTTH in the critical period of the fourth larval instar and the small PTTH in the first critical period of the fifth larval instar.

Determination of the amounts of small and big PTTH in the brain of day 3 final instar larvae and day 1 pupa, whose total activity is increased fivefold, showed that the larval brain contains mostly big PTTH, while the pupal brain contains as much PTTH as the larval and six times more small PTTH (O'Brien *et al.*, 1986) (Table XVI).

3.6.3. Embryonic PTTH

The presence of a PTTH similar to *Bombyx* 4K-PTTH was discovered in embryos of this species. It was also found in eggs and ovaries (Fugo *et al.*, 1987). Another form, with a molecular weight of 20,000–30,000, appears after the differentiation of the central nervous system (Chen *et al.*, 1987). In *Manduca*, a PTTH activity appears in the head of the embryo as early as 24 hr after egg laying, i.e., before brain differentiation (Dorn *et al.*, 1987).

3.6.4. EDNH versus PTTH

The presence in *Bombyx* of a *Samia* PTTH raises the question of whether this substance plays a physiological role in *Bombyx*. If so, it could be a role of the EDNH (egg development neurosecretory hormones) type, confirming the increase in the titer of this substance in the pupa and the adult, namely, at the time of the vitellogenesis.

A comparison between *Manduca* big and small PTTH and EDNH from

Table XVI. *Manduca* Brain Content and PG Sensitivity at Various Ages[a]

	Brain content		PG sensitivity to	
Age of the insect	Big PTTH	Small PTTH	Big PTTH	Small PTTH
Day 3 last instar larvae	+		+	
Day 1 pupae	+	+++	+	++

[a]From the data of Bollenbacher *et al.*, 1984; O'Brien *et al.*, 1986.

Aedes aegypti was made using the *Manduca* PG *in vitro* assay and the *Aedes* ovary *in vitro* and *in vivo* assays. Crude extracts of the adult mosquito heads strongly increased ecdysone synthesis by *Manduca* PG. However, *Manduca* big and small PTTH from pupal brain appeared to be devoid of EDNH and PTTH activity in *Aedes* (Kelly *et al.*, 1986a). Further, the characterization of EDNH and PTTH from *Aedes* pupae and adults revealed two protein fractions with a molecular weight of 11 and 24 kDa. They are present in both pupae and adults, suggesting a possible stimulatory role in steroidogenesis in the PG and ovaries (Whisenton *et al.*, 1987a).

3.6.5. Interspecific Activity of PTTH

We have shown that the PTTH exists in several molecular forms in *Bombyx* and *Manduca*. In *Manduca*, it appears that the active form in the larva is the big PTTH, while the small PTTH is active in the pupa (O'Brien *et al.*, 1986). This is not true of *Bombyx*, in which the larval and pupal PG are activated by the same PTTH, 22K-PTTH. Moreover, the latter is inactive in *Samia*, in which activation is induced by 4K-PTTH.

As for other species, interspecific PG sensitivity and PTTH specificity were investigated in *Bombyx*, *Manduca*, and *Mamestra* (Agui *et al.*, 1983). As shown in Table XVII, cross-reactivity was revealed for all three PTTH, but the sensitivity of the PG and the effectiveness of the PTTH appear to vary among the three lepidopterans.

Manduca crude PTTH was also assayed *in vitro* on the *Sarcophaga* larval ring gland and vice versa (Roberts *et al.*, 1986). Cross-reactivity was observed but only under certain conditions: when the PTTH and PG donors were not of the same age, no dose-response curve could be obtained; on the contrary, when PTTH from *Sarcophaga* prepupae and PG from day 3 fifth instar *Manduca* larvae were used, a typical PTTH-PG dose-response curve was obtained with an ED_{50} of 0.6 brain equivalent.

TABLE XVII. Cross-reactivity of the PTTH of *Manduca*, *Mamestra*, and *Bombyx*[a]

	PTTH activity (ED_{50})		
PG donor	*Manduca*	*Mamestra*	*Bombyx*
Manduca	0.16	0.35	0.17
Mamestra	0.62	0.91	0.41
Bombyx	0.94	1.21	0.34

[a]Modified from Agui *et al.*, 1983.

3.7. PTTH ACTION MECHANISM

It was suggested that the PTTH acted on PG through the intermediary of cAMP (Gersch and Birkenbeil, 1980). Indeed, cAMP is produced by PG homogenates and increases the quantity of ecdysone produced *in vitro* by the PG (Vedeckis *et al.*, 1974). Furthermore, the cAMP level in the PG rises during periods of PTTH release in *Manduca* (Vedeckis *et al.*, 1976). In subsequent investigations on *Manduca* (Smith *et al.*, 1984; Smith and Gilbert, 1986), the role of cAMP as a second messenger in the stimulation of the PG by the partially purified big PTTH was demonstrated *in vitro*. Stimulation is achieved in a dose-dependent manner and substances that increase the intracellular cAMP titer (1-methyl-3-isobutylxanthine, dibutyl cAMP, and forskolin) mimic the stimulation of the PG by PTTH. Modifications of the resting potential of the *Galleria* PG causing changes in permeability were reported during the activation of the PG by the activation factor of *Periplaneta* (Gersch *et al.*, 1976; Gersch and Birkenbeil, 1980).

According to Gundel *et al.* (1982a,b), the resting potential of the *Galleria* PG depends on the potassium ion concentration of the incubation medium. In *Manduca,* however, experiments conducted with larval PG and media of different ionic compositions indicate that the inward current underlying the action potential is carried by calcium (Eusebio and Moody, 1986).

The stimulation of ecdysteroid synthesis by the PTTH (Smith *et al.*, 1985; Smith and Gilbert, 1986) requires the presence of extracellular calcium, although the basal synthetic activity is not calcium-dependent. The action of calcium occurs before that of cAMP according to the following sequence of events:

1. Stimulation of the entry of Ca^{2+} by PTTH
2. Stimulation by Ca^{2+} of the formation of cAMP in a calcium-sensitive adenyl cyclase
3. Activation by the cyclic nucleotide of cAMP-dependent protein kinases responsible for phosphorylations
4. Stimulation of the enzymes involved in the synthesis of ecdysone

The effect of a brain factor on RNA synthesis in the PG was demonstrated by various authors *in vivo* and *in vitro* (Krishnakumaran *et al.*, 1965; Oberlander *et al.*, 1965; Kobayashi *et al.*, 1968; Gersch and Sturzebecher, 1970; Gersch *et al.*, 1973; Weber and Emmerich, 1976; Bräuer *et al.*, 1977). In the preparations of PG of *Bombyx* diapausing pupae, it was shown (Aizono *et al.*, 1986) that the PTTH stimulates the incorporation of labeled uridine in the RNA in a dose-dependent manner. This effect is inhibited by actinomycin D. Studies by electrophoresis show that this essentially involves ribosomic RNA.

3.8. IS PG REGULATED BY FACTORS OTHER THAN PTTH?

3.8.1. Ventral Nerve Cord

We have shown that the PG is richly innervated by the subesophageal ganglion and the anterior thoracic ganglia. A possible role of these ganglia in the regulation of its activity was considered in order to explain the results of experiments in which ligated, decapitated, and decerebrated insects are still able to resume molting, i.e., *Carausius* (Dupont-Raabe, 1952b), *Bombyx* (Kobayashi *et al.*, 1960), *Antheraea* (McDaniel and Berry, 1967), *Manduca* (Judy, 1972; Fain and Riddiford, 1976), and by the demonstration of PTTH activity in purified extracts of the ganglia of the cockroach *Periplaneta* (Gersch and Stürzebecher, 1968).

Gibbs and Riddiford (1977) investigated the presence of the PTTH in the ventral nerve cord ganglia of *Manduca* penultimate instar larvae. Following cervical ligation of the larvae, injections of different nerve ganglia extracts were made into the hind part of the body. While the brain is highly active, the subesophageal ganglion, like the thoracic ganglia and the abdominal ganglia, seems inactive and possibly even inhibitory. Ventral nerve cord ganglia were also assayed in diapausing pupae of *Manduca* (Safranek and Williams, 1980); only the brain was able to initiate wing epidermis apolysis, used as a bioassay.

Malá *et al.* (1977) performed removal of the brain and of the nerve ganglia innervating the PG in *Galleria*. They reported that growth was only disturbed by removal of the brain. The implantation of different ganglia in ligated final instar larvae was also carried out. In these experiments, the time of molting was significantly shortened by brain implants and slightly shortened by subesophageal ganglion implants, while the prothoracic and mesothoracic ganglia are practically without effect. Inhibitory action of the subesophageal ganglion, however, was suggested by Alexander (1970) in the same species. The author reported that immobilization of the larvae in narrow glass tubes delayed pupation, which was restored within normal intervals by removing the subesophageal ganglion.

Recent experiments performed on *Drosophila,* whose ring gland steroidogenesis was determined *in vitro* in the presence of brain and ventral nervous system extracts, showed that both nervous tissues contain PTTH activity, the ventral nerve ganglia being more active than the brain (Henrich *et al.*, 1987).

3.8.2. Nerve Connections

While the disconnection of the PG was reported to have no effect on its secretory activity in the lepidopteran *Papilio demoleus* (Srivastava *et al.*, 1977), in *Periplaneta* the persistence of a connection between the PG and the prothoracic ganglion activated its functioning *in vitro* (Marks and Reinecke, 1965; Richter and Gersch, 1983). Experiments performed *in vivo* also showed that sectioning of the PG nerves acts in two different ways according to the time performed, i.e., accelerating the time of molting or inhibiting the ecdysteroid peak (Richter, 1983). Accordingly, the electrical activity of the PG nerves was demonstrated to vary during the intermolt period with a maximum at the time of the hemolymph ecdysteroid peak (Richter and Gersch, 1983). Moreover, *in vivo* and *in vitro* experiments revealed that a rise in the electrical activity of the PG nerves occurs when the prothoracic ganglion is separated from the subesophageal ganglion (Richter, 1985), with the latter probably exerting an inhibitory role.

3.8.3. Ecdysone and 20-HE Feedback

3.8.3.1. Neurosecretory Cells

The ecdysteroid titer of the hemolymph appears to exert a feedback control both on the ns cells which synthesize PTTH and on the PG.

Removal of the PG and injection of 20-HE into the larvae of *Aeshna* and *Mamestra* induces changes in the appearance of the ns cells (Charlet and Schaller, 1975; Agui and Hiruma, 1977a). These changes are also found in the brains of diapausing pupae incubated with 20-HE (Agui and Hiruma, 1977b). They are accompanied by elevated brain PTTH activity, which is detectable in diapausing PG incubated with pieces of integument. In *Leucophaea,* cultured brains accumulate ns products in the PI and release them if 20-HE is added (Marks *et al.*, 1972). By parabiosing 1-day and 8-day fed *Rhodnius,* in which the ecdysone titer is high, Steel (1978) induced overloaded perikarya after 8 days in the former. This seems to indicate that ecdysone causes inhibition of the release processes of brain ns products. Tritiated ecdysone injections show that this hormone exerts its action at the level of the perikarya, where selective concentration occurs.

20-HE was also shown to stimulate the production of cAMP by the brain of *Manduca* (Vedeckis and Gilbert, 1973), a process which appears to be related to PTTH secretion and to influence the electrical activity of the ns pathway.

In *Periplaneta,* records from the NCCII in an *in vitro* subesophageal ganglion–brain–NCCI–NCCII preparation indicated that 20-HE reduces electrical activity (Richter and Gersch, 1981). On the contrary, in *Rhodnius* ovariectomized adult females, in which bursting activity of the ns cells of the PI and the CC was significantly reduced, the injection of 20-HE reestablished the normal spike frequency.

3.8.3.2. Prothoracic Gland

In addition to its stimulation by PTTH, the PG appears to be directly regulated by the hemolymph ecdysteroid titer in *Platysamia* and *Samia* (Williams, 1952; Siew and Gilbert, 1971).

The existence of a negative feedback exerted by ecdysone and 20-HE on the PG was confirmed by RIA determination of ecdysteroids in *Pieris* (Beydon and Lafont, 1983) and *Mamestra* (Bodnaryk, 1986b). The two hormones ecdysone and 20-HE inhibit the production of ecdysteroids, and 20-HE is much more active than ecdysone in both species.

In *Pieris,* the inhibitory effect of ecdysone and of 20-HE is observed in decapitated pupae, suggesting the direct action of the hormones on the PG.

3.8.4. Juvenile Hormone Feedback

3.8.4.1. Juvenile Hormone and Neurosecretion

Stimulatory action of the JH on the activity of the PI cells in the adult was demonstrated in several insects. In *Schistocerca* and *Locusta,* removal of the CA causes storage of type A ns products in the PI (Highnam *et al.,* 1963), and supernumerary CA implants or the application of JH or JHA clearly stimulates the synthesis and transport of ns products (McCaffery and Highnam, 1975a,b). Allatectomy depresses the synthetic activity of the ns cells of the PI in *Tenebrio* (Mordue, 1965), *Leptinotarsa* (De Wilde and De Boer, 1969), and *Calliphora* (Thomsen and Lea, 1969), while JH or JHA injections or applications activate the ns cells in adult diapausing *Leptinotarsa* (Schooneveld, 1972; Schooneveld *et al.,* 1977) and in *Mamestra* larvae (Hiruma *et al.,* 1978a; Hiruma, 1986).

3.8.4.2. Juvenile Hormone and Prothoracic Gland

Apart from its action on the brain ns cells, JH seems to act directly on the PG by stimulating it. In fact, Williams (1959), Gilbert and Schneiderman (1959), and Ichikawa and Nishiitsutsuji-Uwo (1959) succeeded in terminating diapause and the second pupal molt in decerebrated pupae of the great saturniids by CA implants or extracts, an effect which might be explained by the fact that lepidopterans store their ns products in the periphery of the CA (see Chapter 1). Similar results were obtained by injecting JH or JHA (Krishnakumaran and Schneiderman, 1965; Bhaskaran, 1980; Jones *et al.,* 1986). In addition, in the cockroach *Diploptera,* CA removal caused a reduction of the ecdysteroid titer of the larva (Kikukawa and Tobe, 1986a). Hence it may be possible that the CA exert a dual action on the PG.

Later studies revealed that the developmental effects induced by JH vary according to the time of the experimental procedure, and that PG sensitivity to JH undergoes a switchover during the pupal intermolt period, at least in lepidopterans. Juvenile hormone application before the switchover, namely, at the time of lowest JH hemolymphatic titer, slows development, whereas after the switchover it accelerates metamorphosis. This was observed in *Manduca* (Nijhout and Williams, 1974b; Safranek *et al.*, 1980a), *Mamestra* (Hiruma *et al.*, 1978b), *Spodoptera* (Cymborowski and Stolarz, 1979), *Laspeyresia* (Sieber and Benz, 1980a,b), *Galleria* (Sehnal *et al.*, 1981), and *Trichoplusia* (Newitt and Hammock, 1986).

Hemolymph ecdysteroid determinations confirmed the experimental results in *Galleria* (Sehnal *et al.*, 1981), *Mamestra* (Hiruma and Agui, 1982), *Spodoptera* (Cymborowski and Zimowska, 1984), and *Ephestia* (Lazarovici *et al.*, 1984), showing that the ecdysteroid titer remains very low when JHA is applied before the switchover and rises significantly if the application is carried out after the switchover (Table XVIII).

In *Spodoptera*, for example (Cymborowski and Stolarz, 1979), JHA application to final instar larvae during the feeding period causes a delay in metamorphosis and induces either a prolongation of the feeding period or supernumerary instars. When applied after cessation of feeding it causes metamorphosis to accelerate. The switchover in the sensitivity of the PG was demonstrated to take place at a precise time in the middle of the third scotophase, soon after feeding

TABLE XVIII. Hemolymph Ecdysteroid Titer in Last Instar Larvae of *Spodoptera*, Following Neck Ligation and JHA Application[a]

Insects	ng Ecdysteroids/ml hemolymph for insects ligated at		
	Day-2	Day-3	Day-4
Neck-ligated larvae before switchover	275	490	6620
Neck-ligated larvae before switchover receiving JHA application	6	1	1
Neck-ligated larvae just after switchover	91	880	4404
Neck-ligated larvae just after switchover receiving JHA application	44	1589	
Neck-ligated larvae 6 hr after switchover	627		
Neck-ligated larvae 6 hr after switchover receiving JHA application	6204		

[a]From the data of Cymborowski and Zimowska, 1984.

cessation and heart exposure, and before gut purge, when the larvae reach their maximum weight. It coincides with the ecdysteroid commitment peak. Observation of the PG revealed its regression in the insects treated before the switchover (Cymborowski and Zimowska, 1984) (Table XVIII).

3.8.4.3. Juvenile Hormone Action Site

The JH action site in PG activation is still under debate. Some authors consider that action should be direct, but a number of investigators favor indirect action (Sieber and Benz, 1980a in *Laspeyresia*; Sehnal *et al.*, 1981 in *Galleria*; Rountree and Bollenbacher, 1986 in *Manduca*; Garcia *et al.*, 1987 in *Rhodnius*). In *Manduca,* certain data suggest direct action (Nijhout and Williams, 1974b; Safranek *et al.*, 1980) but other studies show that JH stimulates PTTH release in neck-ligated larvae implanted with brain-CC-CA complexes. If the organ complex originates in JHA-treated larvae, the commitment is delayed, but if it originates in normal larvae prior to PTTH release, the commitment occurs in the normal time. Further, if PTTH is added to *in vitro* incubated PG of different ages, the PG of JHA-treated insects are activated at any time, while the PG of normal insects are activated until day 3 and then become progressively refractory to activation by PTTH (Rountree and Bollenbacher, 1986) (Fig. 33).

The physiological significance of the switchover in PG sensitivity to JH is not quite clear, the more so as it occurs in head-ligated insects. The involvement of factor released from some tissue in the body of the larva looks like an attractive possibility. Furthermore, it may be noted that the periods in which JH inhibits the PG are those in which it is absent from the normal organism, and that the periods in which JH stimulates the PG are those in which it is present in the normal organism.

The sensitivity (competency) of the PG to PTTH varies during the larval-pupal stage (Ciancio *et al.*, 1986) and in diapausing insects (Bowen *et al.*, 1984), day 3 larvae being nearly twice as sensitive to PTTH than day 0 pupae. Thus some changes probably occur at the PG level, regulated by the drop in hemolymphatic JH titer (Watson *et al.*, 1987a).

3.8.5. Hemolymph Stimulatory Factor

In *Manduca,* further investigations showed that JH stimulates the production of ecdysteroids by the PG but not in a dose-dependent manner, confirming that its stimulatory effects may be indirect (Gruetzmacher *et al.*, 1984a). Indeed, incubation of the PG with the incubation medium of the fat body revealed that this tissue exerts a dose-dependent stimulation on the PG. The stimulatory factor from the fat body, also termed the hemolymph stimulatory factor (HSF), was shown to have additive effects to those of the PTTH and to display titer fluctua-

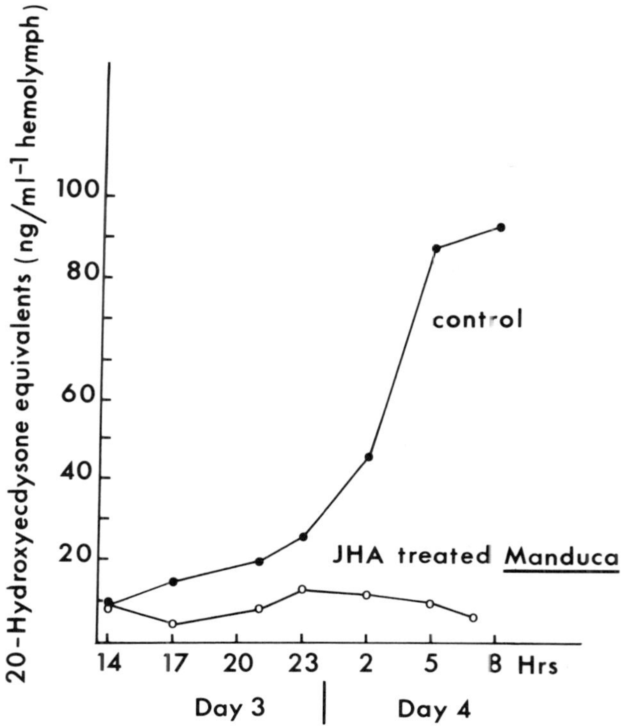

FIGURE 33. Ecdysteroid titer modified by treatment with ZR 512, a juvenile hormone analog (JHA) in *Manduca* larvae. (Modified from Rountree and Bollenbacher, 1986.)

tions in the hemolymph during the final larval instar (Gruetzmacher *et al.*, 1984b) (Fig. 34). The titer of the HSF is low on day 1–3, when the ecdysteroid titer increases only slightly, and increases by threefold on day 7, just prior to the dramatic surge of ecdysteroids (Watson *et al.*, 1985, 1987b).

Unlike PTTH, HSF has been shown to be heat-labile. Its molecular weight was estimated to be in the 30-kDa range (Gruetzmacher *et al.*, 1984b). Its mode of action on the PG differs from that of the PTTH in that it does not involve cAMP, although its maximum activity requires the presence of extracellular calcium. It stimulates the PG within a very short time compared with that of the PTTH (Smith *et al.*, 1986). It could be involved in the transport to the PG of steroid substrates required for steroidogenesis.

The question arises as to whether the existence of HSF can explain the variations in sensitivity of the PG and its different reactions to an input of exogenous JH. Future experiments will perhaps enlighten us on this matter. We

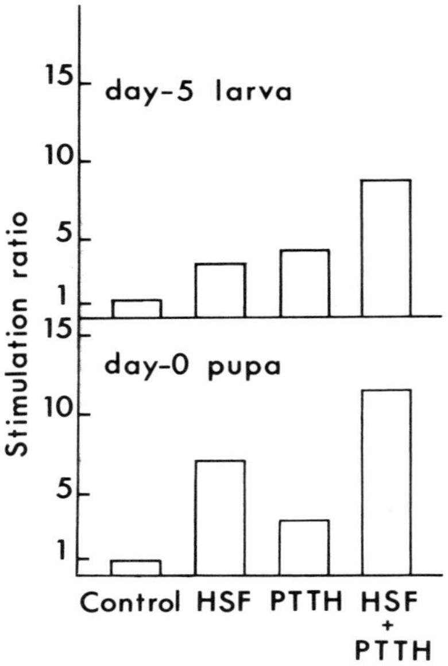

FIGURE 34. Stimulation ratio of PG by PTTH and hemolymph stimulating factor (HSF) in *Manduca.* Note the increase in prothoracic gland activation when both PTTH and HSF are present in the medium. (Modified from Watson *et al.,* 1985.)

already know that its production depends on the JH titer. In the pupae of *Heliothis,* a diapause-terminating factor was demonstrated in the hemolymph and fat body. It is necessary for the functioning of the PG *in vitro* (Gray *et al.,* 1987) and thus could play a similar role to that of HSF.

3.8.6. Prothoracic Gland Activation in Debrained Insects

Although the stimulatory action of the PTTH on the PG and the action of the PG in the production of ecdysteroids are firmly established, a number of results indicate that substitutes may occur and that endocrine processes could be more complex than generally believed.

Prothoracic gland activation by injury or cold was observed in the diapausing debrained pupae of *Antheraea* (McDaniel and Berry, 1967) and *Manduca* (Judy, 1972). Moreover, in the diapause-destined, freshly pupated *Manduca,* neck ligation accelerated the initiation of adult development (Safranek *et al.,* 1986). These paradoxical experimental results might be explained by the occurrence of other stimulatory factors besides PTTH, such as JH or HSF, or even by ecdysteroid synthesis outside the PG.

The continuation of development after PG removal has been reported in several insect species, i.e., *Periplaneta* (Chadwick, 1956; Gersch, 1977), *Galleria* (Sláma, 1983), *Manduca* (Safranek *et al.,* 1986), *Leptinotarsa* (Hsiao *et al.*, 1975), *Tenebrio* (Glitho *et al.*, 1979), and *Dermestes* (Delbecque and Sláma, 1980), and determination of hemolymphatic ecdysteroid titers revealed that ecdysteroid peaks could occur in decapitated insects (Hsiao *et al.*, 1975; Lafont *et al.*, 1977; Delbecque *et al.*, 1978; Delbecque and Sláma, 1980; Sehnal *et al.*, 1981, 1986).

These results have sometimes been explained by an incomplete extirpation of the PG, but it has also been proposed that another tissue, namely, the oenocytes, whose ultrastructural characters recall those of the vertebrate steroidogenetic cells, might be involved in ecdysteroid synthesis (Romer *et al.*, 1974). Moreover, ecdysteroid synthesis by the follicular cells of the ovary has recently been discovered (see Chapter 4).

3.9. CONCLUSION

Prothoracicotropic hormone, which controls PG activity, is the first neurohormone demonstrated in insects. It involves the action of cAMP, induces changes in gland permeability, and triggers ecdysteroid secretion.

Different bioassays *in vivo* and *in vitro* were used, mainly in *Samia* and *Manduca*.

Thanks to RIA determinations of ecdysteroids as well as decapitation and neck ligation experiments performed at different times in the intermolt period, precise chronologies of PTTH release have been established in connection with behavioral and morphological events.

The production and release sites of *Manduca* PTTH have now been identified.

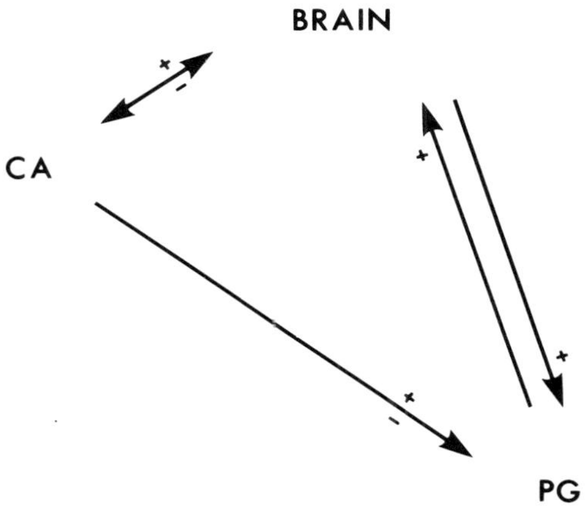

FIGURE 35. Schematic representation of the feedback control mutually exerted by the corpora allata, prothoracic gland, and neurosecretory cells of the pars intercerebralis.

Many investigations were conducted by different teams in an attempt to identify PTTH molecular structure. In *Bombyx* as in *Manduca,* two forms of PTTH have been identified, one big and the other small. The small PTTH of *Bombyx* exists in three different forms of comparable molecular structure. 4K-PTTH I and III are nearly completely identified, whereas 4K-PTTH II has been fully identified.

Certain results suggest that the PG is regulated not only by the PTTH but also by a hemolymph factor. Furthermore, ecdysteroids and JH exert either direct or indirect action on its activity (Fig. 35).

4

Regulation of Corpora Allata Activity and Juvenile Hormone Titer

INTRODUCTION

Besides ecdysones, juvenile hormone (JHs) are the only known insect hormones. They are vitally important in the control of several aspects of male and female reproduction, and regulate the development of the insects, ensuring the maintenance of the juvenile characters. The existence of a juvenile hormone was demonstrated as early as 1936 in *Rhodnius* (Wigglesworth, 1936) and confirmed by many subsequent findings (see Wigglesworth, 1985).

Research was facilitated by the relative ease of ablation of structures with clearly defined contours, like the corpora allata (CA), and by the discovery of JH analogs active by simple application on integuments (JHA).

The understanding of the regulation of CA activity made considerable headway thanks to the introduction of new *in vitro* assays and the development of determinations designed to ascertain the JH titer of different organs and of the hemolymph. The knowledge of the chemical processes involved in JH synthesis and metabolism served for the interpretation of certain experimental data.

On the one hand, the mechanisms regulating Ca activity and JH titers are exerted on the CA. As we shall show, these are inhibitory and stimulatory mechanisms. However, other distinct factors also play an important role because they enable the JHs to be protested by bounding with binding proteins, and also favor their destruction by stimulating the production of catabolizing enzymes.

4.1. JUVENILE HORMONES, THEIR ESTERASES AND BINDING PROTEINS

4.1.1. Juvenile Hormone Molecular Structure

Contrary to what was initially believed, insects use not a single JH but several, i.e., JHI (C_{18} JH) (identified by Röller *et al.*, 1967), JHII (C_{17} JH) (Meyer *et al.*, 1968), JHIII (C_{16} JH) (Bowers *et al.*, 1965; Judy *et al.*, 1973), and JHO (Bergot *et al.*, 1980) (Fig. 36).

The distribution of the various JHs in the different species and during development is still not completely known. JHI, II, and III are present in lepidopterans, whereas JHIII predominates in many other insects, especially in the adult. It appears to represent the earliest form of JH. Juvenile hormones are acylic sesquiterpenoid epoxides. The structure of JHIII, which closely resembles farnesol, a well-known analog, is the simplest of the juvenile hormones, from which the others appear to derive.

4.1.2. Juvenile Hormone Titer

Juvenile hormones are found in several tissues, such as the ovary, accessory glands, gut, and especially the fat body. Variations in titer in the hemolymph appear to match those of biosynthetic activity of the CA, as observed during the

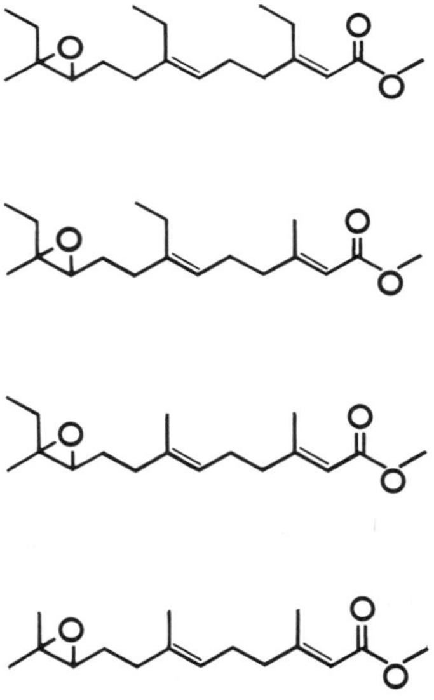

JH O

JH I

JH II

JH III

FIGURE 36. Molecular structure of the juvenile hormones of insects.

larval-pupal development of *Manduca* (Granger *et al.*, 1982), *Leptinotarsa* (Khan *et al.*, 1983), and *Diploptera* (Tobe *et al.*, 1985).

4.1.3. Juvenile Hormone Esterases

The JH titers in the tissues and the hemolymph do not depend exclusively on the synthetic activity of the CA, but on JH degradation by esterases as well. This can occur by several biochemical pathways and by different types of esterases. Carboxyesterases, which are fairly abundant in the hemolymph, the fat body, and the epidermis, carry out ester hydrolysis and convert the JH to JH acid. They have little effect on the JH linked to binding proteins and do not appear to be very important in JH degradation. Juvenile hormone-specific esterases, on the contrary, degrade free JH as well as JHs complexed to carrier proteins (see De Kort *et al.*, 1979).

4.1.4. Juvenile Hormone-Binding Proteins

Once released into the hemolymph, the JHs bind to carrier proteins produced in the fat body, a process that involves recognition by the carrier proteins of the epoxide function, and a polar region corresponding to the ester function, the hydrophobic chain, and the geometry about the double bond and the JH epoxides.

Juvenile hormone-binding proteins were first identified in saturniids (Whitmore *et al.*, 1972) and in sphingids (Goodman and Gilbert, 1978), and later in other species but not all those examined (see De Kort and Granger, 1981). The binding proteins of JHI are different from those of JHIII, but the latter vary with the species (De Kort *et al.*, 1984).

The binding of JHs to proteins or lipoproteins is important for the transport of hormones and their protection against degradation by esterases. In *Diatraea,* an insect exhibiting larval diapause, it was observed that the titers of binding proteins increase in the prediapausing larvae, reach a plateau at the onset of diapause, and remain high throughout diapause (Dillwith *et al.*, 1985), which probably results from continued production in the fat body. In *Manduca,* cyclical fluctuations of hemolymph JH-binding capacities were observed during each larval instar, with low capacities at the beginning and the end of the instar and higher levels during the midpart (see Goodman and Chang, 1985). These data suggest the possible existence of a hormonal regulation of the binding protein synthesis, but this has not yet been demonstrated.

The liposolubility of JHs suggests that these hormones may be stored in lipophilic pools of certain tissues, which would explain the additional methods of regulation.

4.1.5. Juvenile Hormone Analogs

A large number of substances (more than 100) whose action mimics that of the JH and is exerted by simple contact with the integument of the insects have been known since the 1960s (see Sláma *et al.*, 1974; Sehnal, 1983). They are invaluable for the experimental worker, who can use them easily and quantitatively to restore or disturb the hormonal balance. The structure of these JHA, like their activity, is varied, and the best known are farnesol, juvabione (paper factor), and methoprene (altosid or ZR 515) (Fig. 37).

4.1.6. Precocene—Azadirachtin—Antibodies

Other substances, such as the preocenes I, II, and III, derived from plants like *Aegeratum* (Bowers *et al.*, 1976), exert anti-JH action and cause the rapid necrosis of the CA, which stops synthesizing JHs (see Tobe and Stay, 1986; Staal, 1986). Precocenes can be used as experimental tools in cases where ablation of the CA is impossible because their effects mimic the action of this operation and are exerted following simple topical applications. Furthermore, researchers have felt that they could serve as pesticides in the fight against

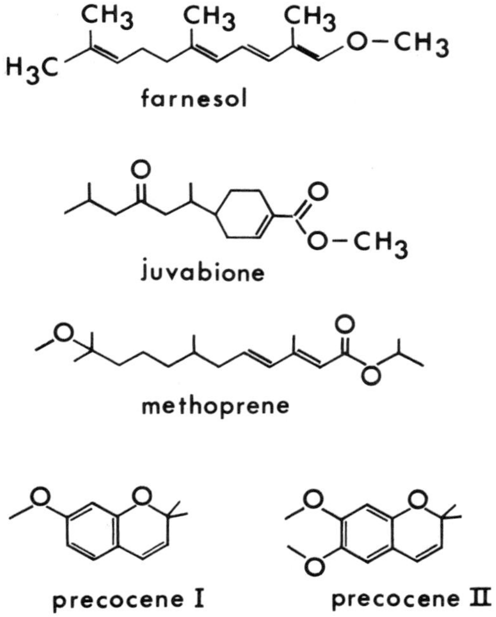

FIGURE 37. Molecular structure of the main juvenile hormone analogs and of precocene I and II.

particularly sensitive insects and especially as proallactocidins. Precocenes have therefore sparked many investigations (Fig. 37).

The action mechanism of precocenes has been debated because for some researchers their effect on the CA involved action on the brain. However, this view is not widely shared. At all events, the ultrastructure of precocene-treated glands did reveal degeneration processes (*Oncopeltus, Locusta, Diploptera*).

The question arises as to whether all species react in the same way to precocene. While this question is still unanswered, some ideas have been derived concerning the way in which precocene acts specifically at the level of the CA cell, where it is oxidized due to the presence of epoxidase, an enzyme of the JH synthetic pathway.

Another molecule, azadirachtin, of plant origin and derived from *Azadirachta indica,* acts as an antifeeder and disturbs the development of the insects. It has been reported to inhibit JH synthesis, but its main effect seems to be through ecdysone synthesis in *Oncopeltus* (Redfern, 1982) and *Locusta* (Sieber and Rembold, 1983; Mordue *et al.*, 1986; Dorn *et al.*, 1986). Its toxic effects have been demonstrated in the PG of *Manduca* (Bidmon, 1986) and *Locusta.* Its mode of action is still unknown.

One method that could be effective is the inhibition of activity in the CA by an antiserum raised against allatotropic brain neurohormones. This has been carried out in *Locusta* (Rembold *et al.*, 1980; Ulrich *et al.*, 1985; Girardie *et al.*, 1987a) (see Section 4.5).

4.2. JUVENILE HORMONE ASSAYS

Juvenile hormone has been identified and quantified for many years, and by many authors, by evaluating the activity of transplanted CA in insects temporarily deprived of JH. Semiquantitative bioassays were later devised for saturniid moths (Schneiderman and Williams, 1954) and *Galleria* (Gilbert and Schneiderman, 1960).

In the *Galleria* assay, the procedure is based on the evolution of the integument in a case of injury followed by covering the injured place with liquid paraffin containing an amount of JH to be determined.

4.2.1. Physicochemical Determination

Juvenile hormones can be determined qualitatively and quantitatively by the physicochemical techniques of mass spectrometry, which are highly sensitive and selective, and which help to distinguish the four JHs. They necessarily include one or more separations, using either a gas chromatograph or a high-performance liquid chromatograph, or a combination of the two, followed by mass spectrometry. This is a rather lengthy procedure which offers very reliable results.

4.2.2. Radioimmunoassay

This highly sensitive and specific method was developed for JHI, II, and III (Baehr *et al.,* 1976; Strambi *et al.,* 1981), and modified by Granger *et al.* (1979). Apart from their advantages, radioimmunological methods present certain drawbacks, such as cross-reactivity to precursors and metabolites. Moreover, the JH molecule is a poor immunogen and must be conjugated to an immunogenic molecule in order for RIA to work.

Using the Baehr method, JH was conjugated via an *N*-hydroxysuccinimide ester intermediate of JH acid to human serum albumin, with iodated histamine conjugates of the JH as tracers. The advantage of this method is that the results shed light on the different JHs. In some cases, however, the values obtained have been considered too high in comparison with physicochemical measurements. This method was recently improved by the use of acetylcholinesterase for coupling to acids from hydrolyzing diols. It is more specific and more sensitive (Baehr *et al.,* 1987).

A modified RIA bioassay for JHI was devised in *Manduca* using a paired gland protocol and incubating the glands in Marks medium (Granger *et al.,* 1979).

Another RIA procedure was developed by Strambi *et al.* (1981), in which antigens and tracers are prepared via the diol derivative of JHI, with JHI, II and III being determined by a single antiserum. The method consists of three main steps, i.e., opening of the 10,11-epoxide ring, succinylation and conjugation to human serum albumin, and acid hydrolysis of the JH to the 10,11-diols. This method offers several advantages due to the high immunoreactivity and the high solubility of the diol derivatives. Its main drawback is its failure to distinguish between the different JHs measured, unless they have been separated previously.

4.2.3. Short-term Radiochemical *in Vitro* Assay

A short-term radiochemical *in vitro* assay was developed to determine the synthetic activity of the CA (Pratt and Tobe, 1974). The incorporation of the methyl group of radiolabeled methionine into the methyl ester group of JH is the basis of this method, in which the CA are incubated for 3 hr with labeled sesquiterpenoid acids and [methyl-^{14}C]methionine. A radioactivity determination of the incubation medium by liquid scintillation spectrometry after thin-layer chromatography is used to obtain the activity level of the CA. Subject to the careful prevention of any pollution by the fat body, whose JH esterases could distort the results, and confirmation that the labeled methionine is incorporated at a ratio of 1:1, this method proves to be highly satisfactory and is currently widely used. A modification of the procedure proposed by Feyereisen (1985) makes it the fastest method, by partitioning the biosynthesized JH between an aqueous and organic phase and suppressing the HPLC separation.

Note that not all of the assays mentioned above have the same utility. In fact, the titer of the hemolymph and the tissues can only be measured by physicochemical methods or RIA, whereas the level of activity of the CA, apart from transplantations, can only be analyzed *in vitro*.

4.3. CORPORA ALLATA INNERVATION AND STRUCTURE

4.3.1. Corpora Allata Innervation

Because CA activity is influenced by different factors, it is important to understand how information is transmitted to these organs. The CA are innervated by two pairs of nerves, the NCA1, which exit from the CC, and the NCA2, which originate in the subesophageal ganglion. While the starting point of the NCA2 is perfectly clear, that of the NCA1 is relatively obscure because they exit from the CC. It is well known that these nerves originate in the brain and generally contain ns and non-ns axons, but in most cases it is unknown whether the ns axons originate in the NCCI arising from the PI, in the NCCII originating in the lateral ns cells of the protocerebrum (pars lateralis), or in the NCCIII originating in the tritocerebrum.

The cobalt chloride technique affords an excellent tool for determining the source of the axons that innervate an organ. However, the evidence at present concerns only a small number of species. In *Schistocerca,* Mason (1973) reported that the axons innervating the CA originate only in the NCCII and NCCIII. In *Locusta,* however, nerve sections, electron microscopy, and cobalt chloride filling show that they originate in both the NCCI and the NCCII (Cassier and Fain-Maurel, 1970b; Rademakers, 1977b) (Fig. 38).

According to Poras (1983), the axons in NCCII, which alone innervate the CA of *Locusta,* arise in 15 ipsilateral cells and two contralateral cells, which also appear to innervate the CC. An important factor for the interpretation of certain experimental results is that the lateral neurons innervating the CA form a loop in the vicinity of the PI, where their collaterals appear to form synapses. Hence cauterizations of PI can easily interrupt their path and lead to the denervation of the CA.

In the cockroach *Periplaneta* (Fraser and Pipa, 1977), the source cells of the NCA1 are found in both the PI and the pars lateralis. Each CA is innervated by 50–60 cells in the contralateral PI, 8–12 cells in the contralateral pars lateralis,

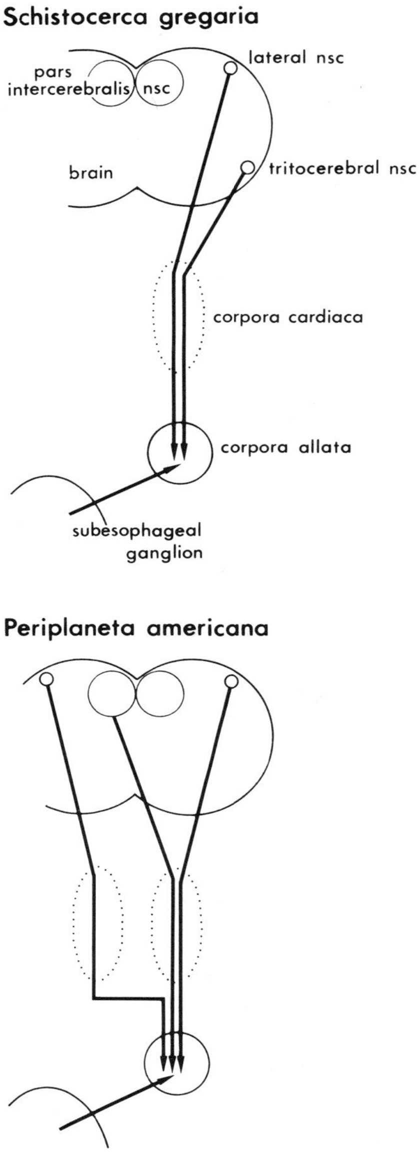

FIGURE 38. Corpus allatum innervation as revealed by cobalt chloride filling in four insects. In addition to their subesophageal innervation, the corpora allata are innervated by lateral and tritocerebral neurons in *Schistocerca* (Mason, 1973), by pars intercerebralis and lateral neurons in *Locusta* (Cassier and Fain-Maurel, 1970b; Rademakers, 1977b), by pars intercerebralis, lateral, ipsilateral, and contralateral neurons in *Periplaneta* (Fraser and Pipa, 1977; Gundel and Penzlin, 1980), and by pars intercerebralis and lateral neurons in *Manduca* (Nijhout, 1975). nsc, neurosecretory cells.

and 16–20 cells in the ipsilateral pars lateralis. Using scanning electron microscopy and nickel chloride backfilling, similar results were obtained for *Diploptera* (about 40 median neurons and 20 lateral neurons) (Lococo and Tobe, 1984).

In the larva and pupa of *Manduca* (Nijhout, 1975), the CA receive axons originating in the PI and in ns cells of the pars lateralis. It appears, therefore, that the CA are often innervated by both the PI and the pars lateralis (Fig. 38).

In *Leptinotarsa* (Khan *et al.*, 1984), the CA are innervated by two pairs of lateral neurons which split into two pathways, one running into the CC and CA, and the other branching as in *Locusta,* in the vicinity of the PI. In this species, NCCI and II are fused.

In those investigations dealing with the NCA2, the source cells have been shown to be four to six medioventral ns cells of the subesophageal ganglion in *Periplaneta* (Pipa and Novak, 1979), *Locusta* (Chalaye, 1967), and *Schistocerca* (Mason, 1973). The NCA2 are probably more primitive than the NCA1 because the origin of the CA is ventral. Moreover, in ephemeropterans, the CA are only innervated by the NCA2 (Table XIX).

Neurosecretory material has been observed histologically in the CA and in their nerves in species belonging to the major orders (see Raabe, 1982). Electron microscopy studies have also been carried out, particularly in *Leucophaea* (Scharrer, 1964a), *Locusta* (Cassier and Fain-Maurel, 1970b) (Fig. 39), *Calliphora* (Normann, 1965), and in many other species (see Cassier, 1979). The results of these investigations indicate that two and sometimes three types of granules are present in the ns material.

4.3.2. Structure and Ultrastructure

Given the importance of the role played by the CA in the physiology of insects, researchers have long attempted to establish correlations between the cyclic variations in the activity of these organs and their appearance. Apart from

TABLE XIX. Nervous, Neurohormonal, and Ovarian Factors Involved in Corpora Allata Regulation

	Nervous		Neurohormonal		Ovarian	
Insect	Stimulation	Inhibition	Stimulation	Inhibition	Stimulation	Inhibition
Diploptera		+	+	+	+	+
Locusta	+	+	+	+	+	
Leptinotarsa		+	+	+		
Manduca	+	+	+			
Galleria		+	+			
Pyrrhocoris	+	+	+	+		

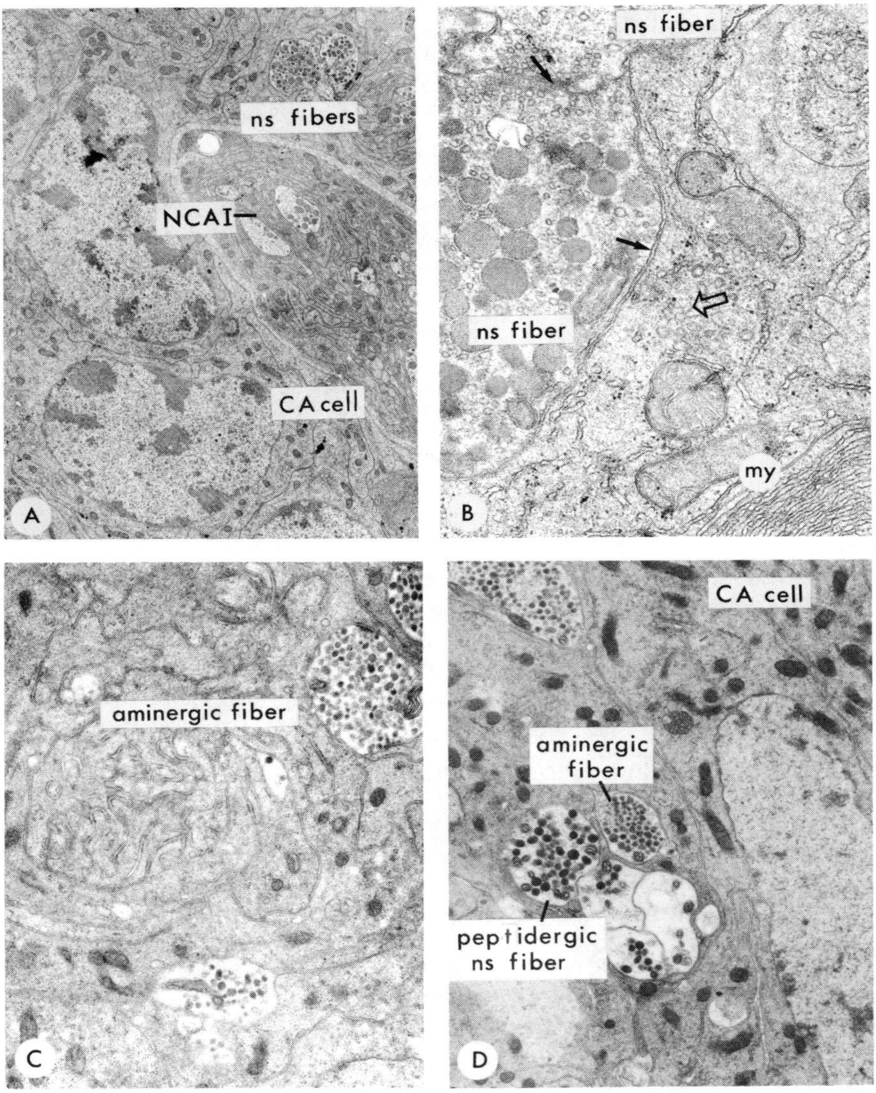

FIGURE 39. Neurosecretory innervation of both peptidergic and aminergic type in the corpora allata. (A) *Locusta.* Inactive corpora allata in a 3-day-old adult female. The NCA1 enter the corpora allata and branch, forming bundles containing both bare axons and axons ensheathed with connective tissue. ×3200. (B) *Locusta.* Corpora allata of a 3-week-old female whose vitellogenesis is terminating. The SER is poorly developed (open arrow) and myeloid figures are observed (my). Two bare peptidergic-type neurosecretory axons are linked to each other and to a glandular cell. Note the abundance of synaptoid vesicules and the

volumetric variations of the CA, histological, and ultrastructural modifications have been identified.

4.3.2.1. Volumetric Modifications

Early investigations on various species suggested that a correlation exists between CA secretory activity and volume. In the adult female blood-sucking bug *Rhodnius* (Wigglesworth, 1936), the blood meal induces a considerable enlargement of the volume of the CA, resulting from an increase in size of the glandular cells, the cytoplasm of which becomes more abundant. Vitellogenesis is triggered at the same time. If the insect is then deprived of food, vitellogenesis is arrested and the size of the CA decreases significantly.

In the phytophagous bug *Oncopeltus,* the CA also changes in size in relation to food intake and ovarian development. In young fasting females, the ovaries show little development and the CA is small, whereas in fed females, the ovaries display considerable growth and the CA increases sixfold (Johansson, 1958).

In the ovoviviparous cockroach *Leucophaea,* the normal reproductive cycle includes a long incubation period during which vitellogenesis is interrupted. The volume of the CA is large during vitellogenesis, diminishes considerably during gestation, and then increases again after parturition, when vitellogenesis resumes (Engelmann, 1957; Scharrer and Von Harnack, 1958). In the adult males, by contrast, no such changes are observed. In the course of larval life, of reproductive activity, and in relation to the polymorphism, a relationship is also observed between the volume of the CA and their level of activity. In some insects, however, this correlation seems not to exist, suggesting that secretion and release processes are not necessarily interrelated (Mordue, 1965) (see Tobe and Stay, 1986).

4.3.2.2. Cell Multiplication

The periodic changes in volume of the CA result both from variations in the nucleoplasmic ratio and from cell multiplication (Scharrer, 1964a in *Leucophaea*; Tobe *et al.,* 1984 in *Diploptera*). In *Diploptera,* the cells numbering 6000 upon emergence increase to 9000 on the fifth day after mating and then fall to 6000. In

---

presence of synaptoid areas at the junction point of the two neurosecretory fibers (filled arrow). ×40,000. (C) Inactive corpora allata during egg care in *Labidura.* The glandular cells are associated to numerous bare neurosecretory fibers loaded with small granules with an aminergic appearance. Note the many points of contact between the granules and plasma membrane. ×16,400. (D) Inactive CA in a virgin adult *Blatella germanica* female. The glandular cells are linked to many bare neurosecretory fibers (ns), both aminergic and peptidergic. ×16,400 (A–D, courtesy of Cassier.)

certain experimental series, however, a lack of correlation has been observed between the synthetic activity of the CA and their cell increase (Tobe *et al.*, 1984). The increases in cell size and JH synthesis appear to be processes that are differently regulated. In *Diploptera,* it seems that mating triggers mitosis and differentiation, while the ovary is responsible for the increase in the JH titer.

4.3.2.3. Structural and Ultrastructural Changes

A large body of research has been devoted to the structural and ultrastructural variations of the CA, especially during the gonadotropic cycle (see Tobe and Stay, 1986 and also, for discussion, Johnson *et al.*, 1985).

The first standard data were provided by Scharrer (1964a, 1978) and Joly *et al.* (1967), and concern both the ovoviviparous *Leucophaea* and the oviparous *Locusta,* in which glandular cells of active CA were observed to have abundant cytoplasm and well-developed cytoplasmic organelles (Golgi bodies, ergastoplasm, mitochondria). In inactive CA, on the contrary, the cytoplasmic volume is small, the cell membranes are folded, and the Golgi bodies, ergastoplasm, and mitochondria show little development.

In active glands, subsequent investigations generally indicated the presence of a thin basal membrane that deeply penetrates between the glandular cells, which are separated by wider intercellular spaces, while remaining bound by gap junctions. The nuclei show greater size and an increase in the number of nucleoles and perinuclear spaces. The Golgi apparatus of the glandular cells undergoes very slight modification, whereas the mitonchondria, which are highly developed and often very elongated in active CA, degenerate in resting CA and are incorporated into lytic structures. Very characteristic elements of the CA placed at rest are autophagic vacuoles, which display various aspects (lysosomes, dense or residual bodies, dense multivesicular bodies).

At the level of the endoplasmic reticulum, the insects exhibit considerable diversity. Some of them (heterometaboles, holometabole larvae) possess little SER and a predominant RER, whereas in others the SER is highly developed, forming large concentric whorls (locusts, *Bombyx, Calliphora*). A parallel development of the SER and the RER may also occur, as in cockroaches, but the reduced role of the SER is often observed.

An autoradiographic analysis performed with labeled farnesic acid in adult *Schistocerca* females revealed that the silver grains are preferentially fixed on the surface of the SER, which accordingly appears to play a predominant role in the synthetic processes in this species. It is nevertheless true that the synthesis of the enzymes indispensable to JH synthesis should take place in the RER.

In a number of cases, ultrastructural variations of the CA have been compared with the hemolymphatic JH titer. A high JH titer is associated with proliferation of the SER, especially in *Labidura* (Baehr *et al.*, 1982) and *Manduca* (Sedlak *et al.*, 1983).

4.3.2.4. Intercellular Coupling

The presence of gap junctions connecting CA cells has been observed. These could be involved in the regulation of JH biosynthesis. Cell coupling was demonstrated by iontophoretic injection of Lucifer yellow and injection current pulses into one cell and recording electrotonic potentials from other cells. From the age of the female studied, it can be inferred that the electrotonic potentials vary inversely with the rate of JH biosynthesis (Lococo *et al.*, 1986).

4.3.2.5. Neurosecretory Innervation

A correlation between the ns content of the CA and their level of activity was reported in *Tenebrio* (Mordue, 1965); ns is abundant in the brain and CA of virgin females, but absent in mated and egg-laying females. In *Periplaneta* (Khan *et al.*, 1978), the amount of ns material contained in the CA decreases when these organs gain in size, both in the larva and in the adult.

Changes in ns axon diameters were correlated with JH synthesis in *Manduca* (Sedlak, 1981; Sedlak *et al.*, 1983). High JH synthesis periods were associated with large-diameter axons, while low JH stages were characterized by axons of much smaller diameter.

The number of membrane-bound ns granules and exocytosis figures was determined in short- and long-day exposed *Leptinotarsa* (Khan and Buma, 1985). This investigation revealed wide differences between the diapausing short-day and the reproductive long-day insects.

4.3.2.6. Conclusion

To conclude, it can be stated that the ultrastructural data on CA are not really uniform and that further research will be needed to elucidate the nature of the organelles associated with JH synthesis. The lipidic nature of the JHs makes their fixation difficult and prevents the observation of their storage or release.

4.4. NEUROHORMONAL VERSUS NERVOUS REGULATION

The CA in larvae and in most adult females exhibit discontinuous activity that is related to specific phases of the insect's life cycle and to external environmental factors. During postembryonic development, the CA display cyclic activity during each intermolt period. In the reproductive phase, the gonadotropic activity of the CA varies during each vitellogenetic cycle, the variations being especially pronounced in viviparous and ovoviviparous species in which the CA are inactive during gestation. Corpora allata activity is also affected by temperature, humidity, photoperiod, and diet, the functional dependence on these external factors occurring through the central nervous system, which receives and integrates the information.

Experiments designed to identify the control mechanisms of the CA consisted essentially of the denervation of these organs, the cauterization of different parts of the brain and particularly of the PI, and the determination of the titer of the JH and of the synthetic activity of the CA at different times and in different conditions, especially by *in vitro* studies.

The results obtained, which are rather complex, will be presented by species investigated, followed by an overall review of the data concerning the nervous and hormonal regulations involving allatostatins and allatotropins, and the feedback of the JH, of ecdysone, and of the ovary.

4.4.1. Cockroaches

The nature and source of the information transiting through the CA nerves in cockroaches were fully investigated by Engelmann and Lüscher (1957) in the ovoviviparous cockroach *Leucophaea.* Meticulous anatomical studies showed that, in addition to the ns axons issuing from the ns cells of the PI, the NCCI contain axons arising from non-ns neurons located in the vicinity of the ns cells. While the bundles of ns axons originating in the right and left ns cells intersect in the posterior part of the brain, the axons originating in the non-ns neurons leave the brain through the corresponding ipsilateral nerve. Selective electrocoagulations of the ns cells of the PI and of the contiguous nerve cells yield opposite results: activation fails to occur in the former, while it occurs in the latter. Experiments involving unilateral severance of the nerve connections also show that the volume of the CA only increases on the operated side. This indicates that the inhibition exerted by the brain on the CA is nervous and not ns, and that it does not originate in the ns cells themselves but in the adjacent neurons.

A series of thorough investigations were carried out on another ovoviviparous cockroach, *Diploptera,* and we also have some information on *Periplaneta.*

4.4.1.1. Role of Innervation

Investigations conducted on *Diploptera* (Stay and Tobe, 1977; Stay *et al.*, 1980; Tobe and Stay, 1980) showed that disconnection of the CA activates them in the virgin female, which agrees with the foregoing results (Fig. 40). Other studies, carried out by Fraser and Pipa (1977) on *Periplaneta,* confirmed the

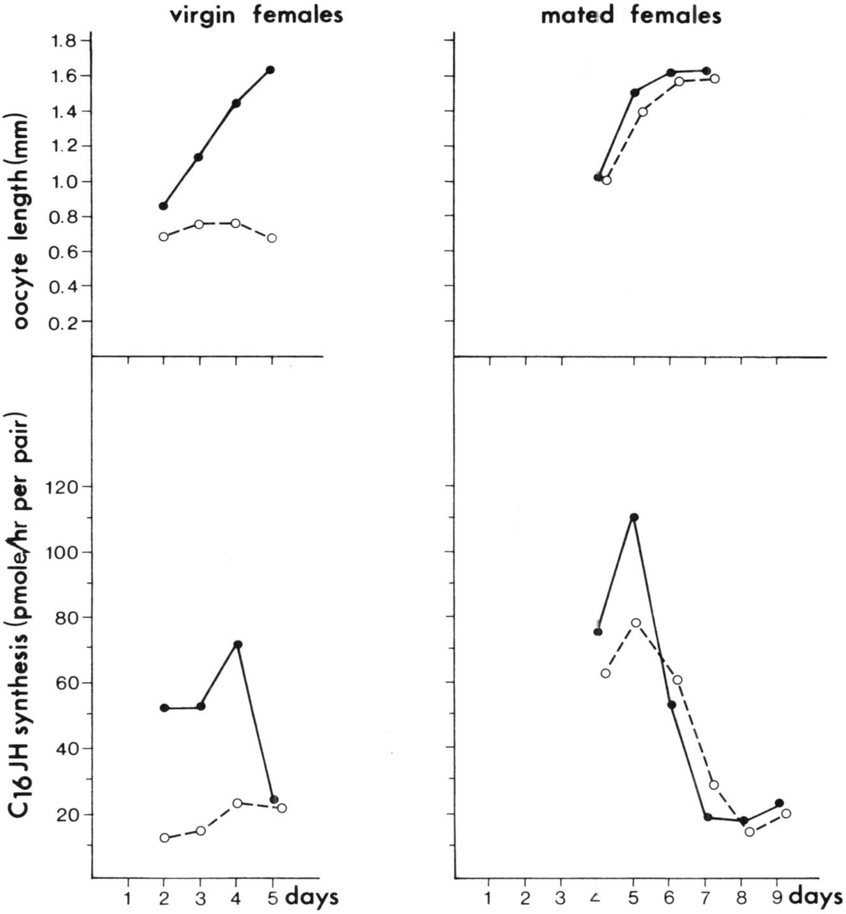

FIGURE 40. Effect of denervating corpora allata in virgin and mated *Diploptera* females on juvenile hormone synthesis rates and oocyte lengths. —, Denervated; ---, sham-operated. (Modified from Stay and Tobe, 1977.)

inhibitory role of the NCA1 in the larva, in which their severance induces supernumerary molts. In *Periplaneta,* however, the comparison of synthetic activity in the CA of females subjected to conditions such as starvation, isolation, and the absence of mating, led Weaver (1984) to question the inhibitory role of the NCA1, while for Pipa (1986), the NCA1 exert an inhibition that is specifically removed by mating.

Although the role of the NCA2 has not been demonstrated in many insects, it appears in *Leucophaea* that its presence is indispensable for activation in the CA (Engelmann, 1957).

Cautery of both median and lateral ns cells was shown to remove brain inhibition on the CA of the *Diploptera* virgin female, causing a stimulation similar to that determined by disconnection, and it was suggested that the lateral ns cells act indirectly in stimulating the release of allatostatin by the median ns cells (Ruegg *et al.*, 1983) a hypothesis supported by the data of Poras (1983) in *Locusta* (see Section 4.3.1).

4.4.1.2. Nervous or Neurohormonal Control

Experiments with unilateral allatectomy and supernumerary CA implantation show that the regulation of the activity of these organs is at least partly hormonal. In fact, some compensation occurs in both cases, whether or not the CA are innervated (Tobe and Stay, 1980).

Furthermore, the transplantation of larval CA in adult females causes a five- to sixfold increase in their synthetic activity, while the transfer of the CA of the adult female to a larva results in a drop in activity. These two opposite effects of stimulation and inhibition are thus subject to humoral regulation (Szibbo and Tobe, 1983) (Fig. 41). Humoral regulation is confirmed *in vitro,* in *Diploptera,* by the reduced synthetic activity of the CA placed in the presence of a brain extract (Rankin *et al.*, 1986; Paulson and Stay, 1987) and by the variations in allatotropic activity of the hemolymph during the genital cycle of *Nauphoeta*. In

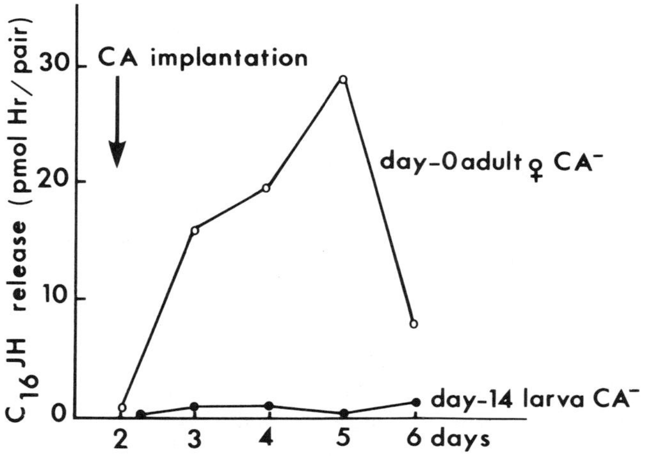

FIGURE 41. JHIII release in allatectomized day 14 last instar larvae and day 0 adult *Diploptera* females implanted with day 14 larval corpora allata. Note the absence of stimulation in the larvae. (Modified from Szibbo and Tobe, 1983.)

this species, the hemolymph of females in vitellogenesis or in gestation is strongly active, whereas the hemolymph of females in previtellogenesis, in chorionization, or in ovulation is inactive (Wilhelm *et al.*, 1987). The origin of the factors involved still needs to be elucidated.

However, it is not inconceivable that brain factors, which may take the hemolymphatic pathway, particularly in experimental conditions, are released in the CA themselves at the level of the neuroeffector junctions, whose presence has often been reported (see Chapter 1). It was shown recently that CA extracts *in vitro* exert allatostatic action in *Diploptera* (Paulson and Stay, 1987).

4.4.1.3. Role of the Ovary

We have seen that denervated CA from mated females retain cyclic activity comparable to that of control insects, indicating that their functioning is also humorally regulated. Hence, in addition to inhibitory nervous control, the CA are subject to stimulatory humoral control (Stay *et al.*, 1980). Surprisingly, the origin of the humoral factor stimulating JH biosynthesis by the CA in *Diploptera* was identified in the ovary (Stay *et al.*, 1980, 1983) (Fig. 42), whose extirpation causes the cyclic activity of the CA to disappear, and whose reimplantation restores it. This action is exerted directly on the CA in the absence of any innervation, a result confirmed in *Periplaneta* (Weaver, 1981) and *Nauphoeta* (Lanzrein *et al.*, 1981).

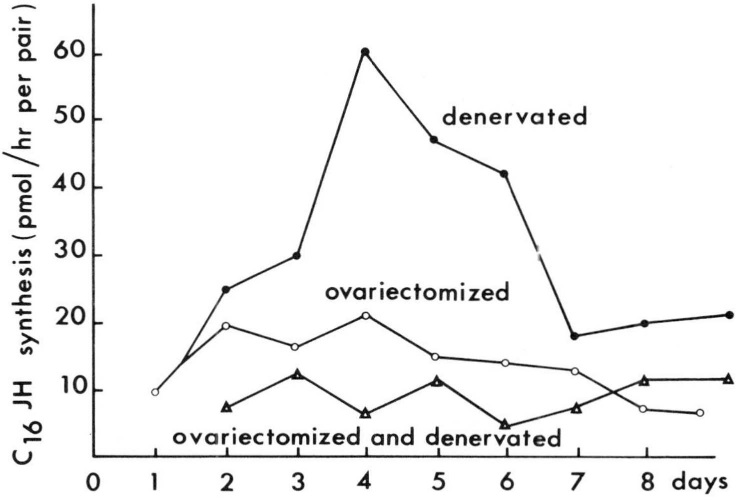

FIGURE 42. Effects of denervation and ovariectomy on JHIII synthesis in an adult *Diploptera* female. (Modified from Stay *et al.*, 1983.)

Tests of different parts of the ovary on ovariectomized females show that only the terminal oocytes are active on the CA. The pre- and end-vitellogenetic oocytes are inactive, while activity is at a maximum in the vitellogenesis oocytes (Rankin and Stay, 1984). Implantation experiments also suggest the action of not one but two ovarian factors, one inhibitory and the second stimulatory (Rankin and Stay, 1985).

4.4.1.4. 20-HE

Recent knowledge about ecdysteroid synthesis in the ovaries suggests that their action on the CA could be due to ecdysone. It has effectively been shown that 20-HE mimics the action of the ovary (Friedel *et al.*, 1980b; Stay *et al.*, 1980) and that castration suppresses the hemolymphatic ecdysone peak of the first reproductive cycle (Weaver *et al.*, 1984) in *Periplaneta.* In *Diploptera,* however, the situation is different because the hemolymphatic ecdysteroid titer proves to be the same in control females and ovariectomized females, precluding the hypothesis of purely ovarian origin, and implying that the allatotropic and allatostatic ovarian factors cannot be identified with this hormone (Stay *et al.*, 1984).

4.4.1.5. Mitoses

We have discussed above the importance of cell multiplication in the activation of the CA. It has been shown that the number of cells in the CA is not necessarily correlated with the synthesized JH titer (Stay *et al.*, 1983; Ruegg *et al.*, 1983; Tobe *et al.*, 1984), although the increase in JH titer requires cell multiplication. Hence it is not inconceivable that different regulation mechanisms exist for cell multiplication and for JH synthesis, especially since these are totally different processes. It should also be pointed out that the existence of the variations in the competence of the CA has been demonstrated during the genital cycle in *Nauphoeta* (Wilhelm *et al.*, 1987) and during development in *Diploptera* (Paulson *et al.*, 1987).

4.4.2. Locusts and Other Orthopterans

Severance of the NCA1 does not have the same effects in locusts as in cockroaches. Instead of activating the CA, this operation inhibits their functioning. In a histological and experimental investigation carried out on *Locusta,* Girardie (1966b, 1967) thoroughly examined the mechanisms of cerebral control of CA activity in these insects.

The histological and ultrastructural study revealed three distinct types of ns cells in the PI: the numerous, small type A and B cells located in the central regions of the PI, and the larger type C cells located around the A and B cells.

Electrocoagulation performed with 15- to 40-μm electrodes allowed selective removal of A-B and of C ns cells. These two experiments produced contrasting effects. Removal of the A-B cells is equivalent to an implantation of CA and causes supernumerary molts, the formation of green pigment in gregarious insects, and the stimulation of vitellogenesis. Removal of C cells, on the contrary, has the same effect as removal of the CA. It induces the appearance of precocious metamorphosis, depigmentation, and the suppression of vitellogenesis, sexual behavior, and the yellow coloration of mature males. Hence the CA are under twofold cerebral control: inhibiting control exerted by the A-B cells, and stimulating control exerted by the C cells. Both of these are exerted by a humoral pathway as demonstrated by implantation experiments. It seems, however, that normal innervation favors the functioning of the CC-CA in *Schistocerca* and *Locusta*. In *Schistocerca,* implanted CA were unable to induce sexual receptivity or ovarian development in allatectomized females (Highnam *et al.*, 1963; Pener, 1965, 1967; Strong and Amerasinghe, 1977). In *Locusta,* positive effects were obtained only by implanting large numbers of organs, and the integrity of nervous connections between CA and CC was shown to be of great importance for this activity (Cazal *et al.*, 1971; Couillaud *et al.*, 1987).

According to Retnakaran and Joly (1976), the inhibitory activity of the A-B cells is not exerted on the CA but on the JH itself. The secretion of the A-B cells appears to inactivate the hemolymph JH by increasing the titer of specific esterases of this hormone. Ultrastructural examination of the CA after removal of the C cells substantiates this hypothesis since they display a normal appearance.

As to the lateral cells of the protocerebrum, which, as we have seen above, innervate the CA exclusively, and which collaterals synapse onto the PI ns cells their electrocoagulation and reimplantation in larvae and adults (Girardie, 1974) yield results that elude interpretation. They exert approximately the same action as the type C ns cells of the PI, but it cannot be positively asserted whether their effect is humoral, and whether it is direct or exerted on the cells of the PI, as recently suggested in a cockroach.

Other investigations have attempted to provide further details on the role of the lateral cells that innervate the CA in *Locusta,* but unfortunately the results contain contradictions. Thus, for Poras *et al.* (1983), performing experiments on the severance of the NCA1 and cautery, the lateral ns cells appear to inhibit the functioning of the CA in the adult diapausing female but stimulate the CA in the larval instar. This was confirmed on the CA of larvae and adults *in vitro* (Couillaud *et al.*, 1984). In fact, both in the adult female and in the larva, severance of the NCA1 causes a decrease in CA activity. In the vitellogenetic female, electrocoagulation of the PI does not alter the activity of the CA, implying that two different allatotropins may occur, one arising in the PI and the second in the lateral ns cells, with only the second participating in the regulation of the CA during vitellogenesis (Couillaud and Girardie, 1984).

Apart from the neuroendocrine control exerted on the CA, regulations appear to occur, in *Locusta,* arising in the reproductive system. Thus after ovariectomy in a young female, the *in vitro* synthesis of JHIII is decreased with the persistence of small activity peaks (Couillaud and Girardie, 1984). In the male of this species, moreover (Couillaud, 1986), ablation of the complex made up of the testicles, seminal vesicles, and accessory glands also causes a decrease in JH synthesis.

Another regulator of CA activity seems to be the hemolymphatic JH titer, which probably acts via the brain, because the addition of JHIII to CA *in vitro* does not change their functioning (Couillaud and Girardie, 1985).

Do the mechanisms revealed in *Locusta* exist in other locusts? Electrical stimulation of the PI in *Anacridium* reactivates the disconnected CA and terminates diapause, thus confirming the stimulatory and humoral function of the PI (Girardie *et al.,* 1974).

The situation seems to be slightly different in *Schistocerca.* We have already seen that severance of the NCA1 produces the same results as severance of the NCCII. This suggests that the lateral protocerebral ns cells play a role in the regulation of CA activity. According to Strong (1965a,b), these cells are allatotropic and their action is exerted by the ns material transported from the brain toward the CA via the NCCII. Mason's investigation using cobalt chloride supports this view since in *Schistocerca,* the CA receive axons from the NCCII only.

The imaginal reproductive diapause of *Tetrix* (Poras, 1982) can be broken by the implantation of CA, provided that they arise in diapausing females and not in 1-day females, and that they are disconnected from the brain–CC complex. This implies an inhibition of the CA during diapause, originating in the PI.

In gryllids, the occurrence of both an inhibitory and a stimulatory factor was demonstrated; the brain and subesophageal ganglion extracts are inhibitory while those of the CC are simulatory (Wicker, 1987, in *Acheta*). Moreover, in *Teleogryllus,* whose JHIII synthesis and ovogenesis closely depends on mating and the disposal of a favorable substrate for egg deposition, the ovary appears to play a humoral inhibitory role (Loher *et al.,* 1987).

4.4.3. *Anisolabis* and *Labidura*

To estimate CA activity in the dermapteran *Anisolabis,* several operations were performed in pupae and adults (removal of the brain, the subesophageal ganglion, and the prothoracic ganglion) followed by the implantation of the CA into allatectomized pupae, in order to estimate their activity from their ability to induce supernumerary molts. It appears that the activity of the CA of decerebrated insects is greater than that of the intact insects (Ozeki, 1980).

In *Labidura,* where the period of vitellogenesis is followed by a period of egg care during which the maturation of the oocytes is arrested, the hemolymphatic JH titer was analyzed at the same time as the examination of the volumetric variations of the CA and the study of their ultrastructure. These data show that the CA are inactive during the egg care period and that cautery of the PI is followed by a rapid decrease in the JH titer. This indicates that the PI exerts a stimulatory action on the CA, while the nature of this action remains to be elucidated. Note that ovariectomy causes hyperactivity of the CA, as in cockroaches.

4.4.4. *Leptinotarsa*

In *Leptinotarsa,* insects raised in short days enter into reproductive diapause. The existence of hormonal controls of the functioning of the CA was demonstrated by the implantation of CA into allatectomized females, resulting in a stimulation or an inhibition of vitellogenesis, depending on the photoperiod conditions to which the insect was subjected (De Wilde and De Boer, 1969). These findings were confirmed by experiments on the unilateral ablation of the CA, the implantation of supernumerary CA, and the application of JHA, whose effects were observed *in vivo* on the volume of the CA and the hemolymphatic JH titer, and *in vitro* on CA activity (Schooneveld *et al.,* 1979).

By cautery of the brain centers innervating the CA and the transfer of short-day feeding insects to long-day starvation conditions and the estimation of CA activity, it was shown that the CA were activated only by destruction of the pars lateralis (Khan *et al.,* 1986).

As to nervous regulations, females raised in short-day photoperiods enter an ovarian diapause that cannot be terminated by disconnecting the CA. Other experiments, such as the denervation of the CA of short-day females, the implantation of CA of long-day females into short-day females, and the transfer of CA from short- and long-day individuals, show that the CA can be activated or inhibited in the absence of any innervation. Nevertheless this plays a role in normal physiology because the synthetic activity of the CA *in vitro* is not modified, as *in vivo,* by the addition of JHIII. This suggests an indirect action of the JH via the nervous system, made even more likely because the normal decrease in CA activity during starvation is no longer observed after denervation (Schooneveld *et al.,* 1977, 1979; Khan *et al.,* 1982, 1983).

4.4.5. Moths

In *Manduca,* the presence of the brain is unnecessary for the decrease in the hemolymphatic JH of the final instar larva (Safranek *et al.,* 1980), a decrease that is of crucial importance for the onset of metamorphosis. In younger larvae, however, a transplanted brain increases the functioning of the CA and causes supernumerary molts (Bhaskaran *et al.*, 1980).

The origin of allatotropin, investigated by cautery of the median and lateral ns cells, appears to lie in the median II group (Buys and Gibbs, 1981). Determination of the synthetic activity of the CA *in vitro* shows that the activation of isolated CA is different from that of CA left connected with the CC and the brain. Thus, besides a brain stimulatory humoral control, a nervous or neuroeffector brain control also exists (Granger *et al.*, 1982) (Figs. 43 and 44). Furthermore, a specific allatostatin of the JHI was identified in the brain (Granger and Janzen, 1987). It acts very rapidly (1 min) in a dose-dependent manner. Its presence is observed in particular on day 4 of the final larval instar; in other words, at the time when JH must drop in order to allow pupal commitment. The synthesis of JHIII appears to be controlled by a different brain allatostatin which agrees with the existence of fluctuations in the respective synthesis titers of JHI and JHIII, during development (Granger *et al.*, 1982, 1984) (see also Baker *et al.*, 1987).

In *Manduca,* as in other species, 20-HE influences the activity of the CA in a dose-dependent manner, and this influence is indirect and passes via the brain. The CA of the final instar larvae are in fact stimulated by 20-HE, but only when they are normally connected with the brain–CC complex (Whisenton *et al.*, 1985, 1987b; Watson *et al.*, 1986). It is interesting to note that, once briefly stimulated, the CA remain active for as long as 11 hr (Table XX).

The experiments of Granger and Sehnal (1974) and Sehnal and Granger (1975) on *Galleria* also revealed the existence of a twofold cerebral control on the CA: neurohemal stimulatory control and nervous inhibitory control. The

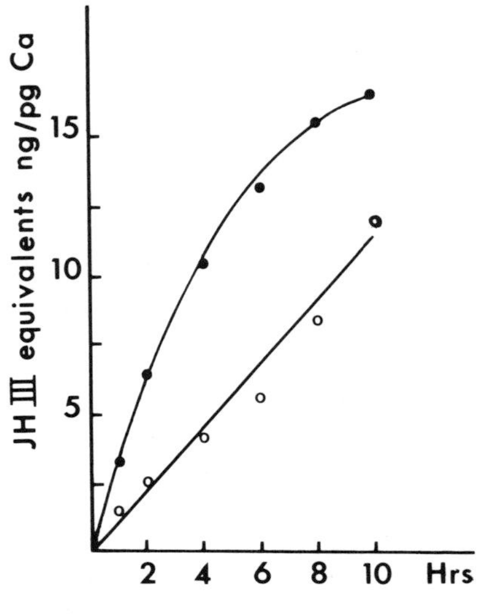

FIGURE 43. *In vitro* JH biosynthesis by the corpora allata of day 7 final instar *Manduca* larvae. Corpora allata incubated alone (○) or in the presence of a brain–CC complex (●) which stimulates corpora allata activity. (Modified from Granger *et al.*, 1984.)

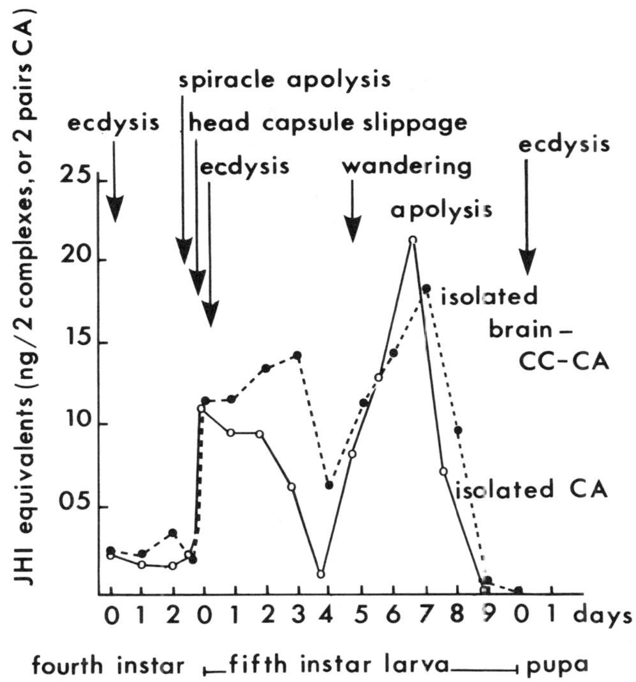

FIGURE 44. *In vitro* JHI biosynthesis by *Manduca* isolated corpora allata (●) and brain–corpora cardiaca–corpora allata complexes (○). (Modified from Granger *et al.*, 1982.)

experiments were conducted on final instar larvae. A brain implant during the first half of the instar induces supernumerary larval molts, indicating activation of the CA. Allatotropic activity seems to be located in the PI because brains deprived of this region are no longer active. In the second half of the larval instar the CA becomes insensitive to cerebral action and do not react to brain implants,

TABLE XX. Stimulation of Juvenile Hormone Synthesis by Brain–Corpora Cardiaca and Role of 20-HE[a]

	ng RIA equivalents		
Tissue	20-HE	JHI	JHIII
Brain–CC–CA	0	0.19	0.91
Brain–CC–CA	+	0.47	1.68
Brain–CC + CA	0	0.53	1.06
Brain–CC + CA	+	0.53	1.10

[a]Modified from Granger *et al.*, 1987.

but this insensitivity is removed by severance of their nerves. Hence it appears that, apart from its activating function, the brain has an inhibitory nervous function. It has also been shown that the activity of implanted brains depends on the stimuli perceived. Thus brains from chilled or injured insects are particularly active, and this activity can be seen after implantation (Bogus and Cymborowski, 1984; Bogus *et al.*, 1986). The occurrence of a stimulatory factor from the brain was confirmed by JH determination (Sehnal and Rembold, 1985).

In *Deilephila,* the CA of the larva secrete *in vitro* at the same rate throughout the instar, but a progressive decrease occurs in brain–CC complexes, which mimics the hemolymphatic titer (Retnakaran *et al.*, 1985).

4.4.6. Bugs

In *Pyrrhocoris* (Hodkova, 1976, 1977, 1979), reproductive activity is dependent on the photoperiod, an ovarian diapause occurring in short-day females. Reproductive diapause is broken by severance of the NCA1 and also by implantation of the CA, but implantation of the brain–CC–CA complex is far less active. Thus the brain appears to exert an inhibitory action on the CA. Moreover, the implantation of the CC–CA complex into females deprived of their own CA is far more effective in long-day females than in short-day females, as though the former were capable of stimulating the implant. The hypothesis of humoral stimulation of the CA may also be inferred from the NCA severance experiments performed on long-day females, which failed to cause any decrease in fecundity.

A stimulatory and an inhibitory control exerted by the brain on the CA were also demonstrated in *Rhodnius* (Baehr, 1973, 1976), in which cautery of the PI causes two opposite effects—the appearance of adultoids when performed during the larval stage and interruption of vitellogenesis in the adult. Note, however, that the allatotropic action is a very early one, disappearing within 24 hr after the blood meal. Furthermore, severance of the NCA1 stimulates vitellogenesis in fasting females and thus confirms the inhibitory functions of the brain on the CA.

4.5. PURIFICATION OF ALLATOTROPINS

We still know very little about the molecular structure of allatotropic and allatostatic factors. In *Locusta,* a partial purification performed from the brain and CC of the adult female (Ferenz and Diehl, 1983; Gadot and Applebaum, 1985; Gadot *et al.*, 1987; Rembold *et al.*, 1986) revealed the existence of a small, thermostable, allatotropic peptide with a molecular weight of about 700–2000. Its presence is observed in the CC, the brain, and the subesophageal ganglion, extracts of which stimulate JHIII synthesis at the respective rates of 45.9, 26.9, and 29.4 pmol/gland/hr (Gadot *et al.*, 1987).

Furthermore, the purification and separation of PI and CC proteins was thoroughly investigated by the Girardie group (Girardie *et al.*, 1985). Three proteins with a disulfide bridge were isolated. They are distributed in the A1 type ns cells (neuroparsin A), in their storage lobe (cardiacin), and in the glandular cells of the CC (neuroparsin B), respectively (Girardie *et al.*, 1987a). Antibodies were raised against neuroparsin A. When injected into locusts, they mimic the effect of PI electrocoagulation, i.e., an increase in JH activity, suggesting that neuroparsin A intervenes in JH inhibitory regulation. Its site of action may lie outside the CA, since no modification in the CA synthesis rate was observed in the injected insects (Girardie *et al.*, 1987b).

In the cricket *Acheta,* a small peptide, possibly the CA inhibitory factor from the CC, was identified (Wicker and Wicker, 1987) with the following molecular structure:

Ala-Ala-Ala-Pro-Phe

Two thermolabile proteinic factors were isolated in *Manduca,* with molecular weights of 4 and 6.8–13 kDa, respectively. They display specific antagonistic dose-dependent effects. The former is allatotropic and stimulates only JHIII synthesis, and the latter acts only on JHI synthesis (Granger *et al.*, 1984; Granger and Janzen, 1987).

The action mechanism of brain allatostatin from *Diploptera* was shown to be calcium-independent and to involve cAMP as a second messenger (Meller *et al.*, 1985; Aucoin *et al.*, 1987).

4.6. RECAPITULATION

The control of the functioning of the CA was investigated in several insect species, particularly in cockroaches, locusts, and *Manduca.* Classic methods of implantation and transplantation were used besides other, more modern methods, such as JH titer determination and measurement of JH *in vitro* synthesis. The histological and ultrastructural analyses were also useful because they helped to confirm the periods of CA activity, which are characterized by cell multiplications, volume increases, and cytological changes, especially in the SER. The identification of neurons whose axons innervate the CA, achieved in particular by the $CoCl_2$ method in a number of species, allowed the performance of severances and selective nerve cauteries whose interpretation is clear. However, regulation of the functioning of the CA is not yet clearly understood. Although we know that both nervous and humoral stimulatory and inhibitory controls from the brain are involved, other factors originating from the ovary are also important, as is hormonal feedback.

4.6.1. Nervous Regulation

4.6.1.1. Nerve Severance (see Table XIX)

Severance of the NCA1 in the cockroaches *Leucophaea* and *Diploptera* and in *Locusta* yielded the same results as severance of the NCCI, suggesting that the axons contained in the NCCII play no part in control of the CA. In *Diploptera,* severance of the NCCII substantiates this hypothesis because it has no effect on vitellogenesis. In *Schistocerca,* however, Strong (1965a,b) reached a different conclusion: severance of the NCCII had the same effect as that of the NCA1.

Severance of the NCA1 sometimes results in a volume increase in the CA and in their physiological activation, which can be detected by juvenilizing or gonadotropic effects and, above all, by JH determination. These effects point to the inhibitory function of the brain on the CA. In other cases, on the contrary, this operation arrests the activity of the CA, as in *Locusta* and *Schistocerca,* the earwig *Anisolabis,* and the mealworm *Tenebrio.* One is thus led to conclude that in these species the brain exerts a stimulatory rather than an inhibitory control on the CA. Occasionally, severance of the NCA1 appears to have no effect on the CA, as in *Leptinotarsa* and *Manduca.*

4.6.1.2. Cautery of the Pars Intercerebralis

Like severance of the NCA1, cautery of the PI can cause stimulation of the CA, although it generally inhibits the activity of these organs. This operation has been carried out chiefly on the adult female, whose responses were studied in relation to vitellogenesis. As we shall see later (Chapter 6), in some species the PI exhibits gonadotropic action independent of the CA and thus complicates the interpretation of the results observed after its cautery. In fact, while cautery that induces the stimulation of vitellogenesis indicates the presence of a factor inhibiting the CA in the ns cells and the absence of an independent gonadotropic factor, on the other hand, cautery that inhibits vitellogenesis cannot be directly interpreted, as this may be due to the suppression of a gonadotropic factor or that of an allatotropic factor.

The precise determination of what actually occurs requires additional experiments involving the implantation of CA into parsectomized insects and the measurement of CA activity. Furthermore, the results obtained by cautery of the PI do not necessarily indicate that the median ns cells are involved because it has been shown in *Locusta* that the lateral cells that innervate the CA form loops at the level of the PI, probably causing their severance during the cautery of the PI.

The role of the nerves originating in the subesophageal ganglion (NCA2) has rarely been investigated (Table XII). Some authors have reported that the NCA2 take part in controlling CA activity and that they have a stimulatory role, as in *Leucophaea,* or an inhibitory role, as in *Leptinotarsa.* However, their severance has no effect on the activity of the CA in *Schistocerca* and *Locusta.* In *Diploptera* and *Periplaneta,* severance of both the NCA1 and the NCA2 has the same effect as severance of the NCA1 alone, thus implying that the NCA2 are not of fundamental importance.

4.6.2. Allatotropins and Allatostatins

Many denervation and cautery experiments on different parts of the brain have been conducted on several species. The results reveal inhibitory or excitatory actions, but it is often difficult to state whether they are humoral or nervous. In several cases, however, the existence of neurohormonal regulation was demonstrated, either inhibitory and stimulatory (*Diploptera, Locusta, Leptinotarsa, Pyrrhocoris*), or only stimulatory (*Galleria* and *Manduca*).

A number of studies were carried out to identify allatotropins and allatostatins, and encouraging results were obtained concerning *Locusta, Acheta,* and *Manduca.*

The origin cells of allatostatins and allatotropins appear to lie in the lateral ns cells of the brain in *Leptinotarsa.* In cockroaches and locusts, both median and lateral ns cells are involved, but not necessarily in the same way, since the lateral ns cells may only exert a stimulatory role on the median ns cells. Apart from the brain, the subesophageal ganglion of *Acheta* and *Diploptera* (Wicker, 1987; Rankin and Stay, 1987) displays an allatostatic activity, as do the other ventral ganglia (Rankin and Stay, 1987).

The neuropeptide release mode involved in the regulation of the functioning of the CA raises problems, considering the presence of many neuroeffector junctions in these organs. The normal site of allatostatin and allatotropin release probably lies in the CA themselves, although in certain species these substances appear to act remotely.

4.6.3. Role of the Ovary

The role of the ovary in regulating CA activity was revealed in *Diploptera* and other cockroaches (Stay and Tobe, 1978; Stay *et al.,* 1983; Lanzrein *et al.,* 1981), and it was shown that it involved a humoral stimulatory action, the effects of castration being inhibited by the reimplantation of the ovary. The role of the ovary has also been observed in other species such as *Polistes* (Röseler *et al.,* 1980), *Acheta* (Strambi *et al.,* 1981), *Aedes* (Rossignol *et al.,* 1981), *Labidura* (Baehr *et al.,* 1982), and *Locusta* (Couillaud and Girardie, 1984).

Besides its stimulatory action on the CA, the ovary also appears to exert inhibitory action at the end of the gonadotropic cycle in *Nauphoeta* (Lanzrein *et al.*, 1981) and *Aedes* (Rossignol *et al.*, 1981). This probably explains the histological and volumetric aspect of the CA of castrated insects (Scharrer and Von Harnack, 1961).

The action mechanism of ovarian factors on the CA has not yet been elucidated. An action via the brain is improbable, at least in *Diploptera*, because the cycle of JH biosynthesis takes place in the absence of the head.

It was suggested that ovarian ecdysone could be responsible for the regulatory effects of the ovary on the functioning of the CA, especially since the active follicles are those containing mature ovocytes, which are known to contain ecdysteroids. Yet this does not appear to be the case in *Diploptera* (Stay *et al.*, 1984) or in *Locusta*, in which the addition of 20-HE to the CA incubation medium does not cause any change in activity (Couillaud and Girardie, 1985).

4.6.4. Ecdysone and 20-HE

Even if the ovary does not appear to act on the CA via ecdysteroids, the latter influence the synthetic activity of the CA. Thus ecdysteroid injection enhances nuclear RNA synthesis in the CA of the pupae of *Philosamia* and *Hyalophora* (Siew and Gilbert, 1971), and causes an increase in the JH titer of the hemolymph (Joly and Joly, 1975) and an elevation of CA synthetic activity both *in vivo* and *in vitro* in *Diploptera* (Friedel *et al.*, 1980b), *Nauphoeta* (Lanzrein *et al.*, 1981), *Manduca* (Safranek *et al.*, 1980; Whisenton *et al.*, 1985), and *Aedes* (Kelly *et al.*, 1981b; Rossignol *et al.*, 1981; Birnbaum *et al.*, 1984). It is not yet known whether the action of 20-HE on the CA is direct or indirect. In *Diploptera* and *Manduca*, it is not exerted *in vitro* on isolated CA suggesting indirect action. Recent investigations in *Manduca* seem to indicate that its action is reversed during the development of the insect (Granger *et al.*, 1987).

4.6.5. Juvenile Hormone

The hemolymph JH titer affects the activity of the CA. The implantation of supernumerary CA reduces the activity of the host's CA, which becomes smaller in several insect species (see Raabe, 1982), and the same effects are produced by the injection or application of JH or JHA in *Diploptera* (Tobe and Stay, 1979), *Leptinotarsa* and *Manduca* (Staal *et al.*, 1981; Edwards *et al.*, 1983). It should nevertheless be pointed out that the active doses are much higher than the physiological doses.

Besides this negative feedback from the JH, a positive feedback also exists, observed in *Philosamia* and *Hyalophora* (Siew and Gilbert, 1971) and in different cases of compensating hypertrophy resulting from the unilateral ablation of a CA in *Leucophaea* (Engelmann and Lüscher, 1957), *Diploptera* (Stay and Tobe,

1978), and *Leptinotarsa* (Schooneveld *et al.*, 1979). One may inquire whether this is due to a direct response of the CA or to a cerebral indirect response. By unilaterally removing one CA, Cassier showed that a cerebral response is involved because compensating hypertrophy is observed only when innervation is intact (*Locusta,* Cassier, 1966). Positive feedback exerted by the CA on the ns cells of the PI has also been demonstrated in other species. In *Leptinotarsa,* as in *Locusta,* they appear to be indirect.

4.6.6. Esterase Regulation

The JH esterase (JHE) titer of the hemolymph and the tissues varies during development and reproduction (see De Kort and Granger, 1981; Jones *et al.*, 1982; Tobe and Stay, 1986). At the level of the hemolymph, a first peak was observed in several lepidopterans at the time of larval-pupal commitment, before gut purge and the onset of wandering behavior (Jones *et al.*, 1982), and a second peak was observed before molting. The data concerning the regulation of esterase synthesis indicate a possible role of the JH on the one hand and of a neurohormone on the other. Our knowledge is extremely limited in both cases.

It has been suggested that esterase synthesis, which takes place in the fat body, is influenced by the JH. In fact, if the fat body of *Hyalophora* is kept *in vitro* in the presence of JH, an increase in the production of esterases is observed (Whitmore *et al.*, 1972). *In vivo,* moreover, JH treatment yields the same result in diapausing adult females of *Leptinotarsa* (Kramer, 1978). However, at the larva-pupa transition, the effect of the JH on JHE production is modified; thus in *Galleria,* JH treatment of the pupa produces an elevation of the JHE titer, while in the larva, by contrast, a decrease in the JHE titer is observed (Reddy *et al.*, 1979). Similar findings were made in *Tenebrio* (Reddy and Kumaran, 1985). In *Manduca* pharate pupae, JH causes an increase in JHE (Sparks *et al.*, 1983).

The involvement of neurohormones in the regulation of JHE synthesis is suggested by some data but not yet positively established. In *Locusta,* it is stimulated by the type A-B ns cells of the PI (Retnakaran and Joly, 1976), and in lepidopterans an increase in JHEs is observed in ligated larvae after brain or subesophageal ganglion implantation (McCaleb and Kumaran, 1980 in *Galleria*) or after the injection of homogenates (Jones *et al.*, 1981 in *Trichoplusia*). In this species, the activity is not uniformly distributed in the nervous system, the thoracic ganglia and the abdominal ganglia being inactive. It also varies during development and disappears at the end of the fifth instar and in the pupa.

4.7. CONCLUSIONS

A large body of data is available on the regulation of the functioning of the CA and of the hemolymphatic JH titer because many researchers have focused on these important problems. Nevertheless, a number of contradictions subsist. The regulation of CA activity appears to be highly complex, and this could be related to the diversity of the factors that influence the functioning of these organs.

The fact that a highly complex regulation exists is not surprising, given the role played by the CA in insect physiology. It is probable that during evolution the mechanisms diversified in accordance with living conditions, with certain controls subsisting and others disappearing. It can be stated as a general rule that the CA are controlled by brain neurohormones, the allatotropins, which are stimulatory, and the allatostatins, which are stimulatory, and the allatostatins, which are inhibitory (Fig. 45). Moreover, ecdysone, JH, and the ovary exert

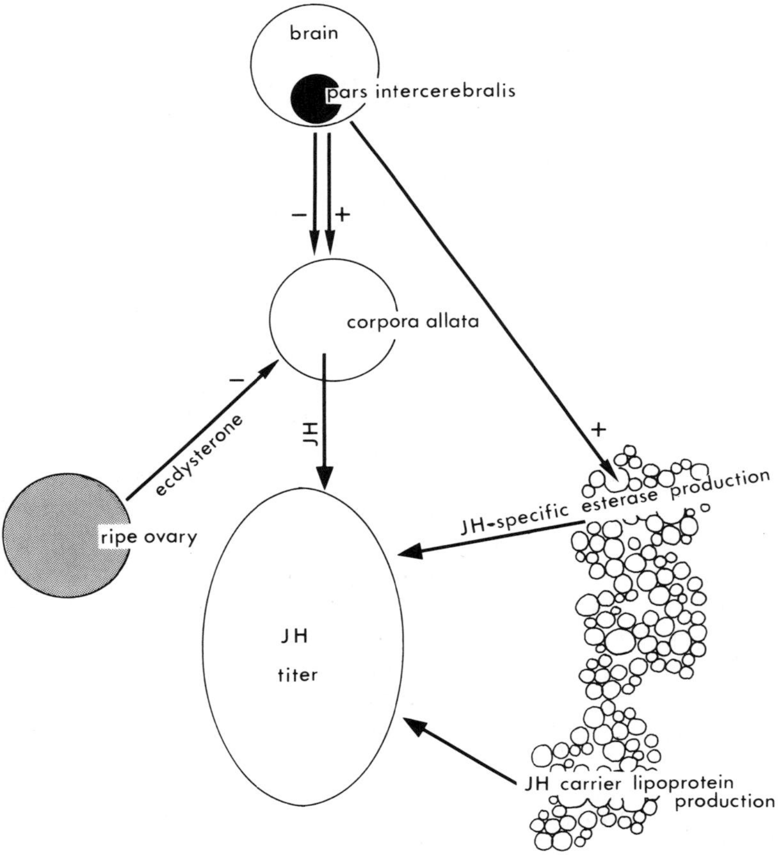

FIGURE 45. Schematic representation of juvenile hormone (JH) titer regulation.

feedback effects, while the brain also exerts nervous or neuroeffector control on the CA, usually with inhibitory effects.

Recent data suggest that different neurohormones could be involved in the synthesis of the different JHs at different stages of development. The regulation of cell multiplications in the CA could also differ from the regulation of the functioning of these organs. The notion of CA competence has also been demonstrated, adding to the data concerning the PG.

Juvenile hormone was recently identified in the embryo of several species (Bergot *et al.*, 1981), and this hormone, which is secreted by the serosa in early stages (Hartmann *et al.*, 1987), appears to play a role in the control of embryonic differentiation (Brüning and Lanzrein, 1987). This suggests that allatotropic and allatostatic neurohormones may already exist in these early stages. This point has not yet been seriously investigated.

Research is currently under way on the identification of the allatostatins and allatotropins in *Locusta* and *Manduca,* and allatotropic factors have been separated in both species. The former, with molecular weights of 700–2000, are thermostable, while the latter, with much higher molecular weights of 4000, 6800 and 13,000, are thermolabile. In *Acheta,* a small pentapeptide, which could be an allatostatin, has been identified.

5

Diapause

The annual succession of the seasons is accompanied by significant changes in temperature, photoperiod, and also food, which may become unfavorable to the insect's survival. It manages to withstand these difficulties by means of diapause, which can occur at different developmental stages, depending on the insect species, i.e., the adult, pupa, larva, and embryo, suggesting the possibility of a different hormonal condition. These diapauses may be intense to varying degrees, and obligatory or facultative. While larval and pupal diapauses exhibit some similarity, those of the embryo and the adult are markedly different.

Diapause is characterized by the arrest of development or reproduction; a low metabolism; a fall in water content, respiratory metabolism, and cytochromes b and c activity; and a low rate of DNA, RNA, and protein synthesis. It is often accompanied by arrest of feeding and changes in behavior and pigmentation. In fact, it is essential to distinguish not one but three different physiological states, corresponding to the period of prediapause, at the end of which diapause is initiated, the period of diapause maintenance, and finally the period of diapause interruption.

Diapause is usually preceded by an accumulation of reserve materials, namely, lipids and glycerol, which implies that the hormonal status of the prediapause insect is modified. It is triggered by changes in environmental conditions, temperature, photoperiod, humidity, and diet, the effects of which are exerted for periods of variable length, sometimes from one generation to the next. It occurs mostly at the onset of winter in temperate zones and in the dry season in the tropics.

The photoperiod is perceived by the ocelli or directly by the brain, which is still capable of responding even if transplanted into the abdomen (*Antheraea*, Williams and Adkisson, 1964). In *Pieris*, the clypeus, which covers the brain, becomes much thinner in the period in which photoperiodic induction takes place (Claret, 1966). Light can also be perceived through the integument by means of epidermal pterobilin (*Pieris*, Vuillaume *et al.*, 1971). We know far less about the action mechanisms of other factors, but it is probable that all the stimuli are integrated in the brain.

Once they are integrated in the brain, the stimuli which initiate diapause determine the nervous and neuroendocrine processes which first initiate the diapause then maintain and finally break it. The ecdysteroids, JH and neurohormones, including PTTH, allatostatins and allatotropins, and diapause hormone act as we shall show below. The determinism of diapause has been the subject of several reviews to which the reader can refer for greater detail (Chippendale, 1983; Denlinger, 1985; Lavenseau *et al.*, 1986).

5.1. IMAGINAL DIAPAUSE

Imaginal diapause is a reproductive arrest and, more specifically, the arrest of vitellogenesis. Its endocrine determinism has been demonstrated in many cases, of which the best known is that of the Colorado beetle, *Leptinotarsa*. Diapause in this insect is initiated by short days, light stimuli being transmitted to the brain which regulates CA activity (see Chapter 4). Diapause is preceded by a prediapausal stage, in which feeding increases while reproduction ceases. Significant metabolic changes take place in the proteins and glucides which accumulate in the fat body. On termination of this stage, the insect burrows into the ground for a period of several months, during which it remains motionless.

Imaginal diapause stems from JH failure (De Wilde *et al.*, 1968). This is confirmed by determinations of the titer of JH and its esterases during adult life, and determinations of the synthetic activity of the CA of the insects in long and short days (De Kort, 1981).

In other insects also, the absence of JH gives rise to reproductive diapause, which is terminated in both sexes by active implants in *Oedipoda* (Broza and Pener, 1969) and *Tetrix* (Poras, 1982), and by treatment with JH or JHA (Bowers and Blickenstaff, 1966 in *Hypera postica*; Schooneveld, 1972, and Schooneveld *et al.*, 1977 in *Leptinotarsa*; Burov *et al.*, 1972 in the bug *Eurygaster integriceps*; Kamm and Swenson, 1972 in *Draeculacephala crassicornis*; Hodek *et al.*, 1973 in ladybirds; Kambysellis and Heed, 1974 in *Drosophila*; and Numata and Hidaka, 1984 in the bug *Riptortus clavatus*).

The inactivity of the CA in diapausing insects is due to cerebral inhibition both nervous and humoral (see Chapter 4). Various investigators and especially Siew (1965), Geldiay (1967), and Schooneveld *et al.* (1977) reported the modifications in the storage pattern of certain ns cells in diapausing insects.

In addition, cauterization of the PI reveals different aspects of the diapause syndrome (De Wilde and De Boer, 1969), and electrical stimulation of the PI of *Anacridium* causes an increase in the secretory activity of the CA and terminates diapause (Girardie *et al.*, 1974).

The ecdysteroids, whose role we examined in the regulation of CA activity, could also affect the onset of diapause in the adult. In fact, it has been shown in

Leptinotarsa that the hemolymph ecdysteroid titer is much higher in adults raised in short days and diapause-destined than in those raised in long days (Briers and De Loof, 1981).

5.2. PUPAL DIAPAUSE

Diapause in the pupal stage, as in the larval stage, results from an ecdysone deficiency due to the lack of secretion and release of the PTTH. It is terminated by cerebral reactivation, as demonstrated on the giant saturniids *Platysamia* and *Antheraea* (Williams, 1946, 1947, 1952). Diapause in these species begins at the end of summer and lasts until spring, when it is interrupted by the surrounding warmth. In experimental conditions, the same result can be achieved by keeping the insect for a short period at low temperature, followed by a return to moderate temperature. Implantation of the brains of reactivated pupae into diapausing pupae induces the termination of diapause. The same endocrine process was found in many species, in which diapause was interrupted by the implantation of activated brain of PG, or by the injection of 20-HE. Regulation occurring in this case is identical to growth regulation (see Chapter 3).

Histological investigations of the PI of the brain reveal an accumulation of neurosecretion during diapause in many species. This was observed in particular in the PI of *Mimas tiliae* (Highnam, 1958) and *Pieris* (Kono, 1973) and in the lateral ns cells of *Manduca* (Borg and Bell, 1977) and also in the frontal ganglion of *Manduca* (Borg *et al.*, 1973).

As surprising as this may appear, experiments designed to interrupt diapause by brain implantation showed that the action of the brain is not always exerted at the same time. Thus, in certain species, the brain appears to exert very early action because its presence stops being necessary fairly soon after the onset of diapause, i.e., *Pieris* (Kono, 1977), *Heliothis zea* (Meola and Adkisson, 1977), and *Acronycta rumicis* (Kind, 1978). A number of lepidopterans, such as *Bombyx, Antheraea, Pieris,* and *Manduca,* are capable of escaping diapause, despite their decerebration (Kobayashi *et al.*, 1960; McDaniel and Berry, 1967; Maslennikova, 1970), even if the CC are also removed (Judy, 1972). This is probably due to the fact that the PTTH is stored in the CA of these species (see Chapter 3). It is interesting to note that the implanted brain/ring gland complex may function according to the rate of its donor (Giebultowicz and Denlinger, 1986 in *Sarcophaga*) but may also undergo activation due to its transplantation (Cloutier *et al.*, 1962 in *Ostrinia*).

As for the ecdysteroids, the determination of their hemolymph titer during diapause has shown that it was very low in different insects, i.e., *Sarcophaga peregrina* and *Sarcophaga crassipalpis* (Ohtaki and Takahashi, 1972; Walker and Denlinger, 1980) (Fig. 46), *Hyalophora* (McDaniel, 1979), *Heliothis* (Loeb

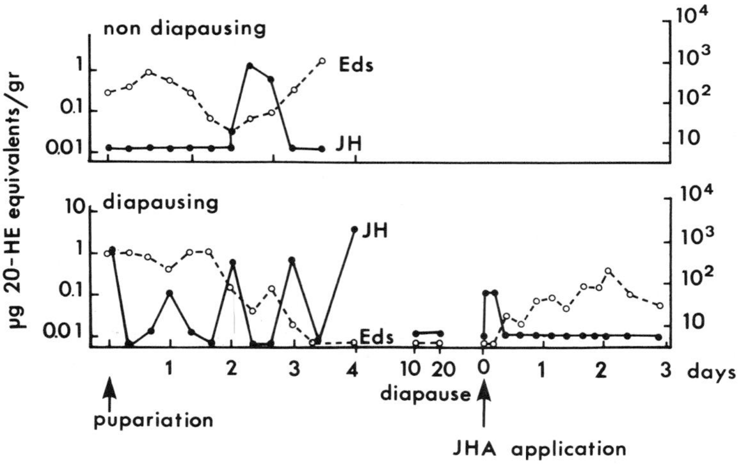

FIGURE 46. Juvenile hormone and ecdysteroid hemolymph titers in diapause- and non-diapause-programmed *Sarcophaga crassipalpis* pupae. (Modified from Walker and Denlinger, 1980.)

et al., 1982), and *Manduca* (Bowen *et al.*, 1984) (Fig. 47). The ultrastructural aspect of the PG of diapausing insects is that of great inactivity (see Herman, 1967).

Very few investigations have been conducted during the prediapausal period, which determines the entry into diapause. In *Manduca,* it appears that the hemolymph ecdysteroid titers are similar in short-day, diapause-destined pupae and long-day, non-diapause-destined pupae, but the prepupal peak is only present in the latter category (Bradfield and Denlinger, 1980; Bowen *et al.*, 1984). At the break of diapause, the hemolymph ecdysteroid titer increases, but less sharply than in the nondiapausing insect. It remains high for a few days, unlike the second ecdysteroid peak of the nondiapausing pupa. In *Sarcosphaga argyrostoma* the ecdysteroid titer is very low in the larvae reared in diapause-inducing conditions (Richard *et al.*, 1987).

A number of experiments of PG ablation and reimplantation (Williams, 1947, 1952; Meola and Adkisson, 1977) illustrated the role of the PG in the break of diapause. Most of the experiments on the injection of 20-HE were conducted in *Aeshna,* solitary bees, *Manduca, Mamestra,* and *Sarcophaga* (Fraenkel and Hsiao, 1968; Hsiao and Hsiao, 1969; Schaller *et al.*, 1974; Žďárek and Denlinger, 1975; Gibbs, 1976; Bodnaryk, 1977; Bradfield and Denlinger, 1980; Bowen *et al.*, 1984). On the whole, they achieved diapause termination, but their effectiveness differed during diapause, undoubtedly due to modifica-

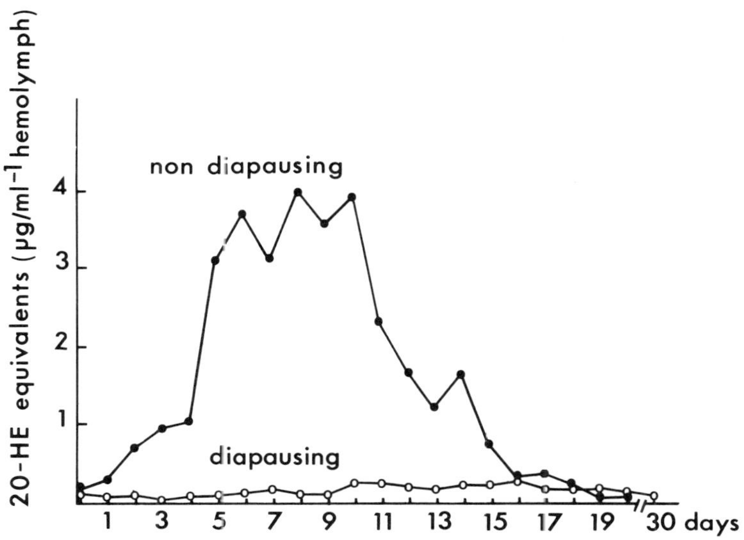

FIGURE 47. Ecdysteroid titer in nondiapausing and diapausing *Manduca* pupae. (Modified from Bowen *et al.,* 1984.)

tions of the sensitivity of the PG. Moreover, injection of large doses of 20-HE was observed to prevent diapause in flies (Žďárek and Denlinger, 1987).

Juvenile hormone is active in the regulation of pupal diapause, at least in certain species. Thus, in *Sarcophaga* the determination of the whole-body JH (by *Galleria* assay) revealed the occurrence of JH pulses every 24 hr, whereas in nondiapausing insects, only one peak was observed (Walker and Denlinger, 1980; Denlinger *et al.,* 1984). However, this was not observed in *Manduca* (Bradfield and Denlinger, 1980), in which the hemolymph JH titer of prediapausing insects does not appear to differ from that of insects destined for development without diapause (Bowen *et al.,* 1984).

The question arises whether cyclic nucleotides play any role in the processes associated with diapause break. In the brain, a pulse in cAMP activity was described in the PI of diapausing pupae at the moment of diapause break (Rasenick *et al.,* 1976), but such pulses were not observed in *Mamestra* (Bodnaryk, 1978) or *Sarcophaga* (Denlinger, 1976) for cAMPc and for cGMP. Furthermore, the injection of a cAMP derivative terminates diapause in silk moths (Rasenick *et al.,* 1976) and prevents its appearance in *Sarcophaga.* Moreover, cAMP analogs were shown to mimic PTTH action on the ring gland *in vitro* (Richard and Saunders, 1987). In *Mamestra,* cAMP and cGMP appear to play opposite roles in diapause maintenance, diapause break being caused by cGMP only (Bodnaryk, 1987).

5.3. LARVAL DIAPAUSE

Like pupal diapause, larval diapause depends on the hemolymph titers of PTTH and 20-HE. It also depends on the JH as demonstrated in several lepidopterans, i.e., *Chilo* (Fukaya and Mitsuhashi, 1957), *Plodia interpunctella* (Waku, 1960), *Diatraea grandiosella* (Chippendale and Yin, 1973; Yin and Chippendale, 1973), *Ostrinia nubilalis* (Yagi and Akaike, 1976), *Monema flavescens* (Takeda, 1978), and the beetle *Trogoderma granarium* (Nair, 1974).

The first investigation, conducted on *Chilo,* showed that the brain and PG definitely caused termination of diapause, but this resulted from the absence of the CA (Fukaya and Mitsuhashi, 1957). The presence of JH was also demonstrated in the diapausing larva.

Later investigations confirmed the functions of the CA in larval diapause, a role of initiation and maintenance. Diapause was triggered by the application of JHA in *Diatraea* (Yin and Chippendale, 1973) and *Chilo* (Yagi and Fukaya, 1974), and its termination was caused by removal of the CA in *Chilo, Ostrinia, Monema,* and *Diatraea* (Yagi and Fukaya, 1974; Yagi and Akaike, 1976; Takeda, 1978; Yin and Chippendale, 1979).

Determination of the hemolymph JH titer using *Galleria* test in diapausing larvae confirms the existence of a high titer in comparison with developing insects (Yin and Chippendale, 1976 in *Diatraea*; De Loof *et al.,* 1979 in the wasp *Nasonia vitripennis*), especially in early diapause (Yagi and Fukaya, 1974 in *Chilo,* Sieber and Benz, 1980b in *Laspeyresia*), and in mid-diapause (Takeda, 1978 in *Monema*).

The hemolymph JH titers of diapausing and nondiapausing *Diatraea* larvae before and during diapause reveal interesting changes. Before and immediately after the onset of diapause, the JH titer is high (4300 *Galleria* units/ml, 1230–2610 *Galleria* units/ml). It is maintained at 700–1500 *Galleria* units/ml up to day 110 and then starts to decline: 200 *Galleria* units/ml on day 130, 70 *Galleria* units/ml on day 190 (Yin and Chippendale, 1979). In nondiapausing larvae, the JH titer, which is high before molting, falls to 140 *Galleria* units/ml in 6 hr, and then remains at a very low level.

The situation is different in other insects. Investigations in *Ostrinia* indicate low JH titers during diapause. Hence in this species, it is possible that the JH performs only the function of initiating diapause and plays no part in its maintenance (Chippendale and Yin, 1979). A study performed in *Gryllus* failed to reveal the presence of hemolymph JH during diapause (Ismail *et al.,* 1984).

If the role of JH in the initiation and maintenance of larval diapause has been demonstrated, at least in some species, its action mechanism remains to be definitely determined. It has been suggested that, during larval diapause and also pupal diapause, JH acts by inhibiting the secretion and release of PTTH which, in turn, suppresses the activation of the PG. Thus the break in diapause was achieved by brain implantations in *Platysamia* (Williams, 1946), *Ostrinia* (Cloutier *et al.*, 1962), and *Monema* (Takeda, 1978).

Inhibition of PTTH release by JH was also postulated in nondiapausing species, in which development is lengthened by the injection of JH and shortened by removal of the CA (see Chapter 3). The existence of a relationship between the activity of the CA and the activity of certain cerebral ns cells was also suggested by the aspect of certain ns cells of the PI during diapause in several insect species, and even *in vitro*. Incubation of the brain of *Mamestra* in the presence and absence of JHA (Hiruma *et al.*, 1978a) also showed that JH acts on the cerebral ns cells by inhibiting their release.

The identification of the ns cells, which store material during diapause and which release material at diapause termination, is important because these cells are probably producing PTTH. Thorough investigations on lepidopterans showed that not all the type A ns cells underwent diapause-related modifications. Histological changes related to diapause and to its termination are almost always observed in a single type, usually called A′ (sometimes B or II) in *Chilo* (Mitsuhashi and Fukaya, 1960), *Euproctis flava* and *Dendrolimus spectabilis* (Mitsuhashi, 1963), *Monema* (Takeda, 1972a), *Pieris rapae* and *Pieris brassicae* (Kono, 1973; Kind, 1976, 1978), and *Mamestra* and *Acronycta turnicis* (Kind, 1976; Kind and Vaghina, 1976; Hiruma and Agui, 1977). In *Pectinophora gossypiella,* however, the A cells appear to act like A′ cells (Raina and Bell, 1978). Furthermore, the lateral ns cells of *Mamestra* seem to show the same activity as the A′ cells (Hiruma and Agui, 1977), but not in *Manduca* (Borg and Bell, 1977). In this species, however, the ns cells producing PTTH were recently identified as lateral ns cells (Agui *et al.*, 1979) (see Chapter 3).

The termination of larval diapause by 20-HE or active PG confirms the role of both JH and 20-HE in *Lucilia caesar* (Fraser, 1958), solitary bees (Hsiao and Hsiao, 1969), *Nasonia* (De Loof *et al.*, 1979), *Laspeyresia* (Sieber and Bentz, 1980a,b), and *Calliphora vicina* (Maslennikova, 1977). In some insects, however, when the operations are performed in early diapausing insects (*Chilo* and *Diatraea*), either the diapause is not broken or larval/pupal intermediates are obtained due to the presence of JH (Chippendale and Yin, 1979).

The determination of ecdysteroid titers shows that they are very low at the entry into diapause in the endoparasitic wasp *Pimpla investicator* (Claret *et al.*, 1978), and at the onset of diapause in *Laspeyresia* (Sieber and Benz, 1980a,b) and *Gryllus campestris* (Ismail *et al.*, 1979). In *Ostrinia* (Bean and Beck, 1983; Gelman and Woods, 1983), an ecdysteroid peak is observed in prediapausing insects, but it is 10 times weaker than in nondiapausing insects. During diapause, the titer remains low and no peak is observed.

The histological and ultrastructural data on the PG during larval diapause indicate a state of inactivity, as in diapausing pupae. This is confirmed by PG incubation experiments in the presence of brain fragments, also showing that the

PTTH activity of the brain is much lower in prediapause larvae than in diapausing larvae (Yin *et al.*, 1985).

5.4. EMBRYONIC DIAPAUSE

In the silkworm *Bombyx,* the number of annual generations varies according to strain and external conditions. Some univoltine strains have a single generation per year, and all their eggs enter diapause. Other bivoltine or quadrivoltine strains have two or four generations per year, and the eggs they lay sometimes show rapid development and are sometimes diapausing. Embryonic diapause is determined by the temperature and photoperiod experienced by the mother in the embryonic or larval instar: a long photoperiod and high temperature induce the appearance of embyronic diapause.

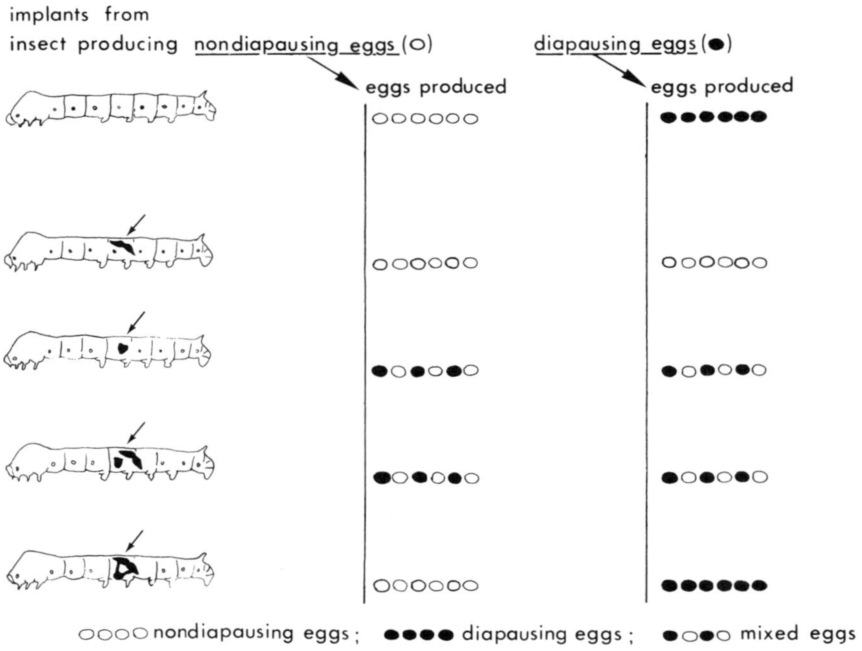

FIGURE 48. Experimental study of the regulation of embryonic diapause in the lepidopteran *Bombyx.* Implantation experiments were performed in larvae with organs taken from larvae determined to produce either nondiapausing eggs or diapausing eggs. Only the complexes from the latter were able to induce diapause. From top to bottom: control, implantation of brain, subesophageal ganglion, both organs disconnected and connected. (After Fukuda, 1963.)

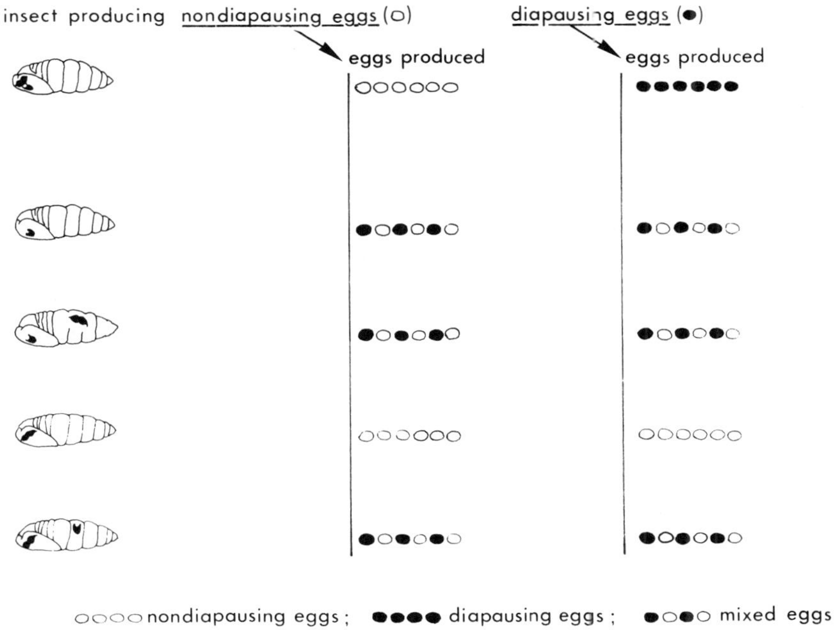

FIGURE 49. Experimental study of the regulation of embryonic diapause in the lepidopteran *Bombyx.* Ablation and implantation experiments were performed in pupae programmed to produce either nondiapausing eggs or diapausing eggs. From top to bottom: control, brain removal, removal and reimplantation of the brain, subesophageal ganglion removal, removal and reimplantation of the subesophageal ganglion. (After Fukuda, 1963.)

The experimental investigations of Hasegawa (1951) and Fukuda (1952, 1963) demonstrated the existence of a diapause factor DH produced in the subesophageal ganglion, whose release is controlled by the brain. The nature of cerebral control appears to be stimulating (Morohoshi and Oshiki, 1969) or stimulating and inhibiting (Fukuda, 1953) (Figs. 48 and 49).

The subesophageal ganglion starts to release diapause hormone immediately after pupation (Hasegawa, 1952), before the onset of ovarian development. However, the command which it receives from the brain at this time results from information stored by the latter at the end of embryonic development, i.e., long before.

The ovary is the effector of the diapause hormone and may produce a mixture of diapausing and nondiapausing eggs, suggesting a diversity of sensitivity of the follicles. The glycogen content of the ovary is high in pupae which produce diapausing eggs. It drops sharply if the subesophageal ganglion is removed and increases considerably if extracts of this organ are injected (Yamashita and Hasegawa, 1966). The action of the diapause hormone on the ovary was confirmed in culture. Apart from its action on glycogen (Yamashita and Hasegawa, 1967) the diapause hormone raises the level in the eggs of triglycerides (Ichimasa and Hasegawa, 1973), cholesterol (Ichimasa, 1976), and 3-hydroxykynurenin, an ommochrome precursor (Hasegawa, 1964). The latter action occurs *in vitro,* making it a convenient assay to estimate the formation of diapausing eggs (Yamashita and Hasegawa, 1967). It is well known that the chorionated oocytes contain large amounts of ecdysteroids, mostly conjugates, which could be converted to free ecdysone and 20-HE during embryonic development, probably determining cuticulogenesis (Lagueux *et al.,* 1981; Hoffmann and Lagueux, 1985). In the diapausing *Bombyx* embryo, the maternal ecdysteroids could play the same role and could release the free ecdysteroids at the end of diapause. In fact, in an Australian locust it was shown that the ecdysteroid level of diapause-destined embryo is three times less than in non-diapause-destined embryos (Gregg *et al.,* 1987).

The cells producing the diapause factor were identified histophysiologically (Fukuda and Takeuchi, 1967) as two medioventral cells located in the subesophageal ganglion. They are unloaded and nearly invisible in pupae and adults producing diapausing eggs, whereas their appearance is normal in the reverse case.

Embryonic diapause does not occur only in *Bombyx.* It is also encountered in other species, such as *Orgyia antica, Adelphocoris lineolatus,* and the flesh fly *Sarcophaga* in which it is also determined by the mother. The source of the diapause factor is found in *Orgyia* in the subesophageal ganglion, the removal of which prevents diapause (Kind, 1969), and in *Adelphocoris* in the thoracoabdominal nerve mass. The ns cells of the subesophageal ganglion implicated in this secretion in *Orgyia* appear to be two medioventral cells located as in *Bombyx.* In *Adelphocoris,* five pairs of lateral cells may be involved (Ewen, 1966).

Fukuda (1963) reported the presence of the diapause hormone only in the subesophageal ganglion. This was confirmed by Hasegawa and Shimuzu (1987). However, recent investigations reveal the occurrence of diapause factor in the brain (Sonobe and Keino, 1975) and in the CC-CA complex of *Bombyx* (Takeda and Ogura, 1976).

The diapause hormone is also found in the subesophageal ganglion of other diapausing species such as *Lymantria, Antheraea yamamala* (Fukuda, 1951; Hasegawa, 1951), and in species without egg diapause but with pupal diapause, i.e., *Antheraea pernyi* and *Phalaenoides glycinae* (Andrewartha *et al.,* 1974) and the cockroach *Periplaneta* (Takeda, 1977). It is also present in the brain, CC, and CA of *Leucania* (Ogura and Saito, 1973), *Periplaneta* (Takeda, 1977), and *Locusta* (Takeda and Girardie, 1985). In the latter species it was demonstrated that a factor with diapause hormone activity is located in the subesophageal ganglion and the glandular lobe of the CC, but not in the PI or lateral ns cells of the brain. It disappears from the glandular lobe of the CC following the disconnection of the latter. Moreover, the CA appear to be necessary for the activity of the CC, a rather surprising observation.

The presence of the diapause hormone in insects devoid of diapause may be understood in different ways. It may be assumed that the diapause hormone displays a different action in nondiapausing insects, such as metabolic action, for example. Moreover, through its molecular structure, it may resemble another factor, which could explain both its presence in nondiapausing species and the discrepancies in results concerning its location.

The question arises as to whether the diapause hormone acts exclusively in embryonic diapause, or whether it plays a part in larval and pupal diapauses. According to Fukuda (1963), the subesophageal ganglion does not act in pupal diapause: it is incapable of initiating diapause in *Philosamia* pupae, and its removal from diapausing pupae does not interrupt diapause. Nevertheless, some investigators, including Maslennikova (1973), support a different view as the outcome of experiments performed on *Pieris*. It seems difficult to discount the existence of a diapause factor transmitted by the egg when larval or pupal diapause is determined by the conditions experienced by the mother.

5.5. PURIFICATION OF THE EMBRYONIC DIAPAUSE FACTOR (DH)

Purification of the hormone of the subesophageal ganglion of *Bombyx* was attempted by Sonobe and Ohnishi (1971), as well as by Hasegawa *et al.* (1972, 1974), Isobe *et al.* (1973; 1975), and Kubota *et al.* (1976). Using 2 million heads, 0.012 g of a substance was obtained, of which 6 μg corresponded to 1 DH unit. This unit is the quantity required for the production of 40% diapausing eggs in a female conditioned to produce nondiapausing eggs. The isolated substance is stable in bases and weak acids. It exists in two forms, DHA and DHB, and DHB is three times more active than DHA.

A thorough analysis of DHA and DHB (Isobe *et al.*, 1975) showed that they are destroyed by proteolytic enzymes. DHA contains 14 different amino acids and two amino sugars, glucosamine and galactose amine, and DHB contains 14 amino acids and no amino sugars.

The molecular weight of DHA is estimated at 3300 ± 40 and that of DHB 2000 ± 200 (Kubota *et al.*, 1976). By using different extraction methods, Sonobe and Ohnishi (1971) also obtained two substances, both thermostable, of higher molecular weight, the smaller having a molecular weight of 5000–10,000, which are destroyed by proteolytic enzymes.

5.6. CONCLUSIONS

Investigations into the regulation of diapause reveal considerable diversity, which is not surprising considering that diapause occurs not only in all stages, but also at different moments of these stages, in which the hormonal states are far from similar. Thus, for example, JH disappears in the pupal stage to allow the metamorphosis and development of the imaginal structures, which is not the case during the larval stages. Moreover, the species differ among themselves for the timing of certain processes, such as spermatogenesis and oocyte maturation. Hence it is probable that the mechanisms regulating the initiation and termination of diapause have evolved independently of each other, although a number of common fundamental factors are implicated, such as ecdysone and PTTH.

Imaginal diapause is a reproductive diapause. It is caused by JH failure due to brain inhibition of the CA. Larval and pupal diapauses result from the absence of PTTH, causing inactivity of the PG, which is reactivated at the end of diapause by PTTH release. Juvenile hormone also plays a role of definite but varying importance. In certain pupae, and especially in larvae, a high JH titer induces and maintains diapause by exerting a negative feedback on the PG and/or the brain, and also by exerting a metabolic role which favors the accumulation of lipid and protein stores. The activity of the CA is regulated by the brain and its neurohormones, allatotropins, and allatostatins. In addition, neurohormonal regulation of the brain is exerted on the fat body to induce the release of the JH esterases, which are very abundant in the hemolymph of certain insects at the end of diapause, when they ensure the rapid elimination of the circulating JH.

In the embryo of *Bombyx* and *Pieris,* a maternal diapause factor originating in the subesophageal ganglion acts on the pupal ovary and determines the conditioning of the eggs. It is well known that in the eggs of *Bombyx* it causes accumulation of glycogen, triglycerides, and 3-hydroxykynurenin, but we do not know precisely what relationship exists between these processes and diapause. It is curious to note that the stimuli which condition secretion of the embryonic diapause factor are perceived long before its release. Purification experiments indicate the existence of two polypeptidic substances, the smaller containing 14 identified amino acids.

6

Reproduction

INTRODUCTION

Reproduction exhibits two main characteristics: a great variety of successive and fundamentally distinct processes, and strict dependence on environmental factors. Its regulation therefore involves numerous sensory receptors, nerve transmission, and integration in the brain, which regulates the synthesis of the two groups of insect hormones—the juvenile hormones and the ecdysteroids—and produces its own neurohormones.

Reproduction results in a succession of interdependent steps which are often quite different in nature and take place at various stages of the insect's biological cycle. The main reproductive events are the following: sex determination, gonial mitoses and meiosis, differentiation of reproductive organs, spermatogenesis, previtellogenesis and vitellogenesis, accessory gland functioning, pheromone production, sexual behavior, mating, ovulation, oviposition, incubation of embryos or larvae, and parturition.

Many environmental factors affect reproduction, including the finding of food. Appropriate seasonal and nutritional conditions are essential for the egg to mature, for mating to succeed, and for the offspring to survive, thanks to its emergence at the right time and place. Internal factors like the nutritional state, the activity of the fat body, and the state of development of the oocytes are also of considerable importance. The different reproductive steps obviously depend on mating, which allows fertilization of the oocytes by the spermatozoa and enables the transmission of male active peptides to the female; the latter triggers oocyte development, egg laying, and monogamy. The ovary is at the origin of a feedback system exerted by mature, nonoviposited eggs on the maturation of young oocytes.

Present knowledge of reproduction has progressed variously, depending on the event considered. The best known field is that of vitellogenesis, while the stages least often explored include spermatogenesis and the first steps of ovarian development.

6.1 SEX DETERMINATION

Numerous attempts to demonstrate the reality of hormonal control of sex determination were made. They consisted of castration and transplantation of disks from one sex to the other in species in which pronounced dimorphism exists between male and female. Most of these experiments gave negative results. In *Orgyia antiqua,* however, female disks transplanted into males developed into male wings, suggesting possible hormonal involvement in disk sex determination (Paul, 1937). This hypothesis was confirmed in an extensive study, performed in the glowworm *Lampyris noctiluca* (Naisse, 1966, 1969).

6.1.1. *Lampyris*

The adult sexual dimorphism of *Lampyris* is well known. The apterous females lie on the ground and produce bright green lights by means of photogenic organs. These organs are reduced in the male, which has normal wings and can fly and detect the place where the female is lying.

The study of the early development of both male and female gonads revealed that the first differences between the sexes appear during the fourth larval instar. In both males and females, numerous cell divisions take place at this time, but whereas in the male they mostly concern the apical mesodermal tissue, which develops considerably and degenerates in the next instar, in the female they occur in the basal mesodermal cells destined to become the follicular cells.

Testes of various ages were transplanted into female larvae or nymphs. After the end of the fifth larval instar, they were not able to induce changes in the female, but when transplanted earlier they induced great transformations. Conversely, when ovaries were implanted into males, no feminization of the host was observed, and in some cases the implanted ovaries were even masculinized.

Thus, the ovary does not possess the ability to feminize the male, but the male testis of the fifth instar larva is able to transform the female gonad into a male one. At this stage, the development of the apical tissue is maximal, and it exhibits strong secretory activity, suggesting that it might be the origin of an androgenic hormone. However, the involvement of another tissue in sex determination was suggested by the fact that ovaries transplanted into castrated males display a certain amount of masculinization. The involvement of neurohormones was first investigated by histology, which revealed that PI ns cells do not function in the same way in males and females. Next, male and female larvae were joined in parabiosis experiments. These resulted in the masculinization of the female gonad, even when castrated males were joined to females whose CC had been removed. In other experiments, the CC were removed from females and males of various ages. This operation did not modify the sexual development of either the females or males operated during the fifth or late in the fourth instar, but in the males operated early in the fourth instar, the gonads underwent feminization.

Therefore, at least in *Lampyris,* male sexual differentiation depends on a neurohormone from the PI, which triggers apical tissue differentiation, which in turn produces an androgenic hormone. Several attempts have been made to confirm the *Lampyris* data.

6.1.2. *Leptinotarsa*

The most thorough investigations were made in *Leptinotarsa* (Richard-Mercier, 1982), where sex determination takes place in the embryo. Despite the large number of experiments made, the existence of a humoral control of sex determination was not demonstrated. Cauterization of polar embryonic cells suppressed the germinal cells but allowed male and female gonads to develop. Further, transplanted gonads did not affect the original sex determination of host gonads and the involvement of neurohormones was not demonstrated either; destruction of the presumptive brain area in young embryos led to the appearance of acephalic larvae with male or female gonads, and incubation of male or female gonads in the presence of male or female brain did not reveal any action of the brain on sexual differentiation.

6.2. OOGENESIS

6.2.1. Oogenesis: First Steps

The determination of mitosis, meiosis, and ovariole differentiation was studied in the blood-sucking bug *Panstrongylus megistus* (Furtado, 1979) (Fig. 50). In this insect, ovariole development takes place during the last larval instar and begins 24–48 hr after the blood meal. The first event is a mitotic crisis affecting both germinal and somatic cells. This crisis is followed by the entry of the oogonia into meiosis, and the differentiation of the tropharium and pre-follicular tissue. It lasts until the onset of previtellogenesis in the young adult female.

Observation of the ns cells of the PI during the last larval instar from the time of the blood meal until the imaginal molt showed the presence of four distinct ns cell types (A, A′, B, and C) with different patterns of activity. In type A cells, the ns material is released before the mitotic crisis, whereas in type A′ cells, it is discharged before the onset of meiosis.

Electrocoagulation of the PI was carried out at two different stages before the mitotic phase and before the meiosis phase. In both cases, the operation prevented the development of the ovary and caused its degeneration. Hence the

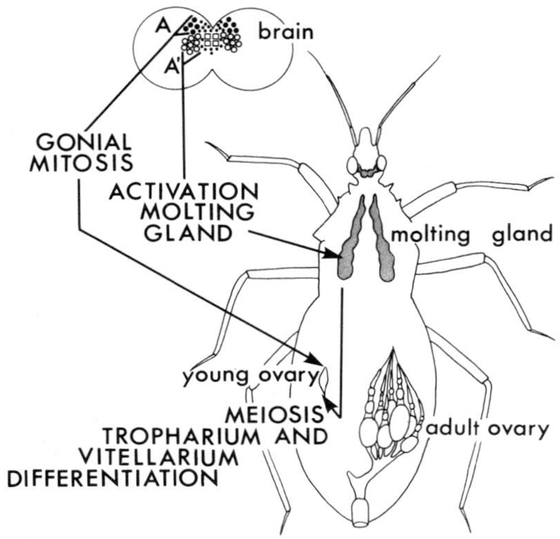

FIGURE 50. Schematic representation of the regulation of gonial mitoses and meiosis in the bug *Panstrongylus megistus*. (From the data of Furtado, 1979.)

ns cells regulate the early development of the ovary in two ways, by helping to trigger both mitosis and meiosis. To achieve better understanding of these mechanisms, brain surgery was continued by destroying only type A ns cells; the results were not identical to those of total removal of the PI, since inhibition of the mitotic crisis was observed but the onset of meiosis was not affected.

Removal of the PG at various times after the blood meal showed that this gland does not trigger the onset of mitosis but is necessary for that of meiosis. Hemolymph ecdysteroid determination demonstrated the existence of two peaks: the first coincided with the onset of meiosis, and the second, which was higher than the first, preceded molting (Furtado *et al.*, 1976). In addition, insects with electrocoagulated PI had their meiosis reestablished by 20-HE injection (Fig. 51).

Ecdysone or 20-HE are therefore responsible for the onset of meiosis and restores normal ovarian development in females deprived of the PI having already completed the mitotic processes. The brain plays a dual role in controlling the initial phases of oogenesis; it acts directly by inducing mitoses and indirectly by stimulating the PG. Juvenile hormone does not seem to intervene, either in early gonial mitosis or in the initiation of meiosis. However, as shown below, it inhibits the processes initiated by ecdysone.

In *Schistocerca* (Jones, 1978), initiation of meiosis in the male also depends on 20-HE. *In vitro* studies showed that the role of this hormone was to stimulate a blood–germ cell barrier which creates a special environment necessary for meiosis onset. In *Locusta* (Lanot *et al.*, 1987), it was shown that the meiosis

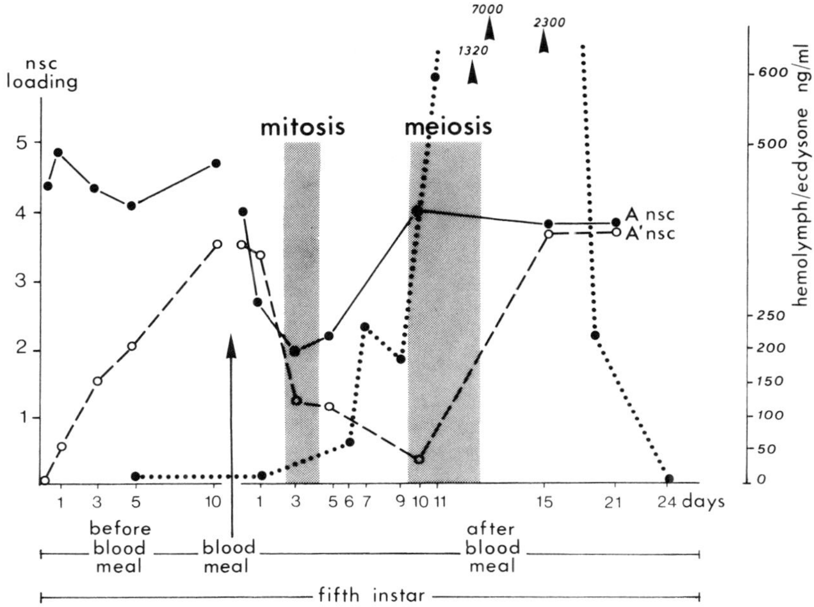

FIGURE 51. Ecdysteroid titer, release of type A and A' neurosecretory products and ovarian development in *Panstrongylus.* (Modified from Furtado *et al.*, 1976.)

reinitiation is preceded by an increase of free ecdysone of the oocytes. *In vitro* the reinitiation of meiosis might be determined by the addition of ecdysone.

6.2.2. Ovariole Differentiation

Not only germ cells but also the mesodermal elements surrounding them have to go through certain steps in order to develop adult structures, and these events are controlled by hormones, JH, and ecdysone.

Early transplantation experiments performed in *Drosophila* (Bodenstein, 1938) and *Iphita limbata* (Ittycheriah and Nayar, 1968) demonstrated that the ovary cannot acquire the ability to develop unless it has been submitted to the particular internal medium of the last larval instar, characterized by the presence of ecdysone and the absence of JH.

Other *in vitro* experiments made on *Tenebrio* confirmed the *in vivo* experiments. In the incubated ovaries, the oocytes began their growth and the follicles became organized as they normally do, during the nymphal instar, provided a sufficient amount of 20-HE was added to the incubation medium. However, as already mentioned, JH suppressed the effects of 20-HE (Laverdure, 1971).

In lepidopterans, similar regulation of ovariole differentiation was shown both *in vitro* and *in vivo* in *Galleria* (Shibuya and Yagi, 1972) and *Bombyx* (Chatani and Ohnishi, 1976). *In vitro,* the growth and differentiation of oocytes, trophocytes, follicular cells, and vitellarium from last instar *Galleria* larvae required the presence of 20-HE, and *Bombyx* silkworm could not initiate ovarian development unless the PG or 20-HE was present.

In the mosquito *Aedes aegypti,* the formation and separation of new follicles in the ovariole was also shown to depend on 20-HE (Beckemeyer and Lea, 1980), and in the pedogenetic insect *Heteropeza pygmaea,* the larval ovaries differentiated one follicle when they were maintained *in vitro* in the presence of 20-HE (Went, 1978).

The development of the oviduct in both *Panstrongylus* (Furtado, 1979) and *Oncopeltus* (Dorn *et al.,* 1986) was found to be dependent on ecdysteroids, 20-HE in the case of *Panstrongylus* and makisterone A in the case of *Oncopeltus* (this development was dose-dependent in the latter). Later, however, during the early adult stage, the presence of JH is also needed (Furtado, 1979). Contrarily to ecdysteroids, JH inhibits the growth and differentiation of the ovary in *Corcyra cephalonica* (Deb and Chakravorty, 1982).

6.2.3. Previtellogenesis

In most cases, previtellogenesis seems to occur independent of the CA, as demonstrated for the *Rhodnius* acrotrophic ovary (Wigglesworth, 1936). However, it was also shown in *Rhodnius* that JH acted on the follicle cells during previtellogenesis, by preparing them to respond to the same hormone during the next step, i.e., vitellogenesis (Abu-Hakima and Davey, 1975). On the other hand, in *Aedes* and *Musca* polytrophic ovaries, JH stimulated previtellogenesis while 20-HE triggered vitellogenesis (Gwadz and Spielman, 1973; Hagedorn *et al.,* 1977; Adams, 1981). Stimulation of previtellogenesis by the CA was also demonstrated in insects with acrotrophic ovaries, such as the bug *Dindymus versicolor* (Friedel, 1974) and the beetles *Dystiscus marginalis* (Joly, 1945) and *Pterostichus nigra* (Ferenz, 1977).

In addition to JH, previtellogenesis is regulated by a neurohormone from the PI in *Tenebrio* (Laverdure, 1972) and *Anacridium* (Girardie *et al.,* 1974).

Consequently, previtellogenesis seems to be regulated in different ways. In ovaries possessing nurse cells, the latter are mainly responsible for previtellogenesis, but the follicle cells also participate, and the oocyte too has definite synthetic properties (Chaminade and Laverdure, 1980). The nurse cells may well be entirely or partly autonomous, while the follicle cells may depend on a brain neurohormone as they do later in vitellogenesis. No information is available as to how oocyte synthetic activity is regulated.

6.2.4. Vitellogenesis

Vitellogenesis is the most important metabolic event in the adult life of the female insect, and numerous studies have been devoted to its morphological and cytological aspects.

Insect eggs are large because of their heavy vitellus. The amount of cytoplasm in the egg is small, and is mostly concentrated in its cortex. It contains glycogen, RNA, and sometimes DNA. Most of the egg is occupied by yolk bodies, chiefly protein yolk, which includes protein–carbohydrate complexes and a little lipid yolk.

Vitellogenesis often occurs in the adult insect but may take place earlier. It may be continuous or cyclic. Sometimes it includes long periods of arrest due to diapause or to embryo incubation. Food stores allowing vitellogenesis to take place are constituted during adult and larval life.

Oocyte growth includes a period of relatively slow growth, i.e., endogenous vitellogenesis and previtellogenesis, and a period of yolk deposition or vitellogenesis, a fast event that mostly occurs in the terminal oocyte of each ovariole only.

The hormonal and neurohormonal control of vitellogenesis varies depending on the insect. Juvenile hormone, ecdysteroids, and neurohormones exert this control in varying degrees as shown below.

6.2.4.1. Cytological Data

The insect ovariole always includes an apical part called the germarium, where the gonial cells divide and enter meiosis: its central and largest portion is the vitellarium, composed of a succession of follicles containing one oocyte, the follicle cells, and, in some ovaries, nurse cells. The follicle cells are separated by interfollicular tissue. Insect ovarioles display two types of organization, i.e., panoistic and meroistic; in the panoistic ovarioles, each oocyte is surrounded by follicle cells only, and no nurse cells are present. In the meroistic ovary, nurse cells or trophocytes contribute to the early growth of the oocyte.

The trophocytes are connected to the oocyte by cytoplasmic bridges. Like the follicle cells, they are often polyploid. The nurse cells are located either in the same follicle as the oocyte, as in polytrophic ovarioles, or in the germarium, where they divide and degenerate, creating a central trophic core connected to the oocytes by long trophic bridges which, during previtellogenesis, supply the oocytes with RNA, proteins, lipids, carbohydrates, and cytoplasmic organelles such as ribosomes and mitochondria. These are known as acrotrophic ovarioles.

The follicle cells display an astonishing variety of functions throughout oocyte development from previtellogenesis to the formation of the egg envelopes and the constitution of resorption body. They are involved in several distinct synthetic processes during previtellogenesis, vitellogenesis, and chorionization. In connection with their varied physiological activities, the follicular cells undergo great changes in their size, morphology, and DNA content (LaPointe *et al.*, 1985).

Vitellogenesis involves the production of specific female proteins termed vitellogenins and their entry into the oocyte. Vitellogenins are mostly produced in the fat body but may also be produced in the ovary. They are transported by the hemolymph, in which their titer is high during vitellogenesis, and accumulate in the oocyte against a concentration gradient 20–100 times their concentration in the hemolymph.

Oocytes do not incorporate proteins indiscriminately but select those which are species-specific vitellogenins. These bind to their receptors and undergo endocytosis (see Röhrkasten and Ferenz, 1985, 1986 in *Locusta;* König and Lanzrein, 1985 in *Nauphoeta;* Raikhel and Lea, 1986; Raikhel, 1987 in *Aedes*).

Insect vitellogenins are lipoglycoproteins with a multimeric structure comprising a variable number of polypeptides with different molecular weights (see Engelmann, 1979). Each insect species possesses one, two, or more vitellogenins, which are often processed in the intercellular spaces surrounding the oocyte.

The lipid yolk bodies which the oocyte contains in addition to protein yolk bodies consist mostly of triglycerides and carbohydrates which originate in material from the nurse cells, follicle cells, and oocyte Golgi apparatus, or else from the fat body.

Vitellogenin penetration into the oocyte during vitellogenesis is made possible by the changes occurring in the follicle cells which retract from the surface of the oocyte and undergo shrinkage and cytoskeletal changes, a process known as patency (Huebner and Injeyan, 1980). The space thus created permits free access to the membrane of the oocyte, which forms long microvilli and displays considerable endocytotic activity, in which clathrin–receptor ligand complexes from the oocyte cortex appear to be involved. In addition, numerous gap junctions between the follicle cells and the oocyte probably contribute directly to the uptake of small molecules which might include ecdysone (Bilinski *et al.*, 1985; Mazzini and Giorgi, 1985).

Once the oocyte has reached its final size, substantial changes occur in its cortical structure. The follicle cells change their mode of activity and produce first ecdysone and then the vitelline envelopes and egg shell. When the oldest oocyte of each ovariole is ripe, the follicular epithelium and the interfollicular tissue open and the egg is ovulated into the pedicel of the ovariole. The follicle cells surrounding the oocyte gather to form what is termed the yellow resorption body, a small mass of tissue in which degenerative processes take place and which disappears within a few days.

6.2.4.2. Role of Juvenile Hormone

In many insects, the CA play a major role in regulating vitellogenesis and control both vitellin synthesis and volume changes in the follicle cells. Corpora allata functioning is in turn controlled by the brain (see Chapter 4) as well as by the ovary, at least in some species.

Vitellogenesis regulation by the CA was first demonstrated in *Rhodnius* (Wigglesworth, 1936) (Fig. 52) and subsequently shown in a number of insects by selective removal and reimplantation methods (see Engelmann, 1979). The pioneering studies in this field concerned *Melanoplus* (Pfeiffer, 1939), *Calliphora* (Thomsen, 1942), *Dytiscus* (Joly, 1945), and *Leucophaea* (Scharrer, 1946), and showed that the CA were necessary for vitellogenesis to start and continue. During the period following these studies, nearly all the insect orders were examined and the role of the CA in triggering vitellogenesis was clearly established. It was also shown that changes in the volume of the CA take place concomitantly with changes in their activity. This was shown in particular in the viviparous cockroach *Leucophaea,* in which vitellogenesis was arrested during the embryo incubation period (Engelmann, 1957, 1959).

It is generally considered that JH intervenes in vitellogenesis by reaching a peak level before the onset of vitellogenesis. Certain species, such as *Melanoplus,* however, mature three sets of oocytes simultaneously in the same ovariole, with each set being exposed to two or three JH pulses (McCaffery and McCaffery, 1983).

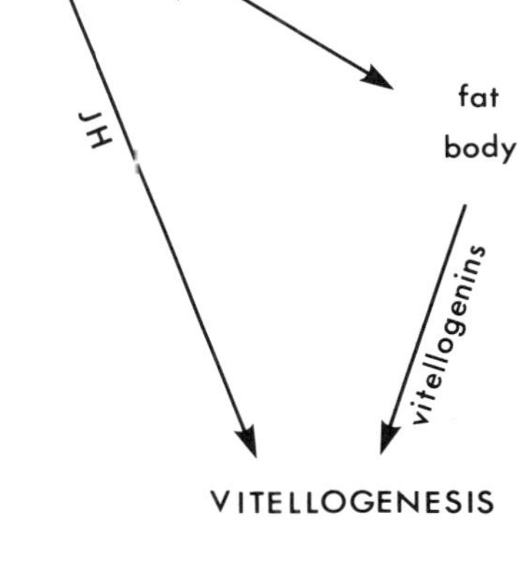

FIGURE 52. Schematic representation of the regulation of vitellogenesis in the bug *Rhodnius,* in which a major role is played by juvenile hormone.

6.2.4.2a. *Juvenile Hormone and Vitellogenin Synthesis.* The role of JH in stimulating the synthesis of proteins, especially vitellogenins, has been demonstrated in numerous intact insects (see Engelmann, 1979; Hagedorn and Kunkel, 1979; Davey, 1984). Juvenile hormone was also shown to trigger vitellogenin synthesis in decapitated insects, in isolated abdomens (Pan and Wyatt, 1971; Brookes, 1969), and in ovariectomized females (Engelmann and Ladduwahetty, 1974). In addition, it triggers this synthesis *in vitro* in fat body cells (Abu-Hakima, 1981) and acellular systems (Chen *et al.*, 1976; Koeppe and Ofengand, 1976). The amount of JH applied to the insect or the number of CA implanted correlate with the amount of vitellogenin produced and vitellogenesis. These quantitative effects were observed in *Byrsotria fumigata* (Bell and Barth, 1970), *Rhodnius* and *Triatoma proctracta* (Regis, 1979; Ruegg and Davey, 1979); in the blood-sucking insects, the number of eggs produced increased for the same amount of blood ingested when supernumerary CA were implanted.

The mode of action of JH on the fat body cells to produce vitellogenins was investigated in *Leucophaea* by Engelmann (1984), who mostly used cell-free systems. Cytosolic JH receptors, probably proteins, were shown to translocate the hormonal message to the nucleus, which in turn transcribed the vitellogenin RNA messengers. These messengers induced the formation of the polysomes involved in vitellogenin synthesis (Della Cioppa and Engelmann, 1980, 1984, 1987; Engelmann *et al.*, 1987).

6.2.4.2b. *Juvenile Hormone and Vitellogenin Entry into the Oocyte.* Besides having a role in vitellogenin synthesis, JH exerts other effects which also contribute to vitellogenesis by acting on the follicle cells. In the previtellogenetic follicle these cells are contiguous, but after the onset of vitellogenesis their shape and volume change, as shown first in *Hyalophora* (Telfer, 1961) and later in other insects, particularly *Rhodnius.* The involvement of JH in the entry of vitellogenins into the oocyte has been proved in several cases and in different ways. *In vitro* the morphological changes in the follicle cells were shown to depend on JH in cockroaches (Bell and Barth, 1971) and in *Rhodnius;* extensive studies of the latter were conducted both *in vivo* and *in vitro* using Evans blue, which stains the interfollicular spaces very clearly (Pratt and Davey, 1972; Davey and Huebner, 1974). The follicle cell changes, however, are not the only events required for the vitellogenins to enter the oocyte (see Telfer *et al.*, 1982). Oocyte penetration by the vitellogenins necessitates, at least in some insects, their combination in the extracellular spaces with a protein produced in the follicle cells and the formation in the cortex of the vitellogenetic oocyte of an endocytotic complex; in *Aedes* this formation was observed to be controlled by the CA (Raikhel and Lea, 1985).

The fundamental part played by the follicle cells was demonstrated with colchicine, which caused the disappearance of microtubules and microfilaments and suppressed the effects of JH (Abu-Hakima and Davey, 1977). These effects were also counteracted by ouabain, an inhibitor of Na^+ and K^+ ATPases, suggesting that the site of action of JH was in the membrane (Abu-Hakima and Davey, 1979; Ilenchuk and Davey, 1982, 1987). Accordingly, JH was shown to stimulate the NaK ATPases of membrane preparations of follicle cells and exhibited saturable binding to microsomal preparations of these cells. By electrophoresis of solubilized membrane preparations, it was possible to show that JH bound to a 43,000-Da peptide, and to a 150,000-Da peptide which also bound to [^{3}H] ouabain; this indicated that JH may bind directly to one of the constituent peptides of the NaK ATPase complex (Ilenchuk and Davey, 1985).

6.4.2.4.2c. *The Dual Role of Juvenile Hormone.* A convincing demonstration of the dual role of JH in vitellogenesis regulation was made in *Periplaneta.* Two vitellogenins were identified which disappeared from the hemolymph after allatectomy but reappeared when CA were reimplanted. When these vitellogenins were injected into allatectomized females, vitellogenesis could not be reestablished, thus demonstrating that JH is necessary for the entry of vitellogenins into the oocyte (Bell, 1969). Further, in *Drosophila grisea* and *Oncopeltus* the application of JHA suppressed the reproductive arrests, showing that the absence of JH was the reason for the nonutilization of vitellogenins because cell patency had not been achieved (Kambysellis and Heed, 1974; Kelly and Davenport, 1976). *Drosophila* mutants underwent similar events (Gavin and Williamson, 1976; Postlethwait and Handler, 1979).

6.2.4.2d. *Oocyte Lipids, Glucides, and Juvenile Hormone.* The oocytes of many insects contain large amounts of lipids, but little is known about the mechanisms controlling their entry. Allatectomy often leads to hypertrophy and lipid accumulation in the fat body, which indicates that the CA intervene in lipid metabolism. Moreover, when ovaries were incubated with active CA, the latter stimulated the uptake of [^{14}C]palmitate from the medium (Gilbert, 1967). According to Schreiner (1977), the first lipid synthesis observed in the *Oncopeltus* ovary did not depend on the CA, but the production of yolk lipid precursors in the fat body was controlled by the CA. Lipid yolk production may also depend on the release of triglycerides from the fat body induced by adipokinetic hormone from the CC (see Chapter 11).

Some of the carbohydrates in the oocyte are bound by proteins, but glycogen is also present in the ooplasm, particularly in its cortex. Although carbohydrate metabolism is mostly regulated by neurohormones, glycogen incorporation by the ovary was shown to depend on the CA, at least in *Oncopeltus* (Schreiner, 1977) (see Chapters 5 and 10).

6.2.4.2e. *Noninvolvement of the Corpora Allata in Vitellogenesis.* Although in many insects vitellogenesis requires the presence of the CA, in others it proceeds quite well in their absence. This was first observed in the small precocious adults of the silkworm *Bombyx* (Bounhiol, 1938) and the saturniid *Platysamia* (Williams, 1952). These species mature their eggs before the adult emergence and store nutritional material during the larval stages. In other lepidopterans which feed and mature eggs at the adult stage, the CA are necessary for oocyte maturation, as demonstrated in *Pieris* (Karlinsky, 1963, 1967) and confirmed later in several other lepidopterans (see Engelmann, 1984).

Worth mentioning is the observation by Röller (1962) in the moth *Galleria* that the effects of CA ablation varied according to the stage at which these organs were removed. Thus the CA might not play an all-or-nothing role but might instead exert additional stimulation of a process which is either autonomous or regulated by other factors. The observation in *Rhodnius* (Davey, 1965) and *Gryllus bimaculatus* (Roussel, 1967) that a certain degree of vitellogenesis and egg laying takes place in the absence of the CA supports this possibility.

In certain insects which mature their eggs at the adult stage, vitellogenesis sometimes take place despite CA removal (Pflugfelder, 1937; Possompès, 1957 in stick insects; Ramamurty and Engels, 1977 in bees; Ejezie and Davey, 1976; Nkouka, 1977a; Langley and Pimley, 1986 in *Glossina austeni, fuscipes,* and *morsitans*).

6.2.4.2f. *Oosorption.* The appearance of oosorption processes is apparently caused by a decrease in the activity of the CA and consequently a reduction in the hemolymph JH titer as shown in *Dytiscus* (Joly, 1945), *Leptinotarsa* (De Loof and De Wilde, 1970), *Periplaneta* (Bell and Bohm, 1975), and *Labidura* (Baehr *et al.*, 1982; Vancassel *et al.*, 1984). Both JH application and CA implantation reduced oosorption in *Schistocerca, Periplaneta, Oedipoda,* and *Drosophila* (Highnam *et al.*, 1963; Bell, 1971; Pener and Broza, 1971).

Oosorption is frequently observed in cases of egg or oocyte retention in the ovaries, two processes which were shown to be regulated by antigonadotropins from the ovary or associated nh tissue (see 6.2.4.5). It is thus possible that these regulatory factors might play a part in the oosorption process itself—the more so as it is generally observed that only certain follicles degenerate while others continue to develop. It was further suggested that a factor from the PI ensures oocyte protection (Lüscher, 1968b).

6.2.4.3. Role of Ecdysteroids

Apart from the few cases mentioned above, in which the CA were not necessary for vitellogenesis, it was firmly believed until recently that JH controlled this process in most insects. Unexpectedly, however, the study of the mosquito *Aedes aegypti* revealed that ecdysone was also involved in vitellogenesis.

6.2.4.3a. *Mosquitoes.* *Aedes aegypti* is an anautogenous mosquito; after emergence, the ovocytes develop for 3 days, during which they complete their previtellogenesis and then enter a period of rest until the blood meal.

In vivo and *in vitro* experiments showed that the CA only played a part during the early period, by stimulating previtellogenesis (Lea, 1969; Gwadz and Spielman, 1973; Fuchs *et al.*, 1980). Later the PI was observed to be indispensable to vitellogenesis because it secretes egg developement neurohormone (EDNH) (Lea, 1972). Lastly, the experiments revealed that 20-HE (Spielman *et al.*, 1971) and the ovary (Hagedorn and Fallon, 1973; Fallon *et al.*, 1974) were also indispensable to vitellogenesis, especially to vitellogenin production.

In addition, various authors have shown that vitellogenesis could be induced in fasting females by 20-HE injection, that *in vitro* the fat body could synthesize vitellogenins if it was brought into contact with an ovary or with 20-HE, and that the ovary secreted ecdysone *in vitro* when brought into contact with the brain of a mosquito or even of another species of insect (Hagedorn *et al.*, 1979).

The following mechanism was proposed: The emergence stimuli reach the brain, which induces secretion of JH, which in turn prepares the fat body for the action of ecdysone (Flanagan and Hagedorn, 1977) and prepares the ovary for the action of EDNH (Shapiro and Hagedorn, 1982), while at the same time inducing previtellogenesis. Later, the blood meal causes the brain to secrete the neurohormone EDNH, which acts on the ovary, where it triggers ecdysone secretion. The latter stimulates vitellogenin secretion by the fat body, thus enabling vitellogenesis to proceed (Fig. 53).

Recent experiments, however, seem to indicate that the mechanism concerned is more complex than that just described because EDNH appears to be released at two different stages of oocyte development (Greenplate *et al.*, 1985) and the same applies to JH (Readio and Meola, 1985). In addition, EDNH release might be induced by a CC-stimulating factor (CCSF) from the ovary (Lea and Van Handel, 1982; Borovsky, 1982) in response to a factor from male accessory glands (MAGF) transferred to the female during mating (Borovsky, 1985b) (Fig. 53). Factors from the fat body and ovary also appear to be involved in behavioral events (seeking of host and egg-laying site) (Klowden *et al.*, 1987).

Comparison of the mechanism regulating vitellogenesis in various species of mosquito shows that they are certainly similar in *Aedes aegypti, Aedes atropalpus,* and *Anopheles albimanus* (Lu and Hagedorn, 1986). Further, it has been suggested that ovarian ecdysteroids and JH act together to stimulate vitellogenin synthesis and vitellogenesis, and that JH is capable of increasing ecdysteroid production by the ovary (Masler *et al.*, 1980; Kelly *et al.*, 1981b; Birnbaum *et al.*, 1984).

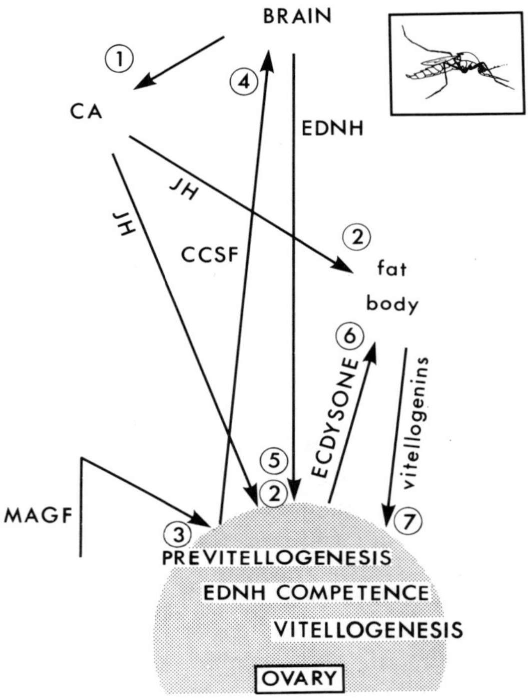

FIGURE 53. Schematic representation of the regulation of previtellogenesis and vitellogenesis in the mosquito *Aedes aegypti*. In this insect, ecdysone exerts a major role in vitellogenesis regulation. In response to brain stimulation by egg development neurosecretory hormone (EDNH) (2), ecdysone is released from the ovary and triggers vitellogenin release from the fat body (6,7). Juvenile hormone acts in the early stage (2); its secretion is induced by brain (1). In addition, a male accesory gland factor (MAGF) stimulates the ovary (3) to release a corpora cardiaca-stimulating factor (CCSF) which triggers EDNH release by the brain. (From the data of Flanagan and Hagedorn, 1977; Hagedorn *et al.*, 1979; Shapiro and Hagedorn, 1982; Lea and VanHandel, 1982; Borovsky, 1985b.)

6.2.4.3b. *Flies.* In *Calliphora vicina,* the role of the brain and CA in vitellogenesis regulation has been known for many years (Thomsen, 1952). The respective roles of 20-HE and of the ovary were also demonstrated recently (Thomsen *et al.*, 1980), when it was found that vitellogenin synthesis was suppressed by ovariectomy and subsequently restored by 20-HE injection.

In *Sarcophaga bullata* and *Musca*, the PI, JH, and ecdysteroids intervene in vitellogenesis regulation, but certain aspects of their action remain unclear. The PI plays an important part in both species, but there is disagreement about whether the effects it exerts through the CA in *Sarcophaga* (Wilkens, 1968; Pappas and Fraenkel, 1978) and *Musca* (Lea, 1975; Sakurai, 1977) are direct or indirect. In both the female and male of these species, 20-HE induces vitellogenin synthesis (Huybrechts and De Loof, 1981; Adams *et al.*, 1985; Agui *et al.*, 1985). Neither the origin of the hemolymphatic ecdysteroids nor the role of the ovary in vitellogenin synthesis has been definitely established. Although the incubated *Musca* ovary was shown to produce ecdysteroids (Adams *et al.*, 1985), as already observed in mosquitoes and *Drosophila* (Rubinstein *et al.*,

1982), an alternative origin for ecdysteroids must nevertheless be considered, since after ovariectomy, hemolymphatic ecdysteroids production still takes place in *Musca* and *Sarcophaga* (Briers and De Loof, 1980; Adams *et al.*, 1985; Agui *et al.*, 1985). Moreover, vitellogenin production was observed in the absence of the ovary in *Musca* (Adams *et al.*, 1985), *Sarcophaga* (Wilkens, 1969; Huybrechts and De Loof, 1981), and *Drosophila* (Postlethwait *et al.*, 1980), as well as in other ovariectomized insects that displayed a hypertrophied fat body and a high level of hemolymph proteins. Uptake of injected labeled ecdysone by the ovary was observed in *Sarcophaga*. The ovary converted this hormone into 20-HE and later into highly polar compounds (Briers *et al.*, 1982b). As regards JH, this was considered to regulate vitellogenesis in *Calliphora* (Thomsen, 1964) and in *Sarcophaga* (Wilkens, 1968, 1969; Engelmann *et al.*, 1971). In *Musca,* however, it seems to be chiefly involved in the early stages of oocyte growth (Adams, 1974). Here again, *Musca* appears to resemble the mosquito more closely than *Sarcophaga*.

In *Drosophila,* 20-HE intervenes in vitellogenin synthesis as well as JH (Postlethwait and Handler, 1979); the latter is required not only for the initiation of vitellogenin uptake and oocyte development but also for their continuation (Wilson, 1982). Whereas 20-HE acts on the fat body, JH acts both on the fat body and on the ovary, which is involved in vitellogenin synthesis in this insect (Jowett and Postlethwait, 1980; Brennan *et al.*, 1982). In addition, only JH permits vitellogenin incorporation into the ovocyte, and therefore vitellogenesis (Postlethwait and Handler, 1979). The latter finding was confirmed by electron microscopy examination of an ap^4 mutant that synthesizes vitellogenins but does not undergo vitellogenesis. The microvillosities and pinocytosis vesicles lacking in mutant oocytes were seen to appear after injection of a JHA (Tedesco *et al.*, 1981). Consequently, although in *Drosophila* the role of 20-HE in vitellogenesis has been confirmed, the process is also dependent on JH in this species, as in certain autogeneous mosquitoes. However, only the synthesis of vitellogenins by the fat body incubated with 20-HE has been obtained (Wu *et al.*, 1987).

6.2.4.3c. *Lepidopterans.* The giant saturniids do not require JH to synthesize vitellogenins or for their eggs to mature in the nymphal stage; moreover, in these insects, ecdysone from the PG has been shown to induce vitellogenesis (Williams, 1952). In two other lepidopterans (*Malacosoma pluviale* and *Bombyx*), whose oocytes mature in the nymphal stage as well, ecdysone was also shown to play a major part in this process. Thus, in isolated abdomens of *Malacosoma,* vitellogenesis was arrested even in the presence of JH, but continued when ecdysone was supplied (Sahota, 1969). The same applied to *Bombyx,* in which ecdysone was reported to act early and briefly on both vitellogenesis and previtellogenesis (Chatani and Ohnishi, 1976; Tsuchida *et al.*, 1987). The steroids involved here originated, at least partly, in the PG (Watanabe and Ohnishi, 1984).

6.2.4.3d. *Other Insects.* Ecdysone has not been shown to have a similar role in other insects. On the contrary, vitellogenesis was inhibited when the PG was implanted in a cockroach (Engelmann, 1957) and a cricket (Thomas, 1964) or incorporated into food (Wright *et al.*, 1971).

Note that the effect of ecdysone on vitellogenesis depends on the doses applied. This was observed in *Aedes* (Beckemeyer and Lea, 1980), *Danaus plexippus* (Herman and Barker, 1976), and *Acheta,* in which the growth of the terminal oocyte was stimulated by 1 μg of 20-HE and inhibited by 10 μg (Chudakova *et al.*, 1982).

Prothoracic gland and 20-HE have also sometimes been shown to act in different ways on young and vitellogenetic oocytes, stimulating the growth of penultimate oocytes but causing the degenerescence of terminal oocytes in *Locusta* (Joly, 1971), *Sarcophaga* and *Phormia* (Pappas and Fraenkel, 1978; Fraenkel and Hollowell, 1979).

6.2.4.3e. *Ecdysone Synthesis in the Ovary.* The discovery that 20-HE and other ecdysteroids are present in numerous adult insects (see De Loof *et al.*, 1984; Lagueux *et al.*, 1984; Hagedorn, 1985) was rapidly followed by the assumption that their production site was in the ovary (Hodgetts *et al.*, 1977; Koolman *et al.*, 1979; Briers and De Loof, 1980; Wentworth and Roberts, 1984). Ecdysone synthesis in the ovary has now been established *in vivo* and *in vitro* in numerous insect species.

The precise site of ecdysone synthesis in the ovary was determined in *Locusta* (Lagueux *et al.*, 1977; Glass *et al.*, 1978; Goltzené *et al.*, 1978) by comparing the synthetic activity of various ovarian components. The results indicated that ecdysone is synthetized in the follicular epithelium. This was confirmed by incubating ovarian follicules with ecdysone-specific antibodies which were selectively bound by the follicular cells (Glass *et al.*, 1978). In addition, isolated follicule cells of *Nauphoeta* were observed to produce ecdysone, most of it 20-HE (Zhu *et al.*, 1983). In another study isolated follicle cells were incubated with tritiated precursors (2-deoxyecdysone and 2, 22, 25-trideoxyecdysone) in order to locate the exact site of their hydroxylation and conjugation, which were both shown to take place in the follicle cells (Kappler *et al.*, 1986).

6.2.4.3f. *Nature of the Ovarian Ecdysteroids.* Whole-body and ovarian ecdysteroids were investigated in *Drosophila, Aedes aegypti, Hyalophora,* and *Galleria* (Bownes *et al.*, 1984; Borovsky *et al.*, 1986; Rubenstein *et al.*, 1986) and a number of other species. In these insects it appeared that most of the ecdysteroids consisted of highly polar conjugates. Moreover, determination of the ecdysteroids in the culture medium of ovaries from *Drosophila* suggested that the ovary produces 20-HE, ecdysone, and highly polar material (Rubenstein *et al.*, 1982; Schwartz *et al.*, 1985). It was shown in the hemolymph of *Aedes* that ecdysone, 20-HE, or both were present in various amounts.

6.2.4.3g. *Chronology of Ecdysone Release.* The timing of ecdysone release in the adult female has been investigated by measuring the amounts of ecdysteroids present in the whole insect, the ovaries, or the hemolymph. There are many such investigations, but the results vary depending on the insect species.

In the autogeneous mosquito *Aedes aegypti,* an ecdysteroid peak appeared in females 10–24 hr after a blood meal, i.e., at the onset of vitellogenin synthesis (Hagedorn *et al.*, 1975), a finding substantiated in other autogeneous mosquitoes *Aedes atropalpus* (Masler *et al.*, 1980), *Aedes casspius* and *detritus* (Guilvard *et al.*, 1984).

In *Locusta,* it was demonstrated that the follicle cells synthesized ecdysone at the end of oocyte maturation during a short period of 8–12 hr (Lagueux *et al.*, 1977; Goltzené *et al.*, 1978). Subsequent research confirmed that ecdysone synthesis occurs at the end of ovocyte maturation, but also showed that it can occur earlier, particularly in ovoviviparous or viviparous species (Imboden *et al.*, 1978; Stay *et al.*, 1984 in cockroaches; Robert *et al.*, 1986a in *Glossina;* see also Hoffmann and Lagueux, 1985; Hagedorn, 1985). In addition, the existence of one ecdysone peak at emergence has often been observed in the young adult.

6.2.4.3h. *Role of Ovarian Ecdysteroids.* Ovarian ecdysteroids appear to exert regulatory function both locally and at a distance. At the follicular level, they enter the oocyte, where they are mostly present as polar conjugates and seem to control the periodic deposition of embryonic cuticula, as suggested by the appearance of distinct ecdysteroid peaks in *Locusta* (Hoffmann, 1980; Lagueux *et al.*, 1984; Hoffmann and Lagueux, 1985). It might also be that they help to regulate chorionization, as proposed below.

As already mentioned, the presence of ecdysteroids in the adult hemolymph is well established. In mosquitoes and flies their major role is to trigger vitellogenin synthesis in the fat body (see Hagedorn, 1985) by acting on vitellogenin gene expression (Gemmill *et al.*, 1986; Racioppi *et al.*, 1986). In insects of other orders, the role of hemolymphatic ecdysteroids is not yet clear.

6.2.4.3i. *Stimulation of Ovarian Ecdysteroid Synthesis.* Ecdysone production was shown to be induced by a brain factor—possibly PTTH itself—which is produced in the PI and released in the CC, as demonstrated in mosquitoes (Hagedorn *et al.*, 1979) and locusts (Charlet *et al.*, 1979). According to some investigators, ecdysone synthesis might also be triggered by JH in *Aedes* (Birnbaum *et al.*, 1984) and *Drosophila* (Schwartz *et al.*, 1985).

6.2.4.3j. *Extraovarian Sites of Ecdysteroid Synthesis.* Although in certain insects ecdysone is produced in the ovary, it is possible that a different tissue might produce it in other insects. Thus, in *Drosophila* (Postlethwait *et al.*, 1980), *Sarcophaga* (Huybrechts and De Loof, 1981), and *Acheta* (Chang and Bradley, 1983), ovariectomy does not affect vitellogenin synthesis, as it does in *Aedes* (Hagedorn and Fallon, 1973; Masler *et al.*, 1980) and *Calliphora* (Thomsen *et al.*, 1980). On the other hand, ecdysteroids were found in both males and females in *Leptinotarsa* (Briers and De Loof, 1981). Lastly, the oenocytes of *Tenebrio* were shown to convert [^{14}C]cholesterol to 20-HE (Romer *et al.*, 1974).

6.2.4.3k. *Vertebrate Steroids.* In connection with the role of ecdysone in reproduction, it is of interest to recall recent studies which revealed the presence of vertebrate steroids such as testosterone, progesterone, and estrogen in *Sarcophaga* (De Clerck *et al.*, 1983), *Leptinotarsa,* and in a number of other insects such as *Periplaneta, Tenebrio, Manduca,* and *Oncopeltus* (Mechoulam *et al.*, 1984), and in *Sarcophaga* (Yokum *et al.*, 1987). Investigations in *Locusta* revealed the presence of small amounts of six vertebrate steroids in the hemolymph of this insect (Novak *et al.*, 1987). The presence of estradiol was demonstrated in *Bombyx* pupa (Ohnishi *et al.*, 1985), in which it was shown that this vertebrate molecule has no effect on vitellogenin synthesis or vitellogenesis, but at high concentration does reduce the number of eggs laid (Ogiso and Ohnishi, 1986).

6.2.4.4. Role of Neurohormones

6.2.4.4a. *Histophysiological Studies.* Many studies have revealed correlations between the amount of type A ns products in the PI and the different phases of the reproductive cycle. They will not be described here, except for the case of *Schistocera.* In this species, egg maturation was greatly accelerated when the females were reared with the males, and the amount of ns products stored in the type A ns cells of the PI gradually increased until oviposition, which was followed by the depletion of ns material. In the females reared without males, the storage of the ns material was more abundant and began earlier. In addition, when the brain or ventral nerve cord was stimulated electrically or the insects were submitted to an enforced activity, the release of ns material and development of the oocytes were observed to occur simultaneously (Highnam, 1961b). Further, [^{35}S]cysteine incorporation experiments confirmed that heavily loaded cells correspond to an inhibition of release while a small load corresponds to an intense process of synthesis and release (Highnam and Mordue, 1970).

6.2.4.4b. *Experimental Studies.* The histophysiological data suggesting that PI ns cells help to regulate vitellogenesis prompted several investigators to examine the effects of removal or electrocoagulation of the PI. This mostly resulted in the arrest of vitellogenesis. PI removal did not perturb vitellogenesis in some insects, including cockroaches, stick insects, *Tetrix, Glossina fuscipes, Oncopeltus,* and *Triatoma,* but the operation suppressed it in at least 20 species belonging to all the main orders (see Raabe, 1982). Nevertheless, these results are still not sufficient to prove that the PI plays a direct part in regulating vitellogenesis, since it is also known to control the functioning of the CA (see Chapter 4). Thus, decisive experiments demonstrating that the PI affects vitellogenesis directly must show that once vitellogenesis has been suppressed by PI removal, it cannot be reestablished by CA implantation or by JH or JHA application.

Corpora allata implantation of JH/JHA application replaced the PI in *Nauphoeta, Leptinotarsa, Manduca,* and perhaps also in *Sarcophaga* and *Musca,* but failed to compensate for its absence in *Locusta* (Girardie *et al.*, 1974; McCaffery, 1976), *Danaus* (Barker and Herman, 1973), *Aedes* (Lea, 1967), and the flies *Calliphora, Sarcophaga,* and *Musca* (Thomsen, 1952; Wilkens, 1968; Lea, 1975). The direct role of the brain was also demonstrated in insects undergoing reproductive diapause, such as *Anacridium* (Girardie and Granier, 1973) and *Leptinotarsa* (Briers *et al.*, 1982a).

Thus direct humoral control of the ovary by the PI is highly probable (Fig. 54). It consists mainly of the triggering of ecdysone production by the follicle

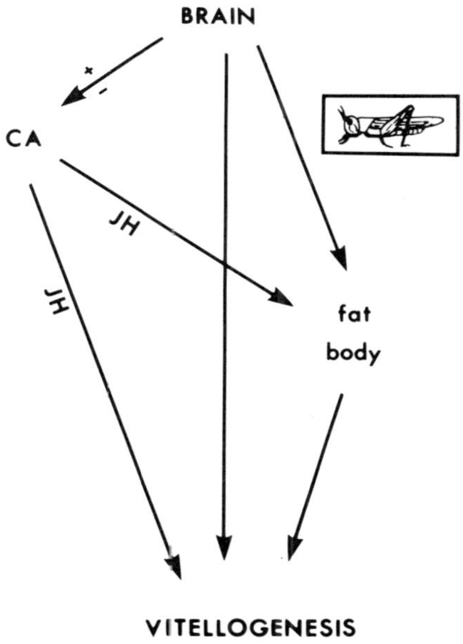

FIGURE 54. Schematic representation of vitellogenesis control in the *Locusta* model. (From the data of Girardie, 1966b.)

cells. In *Aedes aegypti* (Hagedorn *et al.*, 1979), experiments were conducted by incubating the ovary with different nerve ganglia and then determining the quantity of ecdysteroids synthesized. The ventral nerve cord ganglia had no effect, but the brain and CC both proved highly active. Egg development neurosecretory hormone was shown to be present not only in the adult but also in the larva, which suggests that it might be related to PTTH. This might also apply to *Bombyx*, in which PTTH is more abundant in the pupa and adult than in the larva as shown above. In *Locusta* (Charlet *et al.*, 1979), cautery of the PI or removal of the CC suppresses the increase in ovarian ecdysteroids which normally occurs at the end of the oogenetic cycle. Furthermore, the implantation of brain–CC complexes hastened the arrival of the ovarian steroid peak.

6.2.4.4c. *Brain Gonadotropic Neurosecretory Cell.* Investigations designed to identify the brain gonadotropic ns cells were performed in *Locusta* (Girardie, 1966b; Girardie and Girardie, 1967). Selective electrocoagulation of either the A-B cells or the C cells, combined with PI or CA implants, made it possible to demonstrate that all three cell types intervene in vitellogenesis regulation, but in different ways. Electrocoagulation of the A-B and C cells suppressed vitellogenesis, which, however, was reestablished by implanting the entire PI. Corpora allata implants were not effective in insects with cauterized A and B ns cells, but reestablished vitellogenesis in those whose C cells had been removed, thus showing that the C cells act through the CA, which are stimulated by them, and that A-B cells regulate vitellogenesis directly.

6.2.4.4d. *Purification—Interspecific Activity.* The purification of *Aedes* EDNH was investigated by three teams. Different molecular weights were obtained including 3500, 12,000 (Fuchs *et al.*, 1980), 6500 (Hanaoka and Hagedorn, 1980), and 18,700 (Borovsky and Thomas, 1985) for head and whole-body extracts, and the number of active fractions varied from one to two (Masler *et al.*, 1983; Borovsky and Thomas, 1985; Wheelock and Hagedorn, 1985) (Fig. 55).

Interspecies activity was observed among various mosquitoes, including *Aedes atropalpus* and *Aedes taeniorrhynchus,* which responded to purified EDNH from *Aedes aegypti* by synthesizing vitellogenins and maturing eggs (Borovsky and Thomas, 1985).

Moreover, a comparative study of the reactivity of the PG of *Manduca in vitro* and of the ovary of *Aedes in vitro* or *in vivo* showed that the highly purified *A. aegypti* head extracts stimulate ecdysone production by *Manduca* PG in a dose-dependent manner but that the stimulation was not obtained with partially purified EDNH. In addition, neither small nor big *Manduca* PTTH was active in *A. aegypti* (Kelly *et al.*, 1986a).

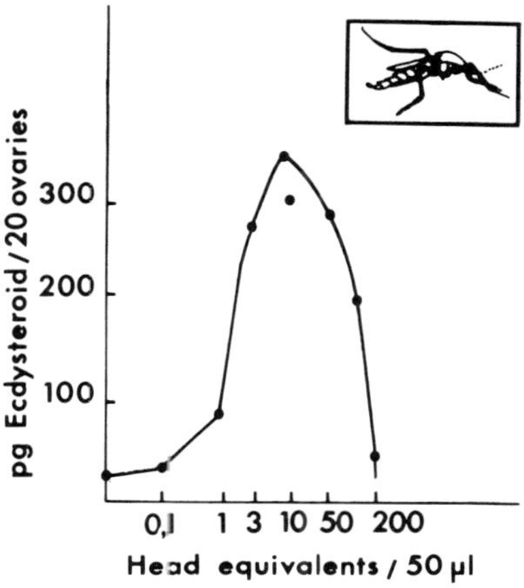

FIGURE 55. Dose-response of ecdysteroid production by the ovary from sugar-fed nonvitellogenetic females of *Aedes* exposed to head extracts purified by reversed phase HPLC. (Modified from Masler *et al.*, 1983.)

6.2.4.4e. *Mode of Action of EDNH.* In *Aedes,* EDNH is released from the CC after a blood meal. Recent parabiosis experiments established that this release was not induced by nerve inputs originating in stretch receptors but rather by a hemolymph-borne factor coming from the ovary, which allows the release of EDNH following the blood meal (Lea, 1972; Lea and Van Handel, 1982; Borovsky, 1982).

The action of EDNH on the *Aedes* ovary involves cyclic AMP because the amount of ecdysone in the ovary is only increased by head extracts if they contain a phosphodiesterase inhibitor (Shapiro, 1983).

6.2.4.4f. *The Ventral Nerve Cord.* The role of the ventral nerve cord has been little investigated in relation to vitellogenesis.

Several histophysiological studies showed a correlation between the load of certain ns cells of the ventral nerve cord and reproductive activity. The first result in this field was the demonstration by B. Scharrer (1955) in *Leucophaea* of the changes occurring in two ns cells of the subesophageal ganglion following castration. This result was confirmed in *Carausius* (Mouton, 1968) and *Teleogryllus* (Dürnberger *et al.,* 1978). In *Acheta,* the same cells released their ns products during the oocyte maturation and egg-laying periods, and stored them when a suitable place to oviposit was lacking or during starvation (Huignard, 1964). In *Locusta,* they were in a storage phase in nonvitellogenic females and were depleted during vitellogenesis (Fréon, 1964). Other, lateral type A ns cells of the ventral nerve cord ganglia were often observed in a storage state during vitellogenesis arrest, e.g., in the viviparous cockroach *Leucophaea* (De Bessé, 1967).

A few experimental data were obtained concerning the role of the ventral nerve cord in regulating vitellogenesis. It was shown that in *Schistocerca,* removal of the last abdominal ganglion prevents vitellogenesis, which may be restored by its reimplantation (Delphin, 1965), and that in young *Acheta* females, precocious and enhanced egg laying followed the implantation of a mature female subesophageal ganglion (Thomas, 1964).

6.2.4.5. Role of Antigonadotropins

Besides the neuroendocrine regulation of the functioning of the ovary, other regulatory mechanisms originating in the ovary itself have been shown to intervene in vitellogenesis and egg laying control in several insect species. These mechanisms involve factors produced in the oocytes and associated tissues (for more details, see Davey, 1984).

6.2.4.5a. *Antigonadotopins.* In many insect species, the growth of the T-1 oocytes is arrested until the T oocytes have been laid, and the latter may be retained in the ovary for long periods in cases of starvation, failure to find a suitable place to oviposit, or other unfavorable environmental conditions.

The injection of extracts of mature oocytes also causes the arrest of T-1 oocyte maturation in *Iphita* (Nayar, 1958), in flies (Adams *et al.,* 1968), and in mosquitoes (Meola and Lea, 1972), but extracts of young oocytes do not have this effect.

In *Rhodnius* the production rate of the first wave of eggs is the same in mated and virgin females but the production of the second wave is completely inhibited in the virgin females, in which the first wave eggs accumulate (Pratt and Davey, 1972). When virgin females are implanted with chorionated eggs, their egg production decreases markedly, suggesting the intervention of an antigonadotropin. The mature eggs retained in virgin females are not stored in the oviducts but in the pedicels of the ovarioles, which are considerably stretched (Huebner and Davey, 1973).

Investigations designed to locate the site of origin of the antigonadotropins were carried out comparing the patency activity of extracts of various tissues located close to ovarioles and pedicels. Both extracts of pedicel and ovariole sheaths were inactive whereas the extracts from the large tracheal mass located at the base of the ovarioles and pedicels were distinctly active (Davey and Kuster, 1981). Histological examination of this region revealed the presence inside the tracheal mass of small abdominal ns organs (the ANO) containing ns cells, peripheral neurons, and ns axons terminals (Kuster and Davey, 1981) (See Chapter 1).

Extracts of isolated ANO revealed strong antigonadotropic activity that affected the JH-induced patency of the follicle cells (Davey and Kuster, 1981). These ns organs are thus certainly involved in the synthesis of the antigonadotropin whose release is caused by the stretching due to egg retention (Fig. 56). Recent investigations showed that it does not compete for JH binding sites and does not involve JHI-sensitive increase in Na-K ATPase (Ilenchuk and Davey, 1987).

In species other than *Rhodnius* such specialized nh organs have not been described and it is not known whether antigonadotropic/oostatic hormone is produced in homologous ANO organs or in mature eggs or embryos, as earlier reports indicated.

Partial characterization of *Rhodnius* antigonadotropin showed that it is a heat-stable, low molecular weight, trypsin-sensitive molecule (Liu and Davey, 1974). In *Aedes aegypti,* a peptide-like molecule which has a molecular weight of 2200 at pH 4.5 but is much heavier at pH 7.0 was isolated from the ovary. It inhibited vitellogenesis and vitellogenin synthesis (Borovsky, 1985a). Moreover, both *Aedes atropalpus* and *Drosophila* ovaries exerted an oostatic effect on *A. atropalpus* (Kelly *et al.,* 1986).

In *Rhodnius* the antigonadotropins were shown to act directly and locally by antagonizing JH stimulation of cell patency, a prerequisite for vitellogenin uptake by the maturing oocytes (Pratt and Davey, 1972; Huebner and Davey, 1973; Davey and Huebner, 1974).

In addition to these local direct effects, antigonadotropins and/or ovary

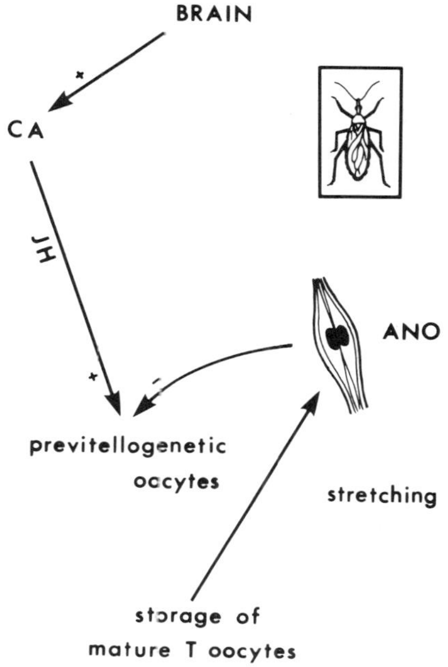

FIGURE 56. Schematic representation of the mechanisms involved in *Rhodnius* in the inhibition of vitellogenesis by an accumulation of mature oocytes. The stored oocytes stimulate the production by the ANO of an antigonadotropic factor which inhibits vitellogenesis. (From the data of Davey and Kuster, 1981.)

factors display indirect effects, which act via the PI ns cells whose load alters after ovariectomy as described in *Schistocerca* (Highnam, 1962), *Euthystira brachyptera* (Müller, 1965), *Calliphora* (Thomsen and Lea, 1969), and *Musca* (Adams *et al.*, 1975). In mosquitoes these indirect effects have been suggested to act on the ovary where they inhibit ecdysteroid production by vitellogenic follicles, e.g., in *A. atropalpus* (Kelly *et al.*, 1986b), and cause the release of a CCSF which triggers EDNH secretion by the PI (Borovsky, 1982). Ovarian factors also act on the CA, as demonstrated in *Diploptera* (Stay and Tobe, 1978; Stay *et al.*, 1983) (see Chapter 4).

6.2.4.6. Oocyte Interactions

The local effect of mature eggs was also shown in the stick insect *Carausius* by the ligation of one oviduct, which caused egg retention, and a vitellogenesis arrest, but only on the ligated side (Thomas, 1979b).

In another stick insect, *Clitumnus extradentatus,* investigations were performed in order to locate the source of antigonadotropin (Mesnier, 1980). After distal sectioning of the lateral oviduct, the eggs accumulated progressively first in the oviduct, then in the pedicels, and finally in their own follicles. When the eggs are retained in the pedicel of the ovariole, the growth of T-1 oocyte is normal, whereas if they are retained in their follicles, they stop growing and degenerate. Thus, an inhibitory factor from the T follicle, which is also present in the T-1 and T-2 follicles, directly regulates the growth of the T-1 oocyte unless it is separated from its follicle by a residual body.

The suppression of this inhibitory factor is, however, not sufficient to stimulate vitellogenesis, since experimental sectioning between the T and T-1 follicles suppressed the growth of the T-1 oocyte. Moreover, section at the base of the T follicle also caused the arrest of growth and subsequent degenerescence of the T oocyte. Therefore, besides the inhibitory effect that the oocytes exert on each other, a stimulatory effect is also exerted on the oocytes by the tissue located below the oocytes. The origin of the local factor stimulating oocyte growth might be the interfollicular tissue, whose appearance during vitellogenesis is definitely secretory (Mesnier, 1980).

6.2.4.7. Conclusion

It is now firmly established that mature oocytes, embryos and ns tissue associated with the ovary produce factors termed antigonadotropin or oostatic hormone, which inhibit the growth of T-1 oocytes. They act directly on the follicular epithelium where they antagonize the effects of JH on follicular cell patency. They also exert feedback effects on the median ns cells of the PI and on the CA. At the level of the follicle and interfollicular tissues, other types of inhibition and stimulation also take place.

6.2.5. Chorionization

Shell formation is another process which has been suggested to depend on brain neurohormones, since after parsectomy or CC removal, intracellular secretion of eggshell material by the follicle cells was perturbed, while the extracellular amalgamation of chorion material was prevented. Moreover, PI or CC implantation into immature females induced early formation of the egg envelopes. The effects of the neurohormonal factors do not seem to involve the CA (Joly *et al.*, 1978). The involvement of ecdysone in oocyte envelope synthesis has also been suggested on the basis of abnormalities observed following 20-HE injection. In *Carausius,* abnormalities of the eggshell capitulum were observed, including a large reduction in size (Mesnier and Thomas, 1981), while *Aedes* displayed abnormalities of the vitelline envelope (Raikhel and Lea, 1982).

6.2.6. Vitellogenesis and Vitellogenin Synthesis in the Male

When ovaries were transplanted into males (see Lamy, 1984; De Loof *et al.*, 1984), both oocyte growth and protein production by the fat body were often observed although the proteic globules of the oocytes of the transplanted ovaries did not always contain specific vitellogenins (Lamy and Karlinsky, 1974 in *Pieris;* Lamy *et al.*, 1975 in *Bombyx*). In other insects, specific vitellogenin production takes place in the fat body of the male, either without a supply of hormones, as in *Galleria* (Lamy *et al.*, 1978) and *Bombyx* (Miwa and Yamashita, 1985), or with such a supply.

In *Diploptera* (Mundall *et al.*, 1979), JH stimulated vitellogenin synthesis, while in *Oncopeltus* (Rankin and Jackle, 1980), *Drosophila* (Postlethwait *et al.*, 1980; Bownes, 1982) and *Sarcophaga* (Huybrechts and De Loof, 1977, 1981), 20-HE was the stimulant. In *Leptinotarsa* and *Leucophaea,* however, 20-HE did not succeed in causing vitellogenin synthesis in males (see De Loof *et al.*, 1984).

It seems, therefore, that in certain insects the male fat body possesses the same fundamental synthetic properties as that of the female. In insects such as *Locusta* (Dhadialla and Wyatt, 1983), a thorough study of vitellogenin synthesis in the male and female throughout their entire development demonstrated that the fat body cells of the male possess JH receptors but are unable to synthesize vitellogenins, since their program is completed at an early stage of development.

6.2.7. Concluding Remarks

Insect oogenesis comprises many stages which are regulated by certain humoral factors, including neurohormones, JH and ecdysone, whose importance and modes of intervention vary depending on the species.

Juvenile hormone plays a fundamental part in oogenesis, in which it controls vitellogenin synthesis and vitellogenesis. In dipterans, however, JH above all helps to ensure previtellogenesis. It inhibits meiosis as well as ovariole and follicle development.

Ecdysone and 20-HE determine the early stages of oogenesis, meiosis, and differentiation. During vitellogenesis, these hormones are secreted by the ovocyte follicle cells in the final phase of maturation. It is not known at present what part they play in insect species in general, but in mosquitoes, flies, and lepidopterans they constitute the chief factor regulating vitellogenin secretion and vitellogenesis.

The brain neurohormones stimulate ecdysone production by the follicle cells, and also intervene in previtellogenesis and gonial mitosis, but it is not known whether one or several factors are involved.

Mature oocytes exert feedback control over vitellogenesis through a neurohormone released from nh organs located in the vicinity of the ovary (ANO); they also respond to antigonadotropins which originate in the oocyte and act locally.

The diversity of the mechanisms controlling oogenis is not surprising, considering the variety of ways in which this process takes place. In many insects, all the terminal ovocytes mature simultaneously thus endowing maturation, ovulation, and egg laying with a cyclic character. In other insects, maturation occurs in the two or three terminal oocytes, which means that each oocyte is subjected two or three times to the same endocrine medium (Weaver *et al.*, 1975) in *Periplaneta;* McCaffery and McCaffery, 1983 in *Melanoplus*). In others again, such as the stick insect or *Rhodnius,* vitellogenesis only takes place in certain ovarioles, and local types of regulation are probably decisive in these species.

Comparison of insects and crustaceans—other arthropods which have been fairly well investigated—reveals similarities between the two classes. Thus, vitellogenesis is stimulated in both cases by a neurohormone and ecdysone. Crustacean ecdysone is produced in the Y organ, which does not degenerate since molting persists in reproducing adults. Neurohormones control the molting gland and ovary in both classes. In both again, the ovary follicle cells display secretory activity and produce a factor stimulating vitellogenesis and vitellin synthesis; in insects it is ecdysone, and in crustaceans it is the vitellogenesis-stimulating hormone.

6.3. SPERMATOGENESIS

During the initial studies of insect endocrinology, several authors observed that removal of the CA caused an anticipated metamorphosis and that the reproductive organs of the anticipated adults were well developed, although reduced in size. This demonstrated that JH was not necessary for the early development of the gonad but did not exclude the possible involvement of other hormones.

6.3.1. Ecdysone

The action of ecdysone on spermatogenesis was first investigated in diapausing pupae of *Platysamia* and *Samia walkeri*. After rupture of the testicular envelope, cysts containing spermatogonia and spermatocytes were isolated in a droplet of the blood of pupating larvae, developing adults, or diapausing pupae. Well-differentiated spermatids were obtained from the pupating larvae and adults, but the spermatogonia and spermatocytes from the diapausing pupae failed to develop. This showed that a factor promoting the multiplication of spermatogonia and maturation of spermatocytes was present in the blood, particularly when 20-HE is released, and suggested that this factor might be ecdysone itself (Schmidt and Williams, 1953). In this connection, testes from the same insect species were shown to develop rapidly when cultured in the presence of ecdysone or active PG (Kambysellis and Williams, 1972).

Other insects, particularly diapausing lepidopterans, were also investigated (see Hagedorn, 1985 for detailed documentation). Testes transplantation from diapausing and nondiapausing larvae of *Papilio xuthus,* as well as implantation of the PG into diapausing insects of this species, confirmed that the PG helps to stimulate spermatogenesis (Nishiitsutsuji-Uwo, 1961). This finding was also confirmed in *Periplaneta* (Blaine and Dixon, 1976). In addition, *in vitro* experiments demonstrated that ecdysone and 20-HE stimulate the testis and determine its growth and differentiation in *Chilo* (Yagi *et al.*, 1969), several species of saturniids (Kambysellis and Williams, 1971a,b), *Monema* (Takeda, 1972b), *Mamestra* and *Spodoptera* (Fukushima and Yagi, 1975). In *Locusta,* too, the addition of 20-HE to spermatocytes significantly increased both the frequency of spermatogonial division and the rate of differentiation (Dumser, 1980).

A very complete study was conducted in the bug *Rhodnius,* whose spermatogenesis progresses slowly during the penultimate and ultimate larval instars and continues in the adult. By various experimental procedures, *Rhodnius* spermatogonial cells were shown to display basal autonomous levels of mitotic activity which was doubled by ecdysone release after feeding, or by injection (Dumser and Davey, 1974, 1975).

The recent demonstration that ecdysone is present in the ovary suggests that it might also be present in the testis. Indeed, investigations in *Calliphora vicina* (Koolman *et al.*, 1979) showed that the male adult testis contains a considerable amount of ecdysteroids, particularly 20-HE, and this finding was confirmed in *Gryllus* (Hoffman and Behrens, 1982), *Sarcophaga* (Briers *et al.*, 1983), *Mamestra* (Shimizu *et al.*, 1985), and *Heliothis* (Loeb *et al.*, 1982, 1984).

The precise site of the testis ecdysteroid production was studied in *Heliothis* last instar larvae (Loeb *et al.,* 1982). Ecdysteroid determination in separate cultures of different parts of the testis and in hemolymph showed that the testis sheath and spermatocysts do contain large amounts of ecdysteroids, so that the testis sheath might be the site at which these substances are released. Subsequent studies performed with antiecdysteroid antibody allowed more precise location of ecdysteroids in the inner layer of the testis sheath and its extensions surrounding the spermatocysts (Loeb *et al.,* 1986). In another lepidopteran, *Mamestra,* the investigation by Shimizu *et al.* (1985) confirmed that in the diapausing pupae the cell sheath of the testis caused the elongation of spermatocysts *in vitro* (see below).

The timing of ecdysone production was examined in the incubated testis of *Heliothis* last instar larvae. It was shown that three ecdysteroid peaks occurred during this instar; the first coincided with the onset of meiotic division, the second preceded spermatocyst elongation, and the third coincided with the fusion of paired larval testes (Loeb *et al.,* 1984). The ecdysteroid peak occurs spontaneously in the incubated testis but might be enhanced by addition of ecdysteroid to the incubation medium (Loeb *et al.,* 1986) (Fig. 57). In post diapausing pupae of *Mamestra,* the maturation of the spermatocyst coincides with rising levels of whole-body 20-HE (Bodnaryk, 1986) and these levels decline after the completion of spermatocyst elongation.

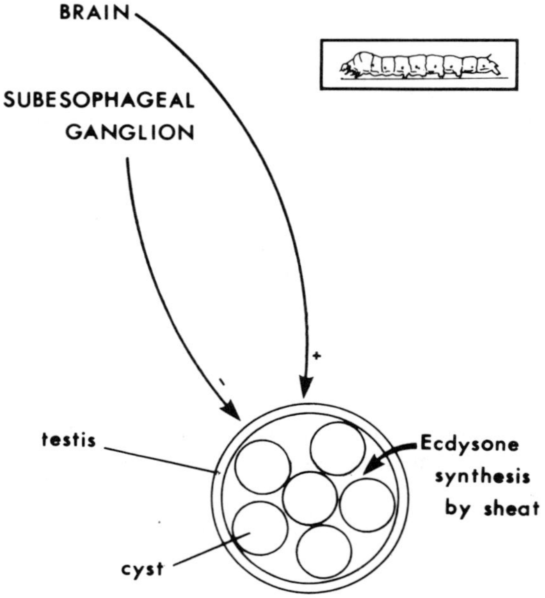

FIGURE 57. Brain, subesophageal ganglion, and testicular sheath involvement in the regulation of spermatogenesis in *Heliothis.* (From the data of Loeb *et al.,* 1982, 1985.)

The stimulation of spermatogenesis by ecdysteroid thus emerges as an established fact. In some cases, however, ecdysteroid involvement in spermatogenesis is less apparent. Thus the activation by 20-HE of incubated *Acheta* testis is slight (Dempster and McFarlane, 1985) and in *Rhodnius* spermatogenesis procedes at a basal rhythm without any ecdysteroid, as shown above. Moreover, it was reported recently that in *Heliothis* both meiosis and elongation took place in the absence of ecdysteroids (Giebultowicz *et al.*, 1987).

20-Hydroxyecdysterone appears to play a part in the renewal of spermatogenesis in postdiapausing larvae, as it triggers eupyrene differentiation but does not exert any stimulatory effects on apyrene spermatogenesis (Friedländer and Benz, 1982). Its role in apyrene-eupyrene determination was also investigated in *Heliothis*. Radioimmunoassay measurements showed that its level was low at the beginning of the pupal instar when the eupyrene spermatocytes elongate and the apyrene spermatocytes are only undergoing mitosis. Ecdysone peaked at the end of the instar during which apyrene spermatozoa differentiate (Loeb *et al.*, 1982). A higher ecdysone titer might thus be necessary for apyrene differentiation. However, it might also be that both types of spermatogenesis are determined at a much earlier stage. Data on this point seem to be lacking.

In the moth *Anagasta Kuehniella,* the proportions of eupyrene and apyrene spermatozoa released appear to vary depending on several factors, particularly the light. In a continuous-light regimen, the number of eupyrene spermatozoa is greatly reduced, in the same way as after 20-HE treatment (Thorson and Riemann, 1982).

Further, an apyrene-spermatogenesis-inducing factor was shown to be necessary in nondiapausing *Laspeyresia* to initiate apyrene differentiation in the bipotential spermatocytes.

6.3.2. Macromolecular Factor

Although ecdysteroids exert stimulatory effects on the testis, it has been shown in the great saturniids that the presence of ecdysone was not sufficient for the development of naked germinal cysts and that *in vitro* the onset of both spermatocyte meiosis and differentiation required the presence of another factor, termed the macromolecular factor. The latter was found to be present in the blood of both male and female pupae, and displayed great changes in activity not connected with diapause termination. In three saturniid genera, this macromolecular factor was observed to be nonspecific, undialyzable, and heat-sensitive. It was suggested to intervene at a later stage than ecdysone, since the latter was assumed to play a permissive role by altering testis wall permeability to the macromolecular factor (Kambysellis and Williams, 1971a,b).

The origin of the macromolecular factor was examined by adding cockroach hemocytes of permanent cellular lineage to cysts from diapausing pupae containing only spermatocytes. Spermatogenetic differentiation proceeded after the completion of meiotic cell division and was followed by spermiogenesis (Landureau and Szöllösi, 1974). Thus, the macromolecular factor might originate in the hemocytes, as proposed by Kambysellis and William (1971a,b). Another possible origin was proposed for this factor on the basis of experiments with *Mamestra,* in which spermatogenesis only succeeded in incubated testis when the spermatocysts were cultured in the presence of ruptured testis sheaths or with the sheath incubation medium. The testis wall was proposed as the site of origin of the macromolecular factor (Shimizu and Yagi, 1978, 1982).

The mechanisms of action of ecdysone and the macromolecular factor were investigated *in vivo* and *in vitro* in *Locusta,* using peroxidase as a tracer. By this method, it was observed that younger cysts containing spermatocytes were permeable, but the permeability of older cysts varied during the intermolt period and sometimes appears to be linked to changes in the hemolymph ecdysone titer (Marcaillou and Szöllösi, 1975). Hydroxyecdysone might therefore intervene in spermatocyte differentiation by modifying cyst permeability to the macromolecular factor as already suggested. It was also shown *in vitro* in *Schistocerca* that ecdysone permitted the formation of a barrier in the testicular follicles (Jones, 1978).

Testicular proteins produced in prepupal, pupal, and adult *Tenebrio* were analyzed by gel electrophoresis and the speed of their incorporation of [^{3}H]leucine was recorded in control insects and in insects injected with 20-HE. This hormone was observed to stimulate [^{3}H]leucine incorporation into specific testicular proteins (Alrubeai and Gorell, 1982).

6.3.3. Juvenile Hormone

As expected, JH was found to play a part in the regulation of spermatogenesis. In *Bombyx* the regulation of the morphogenesis of testes as well as spermatogenesis were investigated by grafting embryonic undifferentiated gonads inside different hosts. It was observed that the growth of the testis and beginning of spermatogenesis took place only in the last instar larvae, in which JH is present in very small amounts (Takeuchi, 1969).

The inhibiting effect of JH on testis development was confirmed in several species, including *Locusta* (Cantacuzène and Şeureau, 1970), *Periplaneta* (Blaine and Dixon, 1976), *Corcyra cephalonica* (Deb and Chakravorty, 1981), and *Spodoptera* (Gelbic and Metwally, 1981). On the whole, in these species, spermatogenesis was delayed and abnormalities in the spermatozoa were observed, especially in supernumerary larvae resulting from the juvenilizing effects of the CA. The high JH titer during larval diapause would thus explain the arrest of spermatogenesis observed during this stage (Yagi *et al.*, 1969). In the moth *Laspeyresia,* JHA application to prediapausing larvae caused spermatocyte lysis, while allatectomy disrupted diapause and stimulated spermatogenesis. It should be noted, however, that diapausing insects do not constantly require a high JH level (Friedländer, 1982).

In *Rhodnius* (Dumser and Davey, 1974, 1975), the application of JHA abolished the 20-HE-stimulated increase in gonial mitoses but did not change their basal autonomous mitotic level. The occurrence of autonomous spermatogenesis was also observed in certain species (see Lender and Duveau-Hagege, 1963).

It should be mentioned that in certain insects unexpected effects were observed after CA implantation or JH/JHA application. Thus, for instance in *Draeculacephala,* JH treatment of diapausing insects slightly hastened sperm maturation (Reissig and Kamm, 1975). In *Eurygaster,* this treatment during prediapause accelerated spermatogenesis, which, however, remained incomplete (Shinyaeva, 1981).

The simultaneous production of normal eupyrene nucleated spermatozoa and of atypical apyrene anucleated spermatozoa is usual in lepideopterans and occurs in other species too. Both eupyrene and apyrene spermatozoa are found in the spermatheca of the female, but only the eupyrene ones fertilize the eggs. The physiological role of apyrene spermatozoa is not clear, though they were suggested to play a trophic role and help the transportation of eupyrene spermatozoa. In some species at least, they are activated before the latter, which suggests that apyrene spermatozoa regulation is different from the regulation of eupyrene spermatozoa.

Testis transplantation, ligature, or JHA application in *Ectomyelois ceratoniae* showed that eupyrene spermatogenesis required a low JH titer but that apyrene differentiation was not related to the decline in the JH level (Leviatan and Friedländer, 1979).

6.3.4. Neurohormones

Prothoracicotropic hormones, which probably triggers ecdysone production in the ovary, could be involved in the production of ecdysone by the testis. Surgical removal of the brain or subesophageal ganglion, nervous section of the ventral nerve cord, ligature of the body, and organ implantation were performed in *Heliothis* early last instar larvae whose spermatocytes were still in the premeiotic stage. The ligation behind the brain or PG did not affect spermatogenesis, but brain removal inhibited spermatocyst elongation, and removal of the subesophageal ganglion triggered the onset of meiosis (Loeb *et al.*, 1985). In addition, injection of extracts had concordant effects, so that neurohormones do seem to regulate spermatogenesis, directly or indirectly. In this connection, it is interesting to note the antagonism between the brain and the subesophageal ganglion. First attempts to isolate the brain factor have shown that it is a thermolabile protein with a molecular weight of 6000–12,000 (Loeb *et al.*, 1987).

6.3.5. Concluding Remarks

The idea that emerges from a number of often conflicting results concerning the regulation of spermatogenesis is that ecdysone plays a major role in this process, as it does in the first reproductive stages of the female. The testis itself, however, seems able to assume its own regulation, perhaps by producing ecdysone, as occurs in the ovary. This implies that regulation by the brain is highly probable, and also that the role of the PG might not be essential. Juvenile hormone appears to play the same inhibitory role in spermatogenesis as in early oogenesis. Other humoral or hormonal factors whose origin is still uncertain are probably also involved in regulating spermatogenesis.

6.4. ACCESSORY GLANDS

6.4.1. Morphogenesis of Accessory Glands and Ducts

The dependency on ecdysone and 20-HE of the organogenesis of genital ducts and accessory glands has been demonstrated in several insect species.

The development of the colleterial glands of *Periplaneta,* which occurs at the end of the last larval instar, requires ecdysone (Dixon and Blaine, 1973) and the development of the male accessory glands of *Plebeiogryllus guttiventris,* which takes place at about the same time, is accompanied by a high ecdysone titer and a low JH level. In *Locusta,* the transfer of the accessory glands in a medium with these hormonal characteristics is necessary for their growth and differentiation (Gallois, 1979). In *Tenebrio,* the differentiation of male accessory glands occurs in the young pupa and includes two periods of maximal mitotic activity. Only the second period requires the presence of 20-HE and coincides with the large ecdysteroid hemolymphatic peak (Szopa *et al.,* 1985).

As far as duct differentiation is concerned, it has been observed to occur only in the presence of ecdysone *in vivo* and *in vitro* (Huet, 1971 in *Tenebrio;* Nowock, 1973 in *Ephestia* and *Samia;* Szöllösi and Landureau, 1977 in *Locusta;* Dorn and Sehn, 1987, in *Oncopeltus*).

However, other tissues of epidermic origin such as the spermatheca and accessory glands escape mitotic stimulation by ecdysone (Szopa and Happ, 1982). These opposite effects of ecdysone might depend on the origin of the tissues, and ecdysone might act as an opener on mesodermal derivatives but as a blocker on ectodermal derivatives.

In *Locusta* and *Schistocerca,* JH treatment during late larval life suppressed the imaginal differentiation of the spermiducts (Szöllösi, 1975). Later on, in the young adult, the situation was reversed. Implantation in young females of additional CA accelerated the development of both the oviducts and colleterial glands

(Lauverjat and Girardie, 1976). In some species, however, duct and accessory gland development, which occur in the adult, seem to be regulated by JH instead of ecdysone (Foster, 1967; Lessman *et al.*, 1982).

Besides the effects exerted by hormones on the reproductive organs, humoral relationships between gonads and ducts have been demonstrated. In *Drosophila,* female ducts were sometimes found to be attached to the gonads, thus causing degeneration (Bodenstein, 1946). In the bugs *Dysdercus* (Pluot, 1979) and *Triatoma* (Regis, 1977), atrophy of one ovary was sometimes observed together with the regression of the upper part of the corresponding lateral oviduct, and experimental section of the oviduct had the same effect in both species, i.e., ovary atrophy due to oocyte blockage in previtellogenesis. This inhibitory effect of disconnecting the ovary and oviduct has still not been explained; it is possible that some factor from the oviduct intervenes in the development of follicle cell competence.

6.4.2. Physiological Role of Accessory Glands

The accessory glands play a large part in the reproductive physiology of female and male insects. Female colleterial glands produce substances destined to embed the eggs or to build a protective envelope or ootheca around the eggs. In tsetse flies, they produce a milky substance which nourishes the larva incubated in the uterus.

The secretory products of male accessory glands (male factors) intervene in several processes taking place in the adult female (see Raabe, 1986b). They were shown to activate oviduct contractions in *Rhodnius* (Davey, 1965), *Locusta* (Lafon-Cazal *et al.*, 1987), and mostly to enhance oviposition as observed in various insects such as locusts, bugs, lepidopterans, mosquitoes, and flies. They also influence the rate of vitellogenesis in several species including *Drosophila* (Merle, 1968), *Acanthoscelides obtectus* (Huignard, 1968; Huignard *et al.*, 1977), *Stomoxys calcitrans* (Morrison *et al.*, 1982), *Aedes* (O'Meara and Evans, 1977), and *Melanoplus* (Friedel and Gillott, 1976). In addition, the male factors also have the role of inducing monogamy, especially in mosquitoes.

The mode of action of the male factors has been proposed to be direct (Leahy and Craig, 1965 in *Aedes;* Riemann *et al.*, 1967 in *Musca*) or indirect, through a factor produced in the spermatheca (Davey, 1965) or through sensory receptors of the genital chamber (Sugawara, 1986) (Fig. 58).

The molecular structure of male factors has been thoroughly investigated in *Aedes, Musca, Drosophila,* and other insect species (see Leopold, 1976; Raabe, 1986b). In crickets, prostaglandins and prostaglandin synthetases might correspond to male factors (Loher *et al.*, 1981). Several studies were recently devoted to their molecular structure and metabolic pathways (Stanley-Samuelson *et al.*, 1986; Wakayama *et al.*, 1986a,b; Murtaugh and Denlinger, 1987). As for pros-

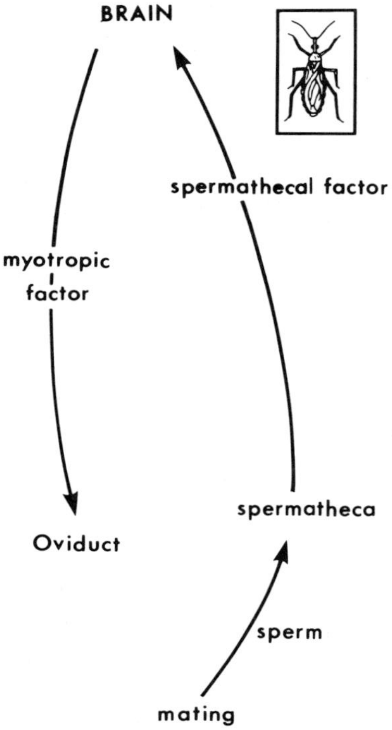

FIGURE 58. Schematic representation of the way in which mating and sperm deposition in the spermatheca induce the release by the brain of a myotropic factor stimulating oviduct contractions in *Rhodnius*. (From the data of Kriger and Davey, 1983, 1984.)

taglandin pathways, it was shown by injecting males with labeled arachidonic acid that the latter was incorporated by testis and accessory glands, transferred first to the spermatophore and then to the female spermatheca and hemolymph (Stanley-Samuelson *et al.*, 1987).

6.4.3. Accessory Gland Regulation

The functioning of the accessory glands and of various glandular cells of the genital tract is mostly regulated by JH, whose synthesis and release depend on brain neurohormones, as already mentioned (see Chapter 4).

The role of JH in regulating the secretory activity of the male accessory glands was discovered in *Rhodnius* (Wigglesworth, 1936) and subsequent studies by several authors showed that in both the male and female, accessory reproductive gland activity was dependent on the functioning of the CA.

In addition to the accessory glands, JH also stimulates the secretory activity of other parts of the reproductive tract such as the oviducts (Quo, 1965) and spermatheca in *Locusta* (Lauverjat, 1969) but not in *Melanoplus* (Gillott and Venkatesh, 1985), male ejaculatory ducts, female bursa copulatrix, and receptacle glands in *Danaus* (Herman, 1975). Juvenile hormone also triggers production by the fat body of certain proteins of the accessory glands (Gillott and Friedel, 1976).

The sensitivity of these glands to hormones differs from one insect to the other; for example, the male *Schistocerca* responds to allatectomy by greatly reducing the secretory activity of its accessory glands, whereas in *Locusta* the secretory activity of these glands following allatectomy continues at a slow rate (Cantacuzène, 1967a,b). As we shall see in Section 6.5.2, these two insects also differ in their sexual behavior after allatectomy.

In cockroaches, the female colleterial glands are very large and diversified. Both the left and right glands help to build the ootheca, the left one by producing a protein and the glucoside of protocatechic acid, and the right one by producing a glucosidase. Their regulation is not the same, however, since only the left gland is regulated by the CA (Bodenstein and Sprague, 1959; Willis and Brunet, 1966); the right one escapes this regulation, as do the male accessory glands of *Oncopeltus* (Johansson, 1958).

In the tsetse flies *Glossina morsitans* and *fuscipes*, removal of the CA or PI, which regulates corpora allata functioning, greatly reduces the production of milk by the colleterial glands. These glands, which are transformed into milk glands, supply food to the larva incubated in the mother's uterus. Because of the reduction of the amount of food ingested, the larva remains small, although it is still able to molt (Ejezie and Davey, 1976; Nkouka, 1977a; Pimley, 1983). The role of JH in milk gland activity in the viviparous cockroach *Diploptera* appears to be completely different since this hormone is necessary to allow the decline of milk gland synthetic activity at the termination of pregnancy (Ter Wee and Stay, 1987).

The role of the ns cells of the brain in male accessory gland activity has not often been investigated. Although in *Gomphocerus* the CA are necessary for the functioning of these glands, the PI also plays an important part which appears to be hormonal, since reimplantation of the brain–CC complex into parsectomized males reestablished accessory gland functioning (Hartmann, 1971). In *Locusta,* however, the accessory glands were not affected by electrocoagulation of the PI or lateral ns cells (Girardie, 1974). In *Rhodnius,* both the CA and brain were found to regulate protein synthesis in the transparent accessory gland (Barker and Davey, 1981, 1983), and the secretory activity of the ns cells of the PI appeared to influence the muscular activity of the spermatheca (Kuster and Davey, 1986).

6.5. MATING

Mating in insects is often preceded by the production of pheromones, which allow the male and female to meet. Other compounds originating from food or from the offspring host plant also play an important role in the meeting of the partners as well as the production of light and sound.

Humoral regulation of pheromone production and sexual behavior both involve JH but not in exactly the same way. As far as neurohormones are concerned, their involvement has been shown in some species.

6.5.1. Cockroaches

In cockroaches, the regulation of pheromone production and sexual behavior appears to vary among species. Female sex pheromone production depends on the CA in *Byrsotria fumigata* (Barth, 1962; Roth and Barth, 1964), but the sexual receptivity of the female is only partly controlled by these organs in *Nauphoeta, Byrsotria, Leucophaea,* and *Blaberus* (Roth and Barth, 1964; Grillou, 1978). The possibility that the PI helps to control mating behavior was suggested by the effects of electrocoagulation of this part of the brain in *Leucophaea;* however, in this species, reimplantation of the PI failed to reestablish normal behavior (Engelmann and Barth, 1968).

6.5.2. Locusts and Crickets

As in cockroaches, the pheromone production and sexual behavior of locusts depend in varying degrees on the CA. In the male, color change and pheromone production clearly cease after allatectomy or sectioning of NCA1, although not of NCA2, as observed in *Schistocerca* (Loher, 1961; Pener, 1965) and *Locusta* (Staal, 1961). The regulation of male sexual behavior itself is more controversial, and it seems that the CA are of greater or lesser importance depending on the insect species. In *Schistocerca* (Loher, 1961; Pener, 1965; Cantacuzène, 1967b) as in *Nomadacris,* males deprived of their CA were not able to copulate. The same was observed in *Locusta* (Girardie, 1966b), but in this species reduced sexual activity was also reported after allatectomy (Pener, 1965), and in most experiments the insects were not considered to be affected by CA removal in terms of their ability to copulate (Quo, 1965; Cantacuzène, 1967b; Strong, 1968). Recent experiments, including precocene application and azadirachtin injection to young larvae of *Locusta,* revealed that the overaged insects resulting from these treatments exhibited sexual behavior despite their degenerated CA (Shalom and Pener, 1986, 1987).

The role of brain ns cells has been also investigated in *Locusta*. Electrocauterization of both the PI and lateral ns cells was performed in young locust males. These operations inhibited the sexual behavior and appearance of the yellow coloring, both of which are processes in which brain ns cells may intervene either directly or by producing allatotropins (Girardie and Vogel, 1966). Further investigations using selective electrocoagulation of PI A-B cells, or C cells, demonstrated that only the C cells intervene in sexual behavior, in two different ways: first, they control the functioning of the CA and thus act indirectly on the appearance of the yellow coloring and sexual behavior, and second, they act directly on sexual behavior (Pener *et al.*, 1972).

In the virgin female of *Gomphocerus rufus,* sexual receptivity is lacking and is replaced by a special behavior of defense against the male, which disappears when the female is ready to mate. After mating, the stretching of the bursa by the spermatophore generates sensory inputs which determine the reappearance of a secondary defense reaction to the male. The removal of the CA from female last instar nymphs or young *Gomphocerus* adults permanently suppressed their ability to copulate, which was reestablished when CA were implanted (Loher, 1966). Corpora allata removal was observed to have similar effects in the grasshopper *Euthystira* (Müller, 1965) but not in *Melanoplus* (Pfeiffer, 1939) or *Chorthippus curtipennis* (Hartmann and Loher, 1974). In *Oedipoda,* however, diapausing insects implanted with active locust CA entered a phase of sexual behavior (Broza and Pener, 1969).

In crickets, the CA do not play a major role in the regulation of mating behavior (Roussel, 1967; Renucci *et al.*, 1985), but female sexual responsiveness is induced by CA implantation (Stout *et al.*, 1976). In *Teleogryllus,* the role of the brain is important, since parsectomy resulted in the arrest of singing and spermatophore production, while abdominal section of the ventral nerve cord did not inhibit periodic singing (Loher, 1974). In this insect, locomotor and stridulatory activities are both regulated by a circadian pacemaker which seems to be located between the PI and the optic lobes (Sokolove and Loher, 1975).

6.5.3. Beetles

In beetles, the role of the CA in the regulation of mating behavior and pheromone production is not firmly established. The CA did not intervene in the regulation of sexual behavior in *Leptinotarsa* (Thibout, 1982) and *Dendroctomus brevicornis* (Hughes and Renwick, 1977b; Bridges, 1982), and the CC seems more effective than the CA in induction of pheromone synthesis in *Ips paraconfusus* (Hughes and Renwick, 1977a).

6.5.4. Lepidopterans

The CA do not seem to participate in the regulation of sexual behavior in lepidopterans (Röller *et al.*, 1963 in *Galleria;* Barth, 1965 in *Antheraea pernyi;* Riddiford and Williams, 1971 in *Antheraea polyphemus* and *Hyalophora;* Tang *et al.*, 1987 in *Lymantria*. In *Platynota sultana,* however, JH seems to cause cessation of calling and onset of oviposition (Webster and Cardé, 1984). The nervous system and neurohormones, on the other hand, appear to control calling behavior in *Hyalophora* and *Spodoptera* (Riddiford and Williams, 1971; Nair, 1981). In *Hyalophora,* both the CC and intact nerve connections with the brain are necessary for this behavior to take place. Moreover, CC-CA homogenates induce a calling response in virgin females (Sasaki *et al.*, 1983; Sasaki and Riddiford, 1984).

Sex pheromone production in *Heliothis* is controlled by a neurohormone from the brain which is present in the hemolymph during the scotophase only. The extractable pheromone content of the female was strongly reduced after neck ligation but increased again when ligated females were injected with brain extracts (Raina and Klun, 1984).

In *Manduca* and *Lymantria,* however, calling appears to depend solely on neural control (Sasaki and Riddiford, 1984; Hollander and Yin, 1985; Itagaki and Conner, 1986).

6.5.5. Dipterans

The sexual behavior of male *Scatophaga stercoraria* depends on the presence of the CA (Foster, 1967), and the female sexual receptivity of *Drosophila* (Manning, 1966; Bouletreau-Merle, 1973) and *Aedes* (Lea, 1968; Gwadz, 1972) can be precociously induced by CA implants or by application of JH or JHA. When *Musca* females were deprived of their CA, they did not copulate but the application of JHA increased the rate of mating (Adams and Hintz, 1969), which depends both on female receptivity and pheromone production in this species.

An important factor in the sexual receptivity of mosquitoes and flies is supplied by the male accessory glands. This factor acts on the female to suppress a second mating. Its injection into virgin females led to permanent inhibition of their receptivity. Another male factor prevents insemination in mosquitoes but does not suppress mating (see Leopold, 1976).

6.5.6. Bugs

In *Pyrrhocoris,* JHA application to larvae resulted in giant supernumerary larvae whose sexual behavior resembled that of adults. The female was receptive and the male tried to copulate, even though this was not anatomically possible because the adult external genitalia were not differentiated (Ždárek and Sláma, 1968). In adult insects undergoing diapause, mating activity was reestablished by JHA application (Ždárek, 1968), as observed in *Draeculacephala* (Reissig and Kamm, 1975).

6.5.7. Conclusion

Studies so far conducted demonstrate that in most cases the CA are needed for mating to take place in cockroaches, orthopterans, and dipterans; in other orders, JH usually seems to stimulate processes which, however, may also proceed without this hormone. It appears to intervene essentially in insects which display interruptions in their sexual life. The most important is diapause, but shorter interruptions related to the search for food, climatic conditions, and so forth may also be important. In this connection it is interesting to note that in the giant saturniids, calling behavior requires hormonal regulation in wild species, whereas in a semidomesticated species like *Antheraea*, no such regulation was observed.

The nervous system is particularly important for mating behavior, which involves the integration of a variety of sensory imputs. Besides the purely nervous regulatory processes, neurohormones might be involved, e.g., 20-HE from the ovary plays a part in pheromone production by *Musca*, the only insect investigated from this point of view.

6.6. OVULATION—OVIPOSITION—PARTURITION

Ovulation, oviposition, and parturition depend on environmental factors and the discovery by the female of an appropriate place for the feeding and growth of its offspring. Internal factors are also of importance, including the state of maturation of the eggs and the presence of bloodborne factors resulting from mating.

Ovulation includes the opening of both the oocyte follicle and interfollicular tissue, and involves contractions of ovarioles, pedicels, and oviducts resulting in oocyte transport. Ovulation may occur just before oviposition, or a longer period of time may elapse between the two processes. Oviposition mostly results from contractions of the genital ducts where the ovulated eggs are stored. Its modalities vary among insects. Some of them, such as stick insects or *Gryllus*, deposit a few eggs each day or night over long periods, but more often egg laying is cyclic and a number of eggs are deposited at regular intervals in batches or inside an ootheca. In the shortlived moths such as *Galleria*, as many as 500–1000 eggs are laid within 2 days of mating. Ovoviviparity and viviparity are not frequent in insects, but several species, including cockroaches and flies, are ovoviviparous and retain their eggs in the genital pouch, often referred to as the brood sac, without supplying them with anything but water. Viviparous insects, on the other hand, supply food to their progeny, either from the modified follicle cells, as in the dermapteran *Hemimerus* and in parthenogenetic aphids, or from the accessory glands, as in tsetse flies and pupiparous dipterans. In both ovoviviparous and viviparous insects, parturition occurs at long intervals.

6.6.1. Overall Cephalic Control

It has often been observed that decapitation or isolation of the abdomen result in immediate oviposition. This effect was considered to be due to the suppression of inhibitory impulses from the head. Although the head inhibits oviposition, it also stimulates this process or makes it possible, since in decapitated insects, egg laying was suppressed. Oviposition or parturition was arrested by the removal of the brain or subesophageal ganglion, whereas ablation of thoracic or anterior abdominal ganglia had no effect on those processes [Thomas and Mesnier, 1973 in stick insects (Fig. 59); Mesnier, 1972a in *Galleria;* Mesnier, 1984 in *Sphodromantis lineola;* Robert *et al.*, 1984 in *Glossina;* Sefiani, 1986 in *Gryllus*].

6.6.2. Role of the Last Abdominal Ganglion

Besides the integrative control exerted by the head, stimulatory regulation by the posterior part of the nervous system was demonstrated in several insects, e.g., *Locusta, Schistocerca* (Quo, 1959; Delphin, 1965), *Galleria* (Mesnier, 1972a; Yamaoka and Hirao, 1971), *Carausius* (Thomas, 1979a), *Sphodromantis* (Mesnier, 1984), and *Gryllus* (Sefiani, 1986).

In *Carausius* (Thomas, 1979a) one egg is continuously maintained between the genital valvae while the others are retained in the lateral oviducts and only pass into the genital pouch when the first egg has been laid. Removal of the posterior abdominal ganglia or section of the last abdominal connectives was found to inhibit oviposition. The posterior operations have the same effect on oviposition as the cephalic operations, but their mode of action is completely different. The egg present in the valvae was blocked after cephalic operations, whereas after posterior abdominal operations, it was ejected but could not be replaced. In the first case oviposition is suppressed by lack of movement of the genital valvae, but in the second it is suppressed by the blocking of the egg's passage from the lateral oviducts to the genital pouch nervously regulated by the last abdominal ganglia in response to sensory inputs generated by the presence of the egg in the egg chamber. The experimental results supported by the demonstration with a scanning microscope that sensory hairs are present in the genital pouch (Thomas, 1979a). In another stick insect, *Clitumnus,* the posterior nervous control is lacking and the genital pouch is devoid of sensory hairs (Thomas and Mesnier, 1973).

The last abdominal ganglion not only stimulates oviposition but also exerts an inhibitory role, at least in *Locusta* (Lange and Loughton, 1985), where the electrical activity of the oviducal nerve was low during oviposition and high when egg laying was stopped and where electrical stimulation of the oviducal nerve did not stimulate egg laying but, on the contrary, resulted in a recoil of the eggs toward the ovary.

6.6.3. Neurohormones

The first evidence of humoral regulation of egg laying was the discovery of the stimulatory effects on oviposition of the hemolymph of the female ready to oviposit. Thus, the hemolymph of mated or ovipositing females induced precocious oviposition in *Bombyx, Iphita, Schistocerca,* and *Hyalophora* (Mokia, 1941; Nayar, 1958; Okelo, 1971; Riddiford and Ashenhurst, 1973). The origin of the active factor was demonstrated to be located in the nervous system in the very first investigations, in which isolated oviducts were brought into contact with brain and CC extracts (Koller, 1954 in *Tenebrio;* Enders, 1955 in *Carausius*). Later, the role of neurohormones in controlling oviposition was demonstrated in a number of insects.

The regulation of ovulation in oviparous insects was only studied in *Clitumnus,* where this process takes place asynchronously in the different ovarioles, a short time before oviposition. Ablation of either the PI or subesophageal ganglion arrested this process, which was reestablished by injection of brain or ventral nerve cord extracts. Neurohormonal regulation may thus be implicated.

Egg laying in stick insects was investigated using various homogenate injections in debrained insects. This revealed that extracts from the brain, subesophageal ganglion, thoracic ganglia, or anterior abdominal ganglia caused oviposition, but extracts from the posterior abdominal ganglia were ineffective. Consequently, a myotropic factor must be present in most of the central nervous system. It acts by triggering oviduct contractions and opening the genital pouch or egg chamber, delimited by the genital valvae (Thomas and Mesnier, 1973) (Fig. 60).

In *Galleria,* the oviposition-stimulating neurohormone (OSH) was shown to be present throughout the entire central nervous system. It stimulates ovipositor movements, oviduct contractions, and exerts modulatory effects on the last abdominal ganglion (Fig. 60).

In some species, such as *Gryllus* and *Sphodromantis,* the presence of OSH is not sufficient to promote oviposition but it causes oviduct contractions and triggers the movements of the ovipositor (Mesnier, 1984; Sefiani, 1986).

The regulation of both ovulation and parturition was investigated in *Glossina fuscipes* (Nkouka, 1977b; Robert *et al.*, 1984) (Fig. 61) and *Glossina morsitans* (Denlinger *et al.*, 1983). These viviparous insects have a very particular physiology, with a long incubation period and very complex behavior in connection with the finding of a suitable place to deposit their progeny. They incubate a single egg and then a larva, which feed on milk gland secretion, as previously noted.

Ovulation control has been studied both in *G. morsitans* (Chaudhury and Dhadialla, 1976; Chaudhury *et al.*, 1981; Gillott and Langley, 1981) and *G. fuscipes* (N'Kouka, 1977b; Robert *et al.*, 1984), where it takes place within 1 min of parturition or mating. The experimental results are to some extent controversial. In *G. morsitans,* injection of mated female hemolymph was able to induce ovulation, but this result could not be obtained in *G. fuscipes*. In addition, PI electrocoagulation and corpus cardiacum removal were reported to arrest this

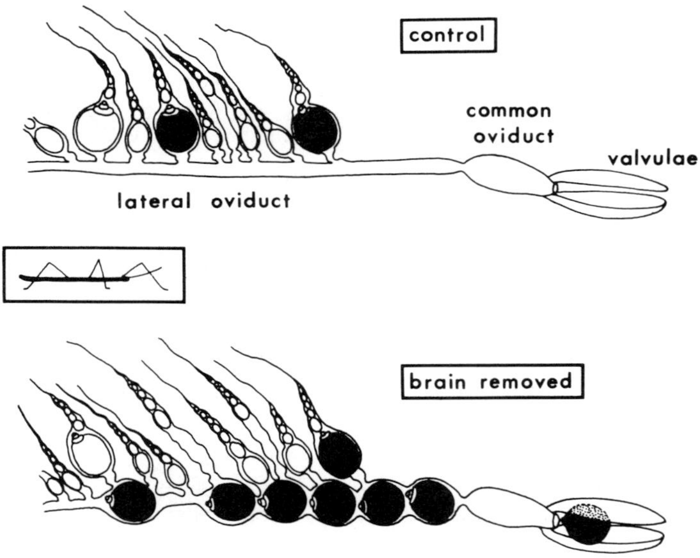

FIGURE 59. Inhibition of egg laying by brain removal in *Carausius.* Note the accumulation of mature eggs in the lateral oviduct of the operated female. (From the data of Thomas, 1979a.)

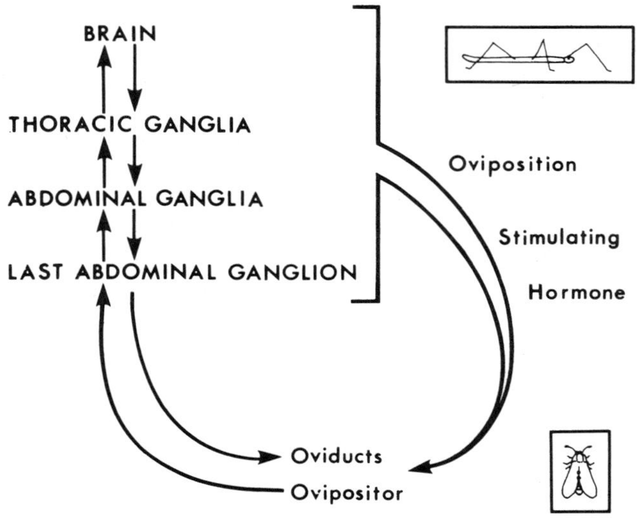

FIGURE 60. Schematic representation of egg-laying regulation in *Carausius* and *Galleria.* Both neurohormonal (open arrow) and nervous (filled arrows) controls are involved. (From the data of Mesnier, 1972a; Thomas and Mesnier, 1973.)

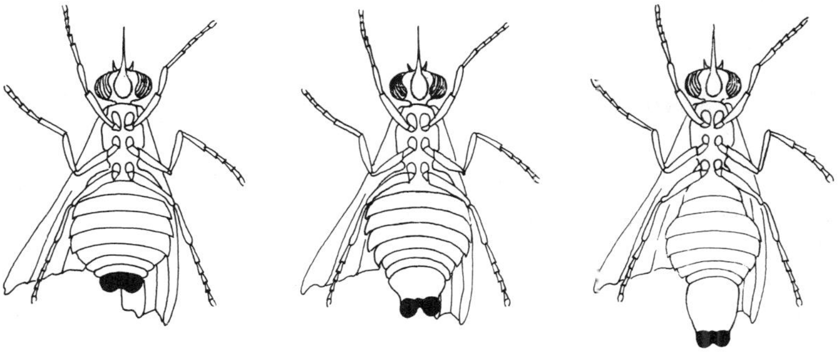

FIGURE 61. The parturition process in *Glossina fuscipes.*

process in both species (Foster, 1974; Nkouka, 1977b; Chaudhury *et al.*, 1981), but neither PI reimplantation (Foster, 1974; Chaudhury *et al.*, 1981) nor injection of varied extracts of nervous and neurohormonal tissues succeeded in doing so in *G. fuscipes* (Robert *et al.*, 1984). This suggests that ovulation in this species is controlled by a nervous pathway, the more so that the muscles of the

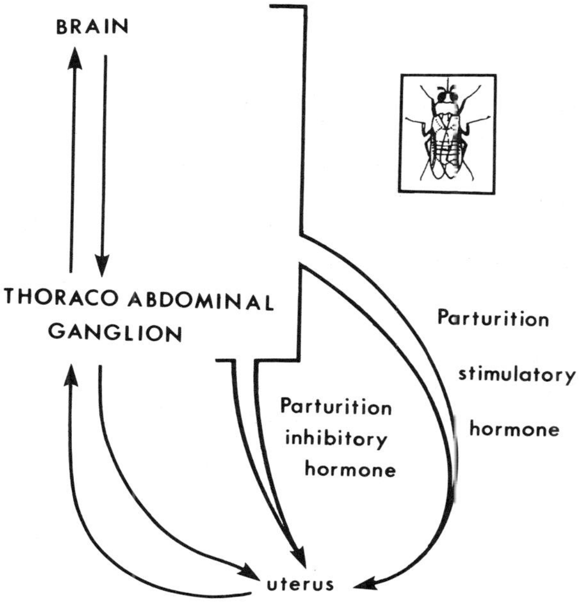

FIGURE 62. Schematic representation of the regulation of parturition in the tsetse fly. Nervous stimuli (→), neurohormonal stimuli (⇒). (From the data of Robert *et al.*, 1984.)

oviduct receive a rich and varied nerve supply of both peptidergic and nonpeptidergic axons (Robert *et al.*, 1984), and ovulation might be regulated by neurotransmitters and neurohormones released at axon endings.

The regulation of parturition was investigated in both *G. fuscipes* (Nkouka, 1977b; Robert *et al.*, 1984) and *G. morsitans* (Denlinger *et al.*, 1983). The results obtained showed that in these insects, a parturition stimulatory hormone (PaSH) is present in both the brain and thoracoabdominal nerve mass, whose extracts reestablished parturition in decapitated *G. fuscipes* (Nkouka *et al.*, 1977b; Robert *et al.*, 1984) and in *G. morsitans* (Denlinger *et al.*, 1983). The presence of PaSH in the blood is of short duration and it probably does not appear until the fly has found a suitable place to deposit its larva (Fig. 62). In addition to PaSH, a hormone inhibiting parturition (PaIH) appears to be stored in the CC and the thoracoabdominal PO (Robert *et al.*, 1984). This factor is probably released throughout the incubation period to avoid abortion. Finally, the peptidergic neuroeffector junctions observed in the extrinsic muscles of the uterus suggest that parturition might be controlled by factors released locally (Robert *et al.*, 1984), the more so as parturition was suppressed by sectioning of the abdominal nerves (Nkouka, 1977b; Denlinger *et al.*, 1983).

6.6.4. Production Sites of the Neurohormones

6.6.4.1. The Brain

The cerebral cells producing OSH, PaSH, and other ovulation/oviposition myotropins were identified in the PI in the bugs *Iphita* (Nayar, 1958) and *Rhodnius* (Davey, 1965). These results were confirmed by the injection of extracts and electrocoagulation performed in *Clitumnus* (Mesnier, 1972b), *Locusta* (Girardie and Lafon-Cazal, 1972; Lazarovici and Pener, 1978; Lange and Loughton, 1985), *Melanoplus* (Friedel and Gillott, 1976), and *Sphodromantis* (Mesnier, 1984).

By means of a plastic "window" inserted into the abdomen of *Rhodnius*, it was possible to observe ovary and oviduct contractions which peaked twice during the gonadotropic cycle, once after the blood meal and the second time just before ovulation/oviposition, two closely connected processes in this insect. No contraction peaks were seen in insects whose PI ns cells had been extirpated (Kriger and Davey, 1982). The origin of the ovulation/oviposition hormone was found to be 10 large cells of the PI, probably type A ns cells, whose extracts induced ovulation/oviposition in mated females deprived of their own ns cells (Kriger and Davey, 1983), and induced myotropic activity in an isolated ovary (Kriger and Davey, 1984). The hemolymph titer of the myotropic ovulation/oviposition hormone was examined during the reproductive cycle. The titer rose after feeding and then, in mated females only, just before ovulation, thus confirming the physiological significance of the hormone concerned (Davey and Kriger, 1985; see Davey, 1984).

It was shown in *Stillbocoris natalensis* (Furtado, 1971) and *Glossina fuscipes* (Nkouka, 1977b), that parturition was controlled by the PI.

6.6.4.2. Ventral Nerve Cord

It has been shown in a number of insects that the OSH was not only located in the brain but also in all the ventral nerve cord ganglia. An attempt to detect the precise origin of OSH was made by Mesnier (1985). Since *Clitumnus* is a nocturnal species which deposits its eggs at the beginning of the scotophase, the activity of nerve ganglia extracts was assayed at the beginning and end of both the photophase and scotophase. These experiments showed that the amount of OSH in the ganglia varied greatly, depending on the time when they were dissected out. Thus, at the end of the dark period and beginning of the light period, the ganglia were heavily loaded, but 1 hr before dark they were almost empty. A histological study of the ventral nerve cord ns cells at four different times in the cycle revealed that the large lateral type A cells exhibited striking load changes. The minimal load was observed 1 hr before the onset of the scotophase, i.e., when the OSH content of the ganglia was very small. In contrast, when the ganglia extracts were very active, these cells were heavily loaded. It therefore seems that the source of the OSH might be in these large, lateral, type A ns cells.

6.6.5. Neurohormone Release Sites

In insects, the sites of neurohormonal release are mostly located in the CC and PO, but in addition, narrow diffuse areas composed of isolated endings apposed to the neural lamella of nerves were observed in electron microscopy and probably constitute other sites of neurohormonal release (see Chapter 1).

It was observed in *Carausius* and *Clitumnus* that CC and PO extracts do not stimulate oviposition as central nervous system extracts do (Thomas and Mesnier, 1973). The neurohemal structures involved in OSH release in *Clitumnus* (Mesnier, 1985) are found in somatic nerves and exibit circadian modifications of their OSH load. Thus, the amount of OSH stored is small in the first hours of the photophase, subsequently increases, and then peaks at the very beginning of the dark phase.

In *Glossina fuscipes,* PaSH is found in extracts of genital nerve branches containing nh areas, but not in the CC or the thoracoabdominal PO. Both nh organs display an inhibitory role.

Thus, in both *Clitumnus* and *Glossina,* the oviposition and parturition-stimulating factors OSH and PaSH are released in nh areas and not in the CC or

PO. However, in other insects, including *Rhodnius* (Davey, 1984), *Hyalophora* (Truman and Riddiford, 1971), *Galleria* (Mesnier, 1972a), *Locusta* (Girardie and Lafon-Cazal, 1972; Lange and Loughton, 1985), and *Melanoplus* (Friedel and Gillott, 1976), CC extracts did stimulate oviposition, and PO extracts did so in *Galleria* (Mesnier and Provansal, 1975).

Moreover, in some species, such as *Galleria,* the PO are active in addition to the nh areas, and the PO of *Clitumnus,* which are inactive in *Clitumnus* itself, are active in *Galleria.* The reasons for these differences in the locations of neurohormonal release sites are not yet clear. It is possible that several factors are involved, or else that the presence of antagonistic molecules mask the presence of the active neurohormone. In addition to being transported by the hemolymphatic pathway, it seems that neurohormones are also transported by nerves and released at the neuroeffector junctions at the muscle level.

6.6.6. Interspecificity of the Neurohormonal Factors

The factors triggering oviposition and parturition display cross-reactions both *in vivo* and *in vitro*. Reciprocal effects were obtained *in vivo* in *Galleria–Clitumnus* and *Sphodromantis–Clitumnus. In vitro* studies were performed on the uterus of *Glossina* (Raabe *et al.,* 1985) and the oviduct of *Gryllus* (Sefiani, 1986). The first effector responded to extracts of the central nervous system of *Clitumnus* while the second responded to *Galleria* extracts. The interspecific activity of the extracts suggests that either the same factor or two related factors are involved which trigger oviposition and parturition (Fig. 63).

6.6.7. Proctolin and Other Myotropic Peptides

The chemical nature of OSH, PaSH, and other ovulation/oviposition myotropins is still unknown. Proctolin, a peptide which has been identified in the

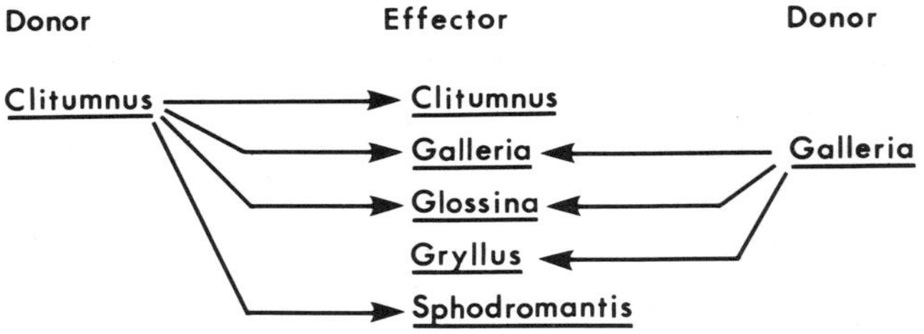

FIGURE 63. Interspecific activity of neurohormones involved in egg laying and parturition control. (From the data of Mesnier, 1984; Raabe *et al.,* 1985; Sefiani, 1986.)

cockroach (Brown, 1975) and exerts a local myotropic effect on the hindgut, was reported by Cook (1981) to act on the oviduct of *Tabanus proximis* at a concentration as low as 10^{-10} M. At this concentration, proctolin increased the tonus and amplitude of oviduct contractions but not their frequency. *In vitro* experiments on the *Gryllus* oviduct also revealed that proctolin stimulated these contractions, but its action was different from that of the central nervous system extracts in that it mostly caused a large increase in their tonus accompanied by a temporary decrease in their amplitude (Sefiani, 1986). It seems, therefore, that proctolin is not the factor involved in activating oviduct contractions, a conclusion also reached by Kriger and Davey (1984). In *Locusta,* however, recent findings indicate that proctolin causes dose-dependent tonic contractions of the lateral oviduct. Its presence was shown by immunocytochemistry in both the oviducal nerve and the innervated regions of the oviduct (Lange *et al.*, 1986). After stimulation of this nerve, its release into the incubation medium was observed. Proctolin was demonstrated in the oviducts of *Periplaneta* as well (Orchard and Lange, 1987).

Two heartbeat-stimulating octapeptides (MI and MII) from the CC, central nervous system, and gut of *Periplaneta* were identified and chemically characterized. Additional myotropic peptides that act on the hindgut were identified quite recently in the head of *Leucophaea* (Holman *et al.*, 1986a–d, 1987a,b; Nachman *et al.*, 1986a,b) (see Chapter 7). Oviposition-stimulating hormone, PaSH, and other ovulation/oviposition myotropins might be among these myotropic factors.

6.6.8. Biogenic Amines

Other substances besides neuropeptides act on oviduct contractions, especially glutamate and biogenic amines, which were suggested to induce ovulation in cockroaches (Hentschel, 1972). On the other hand, serotonin was shown to increase the frequency of the contractions of the oviduct in *Tabanus* (Cook, 1981) and *Gryllus* (Sefiani, 1986) and also stimulated uterine contraction in *Blabera* (Raabe *et al.*, 1985), although not in *Glossina fuscipes*. The *Gryllus* oviduct reacted to serotonin but the strongest effects were obtained with octopamine, which raised contraction frequency by 140–250%; amplitude and tonus remained unchanged. Egg laying in *Gryllus* takes place fairly continuously; this is quite different from the oviposition pattern in *Locusta,* in which inhibitory regulation is well developed (Lange *et al.*, 1984). It involves two dorsal octopaminergic unpaired median neurons (DUMOV) projecting into the basal part of the oviduct where they modulate the neurally evoked contractions. The existence of the DUMOV was revealed by cobalt backfilling, Lucifer yellow injection, and neurophysiological recordings (Lange and Orchard, 1984) (Fig. 64). The effects of various biogenic amines were studied in *Locusta* during

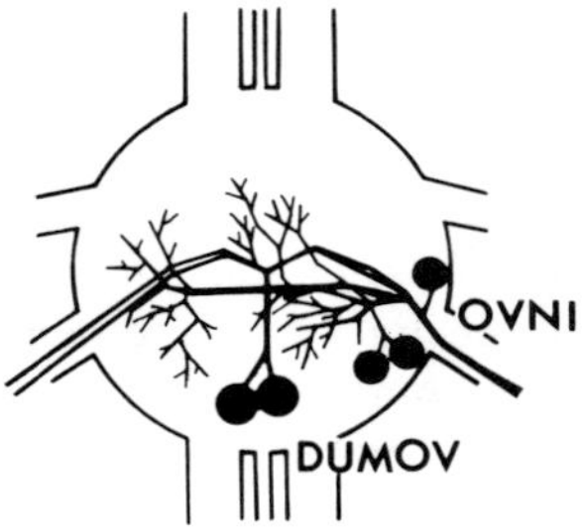

FIGURE 64. Identification, in the seventh abdominal ganglion of *Locusta,* of median octopaminergic neurons (DUMOV) and paired lateral nonoctopaminergic neurons (OVNI) which project into the oviduct. (Modified from Lange and Orchard, 1984.)

electrical stimulation. While norepinephrine and serotonin enhanced the amplitude of neurally evoked contractions and the frequency of myogenic contractions, octopamine reduced the amplitude of neurally evoked contractions (Lange and Orchard, 1984). The presence of octopamine was also discovered in the oviduct of *Periplaneta* (Orchard and Lange, 1987).

Reserpine injections carried out in some insects induce arrest of vitellogenesis and oviposition, which suggests that amines might be involved in the regulation of these processes (Hentschel, 1972), in *Periplaneta.* In *Glossina morsitans,* they prevent ovulation but not parturition (Samaranayaka-Ramasamy and Denlinger, 1985). In *Acheta,* daily injections of haloperidol and benzeracide result in storage of ns products in PI type A ns cells, inhibition of CA activity, and vitellogenesis (Panov and Chudakova, 1987).

6.6.9. Juvenile Hormone and Ecdysone

Juvenile hormone and 20-HE appear to have some influence on oviposition. In *Platynota* the application of JH or JHA to virgin females was reported to elicit oviposition (Webster and Cardé, 1984), but such a finding is exceptional and these hormones did not affect oviduct contractions, at least in *Leucophaea* (Cook *et al.,* 1984). Ecdysone, on the other hand, was demonstrated to have a strong although indirect triggering action in *Rhodnius* ovulation/oviposition. It enhanced the electrical activity of the ns cells of the PI and CC, and stimulated the release of the myotropic ovulation/oviposition hormone from the PI (Ruegg *et al.,* 1981, 1982). Ecdysone exerts this triggering action via an aminergic pathway, since adrenergic receptor antagonists were shown to abolish it (Orchard *et al.,* 1983c) (Fig. 65).

In *Glossina morsitans,* JHA application and 20-HE injection at any time during egg or larva incubation cause abortion (Denlinger, 1975; Meidell, 1982). As regards ecdysteroids, both ecdysone and 20-HE have been shown to affect uterine motility in two opposite ways, i.e., ecdysone enhances it and 20-HE inhibits it (Robert *et al.,* 1986b). These results suggest that ecdysteroids might

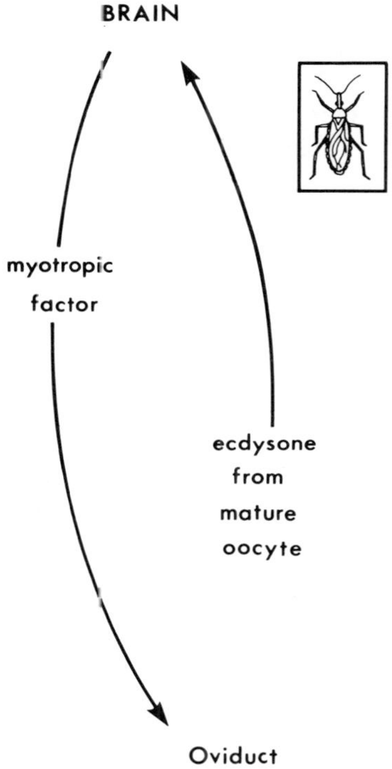

FIGURE 65. Schematic representation of the action exerted by ecdysone from mature oocytes on the pars intercerebralis neurosecretory cells which in turn release a myotropic factor that stimulates oviduct contractions in *Rhodnius.* (From the data of Kriger and Davey, 1983, 1984; Ruegg *et al.,* 1981, 1982.)

play a part in abortion and parturition. The appearance in the hemolymph of an ecdysteroid peak just before parturition supports this hypothesis (Robert *et al.*, 1986a). The site of ecdysone production in the adult *Glossina* remains to be demonstrated. As far as the mode of action of ecdysteroids on the uterus is concerned, it is worth mentioning that their effects are suppressed by tetrodotoxin; this implies that an intermediary nervous tissue is involved in the response to ecdysteroids. The tissue in question might consist of the peptidergic neuroeffector junctions already mentioned.

6.6.10. Mode of Action of the Neurohormones

Oviposition-stimulating hormone and PaSH appear to act directly on the muscles of the oviducts and uterus, the oviducal and vaginal sphincters, and the valvae and ovipositor muscles, as observed *in vitro*. Moreover, active factors are certainly released at the effector level, where peptidergic neuroeffector junctions were observed as well as nonpeptidergic endings (Robert *et al.*, 1984 in *Glossina;* Kiss *et al.*, 1984 in *Locusta*). In *Locusta* both proctolin (Lange *et al.*,

1986) and octopamine (Lange and Orchard, 1984) were identified in certain axons in the oviducal nerve and oviduct, suggesting local release.

The action of hormones regulating oviposition, ovulation, and parturition may involve cyclic AMP, as in the case of most neurohormones, since injection of cyclic AMP into virgin *Glossina morsitans* (Denlinger *et al.*, 1978) and decapitated *Clitumnus* (Mesnier, 1981) stimulated ovulation. In addition, the effects of octopamine on the lateral oviduct of *Locusta* involve cAMP, whose level is raised by the stimulation of the DUMOV (Lange and Orchard, 1986). Application of octopamine agonists and antagonists to the oviduct revealed that octopamine-2 receptors are involved in this stimulation (Orchard and Lange, 1986).

6.7. PARTICULAR MODES OF REPRODUCTION

6.7.1. Social Insects

The main features of reproduction in social insects is the inhibition of reproduction exerted by the queen on the other members of the colony, thus determining the existence of castes. Pheromones and nutrition affect caste determination through the CA, which are inhibited in both primitive social hymenopterans and highly social wasps, bees, and termites.

In *Polistes annularis* topical JH application promoted oocyte maturation in subordinate females and caused changes in their behavior (Bell, 1973; Barth *et al.*, 1975). Moreover, in summer-diapausing females of *Polistes metricus,* JH caused ovarian development (Bohm, 1972). In *Bombus terrestris* (Röseler, 1977; Röseler *et al.*, 1981), the JH titer is low in the queenright workers and high in the queenless workers. In addition, during the prepupal development of workers and queens, both JH and ecdysteroids peak at intervals of a few hours, but the worker peaks precede the queen peaks by about 20 hr (Strambi *et al.*, 1984), suggesting that the effects determined by JH may be time-dependent.

In bees, the role of JH is still controversial and the CA were even considered not to be indispensable either to vitellogenin synthesis (Engels, 1978) or to queen determination from young worker larvae (Rembold *et al.*, 1974). However, JH determination shows that the JH titer is 10 times higher in 3-day-old queens than in workers of the same age (Wirtz and Beetsma, 1972). This is precisely the moment at which the ovarian anlagen differentiated into ovarioles in larvae raised in queen cells, but atrophied in worker cell larvae. The idea that JH triggers reproductive development in bees was also supported by the results of experiments involving JH treatment of larvae reared on worker jelly, which differentiated into either intercastes (Rembold, 1976) or perfect queens (Dietz *et al.*, 1979; Asencot and Lensky, 1984). In other experiments it was reported that

JH might have antagonistic effects, depending on the doses applied (Rutz *et al.*, 1976; Ramamurty and Engels, 1977).

The data relevant to ants also pointed to the CA as triggering sexual development in *Myrmica rubra* (Brian, 1974) and *Pheidole pallidula,* thus inducing the production of eggs determined to develop into reproductives (Passera and Suzzoni, 1979).

In termites, the implantation of CA in pseudergates (a term designating fully grown larvae) resulted in the molting of the receptor insect into supplementary reproductive, showing that the CA help to trigger sexual processes, although they also intervene in the caste determination of soldiers (Lüscher, 1958; Lebrun, 1967).

Few data are available concerning the role of ecdysteroids in reproductive determination in social insects. High ecdysteroid levels were reported in ant larvae destined to develop into workers (Suzzoni *et al.*, 1983), but the opposite situation prevails in bee nymphs, since the future queen pupae have higher ecdysteroid titers than the worker pupae (Lensky *et al.*, 1978). No general conclusion regarding the role of ecdysone can therefore be drawn at the present time.

6.7.2. Unusual Modes of Reproduction

Unusual modes of reproduction are not rare in insects, and may occur sporadically or continuously. They include parthenogenesis, viviparity, and pedogenesis, which are frequently combined. The neuroendocrine regulation of most of these events has never been studied. The only investigations concern aphids and Heteropeza.

6.7.2.1. Aphids

In aphids, the reproductive processes proceed in completely different ways depending on the season and day length. During the long-day summer period, several generations of viviparous parthenogenetic females (termed virginoparae) develop and during the short days of autumn give rise to a generation of sexual males and females, the latter being termed oviparae. The structure of the female reproductive system is different in the two morphs: oviparae form large, yolky, haploid eggs, whereas virginoparae develop, during their larval growth, numerous smaller eggs which evolve parthenogenetically into diploid embryos.

Juvenile hormone influences several features of aphid morphs and acts both on the treated insects and on the progeny incubated in their abdomen (see Chapter 12), causing the latter to develop in parthenogenetic viviparous females (Lees, 1977). Cautery of certain ns cells in the PI, however, abolished the normal response to day length and induced the appearance of oviparous females and males instead of viviparous parthenogenetic females. The existence of a virginopara-promoting substance is therefore probable, as suggested by Steel and Lees (1977). This substance might possibly intervene by stimulating CA functioning.

6.7.2.2. *Heteropeza*

The gall midge *Heteropeza pygmea* displays two modes of reproduction which follow each other in successive generations: first there are the pedogenetic viviparous larvae which incubate their parthenogenetic progeny inside their hemocele; subsequently, as environmental conditions change and overcrowding appears, the mode of reproduction returns to normal, and metamorphosing larvae of both sexes escape from the mother's body, mate, and lay fertilized eggs. In pedogenetic larvae no ducts or accessory glands are present, and the ovary, which lacks ovarioles, contains only follicles, which are released into the maternal inner milieu.

In vivo studies, which involved ligature and hormone treatment, produced somewhat contradictory results, from which it was concluded that JH initiated the development of pedogenetic eggs, whereas 20-HE promoted the production of eggs whose larvae undergo metamorphosis (Pohlhammer, 1968).

In vitro investigations (Went, 1978) showed that JH inhibited follicle formation, which was stimulated by 20-HE, a result similar to that obtained in a mosquito by Beckemeyer and Lea (see Section 4.2.2).

6.8. CONCLUSION

The predominant impression that emerges from the investigations concerning the way in which reproduction is regulated in varied insects is that the mechanisms regulating each reproductive step do not follow the same pattern.

Ecdysteroids and JH play a major role but while JH mostly triggers the functioning of organs, ecdysone or 20-HE is involved in morphogenetic processes (Fig. 66). However, in the regulation of vitellogenesis, 20-HE may replace JH, especially in advanced species like dipterans, and it is necessary to recognize the existence of several different patterns of regulation for the same process.

As regards egg-laying and parturition, JH and 20-HE no longer intervene and their functions are taken over by neurohormones, which also play a part throughout the different reproductive stages, since they regulate endocrine function and also exert a certain amount of direct control over the reproductive processes themselves.

Feedback mechanisms help to regulate reproduction in various ways: these

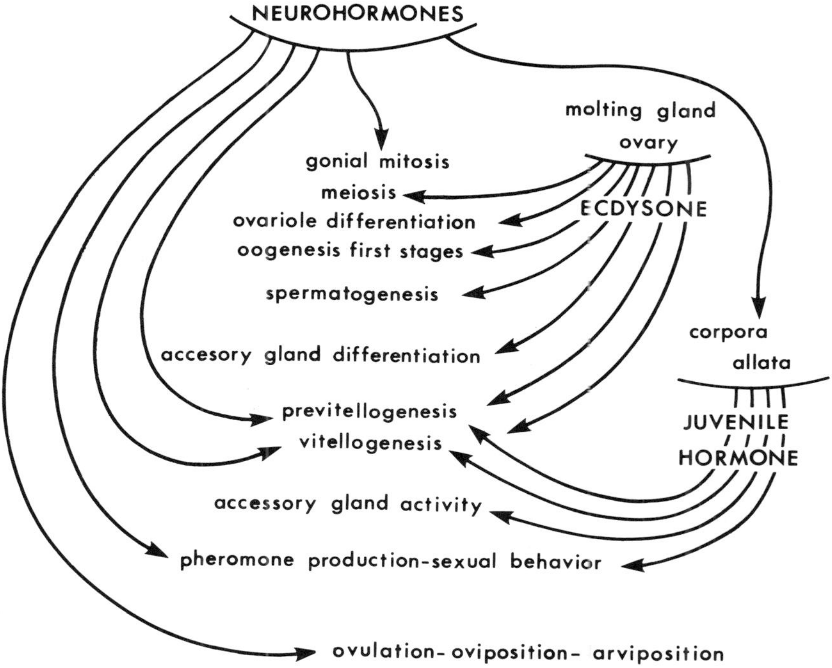

FIGURE 66. Overall schematic representation of the regulation of the various reproductive events.

include the action of mature eggs on the functioning of the CA and ovarian follicle, and the action of 20-HE on the ns cells of the brain. Other chemical regulators which come from the male accessory glands or are produced by the female during copulation intervene actively in female reproduction as they regulate oogenesis, egg laying, and sexual receptivity.

Considerable progress in investigating reproduction has been made since synthetic hormones became available and the titers of the hormones in the hemolymph and tissues could be measured. When in turn it becomes possible to synthesize neurohormones, further advances in this direction may be expected.

The study of the mode of action of hormones on the cell membrane and genes has begun.

7

Muscle Activity

The presence of myotropic neurohormones was revealed some 40 years ago by Koller (1948), who showed that head and brain extracts of different species induced the acceleration of contractions of the Malpighian tubules, oviducts, and gut (Fig. 67). Subsequent investigations dealing with the main visceral organs, and almost always carried out *in vitro* or on semiisolated organs, led to the same conclusion. Most of them dealt with the heartbeat of the cockroach *Periplaneta*. The active factors were discovered in the CC and the central nervous system, and in the effectors themselves.

The effects observed are mainly concerned with the frequency of contractions of the effector, whether it is the heart, the Malpighian tubules, the gut, or the oviducts. Concurrent with this effect, however, increases in the tonus and in the amplitude of the contractions have often been reported.

Visceral muscles in insects contract in a spontaneous, rhythmic manner and display myogenetic functioning, which persists in the absence of any innervation, although this may be highly developed in certain cases. Some skeletal muscles also display a myogenic pattern of contractions.

7.1. VISCERAL MUSCLES (Table XXI)

7.1.1. Dorsal Vessel and Associated Structures

7.1.1.1. Structure and Innervation

The heart, also termed dorsal vessel, is formed of two parts: the heart proper, which occupies the abdominal portion of the body, and the aorta, which extends to the head. The heart consists of ventricles formed of circular muscles and bearing lateral openings or ostioles. One pair of alary muscles is inserted laterally on each ventricle and is associated with connective tissue making up the dorsal diaphragm. The pumping process performed by the heart results from its own contraction and that of the alary muscles, although their contraction rhythms

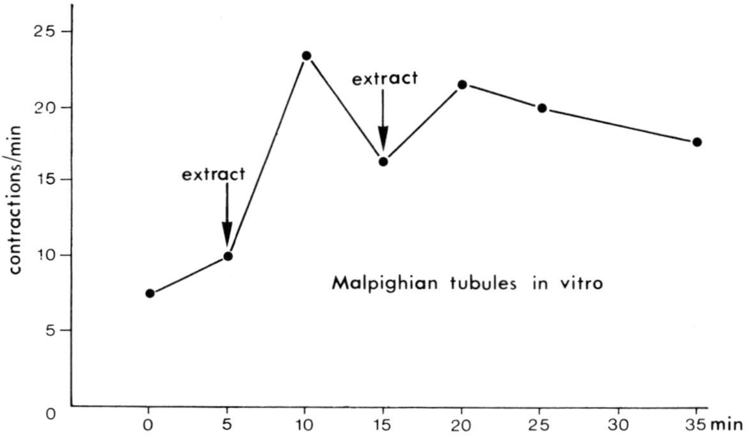

FIGURE 67. Effect of brain homogenates on the contraction frequency of a Malpighian tubule in *Dytiscus*. (After Köller, 1948.)

are different, at least in some species (see Miller, 1985). Accessory pulsatile organs, such as segmental vessels, antennal circulatory organs, and the hyperneural muscle, contribute to hemolymph circulation.

Innervation of the heart in insects has rarely been investigated, and it appears to be highly variable. In the cockroach *Periplaneta* (Alexandrowicz, 1926), the stick insect *Carausius* (Opoczynski-Sembratowa, 1936), and the locust *Locusta* (Roussel, 1972), it includes several components. Two lateral heart nerves originating in the head run along the aorta and the heart. They have been given different names: lateral heart nerves (employed here), cardiac nerves, and lateral cardiac nerve cords. The lateral heart nerves are formed of nerve fibers of different origins. Some of them originate in the CC; others, via the segmental nerves, in the ventral nerve cord ganglia; and still others from nerve cells placed on or near the lateral heart nerves (see Miller, 1985). The general appearance of the lateral heart nerves is that of nh nerves. They constitute, especially in *Periplaneta,* large swellings, with numerous nerve terminals spread against the neural lamella (Johnson, 1966; Miller and Thomson, 1968). Lateral heart nerves and heart neurons have been identified in some species, but they appear to be lacking in others. Recently, a nh network of nerve branches was found to occur in the dorsal diaphragm in eight species of dipterans. It originates in the ventral nerve cord ganglia (Meola and Cook, 1986) and may correspond to the lateral heart nerves described above.

The function of the different nerve components reaching the heart was partly elucidated by investigations conducted on *Periplaneta* (Johnson, 1966;

Miller, 1968a,b,; Miller and Thomson, 1968; Smith, 1969; Miller and Usherwood, 1971). Although the heart of this insect displays a basic myogenic heartbeat, it has a complex regulation, both nervous and neurosecretory. The impulses originating in the central nervous system control the functioning of the heart neurons, which in turn control the contractions of the myocardium. Some of the heart neurons are ns. Moreover, ns axons originate in the ganglia. They belong to two categories, one with small granules, and the second with large granules. Part of their secretory products is released into the lateral heart nerves but, apart from this typical nh functioning, the small-grained fibers innervate the heart neurons and form neuromuscular junctions with the myocardium, which thus displays a double innervation.

The alary muscles also exhibit complex innervation originating in the lateral heart nerves and the segmental nerves. In *Carausius,* and probably in *Periplaneta,* they are also innervated by two neurons located in each segment near the segmental nerves.

7.1.1.2. Heart Myotropic Peptides

Myotropic factors have been identified in the central nervous system by several investigators who assayed the activity of brain, subesophageal ganglion, and ventral nerve cord ganglia extracts on the heart, and sometimes the activity of other organs. According to the data of Ralph (1962b), Rounds (1963), and Rounds and Gardner (1968) on *Periplaneta,* a cardioaccelerator appears to be present throughout the brain and the ventral nerve cord. Gersch (1958) reached the same conclusion concerning *Corethra (Chaoborus) plumicornis* (Fig. 68).

As far as the CC are concerned, the presence of myotropic factors was clearly established in early investigations (Cameron, 1953; Davey, 1961, 1962a). Davey (1963) noted that induced hyperactivity brings about a decrease in the myotropic factor content of the CC and causes the appearance of a powerful heart stimulator in the hemolymph, which is not found in decapitated insects or in those deprived of the CC and subjected to the same treatment. Moreover, the injection of CC extracts causes cardiac acceleration *in vivo* (Hertel, 1971). It was also shown that the CC cardioacceleratory factor is released after electrical stimulation of the brain (Kater, 1968).

In addition to cardioaccelerators, the presence of cardiodecelerators was demonstrated in cockroaches. In *Blaberus giganteus,* central nervous system homogenates, prepared in alcohol-water mixtures in variable proportions, displayed varied activity, and it was observed that antagonistic factors were stored at the same place but not at the same time. In *Periplaneta,* stimulation of the different thoracoabdominal ganglia *in vitro* causes the release of cardioaccelerators from the sixth abdominal ganglion and the release of cardiodecelerators from the thoracic ganglia.

TABLE XXI. Myotropic Activity of Neuroglandular Extracts, Proctolin, and Biogenic Amines on Muscle Contraction[a]

Insect	Homogenates: Brain	Homogenates: CC	Serotonin	Octopamine	Dopamine	Epinephrine	Norepinephrine	Proctolin	Reference
					Heart				
Periplaneta	0	+							Cameron, 1953; Davey, 1961a,b
	+	+	+		+				Unger, 1956; Hertel, 1971, 1975
			+	+	+	+	+		Miller and Metcalf, 1968; Collins and Miller, 1977
Carausius		+							Enders, 1955
Locusta		+							Cazal, 1967
		+	+		+				Hiripi and S.-Rózsa, 1973
Manduca			+	+				0	Platt and Reynolds, 1986
Stomoxys			0	0				0	Cook and Meola, 1983
Corethra	+	+							Gersch, 1958
Calliphora		0							Normann, 1972
					Foregut				
Leucophaea			+					+	Cook and Holman, 1978
Blaberus			+				+		Cook *et al.*, 1969
Periplaneta		−							Cameron, 1953
Locusta		+							Cazal, 1969
			+						Freeman, 1966
Schistocerca			+		−				Klemm, 1972
			−	−				+	Banner *et al.*, 1987
Galleria			+	0	0	+	+		Yamamoto and Fukami, 1980
Corethra	+								Gersch, 1955
					Hindgut				
Leucophaea	+		+		−		−	+	Holman and Cook, 1970; Cook and Holman, 1978, 1979

Periplaneta	+																		Cameron, 1953; Davey, 1962b
				+			+			+							+		Brown, 1975, 1977
				+								0		0					Freeman, 1966; Brown, 1965
Carausius	−	−																	Enders, 1955
Schistocerca				+															Klemm, 1972
Locusta	+																		Cazal, 1969
				+						+		0		0					Freeman, 1966
				+			+										+		Oldfield and Huddart, 1982
Stomoxys																	+		Holman and Cook, 1979
										Malpighian tubules									
Periplaneta	+	+																	Cameron, 1953
Carausius	+	+																	Enders, 1955; Pilcher, 1971
										Oviduct									
			F	A	T	F	A	T	F	A	T		F	A	A	F	A	T	
Leucophaea			−	+															Cook *et al.*, 1984
Blabera			0	0	0														Raabe *et al.*, 1985
Clitumnus			0	0	0														Raabe *et al.*, 1985
Carausius	−	+																	Koller, 1954; Enders, 1955
Locusta	+	+																	Girardie and Lafon-Cazal, 1972
Gryllus	+	+	+			+			+									+	Sefiani, 1986
Locusta			+	+		−	−	−					+	+		+		+	Orchard and Lange, 1985; Lange *et al.*, 1986
Tabanus			+	+	+														Cook and Meola, 1978; Cook, 1981
										Uterus									
Blabera				+															Raabe *et al.*, 1985
Glossina				0															Raabe *et al.*, 1985
										Intersegmental muscles									
Lucilia							+										+		Irving and Miller, 1980
										Extensor tibiae									
Locusta							+										+		Piek and Mantel, 1977; Evans and O'Shea, 1978

[a]A, amplitude; F, frequency; T, tonus; +, increase; −, decrease.

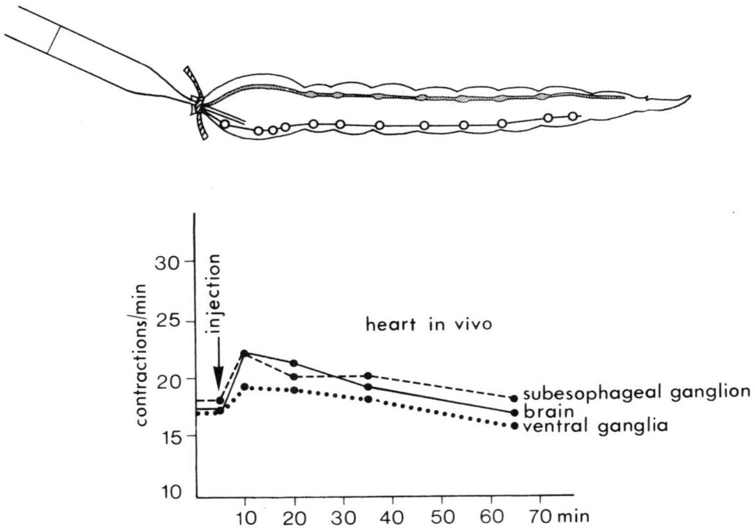

FIGURE 68. Effect of extracts of different parts of the central nervous system on the heartbeat of a decapitated *Corethra* larva. (Modified from Gersch, 1955.)

While the presence of antagonistic factors appears probable, many distinct myotropic factors also seem to exist. In some cases, contradictory results are obtained on different effectors by using the same extract or on the same effector by using different extracts. Hence, for example, brain and CC extracts do not exert the same action on the heart of *Periplaneta.*

The role of PO in cardioaccelerator release was demonstrated in *Periplaneta, Carausius,* and *Locusta* (Raabe *et al.*, 1966) (Fig. 69) and in the freshly molted adult and larva of *Manduca.* In the adult *Manduca,* two cardioacceleratory peptides (CAP1 and CAP2), which elicit a dose-dependent increase in heartbeat *in vitro,* are released from the abdominal PO immediately after eclosion, resulting in a marked rise in heartbeat frequency and facilitated wing inflation. In the larva also, two cardioacceleratory factors (CAF1 and CAF2) from the central nervous system are present in various amounts in the PO, which appear to be more loaded than the CC. The brain and posterior abdominal ganglia are richer than the thoracic ganglia (Platt and Reynolds, 1985).

As for the neurons producing the cardioacceleratory neurohormones, they were not identified except in the adult *Manduca,* where they were found to be located in the mediodorsal part of the abdominal ganglia (Fig. 70). They send their axons throughout the segmental nerves into the PO (Tublitz and Truman,

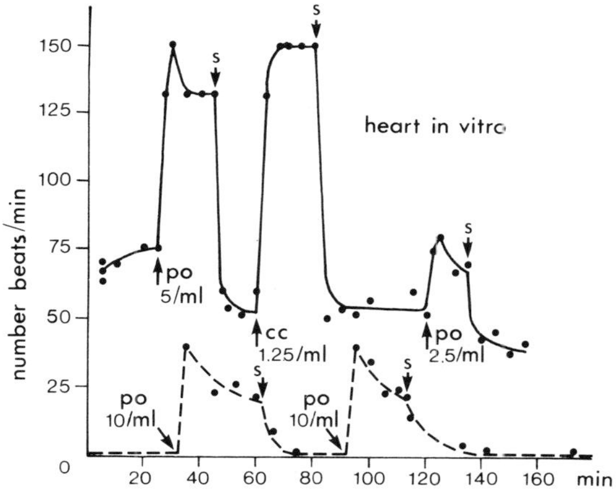

FIGURE 69. Effect of corpora cardiaca and perisympathetic organ extracts on the heartbeat in *Periplaneta.* Dashed lines represent the responses of an arrested heart. cc, Corpora cardiaca; po, perisympathetic organs; s, saline. (Modified from Raabe *et al.,* 1966.)

1985a). Intracellular stimulation of a CAP neuron evokes the release of a certain amount of CAP bioactivity from the PO.

A specific monoclonal antibody was raised against purified CAP1 and CAP2 from the abdominal nerve cord (Tublitz and Evans, 1986). In newly emerged adults, it reduces the CAP activity of the hemolymph and inhibits the increase in cardiac activity.

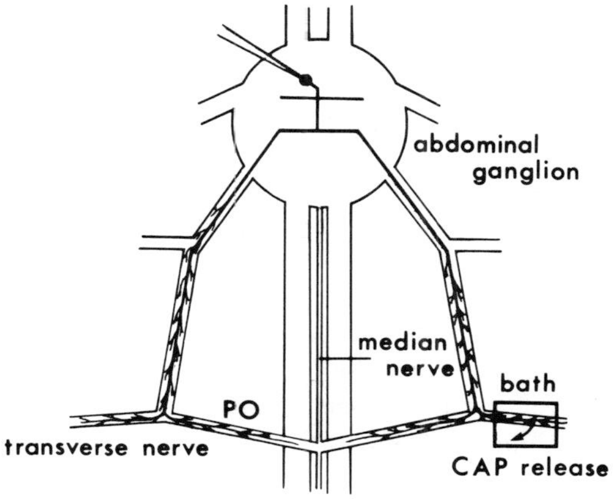

FIGURE 70. Release of CAP activity from the perisympathetic organs of *Manduca* following electrical stimulation of a CAP neuron. (Modified from Tublitz and Truman, 1985a.)

7.1.1.3. Heart and Biogenic Amines

The presence of biogenic amines has been demonstrated in the heart of locusts and cockroaches, the extracts of which contain large amounts of dopamine and serotonin (Hiripi and S.-Rózsa, 1973; Collins and Miller, 1977), while the lateral nerves of the heart of *Blaberus* are rich in dopamine and serotonin (Gersch *et al.*, 1974).

Biogenic amines were shown to display more or less pronounced effects on the heartbeat in *Periplaneta, Locusta,* and some other species, with an acceleration observed at low concentration (10^{-8} M) (see Miller, 1985). However, in certain species, such effects were obtained neither on the heart in *Manduca* (Platt and Reynolds, 1986), nor on the hyperneural muscle in *Periplaneta* (Miller and James, 1976), nor on the alary muscles of the same species (Miller and S.-Rózsa, 1981). In *Periplaneta,* the myotropic activity of biogenic amines upon the heart is as follows:

$$\text{Serotonin} > \text{epinephrine} > \text{octopamine} > \text{dopamine}$$

It should be noted that quite often the increase in the rate of rhythmic contractions induced by biogenic amines occurs only in denervated heart (Collins and Miller, 1977), suggesting the existence of an inhibitory nervous regulation.

With respect to acetylcholine, it was shown that its action is not exerted on either the myocardium or the denervated heart (Miller and Metcalf, 1968), but on the cardiac neurons, of which it increases the spontaneous activity (Miller, 1968a,b; Richter and Stürzebecher, 1969; Richter, 1973).

7.1.2. Gut

The muscles of the digestive tract include the circular and longitudinal muscles, and the contractility of the digestive tract is rather complex. Its different parts—the foregut, midgut, and hindgut—are independent and possess different innervation. The foregut and the anterior portion of the midgut are richly innervated by the nerves from the ganglia of the anterior sympathetic nervous system (frontal, hypocerebral, ingluvial, and proventricular ganglia), but the innervation of the hindgut originates in the last abdominal ganglion. Numerous sensory receptors, both tension receptors and osmotic receptors, are present in the pharynx, foregut, and hindgut.

The digestive tract exhibits myogenic activity. In other words, it displays spontaneous contractions in the absence of stimulation, but it is also dependent on the nervous system, which exerts a double regulation, both nervous and neurohumoral, as we shall see below.

7.1.2.1. Gut Myotropic Peptides

As early as 1955, Gersch showed that the foregut of *Corethra* displays antiperistaltic movements which are stimulated by excitations applied to the ventral nerve cord ganglia, even if the latter are totally disconnected. This suggested the existence of a myotropic humoral factor originating in the nervous system. This idea was also suggested by the work of Koller (1948) in *Tenebrio*, Cameron (1953) and Davey (1962b) in *Periplaneta*, Enders (1955) in *Carausius*, and Cazal (1969) in *Locusta* concerning the hindgut. A study carried out *in vitro* on the brain–CC complex also showed that the latter was able to release a hindgut-stimulating neurohormone into the incubation medium (Marks and Holman, 1974).

Myotropic activity on the gut was demonstrated within the head and the central nervous system, but also in the gut itself and in its nerves (Koller, 1954, Brown, 1967, in *Periplaneta;* Holman and Cook, 1972 in *Leucophaea*). Furthermore, it was shown that the disconnection of the proctodeal nerves results in the inactivation of hindgut homogenates. These important findings are at the origin of the identification of proctolin, the hindgut myotropic factor (see next section).

The effects of hindgut homogenates are very different from the effects of glutamate, the neurotransmitter involved in hindgut nervous regulation (Cook and Holman, 1975a,b). Their activity is maintained after tetrodotoxine treatment, indicating that they exert a direct effect on muscle cells.

7.1.2.2. Gut Myotropic Amines

Serotonin, dopamine, epinephrine, norepinephrine, and octopamine were assayed on both the foregut and hindgut of several insect species. On the whole, the most effective is serotonin, which enhances both foregut and hindgut contractions in all the species investigated but also causes relaxation depending on the doses used, as shown in *Schistocerca* (Banner *et al.*, 1987). As for the other biogenic amines, experiments are less numerous and the results are largely controversial.

Investigations conducted by the histofluorescence method do not reveal any serotonin presence in the midgut of *Aeshna* (Andriès and Tramu, 1985a), but dopamine was found in the foregut of *Schistocerca* (Klemm, 1972). The aminergic axons may originate in the sympathetic nervous system, where biogenic amines were shown to be present in cockroaches and locusts (Klemm, 1971, 1972; Chanussot *et al.*, 1969; Lafon-Cazal and Arluison, 1976; Plotnikova and Govyrin, 1966).

7.1.3. Oviducts

The oviducts, which are not always innervated, have, like most visceral muscles, the inherent ability to contract spontaneously and rhythmically. Their contractions are important for sperm transport to the spermatheca and for ovulation and oviposition.

The injection of extracts of different parts of the nervous system causes contractions of the oviduct, as shown *in vitro* in *Tenebrio* (Koller, 1954), *Carausius* (Enders, 1955), *Locusta* (Girardie and Lafon-Cazal, 1972), *Blabera* and *Glossina* (Raabe *et al.,* 1985), and *Gryllus* (Sefiani, 1986).

7.1.3.1. Myotropic Peptides and Oviducts

Oviduct contractions are also influenced by proctolin, which induces an elevation in the tonus and amplitude of phasic contractions at a concentration of 10^{-10} M in *Tabanus sulcifrons* (Cook and Meola, 1978; Cook, 1981). The oviduct of *Gryllus* also reacts *in vitro* to proctolin, which causes a strong increase in the tonus, an effect different from that of central nervous system homogenates (Sefiani, 1986). The role of proctolin was particularly well demonstrated in *Locusta:* in this species, proctolin causes elevated basal tension, increased amplitude of neurally evoked contractions, and a general stimulation of the myogenic activity of oviduct muscles. The peptide is present in the oviducal nerve, in the innervated parts of the oviduct, and in a few abdominal neurons (Lange *et al.,* 1986). Ultrastructural investigations performed in *Locusta* demonstrated peptidergic neuroeffector junctions in the wall of the oviduct, suggesting a neuromodulatory role for the transported neuropeptides (Kiss *et al.,* 1984).

7.1.3.2. Myotropic Amines and Oviducts

Biogenic amines influence oviduct and uterus contractions, and their presence in these organs was demonstrated in *Locusta* (David and Lafon-Cazal, 1979; Lange and Orchard, 1984) and in the uterus of *Blabera* (Raabe *et al.,* 1985). In *Tabanus,* besides the increase in tonus caused by proctolin, strong effects were obtained with serotonin, which increases not only the tonus but also the frequency and amplitude of the phasic contractions (Cook and Meola, 1978; Cook, 1981).

The denervated oviduct of *Gryllus* reacts to octopamine, dopamine, and serotonin, which increase the frequency and tonus and decrease the amplitude. The activity threshold of octopamine is about 5×10^{-8} M whereas that of dopamine and serotonin is about 2×10^{-5} M (Sefiani, 1986).

In *Locusta,* octopamine reduces the amplitude of the neurally evoked contractions of the oviduct in a dose-dependent manner, inhibits myogenic contractions, and causes release of the basal tonus. Serotonin and norepinephrine increase the amplitude of neurally evoked contractions (Orchard and Lange, 1985) and raise the frequency of myogenic contractions.

The presence of octopamine has been demonstrated in the oviduct, in the oviducal nerves, and in certain neurons (Lange and Orchard, 1984; Orchard and Lange, 1985). The latter are two unpaired median dorsal neurons termed the DUMOV (see Chapter 1). Their electrical stimulation causes an elevation in the cAMP content of the oviduct (Lange and Orchard, 1986) (see Chapter 6). Octopamine, which acts at the postsynaptic level, appears to play a neuromodulatory role.

A number of experiments were performed with the uterus of the ovoviviparous cockroach *Blabera* and the viviparous fly *Glossina fuscipes* (Raabe *et al.*, 1985). The experiments showed a clear effect of serotonin on the uterus of *Blabera,* but this was not observed in *Glossina.* In both insects, however, extracts of the nervous system clearly enhance both the frequency and amplitude of contractions.

7.1.4. Malpighian Tubules

The Malpighian tubules often display considerable motility, owing to the contraction of the thin muscle strips that form spirals around them. They appear to be innervated only in some insects. The role of myotropic factors was only investigated in *Periplaneta* (Cameron, 1953) and *Carausius* (Enders, 1955; Pilcher, 1971). Both brain and CC homogenates as well as epinephrine and serotonin were shown to be effective.

7.1.5. Body Wall Muscles and Blood Pressure

Significant changes in hemolymph pressure take place during insect life, particularly at ecdysis (Sláma, 1976; Provansal *et al.*, 1977). They are due to the contraction of the body wall muscles, which is influenced by 20-HE (Sláma, 1984), but not by homogenates of the PO, at least in *Tenebrio* (Provansal-Baudez and Sláma, 1985). It has also been shown that proctolin acts on the body wall muscles of *Lucilia* larvae, which respond to the peptide by a postsynaptic rhythmic depolarization, an effect also obtained with octopamine and which persists after tetrodotoxin treatment (Irving and Miller, 1980).

7.2. SKELETAL MUSCLES

The activity of proctolin and biogenic amines is not limited to visceral muscles. It was also demonstrated for certain skeletal muscles, such as the locust extensor tibiae (ETi) (Fig. 71).

The extensor tibiae displays a spontaneous contraction rhythm which results from the activity of a small tonic bundle of muscles located at the proximal end of the leg. The myogenic rhythm of the tonic bundle of the extensor tibiae has been used extensively as a sensitive bioassay for myotropic factors and especially proctolin.

7.2.1. Proctolin

Surprisingly enough, when proctolin was assayed on the extensor tibiae myogenic bundle, it was shown to be active and to cause rhythmic contractions and relaxations at nanomolar concentrations (Piek and Mantel, 1977). An increase in the frequency of the contractions was observed at low concentrations, and a permanent contracture at high concentrations. Furthermore, proctolin mi-

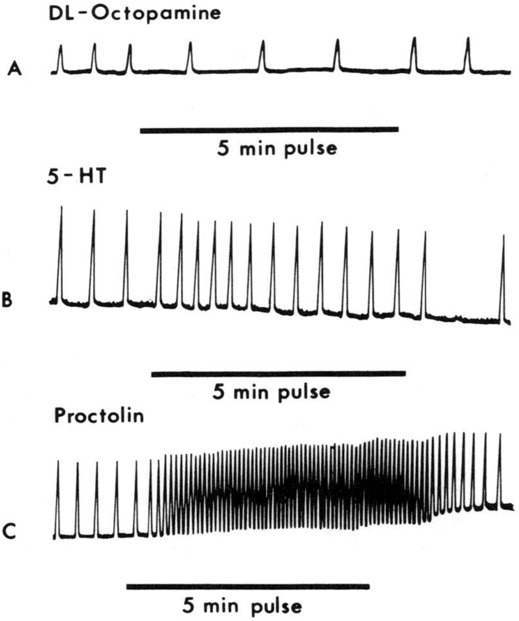

FIGURE 71. Effects of DL-octopamine, serotonin (5-HT), and proctolin on the myogenic activity of the locust extensor tibiae bundle. (Modified from Evans, 1984a.)

croiontophoretic application to the extensor tibiae causes the appearance of depolarizing potentials (May *et al.*, 1979).

7.2.2. Other Peptides

The effect of other insect neurohormones, such as AKH, MI, and MII, on the contractility of the locust extensor tibiae has been reported (O'Shea *et al.*, 1984). In addition, FMRFamide, a molluscan peptide that exerts a modulatory role in the heartbeat of *Aplysia,* was shown to act on the locust extensor tibiae bundle. The FMRFamide effect on this muscle resemble that of octopamine, but it is mediated by a different receptor system and does not involve cAMP (Evans and Myers, 1986).

The occurrence FMRFamide-like neurons, as well as axon terminals in nh organs, were described recently in several insect species (Myers and Evans, 1985) (see Chapter 2).

The C-terminal sequence of FMRFamide is identical to that of two recently identified myotropic peptides, LSK I and LSK II. Hence it may be suggested that these two peptides are responsible for the effects observed in the extensor tibiae.

7.2.3. Octopamine

Dorsal unpaired octopaminergic neurons (DUM) have been identified in locusts (Fig. 72). They are located in the midline of the thoracic and abdominal ganglia, and their axons bifurcate, sending projections into the left and right nerves on each side of the body. In the metathorax, one of these neurons, DUMETi, was shown to innervate the extensor tibiae, where ns endings are observed (Hoyle *et al.*, 1974). Its stimulation inhibits the spontaneous rhythm of the muscle (Evans and O'Shea, 1978; O'Shea and Evans, 1979) and causes the release of octopamine (Morton and Evans, 1984). The perfusion of 10^{-9} M octopamine on the extensor tibiae bundle mimics their stimulation (May *et al.*,

FIGURE 72. Identification of various neurons innervating the extensor tibiae bundle: GABA common inhibitory motoneurons (CL); octopaminergic dorsal unpaired motoneurons (DUM); fast extensor tibiae (FETi) and slow extensor tibiae glutaminergic motoneurons (SETi). Note the colocation of proctolin and glutamate in the SETi motoneurons. (Adapted from the data of Evans and O'Shea, 1978.)

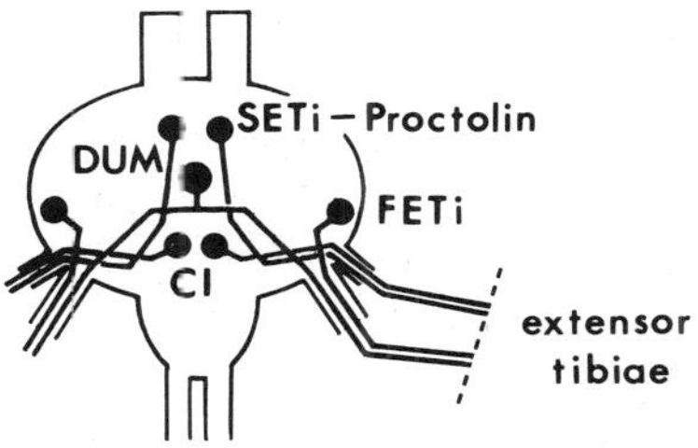

1979). The DUMETi neurons contain about 0.1 pmol of octopamine (Evans and O'Shea, 1978).

In this case, the action mechanism of octopamine is that of a modulatory local neurohormone, which facilitates the amount of neurotransmitter released by the SETi neurons, acting at the presynaptic receptor level (Evans and O'Shea, 1978; O'Shea and Evans, 1979).

7.3. SEPARATION AND IDENTIFICATION OF MYOTROPIC PEPTIDES

The purification and separation of myotropic factors have been attempted by several groups using different methods, so that the data are abundant.

7.3.1. Early Investigations

Since the 1960s/1970s, many investigations have been devoted to the isolation of insect myotropic neurohormones. Most of them used the semi-isolated heart of *Periplaneta* as an assay for active factors and the CC as the donor organ, but the central nervous system, as well as the hindgut, was employed as an effector and also as a donor organ.

Brown (1965) separated several factors from the CC of *Periplaneta,* the first acting on the heart, the second on the hindgut, and the third on both organs. In the CC of the same insect, four cardioaccelerators were discovered distinct from the hypertrehalosemic factors (Natalizi and Frontali, 1966; Natalizi *et al.,* 1970; Traina *et al.,* 1974, 1976). In locusts, the nh and glandular lobes of the CC were fractionated separately (Mordue and Goldsworthy, 1969). Two fractions were identified which displayed both cardioacceleratory and hypertrehalosemic activity.

The factors of the central nervous system have stimulated less research than those of the CC. Ralph (1962b) and Rounds (1963), using aqueous and organic solvents and alcohol-water mixtures in variable proportions, respectively, found several accelerators and also decelerators in the central nervous system of *Periplaneta.* Substances active on the heart have been isolated from the last abdominal ganglion. One appears to be identified as acetylcholine, while another one, which increases the heartbeat amplitude, is a peptide with a molecular weight of 3500–7500, which may be identical to one of the factors that we shall now discuss (Krolak *et al.,* 1977).

Several investigations concerning the factors present in the central nervous system were carried out by Gersch and co-workers (see Section 7.3.3). Three substances active on the heart of *Periplaneta*—factors C, D, and acetycholine—were first found in the central nervous system and in the endocrine glands of this insect. It subsequently appeared that factors C and D each decomposed into two factors (Gersch *et al.*, 1960), giving four factors—factors C_1, C_2, D_1 and D_2—which increased heartbeat frequency but differed by their action on heartbeat amplitude, which was increased by factor D_1, reduced by factor C_1, and remained unaffected by factors C_2 and D_2. Factors C_1 and D_1 were also shown to act not only on the heartbeat but also on pigment movements in the stick insect, and on the spontaneous nerve activity of the phallic nerve in the cockroach (Gersch *et al.*, 1957; Gersch and Richter, 1963).

The results of the foregoing investigations suggest that many factors exist and that there is a correspondence between some of them. Hence Brown's factor P_3 may be identical to Gersch's factor C, factor 2 of Traina *et al.*, and factor B of Mordue and Goldsworthy. Brown's factor P_1 may be identical to Gersch's factor D, factor 1 of Traina *et al.*, and factor C of Mordue and Goldsworthy, which were probably identical to factors identified since 1984. Cross-injection experiments also reveal that distinct species may possess comparable or identical factors. Corpora cardiaca and brain extracts of *Carausius* are active on the contractility of the Malpighian tubules of different coleopterans, and especially of *Dysticus* (Koller, 1954). The CC extracts of *Locusta* stimulate contractility of the different parts of the digestive tract of *Blabera* and *vice versa* (Cazal, 1969). Extracts of PO of *Periplaneta, Carausius, Clitumnus,* and *Locusta* are active on the heart of these four species (Raabe *et al.*, 1966). The different parts of the central nervous system of the coleopteran *Melolontha* stimulate peristalsis of the digestive tract in *Corethra,* and act in the same manner on the heart of *Corethra* and *Periplaneta* (Kirchner, 1962). A substance separated from the head of the honeybee is active on the heart of *Periplaneta* (Natalizi and Frontali, 1966).

7.3.2. Proctolin

7.3.2.1. Discovery

The most thorough purification was finally achieved by using the effector itself, in this case the hindgut, as a source of active factor. This succeeded in identifying a basic pentapeptide, proctolin (Brown, 1975; Brown and Starratt, 1975; Starratt and Brown, 1975), capable of stimulating contractions of the hindgut in the cockroach *Periplaneta*. Proctolin has a molecular weight of 648. It is destroyed neither by chymotrypsin nor by trypsin, nor by carboxypeptidases A and B, but is destroyed by leucine aminopeptidase. Its formula is

Arg-Tyr-Leu-Pro-Thr

It was synthesized in 1977 by Starratt and Brown.

The proctolin unit, defined earlier as the quantity present in the rectum, was finally calculated, and corresponds to 0.86 ng. Its activity threshold on the proctodeal muscle *in vitro* is about 10^{-9} M (Starratt and Brown, 1975) for *Periplaneta*.

Proctolin causes slow-type, graded contractions of the longitudinal muscles of the hindgut, but above a threshold concentration of about 10^{-9} M, it evokes sustained slow-type, graded contractions. Concentrations of 2–8 $\times$ 10^{-10} M potentiates the graded responses and proctolin-induced responses are suppressed by 70–90% by 15-sec preincubation with tyramine at 2 $\times$ 10^{-6} M (Brown, 1975).

In *Leucophaea*, the threshold is even higher (3 $\times$ 10^{-11} M), and two distinct effects can be observed depending on the amount of proctolin used, i.e, an increase in frequency and amplitude of the simple phasic contractions at low concentration, and a tonic effect at higher concentration (Cook and Holman, 1979).

7.3.2.2. Proctolin Effects

Besides its effects on the hindgut, proctolin also modifies the contractility of several other effector muscles. It evokes or increases the contractions of the heart (Fig. 73); the hyperneural muscle (Miller and James, 1976; Hertel and Penzlin, 1986); the antennal heart of *Periplaneta* (Hertel *et al.*, 1985); the body wall muscles in *Lucilia sericata* larvae (Irving and Miller, 1980); the lateral oviduct of *Tabanus* (Cook, 1981), *Locusta* (Lange *et al.*, 1986), and *Gryllus* (Sefiani, 1986); the foregut of *Leucophaea* (Cook and Holman, 1978); and skeletal muscles such as the myogenic bundle of the locust extensor tibiae, and the cockroach coxal depressor. Certain effectors, however, are not influenced by proctolin, i.e., the alary muscle of *Locusta* (Miller and S.-Rozsa, 1981) and the heart of *Manduca* (Platt and Reynolds, 1986) and *Stomoxys* (Cook and Meola, 1983).

7.3.2.3. Proctolin Origin Neurons

As shown above, proctolin was discovered in homogenates of the hindgut in *Periplaneta*. It is also found in the nervous system as shown in RIA studies performed in some species. Proctolin distribution in the nervous system of *Leucophaea, Schistocerca* (Bishop *et al.*, 1981), and *Acheta* (Kingan and Titmus, 1983) is shown in Table XXII. Higher levels are usually observed in the thoracic ganglia and in the last abdominal ganglion than in other abdominal ganglia.

Proctolinergic neurons were identified using intracellular dye injection, RIA, and bioassay with the extensor tibiae preparation for both the soma of the neurons and the perfusate of their terminals at the muscle level after electrical stimulation of the cell body; several neurons were thus identified.

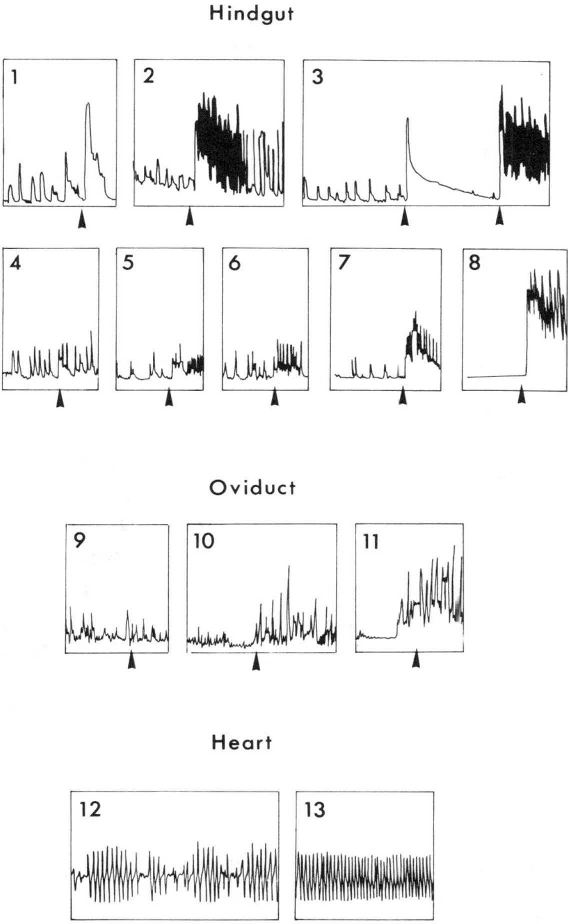

FIGURE 73. *In vitro* responses of visceral muscles to proctolin in *Leucophaea* (1–8, hindgut; 9–11, oviduct; 12–13, heart). (1) A single phasic contraction evoked by the addition of glutamate. (2) Sustained response of the same preparation to proctolin (3×10^{-10} M). (3) Response of the hindgut to sequential treatment with glutamate (1×10^{-4} M, first arrow) and then proctolin (3×10^{-10} M, second arrow). (4–8) Hindgut responses to increasing concentrations of proctolin: 1×10^{-11} M; 3×10^{-11} M; 6×10^{-11} M; 3×10^{-10} M; 2×10^{-9} M. (9–11) Oviduct responses to increasing concentrations of proctolin: 1.2×10^{-10} M; 2.5×10^{-10} and 6×10^{-10} M. (12) Heartbeat control in semi-isolated heart. (13) Addition of proctolin makes the rhythm regular. (Modified from Cook and Holman, 1985.)

TABLE XXII. Proctolin Concentration in the Central Nervous System of Some Insects[a,b]

	Proctolin concentration (pmole/organ)					
Insects	Brain	Th1	Th2	Th3	a 1-5	a6
Leucophaea	0.20	0.45	0.54	0.55	0.11	0.68
Schistocerca	0.47	←	0.37	→	0	0.17
Acheta	1.50	←	0.85	→	0.07	0.74

[a]From the data of Kingan and Titmus, 1983.
[b]a, abdominal ganglion; Th1, 2, 3, prothoracic, mesothoracic, metathoracic ganglia.

Proctolinergic neurons innervating the extensor tibiae were shown to be identical to the SETi glutaminergic neurons as indicated above. Thus proctolin is produced in the same neuron as glutamate and released in the same sites. This important finding reveals a new mode of neuromuscular regulation, i.e., cotransmission (Adams and O'Shea, 1983).

The slow coxal depressor motoneurons (DS) of the skeletal muscles of the proximal segment of the leg of *Periplaneta* are also proctolinergic (O'Shea and Bishop, 1982), as are two lateral white neurons (LW) of the abdominal ganglia of the same species (O'Shea and Adams, 1981) (Table XXIII).

TABLE XXIII. Some Identified Neurons and Their Targets[a]

Target	Identified neurons	Factor involved	Role	Reference
Locust	DUMET, Dorsal unpaired motoneurons to extensor tibiae	O	nm	Evans, 1981, 1984a,b
Extensor tibiae	SETi, Slow excitatory motoneurons	P Glu	nm nt	Evans and Myers, 1986
	FET, Fast excitatory motoneurons	Glu	nt	
	CI, Commun inhibitory motoneurons	GABA	nt	
Locust oviduct	DUMOV, Dorsal unpaired median oviducal neurons	O	nm	Lange *et al.*, 1986
	Protolin-like neurons	P	nm	
Cockroach heart	LW, Lateral white neurons	P		O'Shea and Adams, 1981
Cockroach coxal depressor	DS, Coxal slow depressor	P	nm	Adams and O'Shea, 1983

[a]Glu, glutamate; nh, neurohormone; nm, neuromodulator; nt, neurotransmitter; P, proctolin; O, octopamine.

Electrical stimulation of DS motoneurons causes two types of response (Adams and O'Shea, 1983). The first transient response appears to be due to the activity of glutamate. It causes brief excitatory junctional potentials, followed by a transient increase in force. The second is slower due to proctolin.

Immunocytochemical investigations of the thoracic and last abdominal ganglia of *Periplaneta* demonstrated a large number of proctolin-like neurons (Eckert *et al.*, 1981; Bishop *et al.*, 1981; Agricola *et al.*, 1985). They projected into the proctodeal nerve, in other nerves and in the neuropil. In *Leptinotarsa* (Eckert *et al.*, 1981; Veenstra *et al.*, 1985b), proctolin-like terminals were found in the muscles of the hindgut and foregut.

7.3.2.4. Proctolin Ubiquity

Proctolin has proved to be present in several insect species, i.e., *Periplaneta, Locusta, Gryllus, Musca,* the coleopteran *Sitophilus granarius, Rhodnius, Apis,* and *Malacosoma americana* (Brown, 1977).

It was also identified in crustaceans (Bishop *et al.*, 1984; Schwartz *et al.*, 1984; Siwicki and Bishop, 1985), in other invertebrates (Li and Calabrese, 1985), and in rat brain (Holets *et al.*, 1984). Hence it seems to play an important role in neuromediation.

In crustaceans it plays a modulatory role on the heart (Watson *et al.*, 1983), skeletal muscles (Rane *et al.*, 1984), and in the cardiac ganglion of the lobster (Sullivan and Miller, 1984), but it is also present in the nh pericardial organs (Sullivan, 1979; Schwartz *et al.*, 1984), indicating that it also exerts a neurohormonal function.

7.3.2.5. Proctolin Analogs

Several proctolin analogs were synthesized and their biological activity was investigated on several effectors, i.e., the hindgut (Starratt and Brown, 1979; Sullivan and Newcomb, 1982), the heart (Konopinska *et al.*, 1986), and the locust extensor tibiae (Piek *et al.*, 1979).

The myotropic activity of proctolin appears to have strict structural requirements, each amino acid being necessary for biological activity as well as the free carboxyl group. The presence of the position 2 tyrosine hydroxyl group, however, is not necessary for activity to be present (Starratt and Brown, 1979; Sullivan and Newcomb, 1982). It was also shown that halogenation at position 4 of the tyrosyl residue caused a slight increase effect and prolonged the action of proctolin after washout. This could result from an alteration of proctolin linkage to its receptors (Starratt and Brown, 1979).

Five synthetic proctolin analogs modified in position 2 were investigated on both cockroach and *Tenebrio* for their cardioacceleratory activity. The latter was mostly found to be significantly increased by replacement of the HO group in the Tyr moiety by nitrogen atom-containing groups such as NH_2, NO_2, NMe_2, or OMe. These effects could result from an increase in proctolin resistance against hydrolytic enzymes (Konopinska *et al.*, 1986). The results obtained with the extensor tibiae are different and suggest that more than one class of proctolin receptors might occur (Piek *et al.*, 1979).

As for proctolin antagonists, it was shown that tyramine exerts antagonistic effects at a concentration of 3×10^{-8} to 3×10^{-7} (Brown, 1975; Cook and Holman, 1979).

7.3.2.6. Proctolin Inactivation

Understanding of the role of proctolin requires the knowledge of its degradation modes in the hemolymph. This question was investigated in *Periplaneta* and *Manduca* (Starratt and Steele, 1984; Steele and Starratt, 1985; Quistad *et al.*, 1984).

The half-life of injected proctolin is less than 5 min, which definitely shows that it cannot be transported in the hemolymph without protection. The peptide appears to be inactivated in three distinct pathways, with cleavage at Pro-Thr bond at pH 6.0, Tyr-Leu bond at pH 8.0, and Arg-Tyr bond at both pH (Starratt and Steele, 1984; Steele and Starratt, 1985). These results are of particular interest considering that the pH of the hemolymph of *Periplaneta* varies according to the activity of the insect and that the rapidity of the different degradation modes is different at different pH.

Proctolin metabolism was also investigated in several tissues containing high levels of this peptide (Quistad *et al.*, 1984). The metabolites of proctolin breakdown are mostly [^{3}H]Arg-Tyr. The degradation rate appears to be faster in proctolin-rich tissue, i.e., the proctodeum and the terminal abdominal ganglion. This may be due to different enzymatic cleavages, or to the presence in these tissue of larger enzymatic pools. Most of the proteolytic activity appears to be present in a soluble fraction not membrane-bound.

7.3.3. MI, MII, CC1, and CC2, and Neurohormone D

Several myotropic neurohormones have been identified in recent years, probably corresponding to factors isolated in early investigations. These are factors of the CC of *Periplaneta,* one of which is merely neurohormone D, as well as factors present in the head of *Leucophaea,* the activity of which is exerted particularly on the hindgut (Table XXIV).

The cardioacceleratory neurohormone D from the CC of *Periplaneta,* purified in the 1960s as shown above, was identified by dansyl Edman degradation and liquid phase sequencing (Baumann and Gersch, 1982; Baumann and Penzlin, 1984). The following sequence was proposed:

pGlu-Val-Asn-Phe-Ser-Pro-Asn-(Gly/Trp)

with a molecular weight of 750 –1000 (Baumann and Gersch, 1982).

The molecular structure of two neuropeptides from the CC of the same cockroach was identified at the same time. These were two octapeptides designated MI and MII (O'Shea *et al.*, 1984; Witten *et al.*, 1984), CC1 and CC2 (Scarborough *et al.*, 1984). Their identification involved the following biochemical methods: reversed phase HPLC, enzymatic digestion, amino acid sequence analysis, fast atom bombardment, mass spectrometry, chymotryptic digestion, and comparison with synthetic samples:

MI, CC1	pGlu-Val-Asn-Phe-Ser-Pro-Asn-Trp-NH_2
MII, CC2	pGlu-Leu-Thr-Phe-Thr-Pro-Asn-Trp-NH_2
AKHI	pGlu-Leu-Asn-Phe-Thr-Pro-Asn-Trp-Gly-Thr-NH_2

The characterization of the structure of both neurohormones, with molecular weight estimated at 972 for MI and 987 for MII, revealed that they were not structurally related to proctolin but that they belong to the AKH family and show considerable homology with the prawn red pigment concentrating hormone (RPCH), and with the adipokinetic hormone of locusts (AKH). All these neurohormones are N-protected by a pyroglutamic residue and C-protected by a primary amide ($-NH_2$). MI and MII are relatively hydrophobic and exhibit fluorescent absorbence peaks. A sequence homology was observed between MII and the NH_2 terminal portion of glucagon, which is not really surprising since immunoreactive products to this vertebrate hormone are found in the CC of some insects (see Chapter 2).

Apart from their cardioacceleratory role, the two myotropic neurohormones appear to possess other activities that could possibly result from a similarity of molecular structure, both to one another and to other neurohormones such as AKH. In fact, factors MI and MII, tested for their hyperglycemic activity in *Periplaneta* and for their lipid-mobilizing effects in *Schistocerca,* prove to be effective although less active than AKH. Factors MI and MII also display myotropic activity on the extensor tibiae (O'Shea *et al.*, 1984), their activity thresholds being 10^{-8} (MI) and 5.10^{-9} (MII), respectively (see Chapter 11).

The metabolic degradation of the factor MII was investigated. In the hemolymph the rate of degradation is slow. It is more rapid in contact with Malpighian tubules or fat body homogenates (Skinner *et al.*, 1987).

TABLE XXIV. Effects of Some Isolated Fractions or Molecules on Heartbeat and Hindgut Contraction[a]

Insect, reference	Factor origin	Factor	Molecular weight	Heartbeat	Hindgut contraction
Locusta migratoria and *Schistocerca gregaria*	CC glandular lobe	B		++	
Mordue and Goldsworthy, 1969	CC neurohemal lobe	C		+	
Periplaneta americana	Central nervous system	C1		+	
Gersch *et al.*, 1960		C2		+	
Gersch and Richter, 1963		D1	2000	+	
		D2		+	
Traina *et al.*, 1976	CC or head	1a		+	
		1b		+	
		2a	1795	+	
		2b	1287	+	
Brown and Starratt, 1975	Rectum	Proctolin	500–700		+
Periplaneta americana					
O'Shea *et al.*, 1984	CC	MI CC1	972	+	
Scarborough *et al.*, 1984		MII CC2	987	+	
Periplaneta americana					
Shukle and Judson, 1984	CC	Ic		0	+
		II			+

Periplaneta americana					
Langbein, 1981	Thoracic ganglia				
Leucophaea maderae					
Holman and Cook, 1979; 1983	Head	HSN	400–600		+
Holman *et al.*, 1984					
Holman, *et al.*, 1986a		Leucokinin I	892		+
		Leucokinin II	851		+
Holman *et al.*, 1986b		Leucokinin III	909		+
		Leucokinin IV	905		+
Holman *et al.*, 1986c		Leucopyrokinin			+
Holman *et al.*, 1986d		Leucomyosuppressin	1258		–
Holman *et al.*, 1987a		Leucokinin V	783		+
		Leucokinin VI	917		+
Holman *et al.*, 1987b		Leucokinin VII			+
		Leucokinin VIII			+
Nachman *et al.*, 1986a		Leucosulfakinin I			+
Nachman *et al.*, 1986b		Leucosulfakinin II			+
Manduca sexta adults and larvae					
Tublitz and Truman, 1985b	Central nervous system	CAP_1		+	
		CAP_2		+	
Platt and Reynolds, 1985		CAP_1	4000	+	
		CAP_2	<1000	+	

[a]CC, corpora cardiaca; HSN, hindgut-stimulating neurohormone.

The release of both MI and MII neurohormones is calcium-dependent and occurs in the CC. Their production sites are probably located in the central nervous system.

The presence of peptides identical or similar to MI and MII was investigated in other insects, *Galleria* and *Bombyx,* in a crustacean, the crayfish, and in the rat. In *Galleria* the brain is slightly active but the CC are not, whereas in *Bombyx* both organs are devoid of MI. The results differ for the eyestalk of the crayfish. The brain and hypophysis of the rat are devoid of activity (O'Shea *et al.*, 1984).

7.3.4. Other Factors

Preliminary investigations were conducted in the adult and larva with cardioacceleratory peptides (CAP1 and CAP2, and CAF1 and CAF2) from *Manduca* (Platt and Reynolds, 1985). These factors are thermostable and are destroyed by pronase (CAP1 and CAP2) and α-chymotrypsin, but not by trypsin (CAF1). CAF2 resists both α-chymotrypsin and trypsin. CAF1 and CAF2 are different from serotonin, octopamine, proctolin, neurohormone D, and FMRF-amide. Their molecular weights are 4000 and less than 1000, respectively.

Two small basic peptides termed Ic and II from the CC of *Periplaneta* stimulate hindgut contraction activity, but not the heart (Shukle and Judson, 1984). Their responses with enzymes were different. The activity of both was abolished by chymotrypsin, but trypsin and carboxypeptidase A abolished the activity of Ic only.

7.4 LEUCOKININS, LEUCOPYROKININS, LEUCOMYOSUPPRESSIN, AND LEUCOSULFAKININS

Investigations conducted in *Leucophaea* showed that proctolin was not the only myotropic factor acting on the hindgut, and that a second hindgut-stimulating neurohormone (HSN) occurs in the head and probably in the central nervous system (Holman and Cook, 1979), no proctolin being detectable in head extracts. The properties of the HSN were investigated (Holman and Cook, 1983; Holman *et al.*, 1984) and described as slightly different from those of proctolin. Thermostability and inactivation by enzymes were reported and four distinct factors separated.

In 1986–1987, the molecular structures of 12 different myotropic peptides acting on the hindgut of *Leucophaea* were identified (Holman *et al.*, 1986a–d, 1987a,b; Nachman *et al.*, 1986a,b). These peptides all originate in the head of this cockroach and stimulate hindgut contractions, except for one of them, leucomyosuppressin (LMS), which is inhibitory. These new peptides are called leucokinin I–VIII, leucopyrokinin, and leucosulfakinin I and II.

The methods used for these identifications were HPLC, enzymatic inactivation, amino acid analysis, primary sequence determination, and sometimes fast-atom bombardment mass spectrometry.

The molecular weights of the different factors and the activity thresholds of the natural and synthetic substances are given in Table XXV. Note that the synthetic compounds have approximately the same activity as the natural compounds, and that the activity thresholds of the different principles are of the same order of magnitude.

These substances thus include seven octapeptides and three decapeptides, some of which, such as LI, II, III, IV, V, VI, VII, and VIII, resemble each other closely, whereas others are quite dissimilar. In some cases, a similarity is observed with members of the AKH family. The blockade of the N terminal by pyroglutamic acid, which is frequently found in insect neurohormones, is present here only in leucokinin VI and leucopyrokinin, leucomyosuppressin, and leucosulfakinin II. The other leucokinins are not blocked at the amino terminus. On the other hand, the carboxyl terminal is always amidated as in the AKH family, which does not occur for proctolin:

Leucokinin I (LI)	Asp-Pro-Ala-Phe-Asn-Ser-Trp-Gly-NH_2
Leucokinin II (LII)	Asp-Pro-Gly-Phe-Ser-Ser-Trp-Gly-NH_2
Leucokinin III (LIII)	Asp-Gln-Gly-Phe-Asn-Ser-Trp-Gly-NH_2
Leucokinin IV (LIV)	Asp-Ala-Ser-Phe-His-Ser-Trp-Gly-NH_2
Leucokinin V (LV)	Gly-Ser-Gly-Phe-Ser-Ser-Trp-Gly-NH_2
Leucokinin VI (LVI)	pGlu-Ser-Ser-Phe-His-Ser-Trp-Gly-NH_2
Leucokinin VII (LVII)	Asp-Pro-Ala-Phe-Ser-Ser-Trp-Gly-NH_2
Leucokinin VIII (LVIII)	Gly-Ala-Ser-Phe-Tyr-Ser-Trp-Gly-NH_2
Leucopyrokinin (LPK)	pGlu-Thr-Ser-Phe-Thr-Pro-Arg-Leu-NH_2
Leucomyosuppressin (LMS)	pGlu-Asp-Val-Asp-His-Val-Phe-Leu-Arg-Phe-NH_2
Leucosulfakinin I (LSKI)	Glu-Gln-Phe-Glu-Asp-Tyr-SO_3H-Gly-His-Met-Arg-Phe-NH_2
Leucosulfakinin II (LSKII)	pGlu-Ser-Asp-Asp-Tyr-SO_3H-Gly-His-Met-Arg-Phe-NH_2

Leucokinins display strong homologies. The amino termini are not blocked except for LVI and the carboxyl terminals are amidated. Moreover, the eight peptides share four amino acids in positions 4, 6, 7, and 8, and Pro_2 in LI, LII, and LVII, Gly_3 in LII, LIII, and LV, and an Asn_5 in LI and LIII.

Leucopyrokinin (LPK) is quite different from the leucokinins. Its carboxyl and amino terminals are blocked. In addition, it contains an amino acid sequence which is also present in factor MI:

LPK	pGlu-Thr-Ser-Phe-Thr-Pro-Arg-Leu-NH_2
MI	pGlu-Leu-Thr-Phe-Thr-Pro-Asn-Trp-NH_2

TABLE XXV. Molecular Weight and Threshold Concentration of Myotropic Peptides[a]

Peptide	Molecular weight	Threshold concentration		Head content	
		Natural molecule	Synthetic molecule	ng/head	M/head
LI	892	2.1×10^{-10} M	2.0×10^{-10} M	0.43	0.48
LII	851	1.8×10^{-10} M	1.6×10^{-10} M	0.30	0.35
LIII	909	9.4×10^{-10} M	1.2×10^{-10} M	0.20	0.22
LIV	905	7.2×10^{-10} M	1.4×10^{-10} M	0.21	0.23
LV	783	4.5×10^{-11} M	5.2×10^{-11} M	0.13	
LVI	917	4.9×10^{-11} M	5.6×10^{-11} M	0.15	
LVII			1.3×10^{-10} M		0.09
LVIII			2.8×10^{-11} M		0.06
LPK		6.2×10^{-10} M	6.0×10^{-10} M	1.27	1.36
LMS	1257.7	8.37×10^{-11} M	7.75×10^{-11} M	0.23	0.23
LSKI		1.9×10^{-10} M	2.2×10^{-10} M	0.5	0.34
LSKII		4.2×10^{-11} M	4.5×10^{-11} M	1.6	

[a]From the data of Holman *et al.*, 1986a–d, 1987a,b; and Nachman *et al.*, 1986a,b.

Leucomyosuppressin, the inhibitory factor, is quite different from factors MI and MII, but interestingly enough it shares its carboxyl terminal sequence with the molluscan cardioaccelerator FMRFamide. This is important because FMRFamide-like neurons were recently demonstrated in insects:

LMS	pGlu-Asp-Val-Asp-His-Val-Phe-Leu-Arg-Phe-NH_2
FMRFamide	Phe-Met-Arg-Phe-NH_2

Both leucosulfakinins have an important property: the possession of a tyrosine sulfate ester. Their molecular structures are closely comparable since they share a blocked C-terminal heptapeptide. Their N terminus is blocked by a pyroglutamyl residue in LSKII only. The synthetic analogs of LSKI and LSKII were observed to be active on the midgut:

LSKI	Glu-Gln-Phe-Glu-Asp-Tyr(So_3H)-Gly-His-Met-Arg-Phe-NH_2
LSKII	pGlu-Ser-Asp-Asp-Tyr(So_3H)-Gly-His-Met-Arg-Phe-NH_2
CCK12	Ile-Ser-Asp-Arg-Asp-Tyr(SO_3H)-Met-Gly-Trp-Met-Asp-Phe-NH_2
GastrinII	Glu-Glu-Glu-Glu-Ala-Tyr(SO_3H)-Gly-Trp-Met-Asp-Phe-NH_2
Coerulin	pGlu-Glu-Asp-Tyr(SO_3H)-Thr-Gly-Trp-Met-Asp-Phe-NH_2
FMRFamide	Phe-Met-Arg-Phe-NH_2

LSKI and II exhibit sequence homology with FMRFamide and with the vertebrate peptides gastrin, CCK, and coerulin. LSKII residues Asp_3, Gly_6,

Met_8, and Phe_{10} are identical to those in positions 5, 8, 10, and 12 of CCK, while the similarity between LSKI and CCK is slightly weaker. The residues $Tyr(So_3H)_5$, Gly_6, Met_8, and Phe_{10} of LSKII are identical to those in positions 12, 13, 15, and 17 of human gastrin. It is thus possible that both these peptides are synthesized in neurons which react with the antibodies raised against G/CCK, and FMRFamide. Such neurons were described in the central nervous system and CC of several species (see Chapter 2).

The activity thresholds of myotropic peptides are low and their amounts vary from 0.21 to 1.6 ng/head, with a maximum level of 1.6 and 1.27 for LSKII and LPK, respectively, and a much lower level for the other peptides. Comparatively, the MI and MII neurohormones are definitely more abundant, but the amount of proctolin in the hindgut is rather small (0.86 ng/rectum).

The large number of identified myotropic neurohormones raises the question of their physiological role. It is possible that they exert several distinct functions, as do vertebrate gastin and CCK. If so, they could regulate fat body activities as well as the secretion of digestive enzymes.

7.5. ACTION MECHANISM OF MYOTROPIC FACTORS

The action mechanism of myotropic factors raises a number of questions concerning their release sites.

7.5.1. Neurohormones

The release sites of myotropic factors were thoroughly investigated and sparked passionate debates in certain cases. This is understandable because knowledge of the release site of the factors helps to determine whether they are neurohormones or substances locally active as neuromodulators or neurotransmitters (see Cook and Holman, 1985).

A wide variety of methods is available for solving these problems. A comparative test can be carried out on the activity of the factors on innervated and denervated muscles (Miller and Metcalf, 1968 in the heart of *Periplaneta;* Miller and S.-Rózsa, 1981 in the hyperneural muscle of *Periplaneta*). It is also possible to investigate the electrical activity of the muscles (Richter, 1973) to identify modifications due to the presence of neuromodulators, or to use agents suppressing nervous transmission such as tetrodotoxin, a sodium channel blocking agent, or specific aminergic antagonists. Electron microscopy and immunocytochemistry also provide images that allow an understanding of the relationships between different axons and target cells.

Miscellaneous data suggest that the myotropic factors, or at least some of them, are conveyed by the hemolymph. The presence of myotropic factors was repeatedly shown in nh organs (CC, PO), which are known to be sites of neurohormone release within the hemolymph. In addition, the release of cardioacceleratory factors from the CC was achieved by electrical stimulation of the brain (Kater, 1968) in *Periplaneta.*

7.5.2. Release at Effector Level

Other data suggest a transmission by nerves up to the muscle cells, and a neurotransmitter or neuromodulator role for the myotropic factors. This is not surprising considering the case of proctolin, whose inactivation in the hemolymph is very rapid. The myotropic activity of the target organs and their nerves was demonstrated for the hindgut of cockroaches (Brown, 1967; Holman and Cook, 1970, 1972) and the oviduct of *Locusta* (Orchard and Lange, 1985). Moreover, numerous neuroeffectory junctions were revealed by electron microscopy in all visceral muscles.

Using immunocytochemistry, proctolin-like nerve terminals were discovered in the hindgut (Veenstra *et al.*, 1985b in *Leptinotarsa;* Agricola *et al.*, 1985 in *Periplaneta*) and dopamine-like terminals were observed in the foregut in *Schistocerca* (Klemm, 1972).

When release of the myotropic factors takes place at muscle level, it is difficult to determine whether the factor involved acts as a neurotransmitter or as a neuromodulator (see Brown, 1975; Cook and Holman, 1979), and whether its action is exerted at the pre- or postsynaptic level. By tetrodotoxin application to the preparation, it was possible to show, in some cases, that a myotropic factor was active at the postsynaptic level, with its effects being maintained despite the application of this molecule (Irving and Miller, 1980, octopamine on body wall muscles of *Lucilia;* Shukle and Judson, 1984, heart of *Periplaneta;* Platt and Reynolds, 1985, CAP on the heart of *Manduca*).

Certain molecules seem to act as neurohormones and also as neuromodulators/neurotransmitters. Thus octopamine, which increases lipid discharge from the fat body, is released from the CC of cockroaches or locusts in certain physiological conditions, such as stress or the onset of flight. At the level of certain muscles, however, octopamine has been demonstrated to perform the function of a neuromodulator, i.e., in the locust hind leg and oviducts. It could also act as a neurotransmitter, as in firefly light organs.

Immunocytochemical investigations demonstrated that certain peptidergic neurohormones, such as proctolin, AKH, and other peptides are present in the neuropil of nerve ganglia (Schooneveld *et al.*, 1983; Agricola *et al.*, 1985; Keshishian and O'Shea, 1985). This confirms that they might play a part in neurotransmission or neuromodulation. In the locust brain, in fact, proctolin causes an increase in adenyl cyclase activity (Hiripi *et al.*, 1979).

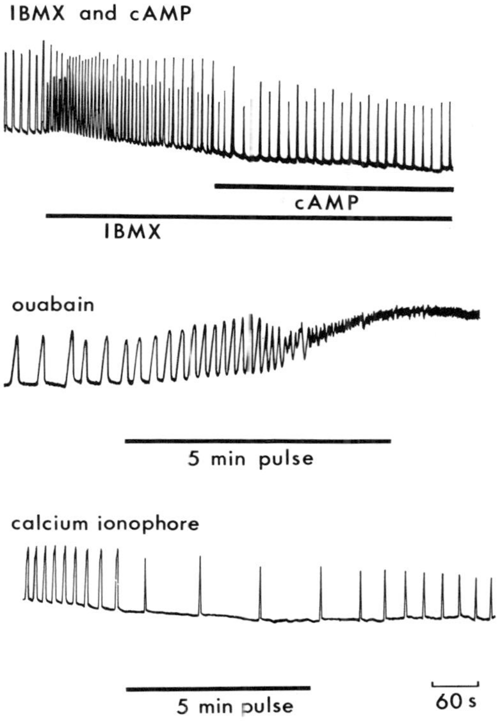

FIGURE 74. Model of action of octopamine on the extensor tibiae bundle. The phosphodiesterase inhibitor IBMX (10^{-2} M) speeds up the rhythm and dibutyryl cAMP (10^{-2} M) slows it. Ouabain, a sodium pump inhibitor, speeds up the myogenic rhythm, suggesting that the effects of octopamine are not mediated by stimulation of the sodium pump. A 5-min pulse of the calcium ionophore reduces the frequency of the myogenic rhythm. The biogenic amine may thus act by increasing the intracellular calcium concentration (closed bars: duration of application). (Modified from Evans, 1984a; Evans and Myers, 1986.)

7.5.3. Cyclic Nucleotides

The involvement of cyclic nucleotides as second messengers in the action of myotropic neurohormones was investigated on several effectors with homogenates and with purified neurohormones.

The potential role of cAMP in the activity exerted by proctolin, serotonin, and octopamine on the myogenic rhythm of the extensor tibiae was investigated, and it was revealed that this nucleotide mediates the effects of proctolin and serotonin but not octopamine (Evans, 1984a,b) (Fig. 74). By contrast, at the oviduct level, octopamine acts through the intermediary of cAMP, of which the concentration rises in the oviduct after stimulation of the DUMOV; its effects are mimicked by a phosphodiesterase inhibitor, 3-isobutyl-1-methylxanthine (IBMX), forskolin, and dibutyril cAMP. Its action appears to be postsynaptic (Lange and Orchard, 1986; Orchard and Lange, 1986).

As for the hindgut, cAMP could play a part in HSN-induced contractions in

Leucophaea (Cook *et al.*, 1975; Cook and Holman, 1979). In fact, phosphodiesterase inhibitors potentiate the hindgut response to head extracts. Moreover, adenylcyclase was found in hindgut homogenates. The proctolin-induced response of the oviduct of *Locusta* do not still depend on cyclic nucleotides (Lange *et al.*, 1987).

However, factors Ic and II from the CC of *Periplaneta* (Shukle and Judson, 1984) do not involve cAMP for hindgut stimulation as well as the central nervous system homogenates for the effect on the Malpighian tubules in *Carausius* (Pilcher, 1971) and the heart in *Periplaneta* (Collins and Miller, 1977).

7.5.4. Ca^{2+}

The need for CA^{2+} was demonstrated for the action of proctolin on the hindgut (Cook *et al.*, 1975), where it causes an elevation of intracellular Ca^{2+} (Dumbar and Huddart, 1981). Ca^{2+} also intervenes at the level of the extensor tibiae muscle (Evans, 1984a,b) (Fig. 74), the heart of *Periplaneta* (Shukle and Judson, 1984), the hyperneural muscle of the same insect (Hertel and Penzlin, 1986), and the oviduct of *Locusta* (Lange *et al.*, 1987).

7.5.5. Receptors

The main investigations concerning receptors were carried out on the myogenic bundle of the extensor tibiae (Evans and O'Shea, 1978; O'Shea and Evans, 1979; Evans, 1981) using α-adrenergic agonists (clonidine and tolazoline) and antagonists (phentolamine and propranolol). These studies showed that three distinct receptor types occur. One of them is located on the terminals of SETi motoneurons, i.e., at the presynaptic level, and the other two receptor types are located postsynaptically on the muscle fibers.

The heart receptors of the myotropic factors CAF1 and CAF2 of *Manduca* are different from serotonin and octopamine receptors. The aminergic responses are blocked by 2-bromolysergic acid diethylamide, but the peptidergic responses are not. The octopamine receptors of *Manduca* share many characteristics with the octopamine receptors of locust leg muscles, while serotonin receptors resemble serotonin receptors located in the Malpighian tubules and the salivary glands (Platt and Reynolds, 1986). They have also been studied in the foregut of a locust (Banner *et al.*, 1987b).

7.5.6. Conclusion

The regulation of the functioning of insect muscles is a complex process which involves neurotransmitters, neuromodulators, and neurohormones.

Several myotropic neurohormones have been identified, both peptides and biogenic amines, which stimulate most of the visceral muscles, including those of the heart, gut, and oviducts. The most active biogenic amines are serotonin and octopamine. As for the myotropic peptides, several factors were recently identified, including the first known insect neurohormone, proctolin, from the hindgut of *Periplaneta,* discovered in 1975.

The molecular structures of two cardioaccelerators from the CC of *Periplaneta* and 12 factors from the head of another cockroach, *Leucophaea,* were discovered in 1984 and 1986–1987, respectively.

Peptidergic neurohormones play an important part in insect neurotransmission, functioning mostly as neuromodulators. In some cases, it has been shown that they are coproduced and cotransmitted in neurons which synthesize classic neurotransmitters, giving rise to the new concept of cotransmission.

The existence of synthetic proctolin stimulated many investigations by RIA and immunocytochemistry, which helped to determine the location of the origin neurons of proctolin and to demonstrate the existence of a peptidergic innervation of certain skeletal muscles. In two cases, that of the extensor tibiae and that of the oviduct, a regulation implying both octopamine and proctolin was demonstrated.

The myotropic peptides synthesized in the ns cells of the central nervous system are released into the traditional nh organs, the CC and the PO. They are also discharged at the effector level, where neuroeffector peptidergic junctions are revealed.

8

Ecdysis and Tanning

The period of ecdysis is critical for the insect's survival. It must successfully shed the old cuticle and survive throughout the period between this process and the completion of tanning. During that time it is extremely vulnerable because its envelope is still soft and fragile. The period between the completion of shedding and that of tanning includes other important steps, particularly the complicated eclosion behavior. This involves the softening of the cuticle, a prerequisite for the expansion that precedes tanning, and requires very precise coordinated regulation that is able to reach all epidermal cells at the appropriate time. Investigations in this field have yielded numerous data and demonstrated the existence of an eclosion neurohormone, a tanning neurohormone called bursicon, and several pupariation factors.

8.1. ECLOSION—MOLTING

8.1.1. Eclosion Hormone

Thorough analysis of the eclosion behavior of the saturniid moth *Hyalophora* showed that it includes several successive steps: a phase of rapid rotational abdominal movements, a rest period, and a second phase of peristaltic movements culminating in eclosion. This is followed by the spreading of the wings, caused by the pressure of the blood that fills them, and is pulsed by contraction of the abdominal longitudinal muscles.

The regulation by the brain of the timing of eclosion was first demonstrated in the saturniid moths *Hyalophora* and *Antheraea,* where it varied for each species and where brain removal caused eclosion to occur at random. Even in isolated abdomens, brain implantation or the injection of brain extracts triggered within 1–2 hr the characteristic behavior which precedes eclosion. When reciprocal implantations were made, the eclosion rhythms corresponded to those of the donor and not of the host (Truman and Riddiford, 1970; Truman, 1971), but each species retained its characteristic behavioral programs. Thus the existence of an eclosion hormone (EH) was demonstrated.

Eclosion hormone acts on the central nervous system, where it triggers the initiation of the species-specific program. This was shown in isolated abdomens of silk moths, in which the action potential of the nerves controlling the muscles involved in pre-ecdysial contractions was recorded. The results differed greatly, according to whether or not the eclosion hormone was present. When it was present, many similar bursts of spikes were first observed in the successive ganglia, but reserved from one side to the other in a way reflecting rotational movements. In the next phase of peristaltic contractions, the impulses were identical on both sides, but phase-shifted from back to front (Truman and Sokolove, 1972). When isolated central nervous systems are subjected to the eclosion hormone they respond by the generation of a program of motor activity which mimics the one expected during the pre-eclosion and eclosion behaviors. These behaviors are initiated by only a few minutes exposure of the central nervous system to EH (Truman, 1978).

The studies concerning EH were extended to *Manduca* (Truman, 1973c). It was demonstrated that in *Manduca,* as in silk moths, an EH was present causing early eclosion when injected into pharate adults and inducing the release of a stereotyped behavior that allows shedding of the old skin. The target sites were shown to be scattered throughout the central nervous system, and a coordinated response was dependent on all the segmentary sites (Truman, 1979).

Investigations in *Manduca* embryos, larvae, and pupae revealed that not only adult eclosion, but also embryonic larval and pupal ecdysis, were under the endocrine control exerted by EH (Truman *et al.,* 1980, 1981), which causes premature ecdysis in larvae and pupae. The amounts of EH in the ventral nerve cord were determine during all larval instars. In final instar larvae, EH increases during feeding period, reaches a plateau, and drops sharply at ecdysis. In the pharate pupae, EH appears in the hemolymph about 30 min before the onset of ecdysis (Truman *et al.,* 1980), and its titer rises sharply and declines steadily during a period of 5 hr (Reynolds *et al.,* 1979).

8.1.2. Eclosion Hormone Assays

Eclosion hormone has been investigated using several bioassays. The first consisted of inducing premature eclosion in pharate adults of *Antheraea* 1–3 hr after the injection of an active extract. Other bioassays were devised in *Manduca* based on the behavioral responses of pharate larvae and pupae (Copenhaver and Truman, 1982).

The amounts of EH may also be indirectly measured using the *Manduca* isolated wing assay, i.e., by measuring the increase in the extensibility of the wing cuticle (Reynolds, 1977).

To determine the titer of hemolymphatic EH, hemolymph was obtained either by decapitation or by introducing a thin tube through the integument into the heart and removing hemolymph from time to time with a syringe (Reynolds *et al.*, 1979).

One EH unit was defined as the amount of biological activity contained in the CC of a pharate *Manduca* adult.

8.1.3. Origin and Release Sites of Eclosion Hormone

In both *Antherea* and *Manduca*, EH gradually increases at the onset of adult development and reaches a plateau, first in the brain and then in the CC; it then diminishes considerably in the CC just before eclosion, while its level in the hemolymph rises sharply (Truman, 1973b).

The time at which eclosion hormone appears in the hemolymph of *Manduca* was determined by isolated wing bioassay. It was demonstrated that this hormone is only released into the blood of the pharate adult once development has been completed, during a restricted period determined by a circadian clock (Reynolds *et al.*, 1979).

In pharate pupae and last instar larvae, no change in the EH level was found in the brain at ecdysis (Truman *et al.*, 1981), but important changes occurred in the ventral nerve cord, whose EH content dropped; this suggests that at these stages the latter is involved in the secretion of EH (Taghert *et al.*, 1980). While the ventral nerve cord content decreased, the EH titer of the hemolymph rose progressively.

Investigations of the neurons involved in EH production using microdissection of ns cell clusters and bioassay of their content indicated that in *Manduca* pharate adults a lateral protocerebral group (Ia) of ns cells was the place of origin of EH (Fig. 75). In this species five large ns cells were stained in the Ia cluster with an antiserum raised against a purified extract of EH. Intracellular recording revealed clear changes in the Ia neurons at about eclosion time; they were electrically inactive on the day before eclosion but exibited a pronounced rate of firing during EH release (Copenhaver and Truman, 1986c). The onset of electrical activity coincided with the appearance of EH in the bath of CC–CA complexes maintained under paraffin oil in a small droplet of saline which was removed periodically for EH bioassay. Iontophoresis of dye into these cells revealed an axon that projected into the CC–CA and a neurite branching pattern to the major neuropilar regions of the brain (Fig. 76). Depolarization of the Ia neurons by intracellular injection of current induced the release of measurable

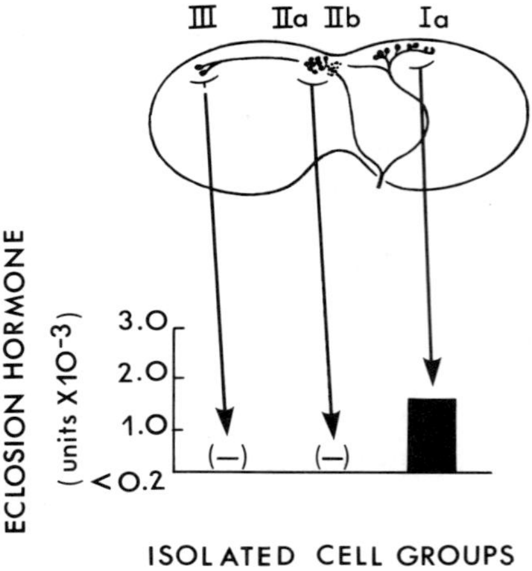

FIGURE 75. Distribution of eclosion hormone activity in the brain of *Manduca*. (Modified from Copenhaver and Truman, 1986a.)

levels of EH activity into a bath surrounding the CC–CA complex (Copenhaver and Truman, 1986a).

8.1.4. Regulation of Eclosion Hormone Release by Ecdysteroids

Experiments in which 20-HE was injected into *Manduca* pharate adults and pharate pupae showed that 20-HE causes a dose-dependent delay in eclosion but that the latter nevertheless proceeded according to a biological clock. Isolated abdomens which were still capable of the preparatory behavior for eclosion were no longer capable of it when injected with 20-HE. Determination of the EH level in the hemolymph confirmed that its release is inhibited by hight ecdysteroid levels. This inhibition was no longer observed before ecdysis. Thus, in pharate adult *Manduca* the timing of EH release is regulated by the declining levels of

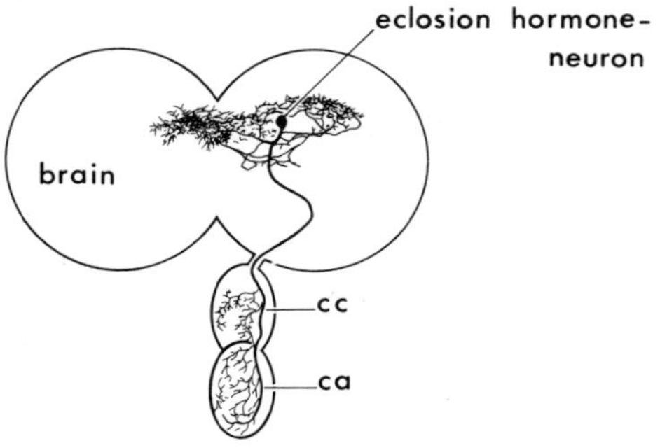

FIGURE 76. Location and pathway of a *Manduca* group Ia neurosecretory neuron. (Modified from Copenhaver and Truman, 1986a.)

hemolymph ecdysteroids, as it is in the pupa and larva. This decline allows the target tissues of EH (i.e., the nervous system and epidermis) to become sensitive to this neurohormone (Truman *et al.*, 1983). The above results are in accordance with the data for *Tenebrio* showing a delay in ecdysis following ecdysteroid treatment (Sláma, 1980). The mode of action of 20-HE decline is not known but it is known to affect not only EH release but other processes as well (see Chapter 9).

8.1.5. Eclosion Hormone and Muscle Degeneration

Besides affecting behavior, EH triggers several processes, including the segmental muscle degeneration which often follows eclosion (Truman, 1973a) (Fig. 77). Eclosion requires certain muscles, particularly the intersegmental muscles which are no longer necessary in the adult stage and atrophy after eclosion.

The role of EH in these processes was demonstrated in *Antheraea* using isolated abdomens injected with EH extracts (Truman, 1973a) or a purified preparation of EH (Schwartz and Truman, 1982). It was also shown that EH did not act on the muscle through the central nervous system, since isolated *Antheraea* abdomens whose central nervous system had been removed still reacted to EH by intersegmental muscle breakdown (Schwartz and Truman, 1984a).

The processes of muscle degeneration do not depend only on EH but are also influenced by the ecdysteroid level. In reality, there are two types of muscle degeneration; a fast process that depends on EH, and an earlier, slower process triggered by the 20-HE decline (Schwartz and Truman, 1982, 1984a). 20-Hydroxyecdysone also enables the muscles to become capable of responding to EH.

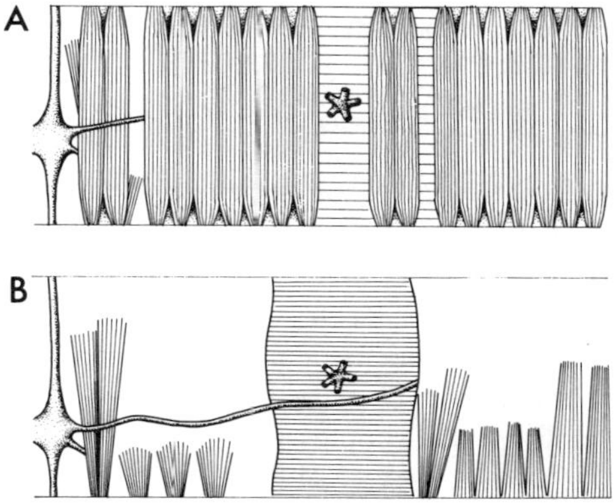

FIGURE 77. Posteclosion breakdown of muscles in *Manduca.* Internal view of the musculature of the right half of the fifth abdominal segment. (A) Moth at eclosion; (B) 3 days after eclosion. The right margin of the drawing is the dorsal midline; the portion of nerve cord on the left marks the ventral midline. Muscles extending to the heart and nerve cord have been omitted. (Modified from Truman, 1973a.)

8.1.6. Neuron Death

At eclosion of the adult *Manduca,* extensive degeneration of abdominal neurons takes place. The first neurons to degenerate are the interneurons, followed by the motoneurons; each neuron dies at a characteristic time, which varies from cell to cell (Truman, 1983). The degeneration of neurons is also observed in implanted ganglia indicating that it depends on a hormonal factor. This degeneration occurs when the level of ecdysteroids declines and an artificially elevated titer of ecdysteroids delays neuron death. It should be noted that muscle degeneration precedes neuron death. The processes of dendritic regression and death were also observed during larval-pupal transformation (Weeks and Truman, 1986).

8.1.7. Various Processes Regulated by Eclosion Hormone

Eclosion hormone is involved in several processes besides muscle and neuron degeneration. These processes are often associated with the ecdysteroid decline preceding EH release. This decline is of importance in endocuticle digestion and melanin production by epidermal cells (see Chapter 9) because it causes EH release and allows the target tissue to become sensitive to this hormone. Eclosion hormone, in turn, triggers wing expansion in the adult and plasticization, and allows the release of two neurohormones which intervene after ecdysis, i.e., bursicon and cardioactive peptides (CAPs) (Fig. 78).

Plasticization is a modification in the mechanical properties of the cuticle which is probably due to small changes in pH (Reynolds, 1977). It makes the cuticle flexible and extensible so that the insect can escape from its old cuticle and later expand its new one. In *Manduca,* EH involvement in plasticization was demonstrated by injecting EH extracts into pharate adults and also into wing veins, where an increase in cuticular extensibility was observed precociously (Reynolds, 1977). Plasticization, however, is also induced later in the molting process by another neurohormone, bursicon.

8.1.8. Presence of Eclosion Hormone among Insects

There is reason to believe that a neurohormone like EH is present in many insect species, although few of them are known at this time. In *Aeshna* larvae, a factor necessary for ecdysis has been demonstrated to occur in the brain; the neurons from which it originates were identified as two pairs of type A ns cells located in the median anterior part of the protocerebrum. These cells were heavily loaded at the time of ecdysis, and empty in permanent larvae, where molting was suppressed by removal of the PG. Elective electrocoagulation of these cells prevented ecdysis (Charlet and Schaller, 1976).

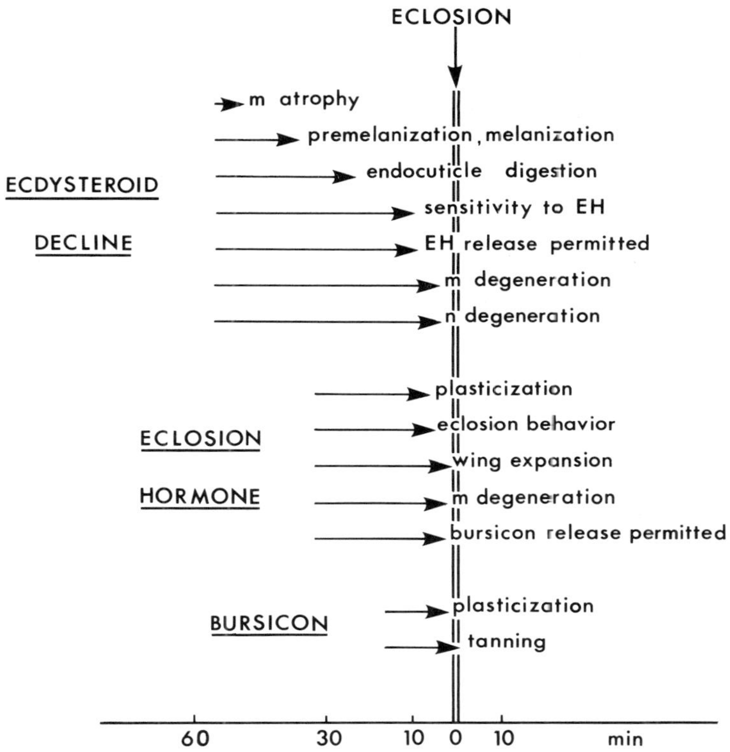

FIGURE 78. Schematic representation of the processes involved in eclosion, tanning, and associated events. EH, Eclosion hormone; m, muscles; n, neurons. (From the data of Truman, 1973a, 1983, 1985; Reynolds *et al.*, 1979.)

An eclosion factor is also present in the protocerebrum, the CC–CA, the thoracic ganglia, and the last abdominal ganglion of *Lymantria* pharate adult (Schnee *et al.*, 1984). In *Bombyx,* EH was identified in the embryos from the moment the neuroendocrine system had developed; its level increased 1 day before hatching and dropped after hatching (Fugo *et al.*, 1985).

8.2. BURSICON

Tanning or sclerotization mostly takes place after molting. However, it should be noted that pre-exuvial tanning processes occur in certain limited areas. The sclerotization of the cuticle, which entails its hardening and darkening, is associated with water loss followed by secretion of the endocuticle and sometimes melanization. In the adult, whose wings need to remain soft in order to be able to unfold, tanning is preceded by changes in the mechanical properties of the cuticle, called plasticization.

The regulation of these different processes has been investigated in several insect species including flies, in which the tanning hormone was first discovered.

8.2.1. The Sclerotization Process

The most frequent tanning in insects is quinonic tanning, involving *o*-quinones which form bonds with the cuticular proteins through the latter's free amine groups and then combine with each other. Another type of fairly widespread tanning is β-sclerotization, in which the protein-quinone bond occurs via the β carbon. In both cases (as in melanization) the precursor is *N*-acetyldopamine.

The source of *N*-acetyldopamine in tyrosine was revealed by Sekeris and Karlson (1962), who pointed out its predominant role in sclerotization. The metabolism of tyrosine generally includes hydroxylation of tyrosine to dopa by a tyrosine hydroxylase, decarboxylation of dopa to dopamine by dopa-decarboxylase, and acetylation of dopamine to *N*-acetyldopamine by *N*-acetyltransferase.

Sometimes, however, the metabolic path of sclerotization is somewhat different, since tyrosine is converted to tyramine, which is then converted to dopa, *N*-acetyldopamine, and *N*-acetyltyramine (Mills *et al.*, 1967). In addition to *N*-acetyldopamine, other substrates for both sclerotization and melanization, i.e., β-alanine and β-alanine conjugates, have been discovered in insects (see Kramer and Hopkins, 1987).

8.2.2. Bioassays

Several bioassays have been used for bursicon, including the very sensitive *Sarcophaga bullata* assay using newly emerged, neck-ligated flies (Fraenkel and Hsiao, 1965).

In *Manduca,* bursicon activity was estimated from the tanning or yellowing of the veins of the wings (Truman, 1973c; Reynolds, 1977). In *Tenebrio,* neck-ligated nymphs 1 day before ecdysis were used as receivers. The responses were estimated 48 hr after exuviation, from the darkening of the cuticle of the abdominal sternites (Grillot *et al.*, 1971, 1976).

8.2.3. Demonstration of Neurohormonal Control of Tanning

When the fly escapes from the puparium, it is small, its cuticle is soft and pale, and its wings are wrinkled. It climbs through the sand to reach the open air, and immediately begins to pump air, expands to its final size, and unfolds its wings. This takes about 20 min; then begins the tanning process, which is completed in 1–2 hr and gives to the fly its final appearance. Experimental investigations into the regulation of tanning in flies were conducted on *Calliphora erythrocephala* (Cottrell, 1962a–c), *Sarcophaga* and other flies (Fraenkel and Hsiao, 1962, 1965).

If a ligature is placed at the neck of a fly during the first 5 min after emergence, tanning is arrested. After this interval, bursicon secretion occurs despite ligation (Fraenkel and Hsiao, 1962, 1965). Hence, secretion of the tanning factor is limited to a very short period. The injection of hemolymph can cause tanning to resume in a ligated fly. The hemolymph, which was inactive on eclosion, becomes active 2–3 min later. In *Sarcophaga,* its activity then rises very rapidly and increases 100-fold within 30 min (Fraenkel and Hsiao, 1962, 1965). One hour after emergence, hemolymph activity gradually begins to diminish.

In addition to flies, the presence of a tanning neurohormone was subsequently demonstrated in cockroaches, locusts, *Tenebrio,* and *Manduca.*

8.2.4. Stimuli Involved in Bursicon Release

It has been shown in several insect species that the release of bursicon depends on sensory inputs. The nerve impulses are transported to the brain, which gives the signal to release bursicon. In fact, if the nervous connections between the brain and the ventral nerve cord are interrupted by severance or ligaturing at the neck, tanning does not takes place, although bursicon is present throughout the central nervous system before emergence. Thus, in both flies and moths (Fraenkel and Hsiao, 1965; Truman, 1973c; Reynolds *et al.,* 1979) expansion and tanning is greatly delayed if the insects are obliged to dig in the sand for a certain time, and in *Manduca* it was also observed that the finding of a suitable place to expand wings was a prerequisite for bursicon release.

By placing ligatures at different times during ecdysis, Srivastava and Hopkins (1975) showed that in cockroaches, the stimuli inducing bursicon release originate in the thorax in the ecdysial split of the cuticle. If the latter is removed or sealed so as to be unbreakable, bursicon is not released into the hemolymph. Severance of the ventral nerve cord at various levels considerably reduces the release of the tanning factor, but a normal release was reestablished when electrical stimulation was applied behind the section.

8.2.5. Production and Release Sites of Bursicon

The distribution of bursicon throughout the central nervous system, CC, and sometimes PO has been investigated in six insect species. The results, summa-

TABLE XXVI. Distribution of Bursicon Activity in Some Insects[a]

		Organ assayed						
Insect	Bioassay	Brain	Corpora cardiaca	sgg	Thoracic ganglion	Abdominal ganglion	Perisympathetic organs	Reference
Periplaneta	*In vitro*	++	+	+	+	+		Mills, 1967b; Vincent, 1972
Leucophaea	*Sarcophaga* bioassay	+	+	+	+	+		Srivastava and Hopkins, 1975
Carausius	*Phormia* bioassay	+	+			+		Vincent, 1972
Locusta	*Phormia* bioassay	+	+	+	+	+		Vincent, 1972
Tenebrio	*Tenebrio* bioassay	+	+	+	+	+	+	Grillot *et al.*, 1971, 1976
Manduca	*Manduca* isolated wing	0			0	+++	+++	Truman, 1973c
			+	0	++	+++		Taghert and Truman, 1982a
Sarcophaga	*Sarcophaga* bioassay	+		0	0	++		Cottrell, 1962b; Fraenkel and Hsiao, 1965

[a]sgg, subesophageal ganglion.

rized in Table XXVI, show that in all parts of the central nervous system, in most of these species, bursicon was found particularly in its abdominal part, which was reported to be six times more active than the brain in *Sarcophaga* and even more active in the pharate adult of *Manduca*. The last abdominal ganglion was also reported to be strongly active in *Periplaneta*.

In *Sarcophaga* the neurons containing bursicon were shown to be located in the PI, whereas in the aphid *Megoura viciae* the source of a tanning factor was found in two ns cells with two processes, one transporting ns material to the CC, and the other along the ventral nerve cord (Steel, 1978).

In *Tenebrio,* comparison of the activity of the different parts of the central nervous system and their nh organs reveals that bursicon is present throughout the central nervous system, in the CC and PO, whose involvement in tanning is indicated by a clear drop in the tanning activity which occurs at the end of the pupal stage, when the bursicon hemolymph level rises. The migration of the tanning factor from the abdominal ganglia into the PO was also shown, by determining the amount of activity in both these organs. Maximal activity was observed in the PO 2 days later than in the ganglion. Thus, the PO appear to be involved in the release of bursicon (Grillot *et al.*, 1971, 1976) (Fig. 79).

As far as the abdominal nerve cord is concerned, bursicon neurons were only identified in *Manduca* larva (Truman, 1973c; Taghert and Truman, 1982b). Cobalt filling of ganglionic neurons from segmentary nerves revealed two dorsolateral paired groups and one median group, but only the former displayed bursicon activity. The axons of the bursicon neurons project through the segmental nerves into the PO.

8.2.6. Timing of Bursicon Synthesis and Release

The timing of bursicon synthesis and release was studied by determining the activity of various extracts at different times and ligaturing the insects at different stages (Mills, 1966; Srivastava and Hopkins, 1975, in cockroaches; Vincent, 1972, in *Locusta;* Reynolds *et al.*, 1979 in *Manduca;* Grillot *et al.*, 1971, 1976, Abboud *et al.*, 1983 in *Tenebrio*).

In most species investigated, bursicon or a bursicon-like neurohormone is synthesized some time before ecdysis but the release takes place very rapidly during molting (e.g., in cockroaches and locusts), or just after it, as in flies and *Manduca*. This is sometimes related to behavioral events such as the finding of a suitable place for wing spreading. In *Tenebrio,* sclerotization is far more time consuming and bursicon activity already appears during pupal life, which probably has some connection with the precocious tanning of certain areas.

In the flies *Sarcophaga* and *Calliphora,* bursicon appears in the hemolymph 3–5 min after emergence or the finding of suitable conditions for expansion; its

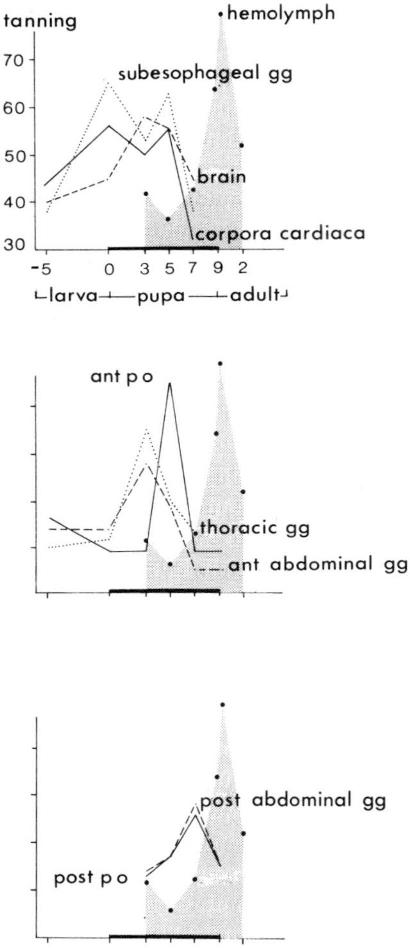

FIGURE 79. The tanning activity of corpora cardiaca, perisympathetic organs, and various parts of the central nervous system in *Tenebrio* during the pupal stage and the beginning of the adult stage. ant, Anterior; gg, ganglion; po, perisympathetic organ; post, posterior. (Modified from Grillot *et al.,* 1976.)

critical period is at about 5 min and by 30 min it begins to disappear from the hemolymph (Fraenkel and Hsiao, 1962; Cottrell, 1962a).

In *Manduca,* the bursicon content of the central nervous system is large before and during eclosion, but 15 min after eclosion it will be reduced by 80%. Bursicon appears in the hemolymph within 2 min of eclosion at the beginning of wing expansion, reaches a maximum by 4–12 min, and then declines progressively. Its half-life is approximately 45 min. Eighty minutes after eclosion, tanning is complete (Reynolds *et al.,* 1979). Bursicon release is stimulated in *Manduca* by EH, which also modifies tissue sensitivity to bursicon. Bursicon and EH therefore appear to be distinct neurohormones, and in the pharate adult their release sites are quite different (Truman, 1973c).

In *Tenebrio* (Grillot *et al.*, 1976), the load of nerve ganglia alters with time and PO loading coincides with the discharge of these ganglia. It is interesting to note that the various ganglia and nh organs release their hormonal content at different times. Thus, release by the anterior abdominal ganglia occurs on day 3 and by the posterior ganglia on day 7, whereas the brain remains active for a longer period. The CC and anterior PO both release their content about 2 days before ecdysis. Other investigations using *Tenebrio* as donor and *Phormia* as receiver confirmed the brief presence of bursicon in the hemolymph at ecdysis, but gave different results for the bursicon content in the ventral nerve cord, where it only diminished during and after ecdysis. This led some authors to suggest the existence of two distinct tanning factors, i.e., the bursicon, released at ecdysis, and another tanning factor, discharged before molting (Abboud *et al.*, 1983). It is also possible that bursicon is released in two distinct pulses, the first being the one demonstrated by Grillot *et al.* (1976) and the second that observed by Abboud *et al.* (1983).

8.2.7. Other Processes Associated with Tanning

Like EH, bursicon appears to regulate some of the processes occurring at ecdysis. In certain cases, however, the effects observed were attributed to factors other than bursicon.

8.2.7.1. Plasticization

Plasticization in flies, which occurs during air pumping by the ptilinum, was shown to be under hormonal control (Cottrell, 1962a). Recent investigations (Fraenkel *et al.*, 1984) demonstrated that several distinct processes take place in the fly during the tanning period, i.e., plasticization and muscle contractions causing pulsing and air pumping. Injection into flies of blood samples from other flies of differents ages shows that pumping is only stimulated by 15-min blood, which suggests that the factor involved is distinct from both EH and bursicon, which are not present in the hemolymph at this time. Hence the existence of several distinct factors is proposed, including EH, a pulsing and pumping factor, a plasticization factor, and bursicon.

In *Manduca,* the plasticization process that precedes ecdysis is dependent on EH. A second plasticization occurs after eclosion. This is quite an important event because it allows the expansion of the new cuticle in the wings (Reynolds, 1977).

The plasticization which occurs in *Rhodnius* after the blood meal must be mentioned here. The active factor in this case is not bursicon. It is present in the brain and is released at undetermined peripheral sites (Maddrell, 1966). The effects of serotonin mimic those of the factor concerned (Reynolds, 1974).

8.2.7.2. Diuresis

Strong diuresis, possibly induced by bursicon, occurs in insects during tanning. In cockroaches, the fractions that cause tanning also cause diuresis. In *Calliphora,* however, the cuticle water associated with amino acid side chains is not displaced by tanning (Vincent and Ablett, 1987).

8.2.7.3. Disaggregation of Wing Cells

In the blowfly *Lucilia cuprina,* the hemolymph contains large numbers of cell fragments, 1 hr after eclosion, as a result of cytolysis of the wing epidermis. This transient process disappears in 1 hr. Like tanning, it is eliminated by early cervical ligature, whereas later ligature allows the cell debris to remain in the hemolymph, while tanning proceeds. A factor stimulating cell disaggregation appears in the hemolymph at the same time as bursicon, i.e., within 10 min of eclosion. Hence it seems plausible that the factor acting on the wing is identical to bursicon. Like bursicon, it is destroyed by proteolytic enzymes and, after filtration and separation, the activities of both factors merge (Seligman and Doy, 1973).

8.2.7.4. Endocuticle Secretion and Resilin Deposition

Another bursicon-stimulated process might be secretion of the endocuticle, which takes place after tanning and depends on ecdysteroids. Like tanning, endocuticle growth is inhibited in *Sarcophaga* by early cervical ligature and is restored by injections of active hemolymph or partly purified bursicon (Fogal and Fraenkel, 1969). In *Locusta,* ligaturing the posterior part of the abdomen at the time of ecdysis diminishes both endocuticle growth and tanning (Vincent, 1971). Moreover, a hemolymph trophic factor which increases cuticular deposition *in vitro* was isolated in *Manduca* (Wielgus, 1983).

Another process which may be controlled by bursicon is resilin deposition in adult insect tendons. In *Calliphora,* it depends on the ns cells of the PI (Sabaratnam, 1974).

8.3. PUPARIATION

Pupariation, which includes different processes preceding sclerotization, differs greatly from tanning, which occurs after ecdysis (Žďárek and Fraenkel,

1972), because it does not involve tanning of an insect which has just molted but that of one which has reached the end of the larval stage without yet having undergone apolysis.

Pupariation includes several distinct processes (Žďárek and Fraenkel, 1972) which were precisely defined in *Sarcophaga* using a new technique (Sláma, 1976) for measuring hemolymph pressure inside the body. By revealing the alterations occurring in muscle activity, this technique allowed separation of the different events (Žďárek *et al.*, 1979). As shown in Figure 80, pupariation includes the following steps: first, an immobilization stage characterized by pressure fluctuations separated by pauses, then the retraction of the anterior segments accompanied by slight regular fluctuations in pressure, a contraction stage characterized by sharp regular peaks of high magnitude and short duration, followed by the start of the shrinkage stage, during which muscle activity gradually disappears, leaving only the small pulses necessary for the smoothing of the cuticle (Fig. 80).

These distinct processes are controlled by neurohormones (Žďárek and Fraenkel, 1969). When *Caliphora* or *Sarcophaga* larvae are ligatured a few hours before pupariation, the anterior part pupariates before the hind part, and the pupariation of the hind part, which is devoid of nerve ganglia, can be accelerated by the injection of hemolymph or nervous tissue homogenate. The active factors are present in the brain, in the ring gland, and also in the fused thoracoabdominal ganglionic mass. In the brain, they are located in the PI, and in the ring gland, they are restricted to the nh area. Outside the central nervous system, active factors appear to be stored at peripheral sites.

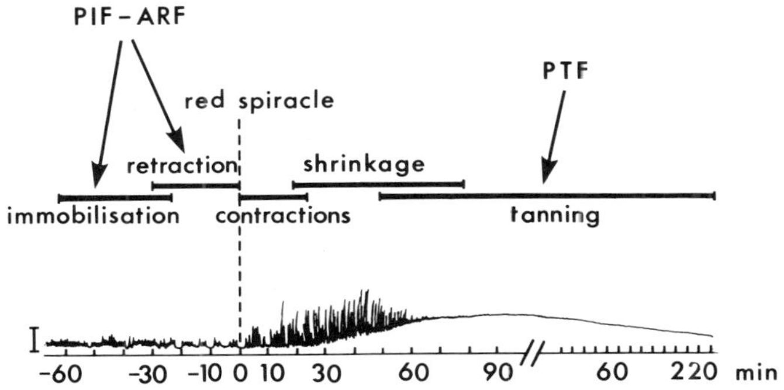

FIGURE 80. Barographic events involved in the pupariation of *Sarcophaga;* immobilization stage (fluctuation of pressure separated by pauses), anterior retraction (slight regular fluctuations of pressure), contraction stage (sharp, regular peaks of very high magnitude and short duration), shrinkage (the pressure baseline starts to rise and continues with small pulses), tanning (the pressure decreases but small pulses persist which allow the smoothing of the surface of the puparium). (Modified from Žďárek *et al.,* 1979.)

Injections of active hemolymph or nervous system extracts which accelerate tanning also accelerate anterior retraction, but these two activities do not occur concomitantly, either in the hemolymph or in the nervous system. This shows that they correspond to distinct factors, i.e., the anterior retraction factor (ARF), and the puparium-tanning factor (PTF). Two other factors are involved in the behavior of prepuparial larvae, i.e., pupariation-immobilizing factor (PIF), and puparium-stimulating factor (PSF) (Žďárek *et al.*, 1981).

The presence of PIF in hemolymph samples injected into wandering stage larvae causes immobilization within 35 min, but this effect disappears within about an hour. Pupariation immobilizing factor occurs in the hemolymph of the pre-red spiracle stage as well as in contracting larvae; it reaches it highest titer 2 hr after the onset of tanning and vanishes when the latter is completed. The insect's sensitivity to PIF changes during pupariation. It is nil when they have just left the food but increases during the 2 hr that precede the onset of the contraction phase. The lateness of the PIF peak suggests that this factor is not only involved in the immobilization process, but might play a part in subsequent processes that also involve the muscle activity which gradually ceases once the shaping of the pupa has been completed.

The effect of PSF is to stimulate the contractions of the somatic muscles of the wandering larvae. This factor can only be detected in the hemolymph during the very short period of strong contractions of the longitudinal muscles. In postfeeding larvae, injection of blood from contracting larvae determines changes in internal pressure (Fig. 81).

The role of ecdysone in triggering puparial tanning processes in flies was known long before the role of neurohormones was discovered (Fraenkel, 1935; Shaaya and Karlson, 1965), and ecdysone was considered to regulate dopa decarboxylase synthesis. To obtain a precise idea of the role of ecdysone, it is helpful to examine hemolymph ecdysteroid titers during larval and pupal development. In both *Sarcophaga* and *Calliphora,* pupariation is preceded by a low ecdysteroid titer and a high narrow peak occurs a few hours after pupariation (Žďárek, 1985). Therefore, high ecdysone titers do not trigger pupariation itself but the processes following it, i.e., apolysis and pupal cuticle synthesis. Therefore the initial experimental results could not be explained by high ecdysteroid levels but rather by the low concentration of ecdysone. This recalls the mode of action of ecdysone during eclosion. As far as this mode of action is concerned,

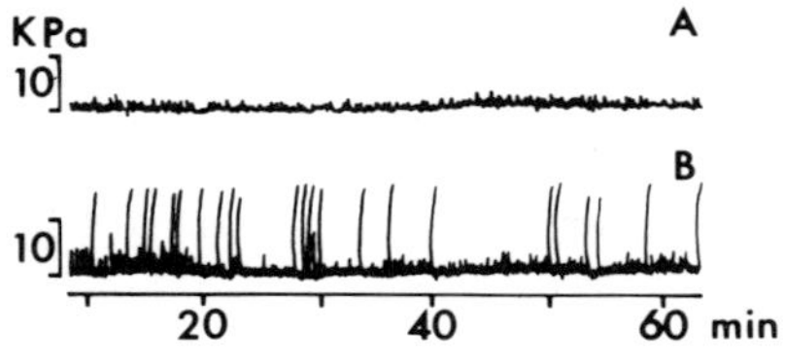

FIGURE 81. Comparative variations in the internal pressure of postfeeding larvae of *Sarcophaga* following injection of hemolymph of larvae at different stages. (A) 1 day before pupariation. (B) Contracting larvae. (Modified from Žďárek, 1985.)

experimental results indicate that it might act through the central nervous system; nevertheless, it is possible that ecdysone has direct effects on tanning, especially the pre-ecdysial tanning of restricted areas (Roberts *et al.*, 1982).

8.4. NEUROHORMONE IDENTIFICATION

8.4.1. Possible Identity of the Active Factors

Research on the mechanisms regulating molting, plasticization, and tanning has revealed the existence of several factors which are active either at different developmental stages, such as larval molting, pupariation, or eclosion, or during the process itself. Several authors compared the physiological activity of these factors in order to determine how many are involved.

Experiments on *Manduca* and *Sarcophaga* showed that bursicon is distinct from EH, as these hormones are present in the hemolymph at different times and their distribution in the central nervous system differs markedly (Truman, 1973c; Fraenkel *et al.*, 1984). Attempts were made to identify bursicon and the plasticization factor in flies (Reynolds, 1976; Fraenkel *et al.*, 1984). In these insects, the factors involved in both plasticization and tanning appear to be distributed in the same parts of the central nervous system. They are released into the hemolymph and disappear from it at the same times, suggesting that they are identical. The dopa decarboxylase inhibitor α-MDH, however, was found to inhibit tanning but not plasticization (Žďárek and Fraenkel, 1972).

In *Sarcophaga*, bursicon was also shown to be different from the pupariation factors ARF and PTF and from other factors involved in processes preceding tanning, i.e., pulsing and pumping (Žďárek and Fraenkel, 1969; Sivasubramanian *et al.*, 1974; Fraenkel *et al.*, 1984). Puparium-tanning factor appears to be different from ARF, but the latter may be identical to PIF (Žďárek *et al.*, 1981).

In short, most of the factors identified appear to be distinct from each other. In the case of hormones such as EH, bursicon, ARF, and PTF, which intervene in sequential processes, this seems necessary, but it is nevertheless surprising that distinct tanning factors are involved in pupariation and eclosion in flies. Bursicon has been compared to other neurohormones, including PTTH, brain gonadotropic factor from flies, and chromactivation factor in *Carausius*. In none of these cases did hemolymph containing bursicon display the same functions as these neurohormones (Fraenkel and Hsiao, 1965).

8.4.2. Purification

8.4.2.1. Eclosion Hormone

In *Manduca*, EH was shown to be a thermostable peptide with a molecular weight of about 8500. In both the brain and ventral nerve cord, the peak of EH activity was located in the same position (Taghert *et al.*, 1980).

Starting from *Bombyx* hemolymph, heads, or embryos, thorough investigations allowed successive purification of EH (from 400- to 2-million-fold) and the following N-terminal amino acid sequence was established by Edman degradation and automated gas phase sequencing (Nagasawa *et al.*, 1983, 1985; Fugo *et al.*, 1984, 1985).

H-Ser-Pro-Ala-Ile-Ala-Ser-Ser-Tyr-Asp-Ala-Met-Glu-Ile

Pure EH was active at a dose of 0.83 ng (10^{-13} M) and the EH content of a *Bombyx* pharate adult head composed two EH units.

8.4.2.2. Bursicon

The properties of bursicon were investigated by Cottrell (1962b) in *Calliphora,* by Fraenkel and Hsiao (1965) and Fraenkel *et al.* (1966) in *Sarcophaga,* and by Mills and Lake (1966) and Mills and Nielsen (1967) in *Periplaneta.*

Cottrell considers bursicon to be thermostable, resistant to boiling for 10 min, to cold down to −15°C, and to desiccation at 120°C. According to Fraenkel and Hsiao (1965), heating destroys a large part of the hemolymph's activity but not that of nerve tissue extracts. Both authors point out that bursicon is not dialyzable and that it is inactivated by alcohol, acetone, and trichloracetic acid, precipitated by ammonium sulfate, and destroyed by the proteolytic enzymes trypsin and pronase.

Electrophoretic separation of bursicon was carried out with *Sarcophaga* hemolymph and various nerve tissues of flies and cockroaches, i.e., PI and abdominal nervous system of *Sarcophaga,* and CC and terminal abdominal ganglion of *Periplaneta.* The results showed the existence of two active fractions: the first is present in the hemolymph and abdominal nerve ganglia of the fly, and in the CC and terminal abdominal ganglion of *Periplaneta,* but it is absent from the PI of *Sarcophaga.* The second active fraction is found in the PI of *Sarcophaga* and CC of *Periplaneta* (Fraenkel *et al.*, 1966).

Bursicon from whole cockroaches was purified 1700-fold (Mills and Lake, 1966). Bursicon is a protein possessing cationic properties. Its molecular weight was estimated to 38–40 kDa in flies and cockroaches, to 20–30 kDa in *Manduca* (Fraenkel *et al.*, 1966; Mills and Nielson, 1967; Taghert and Truman, 1982a). Its molecular structure is not yet known.

8.4.2.3. Anterior Retraction Factor—Pupariation Tanning Factor—Pupariation Immobilizing Factor

As shown above, ARF and PTF, which regulate fly pupariation, are two different substances, both of them distinct from bursicon. They were purified by Sivasubramanian *et al.* (1974), who isolated two protein factors, both of which were destroyed by trypsin and pronase. Anterior retraction factor and PTF are not dialyzable and are denatured by alcohol, acetone, and ammonium sulfate. The ARF was purified 170-fold. Its molecular weight is 180,000, and in blood it appears to contain two subunits of 90,000 each. Puparium-tanning factor, which was purified 200-fold, has a higher molecular weight of about 300,000. In the blood, it appears to contain two subunits of 26,000 each, and in the central nervous system four subunits of 80,000 each. The ARF is thermolabile and PTF is thermostable in the central nervous system but less so in the hemolymph. A similar difference in heat-related behavior has been observed for bursicon. The properties of ARF and PTF confirm that they are not identical, as suggested above. As far as PIF is concerned, its properties are similar to those of ARF (Žďárek *et al.*, 1981).

8.4.3. Interspecific Activity

The physiological effects of EH in the pharate adult of *Manduca* were detected in the hemolymph of several other insect species, i.e., *Calliphora* (Fraenkel and Su, 1984), *Lymantria* (Schnee *et al.*, 1984), *Bombyx* and *Samia* (Morohoshi and Fugo, 1977; Fugo *et al.*, 1984), and in the central nervous system of *Periplaneta, Schistocerca, Melanoplus, Dytiscus, Tipula, Calliphora,* and *Pyrrhocoris* (Truman *et al.*, 1981).

The interspecific activity of bursicon was assayed in flies with the blood and organ homogenates of different species. The blood of several flies exhibited strong interspecific activity (Cottrell, 1962b; Fraenkel and Hsiao, 1962; Reynolds, 1976). In *Sarcophaga* and *Calliphora,* tanning was induced by the blood of the following insects, which had just molted or were about to do so: *Periplanetar* nymphs and adults, *Galleria, Tenebrio, Oncopeltus* (Fraenkel and Hsiao, 1965), *Schistocerca* (Cottrell, 1962b), and *Locusta* (Vincent, 1972). The silkworm *Bombyx* appears to contain no bursicon, either in the adult or in the larva or pupa (Fraenkel and Hsiao, 1965). The location of bursicon has also been compared in the central nervous system and the CC of several species. In newly molted *Periplaneta* larvae and adults, it is present throughout the central nervous system and CC (Fraenkel and Hsiao, 1965). The brain, CC, and ventral nerve cord of stick insects and locusts also contain bursicon (Vincent, 1972).

Anterior retraction factor, PIF, and PTF were found in five genera of cyclorrhaphous flies, and PTF was identified in *Periplaneta* and *Pyrrhocoris.*

8.5. MODE OF ACTION

Like most other neurohormones, EH acts through cyclic nucleotides. Injection of cAMP or cGMP into isolated abdomens of *Hyalophora* produced the same behavioral effects as EH and significantly raised the cAMP content of the central nervous system, but cGMP is twice as potent as cAMP (Truman *et al.*, 1976, 1979). Injection of EH extracts into *Manduca* pharate pupae also quickly raised the cGMP level in the central nervous system, but only just before ecdysis (Morton and Truman, 1985). 20-Hydroxyecdysone appears to regulate the ability of EH to cause this rise. In addition, cGMP appears to serve as a second messenger in the muscle degeneration induced by EH (Schwartz and Truman, 1984b).

The action of cAMP in controlling sclerotization has been pointed out in many species. Its injection, or the injection of dibutyric cAMP, into ligatured insects induced tanning in *Calliphora* (Von Knorre *et al.*, 1972) and *Periplaneta* (Vanderberg and Mills, 1974). In *Tenebrio,* the action of cAMP in tanning processes was demonstrated by RIA of epidermal cAMP during development and after injection of bursicon-rich hemolymph (Delachambre *et al.*, 1979a,b). Epidermal cAMP increases in the pupa simultaneously with blood bursicon, but only in ligatured insects after injection of bursicon. Although bursicon is known to trigger the action of cAMP, the processes affected by that action are not known. Seligman *et al.* (1969) and Delachambre (1976) showed that cAMP activates the formation of tyrosine hydroxylase, thus increasing the rate of tyrosine conversion into dopa, and intervening at the very start of the metabolic chain in *Sarcophaga* and *Tenebrio.* This is surprising since ecdysone, which acts before bursicon, is assumed to stimulate dopa decarboxylase synthesis.

According to Mills *et al.* (1967b), Mills and Whitehead (1970), and Compton and Mills (1982), the role of bursicon is different in *Periplaneta,* in which it activates the membrane adenylcyclase of the hemocytes, thus triggering cAMP synthesis and increasing the permeability of the hemocyte membrane to tyrosine. The latter was reported to enter into contact with the tyrosine decarboxylase in the hemocytes and to be converted to tyramine and then dopamine, without prior conversion to dopa.

In flies, the factors that trigger pulsing and pumping after eclosion also involve cAMP (Fraenkel *et al.*, 1984).

In *Sarcophaga,* PTF also triggers the action of cAMP, but the action of ARF is not mimicked by any nucleotide (Seligman *et al.*, 1977). Injection of cAMP induces early tanning accompanied by incorporation of tyrosine into the cuticle. Confirmation of the part played by cAMP was provided by an experiment involving simultaneous injection of theophyllin and blood diluted to the limit of activity; tanning was completed thanks to theophyllin, which prevents the destruction of cAMP. Experiments were also conducted with actinomycin, which inhibits RNA synthesis and the protein synthesis inhibitors puromycin and cycloheximide, used either alone or combined with cAMP or neurohormones. The results suggest that PTF acts by accelerating the transcription of an mRNA necessary for tanning, while cAMP activates a process depending on the product of the gene derived by transcription of the gene activated by PTF.

8.6. CONCLUDING REMARKS

Investigation of the regulation of the processes connected with molting have demonstrated the existence of several distinct neurohormones: EH, bursicon, ARF, PIF, and PTF, which are found in the hemolymph at well-defined periods.

Eclosion hormone triggers the specific ecdysis behavior necessary for the shedding of the old cuticle and regulates two other processes linked to ecdysis-plasticization and intersegmentary muscle degeneration

Bursicon, which controls tanning, also affects plasticization, endocuticle secretion, disaggregation of wing fragments, and tanning-related diuresis. In *Manduca* the appearance of EH in the hemolymph causes the release of bursicon.

The specific tanning of the puparium of the higher dipterans is not controlled by bursicon but by a different factor, PTF, and puparium formation by retraction of the anterior segments depends on another neurohormone, ARF, which also causes immobilization of the larva.

An interesting finding is that the decline in ecdysteroid levels is of major importance, not only in eclosion, but also in pupariation, because it enhances tissue sensitivity to EH, triggers its release, and causes muscle and neuron degeneration.

The most numerous investigations concern flies and *Manduca,* but other species have also been shown to possess neurohormones that induce molting and tanning.

Bursicon, ARF, and PTF are distributed throughout the central nervous system in *Locusta, Periplaneta,* and *Tenebrio;* however, in *Sarcophaga* and *Manduca,* bursicon is located mostly in the abdominal ganglia. The CC and PO contain bursicon in most of the species investigated (*Periplaneta, Carausius, Schistocerca, Locusta, Tenebrio,* and *Manduca*) and are probably involved in its release. However, experiments on flies suggest that ARF and PTF are released at peripheral sites. As far as EH is concerned, its production and release sites in *Manduca* vary during the development of the insect; at first they take place in the ventral nerve cord and PO, but in the pharate adult they migrate to the brain–CC complex. In this species, EH and bursicon neurons were identified in the brain and abdominal ganglia, respectively.

Part of the molecular structure of *Bombyx* EH has been identified; the molecular weight of this neurohormone is 8500, while the bursicon molecule is distinctly larger, with a weight of 38,000–40,000; the molecular weights of ARF and PTF are larger still, i.e., 180,000 and 300,000, respectively.

cAMP has been demonstrated to be involved in the regulation of eclosion and sclerotization by PTF and bursicon, but the action of ARF does not bring this second messenger into play.

9

Pigment Synthesis and Breakdown—Color Change

The adaptation of insect color and pattern to the environment plays an important role in survival. Such adaptation is achieved both by selective pigment synthesis and breakdown, and by the migration of pigment granules within epidermal or subepidermal cells.

9.1. PIGMENT SYNTHESIS AND BREAKDOWN

Insects display a wide variety of coloration and are capable of altering their pigmentation in accordance with external and internal factors by the synthesis, in variable proportions, of pigments equally varied in color, located in the epidermal cells and the cuticle (Table XXVII).

9.1.1. Factors Involved in Pigment Synthesis

One frequently decisive factor is the color and texture of the environment to which the insects adapt their color and pattern, thus achieving a homochromy that plays a cryptic, protective role.

Maternal influences and gregariousness also play a part in insect pigmentation. The influence of gregariousness on locust pigmentation has been known for a long time. Gregarious locusts have a dark body color tending to black, while solitary insects have a light body color ranging from green to brownish yellow, depending on the background. Similar processes were observed more recently in the lepidopteran *Leucania separata* and in Australian stick insects (Key, 1957). In *Gryllus,* gregariousness influences pigment synthesis, but in a different way, the gregarious insects being lighter than the solitary ones.

Pigment changes may be associated with a particular type of development,

TABLE XXVII. Varied Colors Displayed by Pigments of Integument

Pigment color	Cuticle: Pterins	Cuticle: Melanins	Epidermis: Bile pigments	Epidermis: Carotenes	Epidermis: Ommochromes
Colorless					+
White	+				
Green		+			
Yellow	+	+	+	+	
Orange			+	+	
Brown		+	+		+
Red	+				+
Black		+	+		

such as diapause, or with a developmental stage as in the ontogenic color change of *Cerura*. Certain warning pigmentation playing a protective role against predators are known to develop at certain stages only when the insect is exposed; they are often created by areas of strongly contrasting colors. The O_2 and CO_2 content of the surrounding atmosphere also has a bearing, and possibly plays a part in the color changes associated with gregariousness. In *Locusta*, 1 min daily treatment of the insect with CO_2 induces the appearance of light pigmentation (Fuzeau-Braesch, 1968). Nutritional factors also play a part as shown in *Carausius*, in which ommochrome synthesis requires dietary tryptophan (Bückmann, 1985).

Apart from some cases in which light is directly perceived by the integument (Oltmer, 1968 in *Pieris brassicae*), the compound eye is responsible for mimicry. The removal of the optic lobes (Atzler, 1931) or their disconnection from the brain (Moreteau-Levita, 1972a) makes adaptation to the background impossible. Investigations performed by Bückmann and Düstmann (1962) and Düstmann (1964) showed that blackening or removal of the ventral half of the eye in *Carausius* causes the insect to darken.

Light inputs as well as other inputs from the environment are transmitted to the brain, which triggers the endocrine and neuroendocrine regulations involved in pigment synthesis and lysis. A number of investigations have been devoted to the study of the control of pigment metabolism. These investigations are difficult because cuticle pigmentation often masks the color of the epidermal cells and because several pigments are often contained in the same epidermal cell. Thus a precise determination of the varied pigments involved entails quantitative biochemical determinations.

9.1.2. Pigments

The pigments of the insect integument belong to many biochemical families: the cuticle contains melanins and pterins, the epidermis ommochromes, carotenoids, and bile pigments (insectoverdins).

Pterins are often white (pterin, leucopterin) or yellow (xanthopterin), sometimes red (erythropterin). Their basic structure is a pyrimidal-pyrazinic nucleus with four nitrogen atoms (Fig. 82). Like melanins, they are found in the vicinity of epidermal cells containing ommochromes (Düstmann, 1964 in *Cerura;* Berthold and Bückmann, 1976 in *Carausius*).

Depending on their degree of oxidation, melanins are black, brown, or yellow. They are high molecular weight compounds that are derived from tyrosine through the intermediary of dopa and dopamine, and one important intermediate is dopaquinone (Fig. 83).

Ommochromes are represented by ommines and ommatines, brown xanthommatine, dihydroxanthommatine, and 3-hydroxykynurenin. They are either low molecular weight ommatins or high molecular weight sulfur-containing ommines. These are tryptophan metabolites, often considered as excretion products of certain amino acids, whose hemolymphatic titer increases considerably during the proteolysis preceding metamorphosis (Bückmann *et al.,* 1966 in *Cerura*). Tryptophan and 3-hydroxykynurenin are converted to ommochromes, which are deposited in the epidermis and also in the gut and fat body. Ommochromes display redox properties and change color according to the presence of oxidizing factors. Reduced forms are red (dihydroxanthommatine) while oxidized forms are brown (xanthommatine); 3-hydroxykynurenin is colorless (Fig. 84).

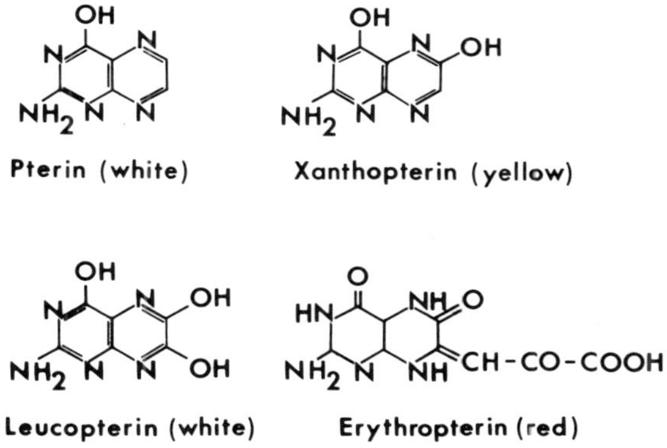

FIGURE 82. Molecular structure of pterins.

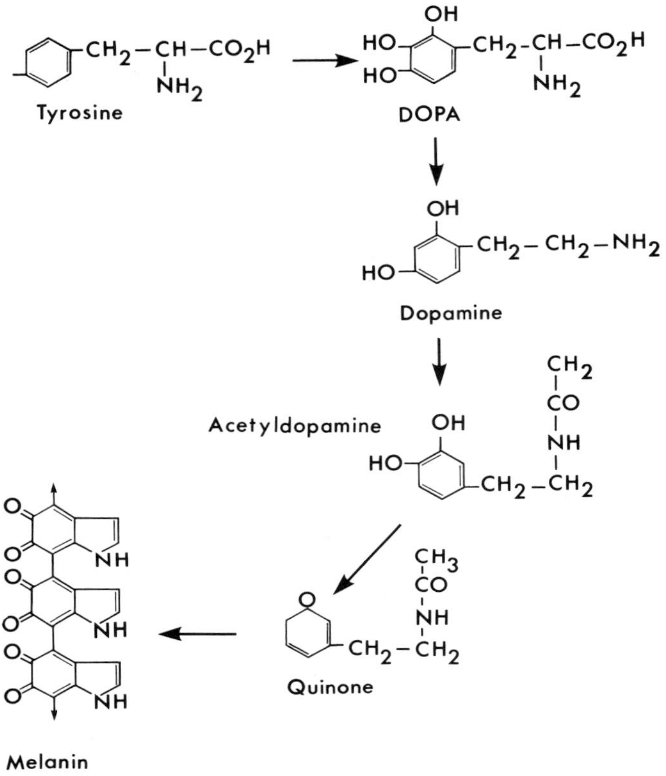

FIGURE 83. Molecular structure and synthetic pathway of melanins.

Bile pigments are blue or green linear tetrapyrrole molecules, termed biliverdins or insectoverdins, found in many insects. They are located in the epidermal cells as carotenoids and ommochromes. When oxidized they turn yellow, brown, or black (Fig. 85).

Carotenoids are yellow and orange. They are often found in epidermal cells containing green or blue bile pigments. They are of nutritional origin (Fig. 86).

9.1.3. Locusts, Grasshoppers, and Stick Insects

Pigment variations and their regulation are closely comparable among locusts, grasshoppers, and stick insects. These insects possess large numbers of variously colored pigments: 16 distinct pigments in *Carausius* and *Locusta*, i.e., several pterins, ommochromes, carotenoids, melanins, and biliverdins (Fig. 87).

In the red locust *Acrida turrita*, the grasshopper *Oedipoda cerulescens*, and

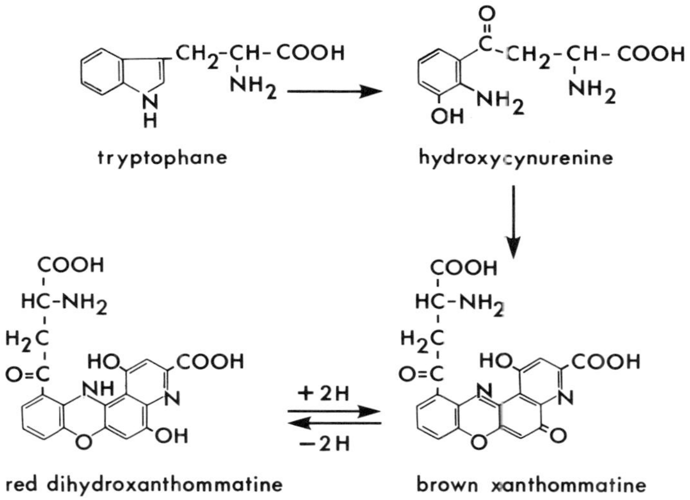

FIGURE 84. Molecular structure and synthetic pathway of ommochromes.

the stick insect *Carausius,* the insect harmonizes its pigmentation with that of the background. The colors range from green to yellow in *Acrida,* but exhibit a broader range in *Oedipoda* (yellow, red, gray, or black) and *Carausius* (green, beige, brown, or black). Pigment synthesis is influenced by JH, ecdysteroids, and neurohormones, whose interactions are not yet perfectly clear.

9.1.3.1. Corpora Allata (Table XXVIII)

Green body coloration depends on the CA in many grasshoppers (Rowell, 1967), in *Locusta* (Joly, 1951; Joly *et al.*, 1956; Staal, 1961) and *Acrida* (Joly, 1952), and in *Carausius* (Raabe, 1961; Düstmann, 1964). In *Acrida* and *Locusta,* the implantation of these organs results within 24 hr in pigmentation of the blood, which changes from yellow to green. The tegumentary change occurs

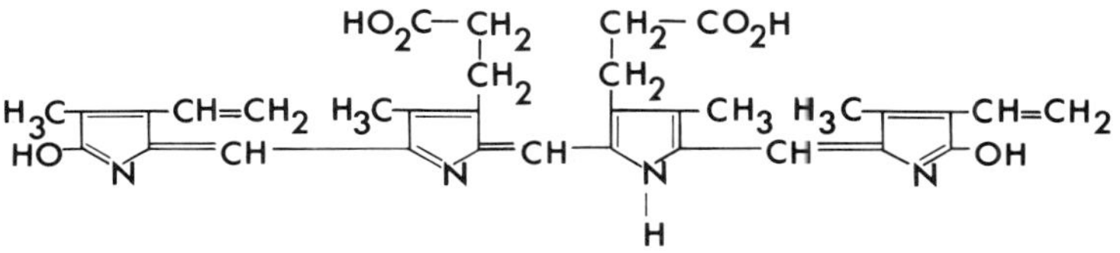

FIGURE 85. Molecular structure of insectoverdins.

FIGURE 86. Molecular structure of carotenoid pigments.

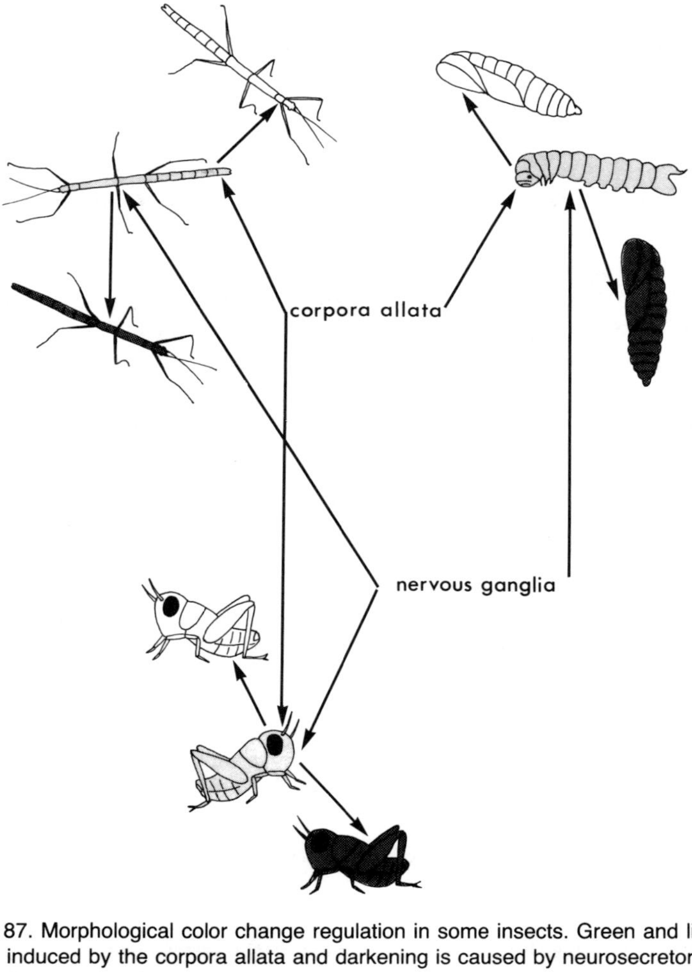

FIGURE 87. Morphological color change regulation in some insects. Green and light body color are induced by the corpora allata and darkening is caused by neurosecretory cells of the central nervous system.

TABLE XXVIII. Pigmentation Changes Induced by JH and Ecdysteroids in Locust and Stick Insects[a]

Insect	Operation	Pigmentation
Acrida	CA−	Green to beige
Carausius	CA−	Green to beige
Locusta	CA+	Beige to green
Carausius	CA+	Green to dark
	PG+	Green to dark

[a]−, Removal; +, implantation.

more slowly and only becomes apparent after molting. It involves two distinct processes: the elimination of melanins and ommochromes, and the synthesis of biliverdins. When the greening of gregarious locusts is induced by CO_2 treatment, the hemolymphatic JH increases markedly (Fuzeau-Braesch *et al.*, 1982).

The maintenance of the green pigmentation also depends on the CA. Their removal is followed after the next molt by the appearance of a gregarious pigmentation in *Locusta* (Joly, 1951) and a beige pigmentation in *Carausius* (L'Hélias, 1955; Raabe, 1961). In both insects, the insectoverdins vanish. In *Carausius,* however, the implantation of CA or the disconnection of these organs leads, paradoxically, to the formation of ommochromes and the appearance of a black body color (Pflugfelder, 1938; L'Hélias, 1955; Raabe, 1961; Düstmann, 1964). This effect is probably due to some interaction between the JH and a darkening neurohormone.

Other correlations exist between pigmentation and CA. In locusts, these organs induce in males yellow body color at sexual maturity (Loher, 1961; Pener, 1965; Pener *et al.*, 1972). In crickets, they are responsible for lightening the body color, which accompanies gregarious life (Roussel, 1966).

9.1.3.2. Prothoracic Gland (Table XXIX)

The role of the PG in pigmentation is not clear. In *Schistocerca,* partial removal of the PG in solitary green larvae gives rise to a yellow body color spotted with black (Ellis and Carlisle, 1961), but the implantation of this gland in *Carausius* causes browning of the light insects by ommochrome synthesis (Raabe, 1961).

9.1.3.3. Neurohormones

The action of neurohormones in the synthesis of dark ommochromes and melanins appears to be established. The source of the active factor, however,

TABLE XXIX. Role of Neurohormones, Juvenile Hormone, and Ecdysteroids in the Synthesis of Melanins and Ommochromes in Several Lepidopterans[a]

	Juvenile hormone		Ecdysteroids		Neurohormones	
Insect	m	o	m	o	m	o
Bombyx	↑	↑			↑	
Cephonodes	↓	↓				
Cerura				↑ ↓		
Hestina						↑
Leucania					↑	
Mamestra						↑
Manduca	↓		↓		↑	
Papilio					↑	
Pieris				↑ ↓	↑	
Spodoptera					↑	

[a]m, Melanins; o, ommochromes.

differs from one species to another. In *Locusta* and *Oedipoda,* the PI and CC contain a darkening factor (Staal, 1961; Moreteau-Lévita, 1972b,c) originating in the C cells of the PI (Girardie, 1967), and in the lateral cells of protocerebrum (Girardie and Cazal, 1965). In *Carausius,* the implantation of the PI or the entire brain is without effect; but the implantation of the subesophageal ganglion causes darkening (Raabe, 1966), and it is possible that the darkening factor originates in two medioventral cells of the subesophageal ganglion, since their histological appearance is quite different depending on body color. In *Schistocerca,* first instar larvae melanization seems to be under the influence of a factor issuing from the metathoracic ganglion (Padgham, 1976a,b).

9.1.4. Lepidopteran Larvae and Pupae

Color changes are frequently encountered in lepidopterans, often occurring in the pupa and sometimes in the larva. They may be associated with diapause or with other developmental stages. Environmental factors such as light, background, temperature, and diet frequently determine the color change, which may also depend on crowding as in *Cephonodes hylas, Leucania,* and *Spodoptera.*

Pigmentation modifications occur either in larvae (*Cephonodes, Cerura, Hestina, Leucania, Mamestra,* and *Manduca*) or in pupae (*Bombyx, Papilio, Pieris*), whose pigmentation depends on the environmental conditions encountered during the last larval stage.

Cerura vinula represents a special case because its prepuparial pigment changes are mostly independent of environmental conditions, except temperature, and correspond to physiological ontogenic stages. The green larvae bearing a light brown dorsal pattern turn uniformly dark red when they stop feeding just before pupation. The red dihydroxanthommatine that is formed results from the reduction of the brown xanthommatine in the dorsal part of the caterpillar and from xanthommatine synthesis in the ventral part, the reactions of both dorsal and ventral parts being independent. Subsequently, the pigment is converted to colorless 3-hydroxykynurenin (Linzen and Bückmann, 1961) and then degraded. At the same time, rhodommatine and ommatine synthesis take place in the gut and the fat body (Fig. 88).

The pigmentation of *Manduca* includes black melanins, pink ommochromes, namely, dihydroxanthommatine, and blue bile pigments. The saturniid larva *Hestina japonica* changes color from green to brown in autumn, after cessation of feeding, the tetrapyrrolic pigments being replaced by an ommochrome (Osanai, 1966). The reverse occurs in spring during the last larval instar.

Pupae of *Papilio xuthus* and *protenor* exhibit a variety of tints ranging from green to brown, which is also related to the type of development (with or without diapause) and to texture, color, and scent of the pupational host plant, which induces a green coloration even in darkness. The pigments are essentially mesobiliverdins, carotenoids, and melanins (Ohnishi and Hidaka, 1957). The coloration of the pupae of *Pieris rapae* and *brassicae* depends on the light intensity and background color perceived during a short period preceding pupation and varies

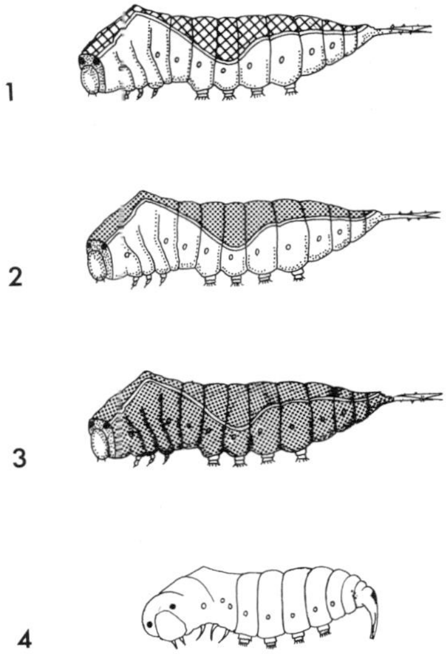

FIGURE 88. The ontogenic color change in *Cerura*. (1) Last instar larva before cessation of feeding. (2) Last instar wandering larva. The brow ommochromes of the dorsal patch had turned red. (3) The reddening spread to the ventral green part of the body. (4) Lightening of the prepupa. (Modified from Bückmann, 1974b.)

from light green to spotted black. The epidermal pigments are mesobiliverdins, yellow carotenoids, and reddish-brown ommochromes, while the cuticle is either transparent or melanized (Fig. 89).

The pupa of *Bombyx,* which varies from black to amber yellow, changes color in response to genetic factors and to the temperature experienced by the larvae from the end of spinning until pupation. The cuticle melanins responsible for the black pigmentation are deposited at low temperature in certain strains; at 30°C they are absent from the integument, which assumes an amber color (Hashiguchi, 1964).

9.1.4.1. Corpora Allata (Table XXX)

Green pigment synthesis and dark pigment degradation appear to depend in lepidopterans, as in orthopterans, on the secretory activity of the CA. Indeed, JH or JHA application stimulates the deposition of green pigments in epidermal cells and inhibits the synthesis of both melanins and ommochromes, whereas CA extirpation causes synthesis of the latter pigments (Hidaka and Ohtaki, 1963; Lhonoré *et al.,* 1980 in *Pieris;* Ikemoto, 1983 in *Cephonodes;* Hori and Riddiford, 1982; Curtis *et al.,* 1984 in *Manduca*).

In some species, however, such as *Leucania, Mamestra,* and *Bombyx,* JH effects are diametrically opposed. In *Leucania,* where the role of a neurohormone, the melanin and red coloration hormone (MRCH), was established for both melanin synthesis and ommochrome synthesis, JH deficiency inhibits melanization (Ogura, 1975). In *Mamestra,* ommochrome synthesis requires both JH and MRCH during the period of new cuticle formation (Hiruma *et al.,* 1984). The implantation of CA or the injection of JH into *Bombyx* larvae gives rise to a change in coloration from light gray to brownish yellow due to ommochrome synthesis (Kiguchi, 1972; Ohashi *et al.,* 1983).

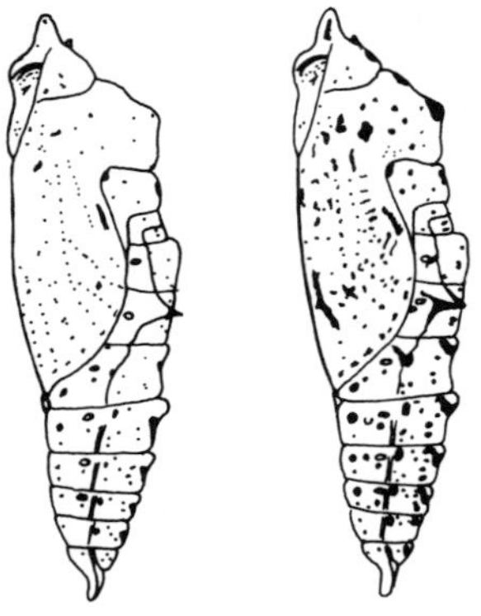

FIGURE 89. Variations in the pigmentation of the pupa of *Pieris brassicae* depending on light intensity and background color. (Modified from Bückmann, 1971.)

TABLE XXX. Origin of the Darkening Factor in Some Insects

Insect	Inducing factors	Color change	Synthesized pigments	Organs involved in pigment synthesis								
				CC	Brain	sgg	th1	th2	th3	ab1	ab2	ab3 . . . ab7
Locusta[a]	Gregariousness	Green to brown-black		O	O							
			Melanins						O			
Schistocerca	Background color	Darkening		O	O	O						
Oedipoda[a]		Blackening										
Carausius	Heat and darkness	Green to black	Ommochromes			O						
Hestina	Diapause	Green to brown	Ommochromes									O-O-O-O-O
Leucania	Crowding	Yellow to brown	Melanins	O—	—O	O						
Papilio	Host plant	Green to brown	Melanins		O—	—O—	—O					
Pieris	Light background color	Green to ±black	Melanins		O—	—O—	—O					
Bombyx	Genetic factors, temperature	Yellow to black	Melanins		O—	—O—	—O—	—O—	—O			

[a]The pigments involved are poorly known; they might be melanins, ommochromes, or oxidized tetrapyrrolic pigments. O, organ involved; O——O, organs involved whose connections cannot be interrupted; ab1, 2, 3 . . . 7, abdominal ganglia 1, 2, 3 . . . 7; CC, corpora cardiaca; sgg, subesophageal ganglion; th1, th2, th3, thoracic ganglia 1, 2, 3.

Investigations were conducted to determine which JH was involved in pigment synthesis. In both *Cephonodes hylas* (Ikemoto, 1983) and *Mamestra* (Hiruma *et al.*, 1984), JH III is much less active than JH I and JH II. In addition, JH I displays maximum activity in *Mamestra*.

9.1.4.2. Inhibitory Factor

A number of studies on *Pieris rapae* (Ohtaki, 1963), and *P. brassicae* (Oltmer, 1968; Bückmann, 1971), *Vanessa io* (Bückmann, 1960), and several nymphalids (Koch and Bückmann, 1984) suggested the existence of a melanin synthesis-inhibiting factor. This hypothesis is based on experiments in which dark ligated insects showed themselves unable to lighten the posterior portion of the body. In *Inachis io* (Maisch and Bückmann, 1987), this factor reduces pupal melanization and increases the carotenoid content. The source of the inhibitory factor is unknown at present. Although the CA might be suspected, it was shown in *Inachis* that neither ecdysteroids nor JH was involved.

9.1.4.3. Ecdysone (Table XXX)

As shown above, color changes encountered in *Cerura* (Bückmann, 1956a,b, 1959) correspond to an ontogenic color change in which ecdysone plays a prominent role. It is interesting to observe that very low doses are necessary for epidermal reddening due to the activation of the conversion of 3-hydroxykynurenin into ommochromes (Bückmann, 1974b), while large doses reached during the last large ecdysteroid peak cause the lysis of ommochromes which precede the pupal ecdysis (Bückmann, 1974a). The time of ommochrome synthesis induction by ecdysteroids is distinct from the critical molting period, which occurs 6 days after the critical period of ommochrome synthesis. It is thus possible that the amounts of ecdysone required for the latter process are less than those required for molting.

The role of 20-HE in pigmentation control was demonstrated in two other lepidopterans. In *Pieris,* the effects induced by ecdysone depend on the amounts of the hormone in the hemolymph (Bouthier and Lafont, 1976; Beydon *et al.*, 1980). When JH is present, low ecdysteroid doses trigger ommochrome synthesis, but larger doses, similar to those reached in the hemolymph of prewandering larvae, induce ommochrome lysis. 20-Hydroxyecdysone is also involved in melanin synthesis, which it postpones in *Manduca.* The decline of hemolymph ecdysteroids correlates with the onset of melanization (Curtis *et al.*, 1984). On the other hand, in *Araschnia levana,* the seasonal morphs of the adult are regulated by the timing of ecdysteroid release (Koch and Bückmann, 1987).

9.1.4.4. Neurohormones

The synthesis of dark brown or black pigments, ommochromes, and especially melanins depends on one or more neurohormones produced or released into different parts of the ventral nerve cord. This is demonstrated by the removal and reimplantation of organs, and the cutting of connectives and ligatures performed on different regions of the body, as shown below.

In *Papilio,* experimental investigations (Hidaka, 1956, 1961) revealed the control mechanisms governing the color change from green to brown. Ligatures were placed on young prepupae after the critical molting period. Wherever the ligature was placed, the green pigmentation was unaffected, but the ligature prevented browning of the posterior part of the body. Hence a darkening factor appears to originate in the anterior region of the body. The removal of different nerve ganglia and the interruption of their connectives shows that the brain, the subesophageal ganglion, and the prothoracic ganglion play a major role in the browning process. By implanting different ganglia into ligatured larvae, it is observed that the brain, the subesophageal ganglion, and the prothoracic ganglion are active only if interlinked. One may thus reasonably ask whether the three ganglia play the same role, or whether some of them only intervene by triggering and transmitting a nerve command while the others secrete and/or release the active factor. If the nerve ganglia are removed at different times, it may be observed that the critical period of browning occurs earlier for the brain and for the subesophageal ganglion than for the prothoracic ganglion, which may be the final source of the active factor.

Green *Pieris* larvae transferred to constant darkness develop into brown pupae. If they are ligated between the thorax and the abdomen, the anterior portion of the body turns brown, while the posterior part remains green. As in *Papilio,* the brain, subesophageal and prothoracic ganglia must be present and have intact connections in order to allow synthesis of the dark pigments (Ohtaki, 1960).

Experiments were also conducted on *Bombyx* prepupae, which are transformed into black pupae when raised at 30°C. In this case, it is the complex formed by the brain, subesophageal ganglion, and the three thoracic ganglia which is necessary for synthesis of the black melanic pigment, which cannot take place if one of these ganglia is removed. Synthesis is restored by the implantation of all five ganglia. By removing one of the five at different times, the role of each can be clarified. It appears that in *Bombyx,* the metathoracic ganglion plays the part performed by the prothoracic ganglion in *Papilio* (Hashiguchi, 1960, 1964).

The results obtained on *Leucania, Spodoptera,* and *Hestina* larvae are slightly different. A melanization factor is present in *Leucania* in the brain–CC complex, and also in the subesophageal ganglion (Ogura, 1975), but the nerve connectives are not indispensable for the functioning of the two nerve ganglia. The subesophageal ganglion is also active *in vitro* (Ogura and Mitsuhashi, 1979). In *Spodoptera,* of which the study is just beginning, the subesophageal ganglion also contains a darkening factor (Yagi, 1980). In *Hestina,* which darkens by the formation of ommochromes, ligatures reveal an active factor released between the third and seventh abdominal segment (Kiguchi, 1972) (Table XXX). Surprisingly, in *Manduca,* neither central nervous system or CC extracts, nor purified MRCH can induce melanization (Curtis *et al.*, 1984).

The data concerning *Papilio, Pieris,* and *Bombyx* show that the release sites of the darkening hormones are distant from their production sites. Ultrastructural investigations carried out in *Papilio* confirm the hypothesis of a migration of the head ns products up to the prothoracic ganglion. In fact, the ns granule distribution is different in the subesophageal and thoracic ganglia. In the subesophageal ganglion, these granules are mostly present in axonal pathways, whereas in the prothoracic ganglion, they accumulate in the neuropil, namely, in larvae destined to become dark (Awiti and Hidaka, 1982).

The functioning of these complexes, which extend to the prothoracic ganglion and even the metathoracic ganglion, can be interpreted in different ways. While it is probable that the brain controls the secretion and release processes, the precise location of these processes is far more difficult to pinpoint. It is conceivable that synthesis and release take place in each ganglion, but it is also possible that the brain alone synthesizes the factor, which is transported along the ventral nerve cord, where it is then released in one or more sites.

9.1.5. Action Mechanism of Hormones and Neurohormones in Pigmentation

9.1.5.1. Melanins

Metabolic processes involved in melanin synthesis were thoroughly investigated in *Manduca* antepenultimate larval instar (Hori *et al.*, 1984; Curtis *et al.*, 1984; Hiruma and Riddiford, 1985, 1986; Hiruma *et al.*, 1985). The methods employed, electron microscopy, determination of phenoloxidase activity in the premelanin granule and dopa decarboxylase (DDC) activity helped to clarify the timing of the appearance of the main steps in the melanization process in allectomized insects and in black mutant insects. This timing was established in relation to head capsule slippage (HCS) (Fig. 90).

In *Manduca,* melanization takes place 3 hr before ecdysis but is preceded by different processes. Thirteen hours before HCS, the premelanin granules containing an inactive phenoloxidase are deposited. Fifteen to sixteen hours after HCS, DDC activation takes place. This enzyme is very important for subsequent processes because it is required for the conversion of dopa to dopamine, the main

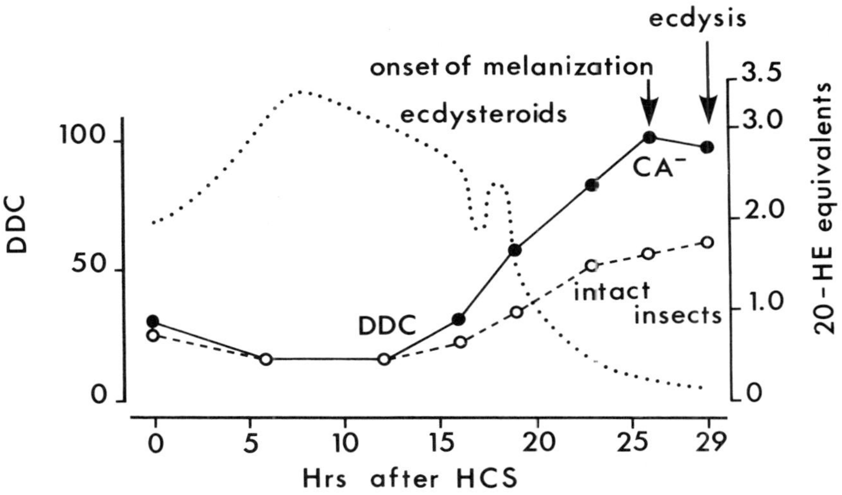

FIGURE 90. Dopa decarboxylase activity in the epidermis of the last instar larvae of *Manduca*. Normal intact insects (○); allatectomized larvae (●). The titer of hemolymphatic ecdysteroids is indicated(· · ·). (Modified from Hiruma and Riddiford, 1985.)

precursor of melanin synthesis in this insect. Twenty-two hours after HCS, the activation of prophenoloxidase occurs. This enzyme is required for the conversion of dopa and dopamine to their corresponding quinones and further melanization.

As shown above, the absence of JH 5–6 hr before HCS is a prerequisite for premelanin granule deposition and DDC activation. Besides, DDC activation, as well as the activation of the diphenoloxidase contained in the cuticular granules, necessitates a drop in the titer of the hemolymphatic ecdysteroids. Just how does this occur?

The study of the ecdysteroid hemolymphatic titers at the time of the second ecdysteroid drop in intact and allatectomized larvae (Fig. 91) reveals that the general pattern, similar in both types of insect, exhibits not a single drop, but a slight fall followed by a short rise and a second large fall. Thus two successive drops occur in ecdysteroid titer, the second more important than the first. If the first is involved in melanization, the second may intervene in later processes, such as sclerotization, which also requires DDC activation (see Chapter 8). In allatectomized insects, the appearance of both drops in ecdysteroid titers are delayed by 4 hr, suggesting that the absence of JH stimulates ecdysone production in some respects. This effect recalls other inhibitory effects of JH on PG, as observed in lepidopterans (see Chapter 3).

20-Hydroxyecdysone may inhibit DDC mRNA transcription since mRNA

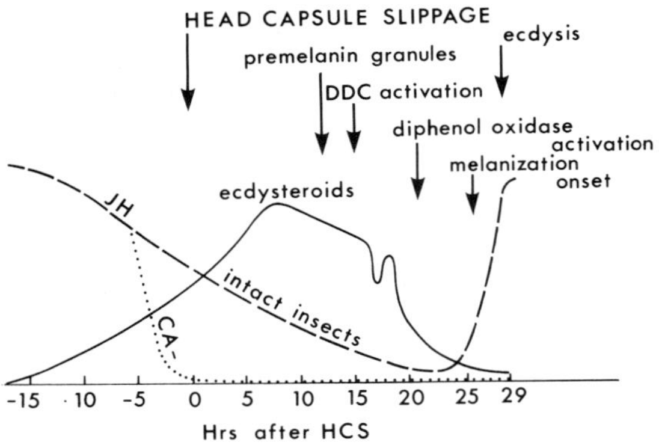

FIGURE 91. Pigment metabolism and endocrine events in the last larval instar of *Manduca*. CA−, Allatectomized larvae. (From the data of Fain and Riddiford, 1975; Curtis *et al.*, 1984; Hiruma and Riddiford, 1986.)

inhibitors prevent the increase of DDC activity. In the embryo of *Drosophila,* in fact, the fall in ecdysteroid titer causes an increase in the DDC transcription. In *Calliphora,* however, it is a rise in ecdysteroid titer that induces DDC mRN synthesis at pupariation.

Ommochrome synthesis is known to depend on the conversion of 3-hydroxykynurenin, which itself results from the conversion of kynurenin. In *Manduca,* it has been shown that the conversion of 3-hydroxykynurenin to ommochromes requires a low level of JH, which permits the activation of kynurenin-3-hydroxylase (Hori and Riddiford, 1982). Ommochrome metabolism also depends on ecdysone titer, as already shown; low levels stimulate their synthesis, while high levels cause their lysis. The action of ecdysone is not exerted on tryptophan, which in its absence is converted to 3-hydroxykynurenin and eliminated, but rather on 3-hydroxykynurenin, which it causes to be converted to ommochromes (Bückmann, 1974a). The action of ecdysone may also occur in another way, with this hormone acting on tissue breakdown, resulting in proteolysis and an increase in the hemolymph tryptophan titer. Furthermore, ommochrome synthesis, which frequently occurs in areas rich in cuticular melanin, may be stimulated by an excess of dopaquinone in epidermal cells after the progress of melanization.

The action mechanism of neurohormones in melanin and ommochrome synthesis is poorly known. Yet it has been observed that cyclic nucleotides, when injected into ligated abdomens of *Leucania* larvae, cause a strong darkening of body color, suggesting that cAMP acts as a second messenger in MRCH stimulation of melanization (Matsumoto *et al.*, 1979).

9.1.6. Purification of MRCH

The earliest attempts to purify lepidopteran melanization neurohormone were made by Hashiguchi *et al.* (1965) using the thoracic ganglia of *Bombyx* pupae. Later, the purification of MRCH was thoroughly investigated (Suzuki *et al.*, 1976; Matsumoto *et al.*, 1985, 1986). Extractions were made from adult *Bombyx* heads according to a 13-step protocol. Three active fractions were demonstrated whose sequencing, begun in 1985, is nearly complete (Matsumoto *et al.*, 1986). The insect used as an assay for MRCH was the abdomen-ligated *Leucania,* whose color change is more spectacular than that of *Bombyx,* with cuticular melanins and red epidermal ommochromes being synthesized according to gregariousness.

Three active peptides were isolated: they are thermostable, with a molecular weight of 6400–8000. The N-terminal amino acid sequences are as shown below. Note that the three molecules display strong similarities and that the strong similarities between MRCHI and MRCHIII are also present in the carboxy terminal sequence, which is also very similar to human insulin-like growth factor II (IGFII).

	1 5 10 15
MRCHIII	H-Pro-Leu -Ser -Glu-Asp-Met-Pro-Ala-Thr-Pro -Ala-Asp-Gln-Glu-Met-Tyr
MRCHI	H-Leu -Ser -Glu-Asp-Met-Pro-Ala-Thr-Pro -Ala-Asp-Gln-Glu-Met-Tyr-Gln-
MRCHII	H-*** -*** -Glu-Asp-Met-Pro-Ala-Thr-Pro -Ala-Asp-Gln-Glu-Met-Tyr-Gln-
IGFII	-Ala -Leu-Leu -Glu-Thr -Tyr -Cys -Ala-Thr-Pro-Ala-Lys -Ser -Glu-OH
	55 60 65

Purified *Bombyx* MRCH causes cuticle melanization in other lepidopterans such as *Spodoptera* (Yagi, 1980), *Leucania,* and *Mamestra*. It has no effect on eclosion and diapause (Matsumoto *et al.*, 1981).

A pupal melanization-reducing factor was purified by Bückmann and Maisch (1987). It is inactivated by pepsin and pronase and its molecular weight is between 10,000 and 5000.

9.2. PIGMENT MIGRATIONS

Physiological color change in insects often displays circadian rhythm. It results from pigment migrations within chromatophores or within epidermal cells.

9.2.1. *Corethra*

In *Corethra,* large chromatophores containing black ommochromes cover the dorsal face of two pairs of anterior and posterior tracheal bladders. These ameboid chromatophores exhibit reversible reactions. On a black background they form a continuous coating of jointed hexagonal cells. On a light background, they contract and gather; the bladder surface becomes dotted with small, dark patchs dispersed on a transparent background (Fig. 92).

Based on ligature and eye-blinding experiments, Teissier (1947) showed that chromatophore function is humorally controlled and ordered by the head. Organ extract injections (Dupont-Raabe, 1949, 1957; Gersch, 1956) and implants (Hadorn and Frizzi, 1949) show that a chromactivating factor is located throughout the central nervous system, causing expansion of the chromatophores. Dilutions of different nerve ganglia extracts and implant experiments demonstrated that the brain is more active than the ventral ganglia and that the CC exhibit lower activity than the brain (Dupont-Raabe, 1949, 1957; Hadorn and Frizzi, 1949).

Hormone release from the ventral nerve cord ganglia was confirmed by ligature (Dupont-Raabe, 1949) and electrical stimulation applied to the abdominal ganglia, separated by a ligature from the anterior region (Gersch, 1956).

Isolation of a segment containing bladders with expanded chromatophores causes contraction of the latter (Dupont-Raabe, 1949, 1957; Hadorn and Frizzi, 1949). Hence it appears that the contraction stage represents a state of rest of the chromatophore, which it reaches in the absence of darkening hormone, and the involvment of a lightening hormone in color change does not seem obvious.

A

B

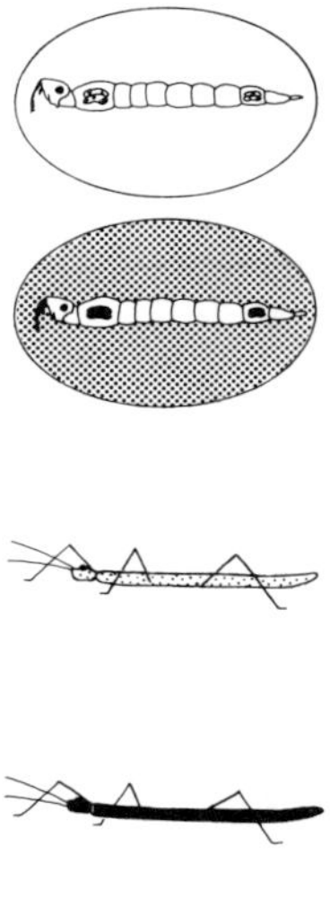

FIGURE 92. Physiological color changes in insects. (A) Contraction and expansion of the chromatophores covering the bladders according to the background in *Corethra.* (B) Circadian color change from grey to black due to pigment migration inside epidermal cells in the stick insect *Carausius.* (Modified from Dupont-Raabe, 1957.)

9.2.2. Dragonflies

Color change in dragonflies depends mostly on temperature. It is partly direct, but a darkening factor from the central nervous system is also involved (Veron, 1973).

9.2.3. Stick Insects

The stick insect *Carausius* displays a circadian color change which affects the entire body and occurs at dusk in varying degrees, depending on the nature and amount of the pigments contained in the epidermal cells. The green insects remain unchanged since the insectoverdins do not move; the beige and gray ones become brown or completely black due to the migration of the dark ommochromes from the base of the epidermal cell to its surface. Orange-yellow pigments spread in the median zone of the cell.

The processes underlying ommochrome migration were investigated by immunofluorescence microscopy using monospecific antitubulin antibodies (Berthold, 1980, 1982). A large system of microtubules attached to both distal and apical cell membrane was demonstrated in *Carausius* epidermal cells, representing a firm skeleton for the movements of the ommochrome granules. The study of the occurrence of Mg^{2+}- and Ca^{2+}-stimulated ATPases that could be involved in the production of the energy necessary for pigment granule migration revealed the association of Mg^{2+}-stimulated ATPases with the microtubular system.

Investigations carried out 50–60 years ago demonstrated that the circadian color change displayed by *Carausius* was humorally regulated by the brain (Giersberg, 1928; Atzler, 1931). Subsequent investigations provided further knowledge of the process (Dupont-Raabe, 1951, 1957; Mothes, 1960).

Like the brain, the subesophageal ganglion was shown to play a prominent role in color change. The removal or simple disconnection of both ganglia eliminates any possibility of darkening, whereas this does not occur when the optic lobes or other nerve ganglia are removed (Dupont-Raabe, 1951, 1957). Injections of organ extracts into insects incapable of darkening by lack of a brain show that an active factor is present throughout the central nervous system with a varied distribution. Its concentration in the brain and subesophageal ganglion is very high, inducing a maximum reaction with an injection equivalent to 1/1000 organ, whereas in the mesothoracic ganglion and the second abdominal ganglion the activity is lost at doses of 1/100 and 1/20, respectively (Dupont-Raabe, 1957) (Table XXXI).

With respect to CC, it was shown that these organs, which are not necessary for the color change to take place, contain abundant amounts of a factor which exhibits medium activity on *Carausius* pigments but high activity on crustacean chromatophores.

TABLE XXXI. Compared Activity of Different Nervous and Glandular Organs[a]

Donor insect	Donor organ	*Corethra* and *Carausius* mean color change with increasing dilutions			
		Corethra chromatophores			
Coretha		1/3	1/30	1/300	
	Periaortic ring (corpora allata + PG)	1.6			
	Corpora cardiaca	3.8			
	Optic lobes	3			
	Brain	4.8	4.7	4.1	
	Ventral ganglia	4.2	1.5		
		Carausius epidermis			
Carausius		1	1/10	1/20	1/100
	Brain	5		4.9	4.9
	Subesophageal ganglion	4.8		5	4.7
	Mesothoracic ganglion			4.3	1.6
	Connectives	4.6		1.7	
	2nd abdominal ganglion	4.2		1.3	
	Corpora cardiaca	3.6	3.1		

[a]The *Corethra* chromatophore expansion and *Carausius* body darkening are estimated according to a five-stage scale which enables us to calculate the mean reaction. 5, maximum expansion in *Corethra* and darkening in *Carausius*; 3, intermediate stage; 1, chromatophore contraction in *Corethra* and lightening of the body color in *Carausius*. (From Dupont-Raabe, 1957, modified.)

9.2.3.1. Intracerebral Location

The location of the brain neurons producing the chromactivating factor was examined by selective removal of different parts of this organ and by the injection of their extracts. The results pointed to the tritocerebrum exclusively (Dupont-Raabe, 1954, 1957). While ablation of the PI, even combined with CC removal, did not interfere with color change, the ablation of the tritocerebral parts of the brain interrupted the circadian darkening of the insects. In addition, only the tritocerebrum homogenates were able to reestablish the darkening of debrained insects. *In vitro* experiments conducted with pieces of integument and purified factors C1 and D1 from the central nervous system of *Periplaneta* (Mothes, 1960) also identified the tritocerebrum for the factor C2 which causes darkening, but indicated that the factor D1 darkening in low concentrations and lightening in high concentration was present in the PI.

9.2.3.2. Release Site

Corpora cardiaca do not release the cerebral chromactivating factor because their removal, even when accompanied by that of the PI, allows normal color change to subsist (Dupont-Raabe, 1957). Furthermore, the determination of factors C1 and D1 in CC showed that they fail to exhibit any circadian change, unlike the brain and the ventral nerve cord ganglia (Mothes, 1960) (see below). The chromactivating factor of the ventral nerve cord is also absent from the PO (Raabe *et al.*, 1966). Thus other nh structures not identified yet must be involved in the release of these factors.

The production and release of factors C1 and D1 were investigated *in vitro* (Mothes, 1960) by evaluating their concentration in the hemolymph and in the different nerve ganglia during the circadian cycle. The hemolymph titer of both factors is low by day, increases rapidly at dusk (the time when hormone release occurs), and remains high during the night. The amount of chromactivating factor contained in the central nervous system also varies during the circadian cycle, but the variations are quite different in the brain and the ventral nerve cord. The brain is loaded by day and reaches a maximum in the late afternoon, just before the time when the hormone titer in the hemolymph increases. In the ventral cord, however, the storage is observed by night, which is difficult to explain.

9.2.4. Interspecific Activity

Carausius organ extracts, which are active on *Carausius,* are also active on *Corethra,* and extracts of *Corethra* heads cause darkening in *Carausius* (Dupont-Raabe, 1949, 1957). It may therefore be supposed that the factors involved might be similar molecules. These factors are also present in species lacking color change (Dupont-Raabe, 1957), such as the cockroaches *Blabera fusca* and *Blatta orientalis,* the grasshopper *Tettigonia viridissima,* the bug *Notonecta glauca,* and the mosquito *Culex apicalis.* Organ extracts from these species cause the expansion of *Corethra* chromatophores and the migration of epidermal pigments in *Carausius.* Thus it is probable that apart from its action on pigments, the chromactivating factor is involved in another physiological regulation. It is interesting to note that the location of the factor is identical among insects. Absent from the PI, it is located in the lateral parts of the protocerebrum and in the tritocerebrum, where ns cells were demonstrated (Raabe, 1963a,c).

9.2.5. Purification of the Neurohormones

The existence of a substance in the heads of many insects that acts on crustacean chromatophores causing concentration of the yellow and red pigments of the prawns and the black pigments of the shrimp *Crangon vulgaris* and also acts on crabs by inducing expansion of the melanophores was revealed 50 years ago (Hanström, 1940). This substance, which is located in the CC (Brown and Meglitsch, 1940; Thomsen, 1943; Dupont-Raabe, 1952a), is found in many but not all insect species. Its source is thus quite different from that of the insect chromactivating factor found in the central nervous system.

Preliminary separation and comparison of the chromactivating factors of the brain and CC of *Carausius* on the one hand, and of the sinus gland of the prawn *Leander serratus* on the other, was carried out by electrophoresis (Carlisle *et al.*, 1955; Knowles *et al.*, 1955). These experiments revealed the existence of two distinct factors termed factors A and C, respectively. Factor A, present in CC and in the sinus gland, causes contraction of red and yellow chromatophores in *Leander,* contraction of the black chromatophores in *Crangon,* but has a slight effect on *Carausius*. Factor C, which has no effect on *Leander,* nevertheless acts on *Crangon* and exerts a strong action on *Carausius*'s black pigments. It is present in the brain of *Carausius* and absent from the CC (Fig. 93).

Recent studies, including the identification of the molecular structure of neurohormones of the AKH family involved in metabolism regulation, revealed that these molecules resemble the molecule of the crustacean red pigment concentrating hormone (RPCH) (see Chapter 11) which explains the reciprocal effects observed. It remains to demonstrate the molecular structure of the brain chromactivating neurohormone and to understand its physiological role in species deprived of a rapid color change. Gersch and co-workers (1957, 1960) conducted thorough investigations in order to purify the neurohormones, as reported above. Neurohormones C and D, and then C_1 and D_1, were separated from the central nervous system of *Periplaneta* and *Carausius*. These factors are active on pigment movements in *Carausius* and *Corethra* (Gersch, 1956; Mothes, 1960) and act on heartbeat, endogenous nerve activity, and crustacean chromatophores. Successive purifications led to the production of crystallized neurohormone D (Gersch *et al.*, 1960) with a molecular weight of 2000.

9.2.6. Summary

The regulation of pigment synthesis and breakdown was extensively investigated in lepidopteran larvae and also in stick insects and locusts. Although our knowledge has advanced, thanks in particular to the identification of the chemical nature of pigments and of their metabolic pathways, many questions remain unanswered.

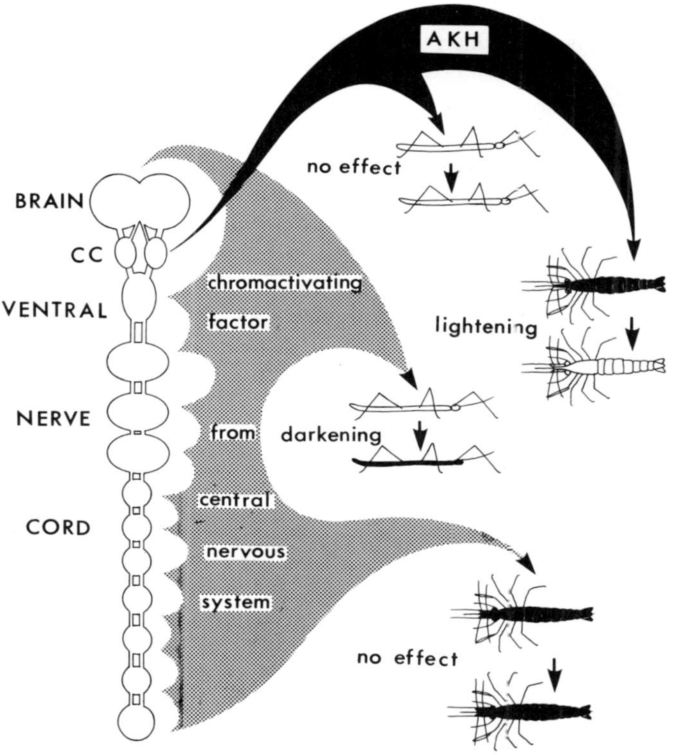

FIGURE 93. Insect chromactivating factor and adipokinetic hormone. Production sites and effects in insects and crustaceans.

Neurohormones play an important role in the stimulation of ommochrome and melanin synthesis. The molecular structure of the melanization and reddish coloration hormone MRCH has been nearly completely characterized. Three distinct peptides were identified displaying a sequence homology with the human insulin-like growth factor II.

The sites of synthesis and release of neurohormones involved in the synthesis of ommochromes and melanins appear to vary. They are limited to the brain or extend along the ventral nerve cord. In lepidopterans, the presence of one or more thoracic ganglia is necessary for the processes to take place. In addition, intact nerve connections between the brain and ventral nerve cord ganglia are always a prerequisite for darkening.

20-Hydroxyecdysone and JH participate in ommochrome, melanin, and bile pigment synthesis. In lepidopterans, the hemolymphatic ecdysteroid titer plays a decisive role. When it is low the pigments are synthesized but when it increases they are lysed. The JH, which stimulates the synthesis and maintenance of bile pigments, plays an inhibitory role in the synthesis of ommochromes and melanins in several species. In certain insects, however, it exerts opposite effects which may result from a positive feedback exerted by JH on the PG.

Rapid color changes, which occur mostly in stick insects in connection with circadian light/dark photophases, are regulated by a neurohormone from the central nervous system, whose production site in the brain is in the tritocerebrum. The release sites of this chromactivating factor, which is neither in the CC nor in the PO, are probably in peripheral nh areas. The chromactivating factor is distinct from AKH but could be identical or comparable to neurohormone D, also called factor MI.

10

Osmoregulation

The water balance in insects depends mostly on whether their mode of life is aquatic or terrestrial, and on the nature of their diet. For herbivorous, sapsucking, and bloodsucking insects the elimination of water is a vital problem, whereas for others there is a constant struggle to find enough water because their small size and intense respiratory metabolism lead to considerable water loss. Insects regulate their water balance by controlling the functioning of the Malpighian tubules and the rectum which act in an opposite way.

From the hemolymph, the Malpighian tubules absorb a liquid which flows through the rectum. They exhibit passive permeability to water and to substances with a molecular weight below 5000, many of which are reabsorbed at the rectum. The latter displays anatomical arrangements such as cryptonephridia and rectal papilla which are designed for reabsorption. Depending on the insect's physiological state, the rectum reabsorbs various amounts of water, but also removes amino acids, sugars, lipids, and certain ions, especially potassium. Activation of the Malpighian tubules causes loss of water, whereas activation of the rectum increases the amount in the insect's body.

Investigations into water metabolism control began in 1956 with research dealing with the bee and the larva of the beetle *Anisotarsus cupripennis*. Subsequent investigations dealt with blood-feeders such as *Rhodnius, Anopheles freeborni,* and *Glossina austeni,* which must lose weight rapidly after a meal to avoid losing their mobility and their ability to fly. Also investigated were species that display different life and feeding modes in relation to water metabolism, including cockroaches, locusts, stick insects, butterflies, and so forth.

10.1. METHODS

Several methods were used to study both water and ion excretion and reabsorption by the Malpighian tubules and rectum, respectively.

10.1.1. *In Vivo* Methods

In vivo experiments were conducted by measuring the ratio initial fresh weight–final fresh weight/dry weight under various experimental conditions designed to alter the osmotic pressure of the hemolymph. Tritiated water was also used to determine the amount of water retained by the insect (Proux and Buscarlet, 1976).

The excretion of the dye amaranth was also used to evaluate the activity of the Malpighian tubules (Mordue and Goldsworthy, 1969). It was injected into locusts whose hemolymph was then punctured every 5 min and examined with a spectrophotometer. Injection of labeled potassium urate was also used. The concentration of radioactive substance present in the rectum denotes the secretory activity of the Malpighian tubules (Mordue, 1969).

10.1.2. *In Vitro* Methods

Most experiments were conducted *in vitro* using several assay methods: the first was Ramsay's method (1954) or one of its variants (Maddrell, 1963, 1972) (Fig. 94). This consists of measuring the amount of urine produced by the Malpighian tubule isolated in a drop of deproteinized serum placed in a small container filled with liquid paraffin. The severed tubules are ligated and cut at one end to allow the urine to escape. The diameter of the drop of urine produced

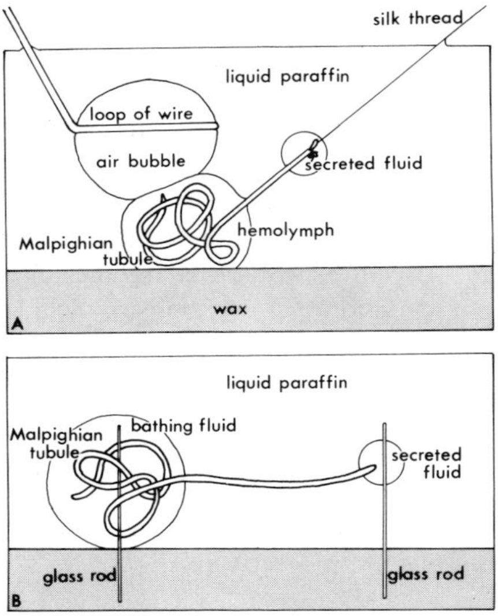

FIGURE 94. Experimental arrangement used to investigate fluid secretion by an isolated Malpighian tubule. [Modified from Ramsay, 1954 (A); from Maddrell, 1972 (B).]

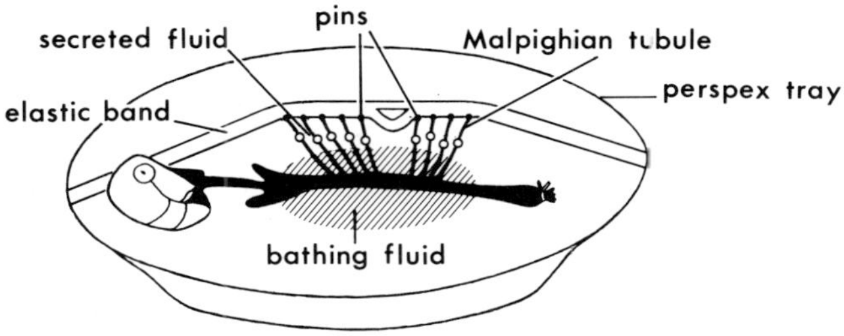

FIGURE 95. Locust Malpighian tubule preparation used to study diuretic activity. (Modified from Morgan and Mordue, 1984.)

is measured with an ocular micrometer. A variant of this method was used by Morgan and Mordue (1984) on semi-isolated tubules (Fig. 95). Other investigators isolated the Malpighian tubules with a short intestinal portion ligated on one side and a capillary tube inserted into the other. Diuretic variations were estimated by changes in the liquid level in the capillary tube (Mills, 1967a; Cazal and Girardie, 1968).

Another method allows measurement of the urine secreted by the use of a dye which is retained by the cells of the Malpighian tubules. The latter are placed, either alone or together with the midgut, in an isotonic dye solution. They are rinsed half an hour later and extracted with an alcoholic solution whose dye content is estimated colorimetrically (Altmann, 1956) or measured with a spectrophotometer (Mills and Nielsen, 1967).

The rate of fluid secretion by isolated Malpighian tubules was also determined by evaluating the modifications of the transepithelial voltage (Maddrell and Klunsuwan, 1973; Williams and Beyenbach, 1983, 1984). Lastly, another technique was devised, based on the increase in the intracellular level of cAMP determined by diuretic hormone activity (Morgan and Mordue, 1985a).

The reabsorption activity of the rectum was evaluated by a technique similar to that used for the Malpighian tubules. The rectum is ligated at one end and a graduated capillary tube inserted at the other. The movement of the liquid in the capillary tube serves to measure the reabsorption activity (Vietinghoff, 1966a; Cazal and Girardie, 1968). In other experiments the rectum operates inside out, reversing the direction of water flow. When immersed in saline, its activity is revealed by its weight increase (Mordue, 1969).

Ion transport at the level of both Malpighian tubules and rectum was investigated using several methods including, mostly, estimation of the electrical gradients (Maddrell, 1977), measurements of transepithelial potentials (Spring *et al.*, 1978; Hanrahan and Phillips, 1985), and measurements of intracellular and intraluminal ion concentrations (Morgan and Mordue, 1983a).

10.2. DIURETIC HORMONE

The first demonstration of the existence of a diuretic hormone was made in *Rhodnius* (Maddrell, 1963; 1964a,b). This insect absorbs a single blood meal representing 10 times its body volume at the onset of each larval instar and immediately excretes a large quantity of urine, whose production is stimulated by a diuretic hormone present in the blood of freshly fed insects (Fig. 96) but absent in fasting insects. By organ removal and extract injection the source of this hormone was revealed into the thoracoabdominal nerve mass. Later investigations confirmed the existence of a diuretic hormone in a number of insects, and interspecific activity of tissue homogenates was often observed (Fig. 97).

10.2.1. Production Sites

10.2.1.1. Brain

Histophysiological studies were conducted on different insects subjected to hydrating and dehydrating conditions. In most cases, significant storage of type A ns products was observed in the PI and CC of insects placed in a dry or dehydrating environment, or injected with saline (Kannan and Prabhu, 1985) (see Raabe, 1982). However, four other median posterior type A ns cells have been shown to exhibit great changes in their load. These changes correlated with

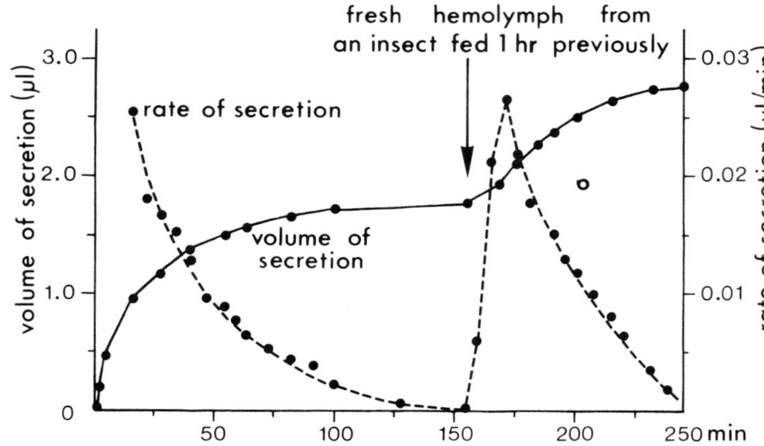

FIGURE 96. Demonstration of the presence of a diuretic hormone in the blood of a freshly fed *Rhodnius*. Adding hemolymph from an insect in diuresis strongly increases the rate of secretion of a resting set of isolated Malpighian tubules. (Modified from Maddrell, 1963.)

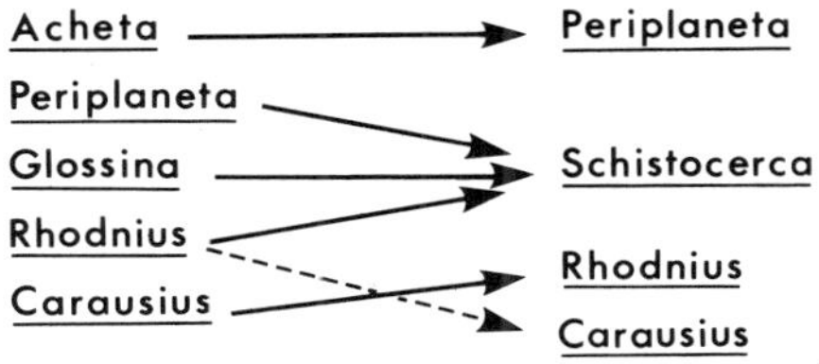

FIGURE 97. Interspecific activity of the diuretic hormones of several insects. (From the data of De Bessé and Cazal, 1968; Maddrell *et al.*, 1969.)

humidity conditions in *Locusta* (Girardie, 1970), *Aeshna* (Charlet, 1974), and *Leucophaea* (De Bessé, 1978).

Experimental investigations using regenerates of the CC storage lobe after NCC sectioning showed that in *Schistocerca*, the diuretic factor of the CC is present in the storage lobe only; it originates in ns cells of the brain and not in the glandular cells of the CC (Mordue and Goldsworthy, 1969; Morgan and Mordue, 1983b). The location of the diuretic hormone was also investigated in *Carausius* by assaying the activity of the different parts of the brain. It was observed that the lateral and tritocerebral parts are devoid of diuretic activity, which is displayed by the central region of the protocerebrum (Pilcher, 1970). The same method of injecting various homogenates confirmed, in several species, the presence of a diuretic hormone in the brain and CC (Table XXXII). As regards the identification of its neurons of origin, the data are controversial; thus, both the PI and posterior median ns cells have been proposed as possible candidates. Removal of the PI induced high water retention in many insects (Table XXXIII), but electrocoagulation of the median posterior ns cells also gives rise to an increase in fresh weight or a swelling of the insect, and reimplantation of these cells in insects in which they have been destroyed restores a normal situation (Charlet, 1974 in the dragonfly *Aeshna;* Proux, 1979; Proux *et al.*, 1980 in *Locusta;* De Bessé, 1978 in *Leucophaea*).

The ventral nerve cord also displays clear changes in the load of certain ns

TABLE XXXII. Water Retention Caused by Removal of Pars Intercerebralis

Insect	Reference
Gryllus domesticus	Girardie, 1966a
Schistocerca gregaria	Highnam *et al.*, 1965
Locusta migratoria	Cazal and Girardie, 1968
Melanoplus sanguinipes	Elliott and Gillott, 1977
Leptinotarsa decemlineata	De Wilde, 1966
Anisotarsus cupripennis	Nunez, 1956
Calliphora erythrocephala	Thomsen, 1952

TABLE XXXIII. Location of the Diuretic Factor

	Compared activity						
				VNC			
Insect	Brain	CC	CA	sgg	th gg	ab gg	Reference
Periplaneta	0	0	+	0	0	+++	Mills, 1967a
	+++		+++				Wall, 1966
Carausius	+++	+++	0	+++	—— +	——	Pilcher, 1970b
	+++				+++	+++	Vietinghoff, 1966a[H],b[H]
Locusta	+++						Cazal and Girardie, 1968
					+	+	Poux, 1978a,b
	+	++		+	—— +	——	Morgan and Mordue, 1984
Schistocerca	+++	+++					Highnam *et al.*, 1965
	+++	+++	++				Mordue, 1969
Melanoplus	+++						Elliott and Gillot, 1977
Gryllus	+++						Girardie, 1966a
Acheta	+	+++		+		+	Spring and Hazelton, 1987
Apis			+++				Altmann, 1956
Anisotarsus	—— +++	——		0			Nunez, 1956
Leptinotarsa	+++						De Wilde, 1966
Onymacris	+	+			+		Nicolson and Hanrahan, 1986
Pieris	++	+++		+	+	0	Nicolson, 1976[HH]
Danaus	+++	—— ++	——	————	++	————	Dores *et al.*, 1979
Calliphora	+++	+++					Thomsen, 1952
	+				+++		Schwartz and Reynolds, 1979
Glossina	+++				—— +++	——	Gee, 1975
Corethra					0	+++	Gersch, 1967
Anopheles freeborni, adult	+++				+		Nijhout and Carrow, 1978
Aedes taeniorhynchus, larva	+++				+		Maddrell and Phillips, 1978
Rhodnius	+	0	0	0	0	+++	Maddrell, 1963, 1964a
Dysdercus	+++	+	0	0	—— +	——	Berridge, 1966

[H], chromatographed extracts; [HH], boiled extracts; 0, +, ++, +++, zero, weak, middle, strong activity; CA-, CC-, corpora allata, corpora cardiaca, ab gg, abdominal ganglia; sgg, suboesophageal ganglia; th gg, thoracic ganglia; VNC, ventral nerve cord.

cells according to humidity conditions (see Raabe, 1982). In *Rhodnius,* in which only the posterior part of the thoracoabdominal nerve mass contains diuretic hormone activity (Maddrell, 1963), a thorough study of the various ns cells was made in connection with diuresis (Baudry, 1968). The presence of three distinct types of lateral ns cells was demonstrated, i.e., six A cells, 12 B1 cells, and 10 B2 cells. Their appearance during feeding and diuresis suggests that the B1 cells are responsible for synthesizing the diuretic factor because they release their ns products in the minutes that follow the meal (Baudry, 1969). Investigation of the diuretic activity of isolated neurons located in the same region before and after feeding (Berlind and Maddrell, 1979) demonstrated active cells which corresponded to Baudry's B1 cells (Fig. 98).

The ventral nerve cord was experimentally investigated in many species (*Periplaneta, Carausius, Locusta, Pieris, Glossina austeni, Corethra, Anopheles freeborni, Aedes taeniorhynchus, Calliphora vicina,* and *Rhodnius*) (Table XXXIII). These studies revealed the frequent presence of a diuretic factor in some or all of the ventral nerve cord ganglia.

Investigations in *Locusta* aimed at identifying the neurons of origin of the diuretic hormone of the ventral nerve cord revealed two medioventral ns cells in the subesophageal ganglion which were shown to be immoreactive to vasopressin antibodies (Rémy *et al.*, 1979; Rémy and Girardie, 1980) (see Chapter 2). Since vesopressin-like peptides are found in the hemolymph and their titer increases with diuresis, it has been suggested that they could be involved in the regulation of this process (Proux and Rougon-Rapuzzi, 1980).

In the subesophageal ganglia of *Acheta,* the concentration of a vasopressin-like factor has also been shown to vary with humidity, and is lower in insects raised in a moist environment (Strambi *et al.*, 1978).

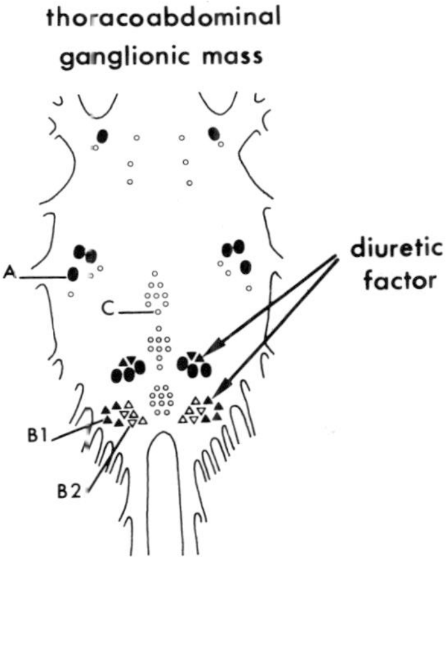

FIGURE 98. Identification of the neurosecretory cells involved in diuretic hormone secretion in *Rhodnius.* (Modified from Baudry, 1968; Berlind and Maddrell, 1979.)

It is interesting to note that the diuretic activity in nerve tissue varies with the hygrometric conditions as shown in *Periplaneta* (Ralph and McCarthy, 1964; Wall, 1965) and *Locusta* (Cazal and Girardie, 1968). The sensitivity of the effectors also varies according to the degree of dehydration, e.g., in the rectum of the cockroach (Steele and Tolman, 1980).

10.2.2. Release Sites

At least in blood-feeders, diuretic hormone can be released very rapidly. In *Rhodnius,* it only takes 3 min (Maddrell, 1963) and in *Anopheles* 1 min (Nijhout and Carrow, 1978). The diuretic factor of the PI is transported by the NCCI in *Schistocerca* and stored in the CC (Goldsworthy *et al.,* 1972b). As far as the ventral nerve cord release sites are concerned, they have been revealed in certain insect species using potassium-rich solutions that cause depolarization of the ns endings and subsequent hormone release into the incubation medium. In *Rhodnius* and *Glossina,* release sites were demonstrated to be present in the proximal part of the first three pairs of abdominal nerves (*Rhodnius*) and the abdominal nerves (*Glossina*). These nerves have characteristic ultrastructural features (Maddrell, 1966; Maddrell and Gee, 1974) and may be considered as nh areas. In *Calliphora,* diuretic hormone is probably also released from the nerves, whose extracts are more potent than those of nerve ganglia (Schwartz and Reynolds, 1979).

The pathways and release sites of the two medioventral vasopressin-like ns cells of the subesophageal ganglion of *Locusta* were revealed by immunocytochemistry, as already mentioned (Chapter 2). The processes of these cells run all through the central nervous system, enter the segmentary nerves, and ramify extensively at their periphery, thus forming nh areas (Rémy and Girardie, 1980).

Severance of the ventral nerve cord or decapitation causes water retention in many species, including the beetles *Anisotarsus* (Nunez, 1956), *Leptinotarsa* (De Wilde, 1966), the mosquito *Anopheles* (Nijhout and Carrow, 1978), *Locusta* (Proux, 1978b), and *Glossina* (Gee, 1975). This suggests that information concerning water metabolism is integrated by the brain. The nature and source of the information are poorly known, but we do know that feeding triggers diuretic hormone release in locusts (Highnam *et al.,* 1966) and in *Dysdercus* (Berridge, 1966). By various experiments performed on *Rhodnius,* Maddrell (1964b) demonstrated that the mechanism triggering diuresis was abdominal distension. In *Locusta,* interesting results were obtained concerning the vasopressin-like diuretic factor. Using electrocoagulation, electrical stimulation, and injection of homogenates followed by RIA determination of the hemolymph titer, it was shown that both the PI and the median posterior ns cells control the release of this diuretic factor and play antagonistic roles. The PI inhibits the release of the diuretic factor while the median posterior ns cells stimulate this release (Proux and Girardie, 1982). It should also be noted that the physiological effects of this vasopressin-like factor are only clearly apparent in insects deprived of their brains' diuretic ns cells (Proux, 1978a).

10.2.3. Diuretic Hormone Titers

Diuretic hormone can often be revealed in the blood (*Periplaneta, Carausius, Locusta, Glossina, Rhodnius, Dysdercus*). Its titer varies with water conditions and in some species it is undetectable (*Pieris, Anopheles*).

To provide an idea of the quantity of circulating diuretic hormone, attempts were made to measure the effective doses initiating fluid secretion. In *Rhodnius,* the third part of the ganglionic mass is sufficient to induce strong secretion (Maddrell, 1964a). Diuretic hormone levels were found to be similar in *Carausius* (Pilcher, 1970) and *Dysdercus* (Berridge, 1966).

In *Schistocerca,* the breakdown rate of the CC diuretic hormone was estimated at 0.05–0.15/CC/hr (Mordue, 1969). In *Rhodnius* (Maddrell, 1964a), diuretic hormone, which is abundant in the hemolymph after a meal, disappears in about 2 hr at the end of diuresis. This appears to be due to catabolic processes exerted by the Malpighian tubules. In *Dysdercus* (Berridge, 1966), *Carausius* (Pilcher, 1970), and *Glossina* (Gee, 1975), but not in *Calliphora* (Schwartz and Reynolds, 1979), the Malpighian tubules destroy diuretic hormone and can thus control its hemolymph titer. The inactivation of neurohormone D by Malpighian tubules was investigated in *Periplaneta* (Baumann and Penzlin, 1987); it was shown that diuretic hormone is destroyed in the cells of the tubules by soluble proteinases which appear to be metalloendopeptidases with a molecular weight of 45,000. In addition, diuretic hormone can be destroyed by an enzyme of the central nervous system (Gee, 1975). Lastly, the breakdown of the diuretic hormone in *Onymacris* was observed to occur in the hemolymph, without involvement of the Malpighian tubules (Nicolson and Hanrahan, 1986).

10.3. BIOGENIC AMINES

As in other physiological processes, biogenic amines help to regulate diuresis in addition to neurohormones. Serotonin was shown to stimulate the activity of the Malpighian tubules in many species, including *Carausius* (Maddrell *et al.*, 1969), *Papilio* (Nicolson and Millar, 1983),*Aedes taeniorhynchus* (Maddrell and Phillips, 1978), *Calliphora* (Schwartz and Reynolds, 1979), and *Rhodnius* (Maddrell *et al.*, 1969) but not in the locusts (Mordue, 1972; Maddrell and Klunsuwan, 1973; Rafaeli *et al.*, 1984). Fluid absorption by the midgut is also enhanced by serotonin in *Rhodnius* (Farmer *et al.*, 1981). Further, epinephrine was reported to act as a stimulant in locusts, but not octopamine

(Rafaeli *et al.*, 1984), while in *Papilio* norepinephrine and dopamine increase fluid secretion by tubules, but less than serotonin (Nicolson and Millar, 1983). Serotonin inhibitors such as tryptamine and tyramine diminish Malpighian tubule secretion in *Rhodnius,* but cyproleptadine and the adrenergic blockers phentolamine and propranolol do not suppress the increased secretion by *Papilio* tubules in response to serotonin. This suggests that serotonin does not act through β-adrenergic receptors.

It is well established that serotonin is not identical to diuretic hormone in *Rhodnius* (Maddrell *et al.*, 1969, 1971; Aston and White, 1974), since in this species diuretic hormone is not present in the fluid secreted by the tubules, whereas serotonin is present in this fluid. The pH and temperature requirements for diuretic hormone conservation are different from those for serotonin.

10.4. ANTIDIURETIC HORMONE

The existence of antidiuretic hormone has been demonstrated in cockroaches, stick insects, locusts, bees, and *Corethra*, chiefly by *in vitro* experiments using isolated Mapighian tubules or rectum (Table XXXIV). In some species this hormone is found in the brain, and its presence in the ventral nerve cord was reported whenever this location was investigated. It is also present in the CC and PO.

Using an everted rectum *in vitro* preparation it was shown that distinct antidiuretic factors were contained in the storage and glandular lobes of the CC in *Locusta* (Proux *et al.*, 1984; Herault *et al.*, 1985). The glandular lobe factor stimulates fluid reabsorption in a dose-dependent manner, and octopamine from the lateral ns cells of the brain is involved in the control of its release.

10.5. ION METABOLISM

Insects are able to regulate the osmotic and ionic concentration of their hemolymph by selective ion reabsorption. Recent investigations of this regulation have shown that the CC were involved in controlling these processes.

Ionic reabsorption in *Rhodnius* occurs in the lower part of the Malpighian tubules. It is stimulated by serotonin (Maddrell and Phillips, 1978) and by a substance which is distinct from diuretic hormone although it is located in the same places (Maddrell and Phillips, 1976).

In the rectum of *Locusta,* it has been shown that a hormone from the CC that stimulates chloride transport (CTSH) increases short-circuit current and the transepithelial electropotential difference, thus enhancing an electrogenic K-stimulated chloride transport (Spring *et al.*, 1978; Spring and Phillips, 1980).

TABLE XXXIV. Location of the Antidiuretic Factor

					VNC			
Insect	Reference	Brain	CC	CA	sgg	th gg	ab gg	PO
Periplaneta	De Bessé and Cazal, 1968		+++					+++
Blabera	Wall and Ralph, 1962; Wall, 1965	+++	+++	++	+++	++	+	
Carausius	Vietinghoff, 1966a,b[H]	+++	+++	+++				
Clitumnus	De Bessé and Cazal, 1968		+++					+
Locusta	Cazal and Girardie, 1968	+++	+++					
	De Bessé and Cazal, 1968							+++
	Proux *et al.*, 1984		++					
Schistocerca	Mordue, 1970		+++					
Apis	Altmann, 1956		+++					
Corethra	Gersch, 1967, 1969					+++		

[H], chromatographed extracts; +, ++, +++, weak, medium, strong activity; CC, corpora cardiaca; CA, corpora allata; ab gg, abdominal ganglion; PO, perisympathetic organs; sgg, subesophageal ganglion; th gg, thoracic ganglia; VNC, ventral nerve cord.

Corpora cardiaca homogenates were observed to have a similar role in raising urine ion concentration in the rectum of the grasshopper *Romalea microptera* (Spring, 1986) and the Malpighian tubules of the cricket *Acheta* (Spring and Hazelton, 1987), where a neurohormonally induced increase in K^+ and more especially in Na^+ was observed.

10.6. JUVENILE HORMONE AND ECDYSONE

Some investigators have reported that CA extracts display an activity comparable to that of CC extracts. Although this may be due in some cases to contamination by the ns products contained in the CC, it was also observed that removal of the CA induces swelling in the locust and the blowfly (Day, 1943; Mordue, 1969). It is difficult to state at present whether this effect is direct or whether CA removal causes retention of diuretic neurohormones.

Ecdysone may also be a factor in water metabolism. Ryerse (1978) showed that the functioning of the Malpighian tubules of the pupa of *Calpodes ethlius* normally stops just before pupation and continues *in vitro* or *in vivo* after an ecdysone injection. In the young tsetse fly, moreover, the addition of ecdysone or 20-HE to the incubation medium of Malpighian tubules containing a diuretic extract increases the excretion rate (Gee *et al.*, 1977). The same was observed in *Schistocerca* (Rafaeli and Mordue, 1982). In *Manduca,* a correlation was found between water loss and ecdysone titer (Jungreis *et al.*, 1982) but in *Papilio demodocus* the rate of secretion by the Malpighian tubules was not increased by 20-HE (Nicolson and Millar, 1983).

10.7. PURIFICATION OF HORMONES INVOLVED IN OSMOREGULATION

10.7.1. Diuretic Hormones

The peptidergic character of diuretic hormones was demonstrated in several species (see Table XXXV). Most diuretic hormone homogenates are unstable at room temperature but boiling increases their stability, doubtless owing to the destruction of the proteolytic enzymes contained in the organ extracts.

10.7.1.1. *Periplaneta*

In *Periplaneta,* Goldbard *et al.* (1970) used the terminal abdominal ganglion as a source of active factors and the rectum as *in vitro* effector. Two factors were separated, a diuretic factor with a molecular weight above 30,000, and an antidiuretic factor with a molecular weight of 8000–10,000.

TABLE XXXV. Some Properties of Hormones Regulating Osmoregulation

Insect	Number active fractions	Stability at 100°C	Solubility in methanol	Molecular weight	Peptidergic nature	Reference
			Antidiuretic hormones			
Periplaneta				8000		Goldbard *et al.*, 1970
Locusta	2	−/+				Herault *et al.*, 1985
Apis		+				Altmann, 1956
			Diuretic hormones			
Periplaneta	2			>30,000		Goldbard *et al.*, 1970
Schistocerca	2				+	Mordue and Goldsworthy, 1969
Locusta			+			Rafaeli and Mordue, 1982
	2	+ 70°		1000–2000	+	Morgan and Mordue, 1983
				6000–7000		Morgan *et al.*, 1987b
	2			2500		Cupo and Proux, 1983
Apis		+				Altmann, 1956
Danaus		+	0	3000	+	Dores *et al.*, 1979
Aedes	1	+		1900–2700	+	Petzel *et al.*, 1986
Anopheles		+				Nijhout and Carrow, 1978
Glossina		−	−	1200	+	Gee, 1975
Calliphora		+	+/−		+	Schwartz and Reynolds, 1979
Dysdercus		+				Berridge, 1966
Rhodnius		+				Maddrell *et al.*, 1969
	2		0	>60,000 <2000	+	Aston and White, 1974; Aston, 1979
			Chloride transport-stimulating hormone			
Schistocerca		+			−/+	Phillips *et al.*, 1980

The activity of factors C1 and D1, isolated from the ventral nerve cord of *Periplaneta* by Gersch and co-workers, was assayed *in vitro* on the rectum of *Carausius* (Vietinghoff, 1966b, 1967) and on the Malpighian tubules of the same species (Unger, 1967). Factor C1 was shown to be antidiuretic to the Malpighian tubules and diuretic to the rectum, while factor D1 was diuretic to the rectum and antidiuretic to the Malpighian tubules.

10.7.1.2. *Rhodnius*

In *Rhodnius* (Aston and White, 1974), a search was made for diuretic hormone in extracts of the thoracoabdominal nerve mass, in the K^+-rich incubation medium, the blood, and in extracts of the head and thorax. After chromatography, two similar activity peaks appeared in the extracts of the head and thoracoabdominal nerve ganglion. This involves a substance of molecular weight above 60,000 and another substance of molecular weight below 2000. The latter is the only one present in the blood and in the potassium-rich incubation medium of the nerve ganglion, suggesting that it is the active factor. Further investigations by Aston (1979) and Hughes (1977, 1979) demonstrated the fragility of the diuretic hormone and the difficulties encountered in its extraction, storage, and purification. They confirmed the existence of two peaks of high and low molecular weight.

10.7.1.3. Locusts

The activity of diuretic hormones was also investigated in the CC of locusts (Mordue and Goldsworthy, 1969; Morgan and Mordue, 1983b; Morgan *et al.*, 1987). By gel permeation chromatography and reversed phase HPLC, the diuretic activity from the storage lobe of the CC was isolated. It formed two distinct peaks located on each side of the large AKH peak. Their molecular weight was between 1000 and 2000. Further investigations showed that besides the small molecular weight peptide (DP2), another peptide (DP1) was present with a molecular weight of 6000–7000. Using high-performance size exclusion chromatography followed by reversed phase HPLC, the factor DP1 was isolated as well as another factor (LCCP) which was not involved in diuresis. Their sequencing was partly elucidated, with the sequences 1–11 for DP1 and 1–40 for LCCP, as follows:

DP1 ?-Gly-?-Gly-Ile-Glu-Ala-?-Val/Met-Tyr-Lys

1 10 20

LCCP Asn -Pro-Ile-Gly-Ala-Asn-?-Val-Val-Asp -Leu-Thr-Arg-?-Glu-Tyr-Gly-Asp-Val-Thr-?-Phe-Phe-Gly-Arg-Lys-

30 40

Val-?-Ala-Lys-Gly-Pro-Gly-?-Lys-?-Gly-Gly-Pro-Tyr-Glu-Leu-Gly-Lys-?-Val

The amounts of both peptides in the CC are very different as the latter contain 17 times more LCCP than DP1. However, dose-response data for DP1 have shown that concentrations of 10^{-9} M are sufficient to induce a diuretic response.

Other investigators in the field of locust diuretic hormones (Cupo and Proux, 1983; Proux *et al.*, 1982, 1987; Schooley *et al.*, 1987) found that the diuretic factors from the subesophageal ganglion of *Locusta* resembled arginine-vasopressin (AVP). Their molecular weight was estimated at about 2500 and their isoelectric point 7.5, so that they differed slightly from AVP but exhibited the same C-terminal tetrapeptide. These vasopressin-like diuretic factors occurred in two forms, a free form and a high-weight form. The latter is probably a precursor molecule present in the pellets only and is not spontaneously immunoreactive with AVP antibodies. Investigations showed that it is found throughout the nervous system like the free factor.

One wonders whether these vasopressin-like factors are similar to the diuretic factors isolated by Morgan and Mordue. According to the data of these authors, AVP, lysine-vasopressin, and vasotocin elute with shorter retention times than the diuretic hormones from the CC. In addition, the vasopressin-like factors are absent from the CC, as shown by Proux. It is therefore highly probable that distinct molecules from both the subesophageal ganglion and the CC are involved in the control of diuresis in locusts.

10.7.2. Chloride Transport-Stimulating Hormone

The chloride transport-stimulating hormone (CTSH) was investigated in locusts and purified more than 100-fold. It is trypsin-sensitive and has a molecular weight of about 8000. It is destroyed at low pH and possesses a negative charge (Phillips *et al.*, 1980). Another molecule with a molecular weight above 30,000 was also found in the locust.

10.8. MODE OF ACTION OF HORMONES REGULATING OSMOREGULATION

Thorough investigations have made it possible to establish that urine secretion is mediated by an apical electrogenic K^+ pump and is accompanied by an increase in the transepithelial electropotential across the Malpighian tubules or rectal sacs (Maddrell, 1977 in *Rhodnius* Malpighian tubules; Proux *et al.*, 1984 in the rectum of *Locusta*).

The role of K^+ ions is crucial in the functioning of both the Malpighian tubules and rectum, but diuresis and reabsorption are also dependent on Na^+ and Cl^- ions (Morgan and Mordue, 1981). Na^+ ions are particularly important in bloodsucking insects as they have to excrete these ions which are supplied in large amounts by the ingested blood, e.g., in mosquitoes (Petzel *et al.*, 1987). They also play a part in rectal reabsorption (Tolman and Steele, 1980 in *Periplaneta*). As regards Cl^- ions, studies by ion-selective intracellular electrodes and electrical radiotracer fluxes demonstrated that in the rectum of *Locusta*, the presence of a Cl^- pump was responsible for transport processes (see Phillips, 1983). Ca^{2+} ions, on the contrary, are not necessary for the functioning of the Malpighian tubules (Maddrell, 1977), and rectal Cl^- absorption stimulated by CC extracts is not dependent on exogenous Ca^{2+}.

The amino acid proline was shown to be an important fuel for the reabsorption activity in the rectum of *Locusta* (see Phillips, 1983).

The involvement of cAMP as a second messenger in the hormonal effects of diuretic hormones, antidiuretic hormones, CTSH, and biogenic amines was investigated. In most of the insects studied, cAMP was found to mimic the effects of the diuretic hormones both *in vivo* and *in vitro* (Steele and Tolman, 1980 in *Periplaneta;* Maddrell and Klunsuwan, 1973; Anstee *et al.*, 1980; Rafaeli and Mordue, 1982 in locusts; Nicolson, 1980; Nicolson and Millar, 1983 in lepidopterans; Maddrell and Phillips, 1978 in *Aedes taeniorhynchus*; Gee *et al.*, 1977 in *Glossina;* Maddrell *et al.*, 1969 in *Rhodnius*). In addition, cAMP was shown to mimic the effects of antidiuretic hormones on the locust rectum (Herault and Proux, 1987) and to increase the rate of fluid secretion in the midgut of *Rhodnius* (Farmer et al., 1981). In *Locusta,* it also mimicked the effects of CTSH (Spring and Phillips, 1980).

Aminophylline and other phosphodiesterase inhibitors display the same effects on *Locusta* and *Aedes* as diuretic and antidiuretic hormones (Hanrahan and Phillips, 1985; Herault and Proux, 1987 in *Locusta;* Petzel *et al.*, 1987 in *Aedes*).

The role of cAMP in mediating water and ion transport was definitely proved by determining the amount of cyclic nucleotide present in the tissues after hormonal action. The level of cAMP was reported to rise rapidly in the Malpighian tubules after they were brought into contact with diuretic hormone (Aston, 1975 in *Rhodnius;* Rafaeli *et al.*, 1984; Morgan and Mordue, 1985a in locusts; Spring and Hazelton, 1987 in *Acheta;* Petzel *et al.*, 1987 in *Aedes*). The same effect was also observed in the rectum with the antidiuretic hormone (Spring and Phillips, 1980; Herault and Proux, 1987 in locusts).

In a few cases, however, it was observed that certain diuretic fractions from the CC of locusts or from the head of mosquitoes increase fluid secretion in the Malpighian tubules without the involvement of cAMP (Morgan and Mordue, 1985b; Petzel *et al.*, 1986). In *Locusta,* both cGMP and Ca^{2+} are involved in the action of one of the diuretic factors although the Malpighian tubules mostly function in the absence of Ca^{2+} (Maddrell, 1977).

Serotonin, at least in some species, appears to operate through a mechanism

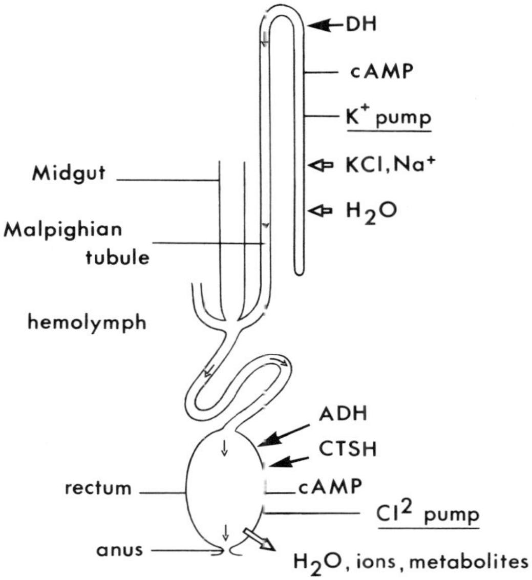

FIGURE 99. Schematic representation of fluid movements in the insect excretory system. Fluid is transported into the Malpighian tubules where many hemolymph solutes diffuse across the walls into the lumen. Much of the water, ions, and other useful substances is then reabsorbed in the anterior hindgut and rectum. Water, ions, metabolites (⇒); hormones (→). (Modified from Phillips, 1983.)

distinct from that of diuretic hormone, persumably involving different receptors (Maddrell and Klunsuwan, 1973; Morgan and Mordue, 1985b).

It still seems that in most cases neurohormonal action triggers the elevation of intracellular cAMP, which, as shown above, is followed by the activation of the cationic pump in the apical membrane and an increase in transepithelial K^+ permeability (Fig. 99).

10.9. CONCLUSIONS

Investigations into the regulation of water metabolism show that all insects possess a diuretic hormone, but only some of them seem to have an antidiuretic hormone as well (Table XXXIV). The presence of a chloride transport-stimulating hormone has also been demonstrated.

These hormones are produced in the brain, the glandular lobe of the CC, and the ventral nerve cord ganglia. The release sites of brain diuretic and antidiuretic hormones are located in the CC. As regards the ventral nerve cord, the PO of cockroaches, stick insects, and locusts play a part in the release of antidiuretic hormone, while in *Rhodnius* and *Glossina*, diuretic hormone is released in distal neurohemal areas.

The mode of action of diuretic hormones on the Malpighian tubules has been thoroughly examined. Like serotonin, these hormones act through cAMP, which activates the cationic pump of the apical membrane of the Malpighian tubules. Cyclic AMP is also involved in the action of antidiuretic factors and CTSH.

Diuretic and antidiuretic hormones were purified from the subesophageal ganglion and CC in several species, including *Rhodnius, Glossina, Aedes,* and *Locusta*. More than one active factor with high and low molecular weight, respectively, was demonstrated to be present within the same species.

11

Metabolism

Insects use energy from two main sources: trehalose and diglycerides. Depending on the species and developmental stage, the relative importance of sugars and lipids as energy providers varies. Some insects also use the amino acid proline as fuel for flight. The main tissue which maintains homeostasis by storing and supplying metabolites is the fat body which also plays a major role in the processes of protein synthesis involved in growth and differentiation.

11.1. LIPID AND CARBOHYDRATE METABOLISM

Lipids are extremely important in the energy metabolism of insects, especially during flight (see Goldsworthy, 1983; Beenakkers *et al.*, 1985). While flight consumes carbohydrates in dipterans and amino acids such as proline in the tsetse fly and Colorado beetle, many other insects consume lipids which they sometimes convert from alimentary carbohydrates. Locusts use carbohydrates as their energy source during the first 20 min of flight, but these are then supplanted by lipids, especially diglycerides, whose hemolymph level rises by a factor of 5 during flight (Beenakkers, 1965).

Most insects often use one major fuel and one or several minor fuels, as is the case for *Calliphora* and *Danaus* (Vejbjerg and Normann, 1974; Dallmann and Herman, 1978). They also use one fuel in the larval stage and another when they become adults (Ziegler, 1984). Lipids also perform an important function in the adult female because triglycerides represent one of the major constituents of the vitellus.

Lipids are stored in the fat body, mainly in the form of triglycerides, but the fat body also contains small amounts of free fatty acids and mono- and diglycerides. The triglycerides are converted by a lipase into diglycerides which are released into the hemolymph where they bound to specific transporting lipoproteins. The diglycerides are then oxidized in the muscles.

The hemolymph lipid concentration varies during flight, locomotor activity, and, in the adult female, as a function of reproduction. It decreases at molting in *Locusta* (Gäde and Beenakkers, 1977; Lafon-Cazal and Morandini, 1982) but increases in fasting insects (Jutsum *et al.,* 1975).

In insects, the most readily accessible and most frequently used carbohydrate energy material, especially for flight, is trehalose, the most abundant carbohydrate transported by the hemolymph. Glucose is only present in small quantities in the blood, although it is used by the muscle, where trehalose is hydrolyzed to glucose by the action of a trehalase. Trehalose is stored in the fat body and other tissues in the form of glycogen. Specific glycogen phosphorylase converts glycogen into trehalose.

In some species, the fuel used for flight is not trehalose but another carbohydrate, such as glucose or fructose. Furthermore, the source of flight metabolites may be the muscle and not the fat body (Downer and Parker, 1979).

The hemolymph trehalose concentration varies during postembryonic development. It increases during muscular exercise and in circumstances such as stress (Downer, 1979) or injury (Wyatt, 1961) (Fig. 100C). It also undergoes regular variations during the circadian cycle. During insect growth, histogenetic periods require a larger energy supply which is obtained from carbohydrates and especially from glucose. Hence it is not surprising that operations which inhibit development affect sugar metabolism.

11.1.1. Adipokinetic and Hyperglycemic Hormones of the AKH Family

11.1.1.1. Adipokinetic Hormones

The existence of a hormone exerting adipokinetic effects was first demonstrated in the CC of *Schistocerca* and *Locusta;* it causes the release of fat body diglycerides during flight, which in turn raises their hemolymph titer (Beenakkers, 1969a; Mayer and Candy, 1969; Goldsworthy *et al.*, 1972a). This hormone is known as the adipokinetic hormone (AKH) (Mayer and Candy, 1969) (Fig. 100A,B).

Adipokinetic hormone is present in large quantities in the hemolymph of the adult locust during flight. Its release begins in 10 min of flight and proceeds until 30 min. The quantity released by one locust corresponds to 1/400 of one glandular lobe (Goldsworthy *et al.,* 1972b), and it causes a 300-fold increase in hemolymph lipids. If the hemolymph of a locust which has just flown is injected into a resting insect, the diglyceride content of its hemolymph rises.

Adipokinetic hormone acts on incubated fat body, from which it causes diglyceride release *in vitro* (Mayer and Candy, 1969). It appears not only to induce conversion of the triglycerides and the fatty acids of fat body to di-

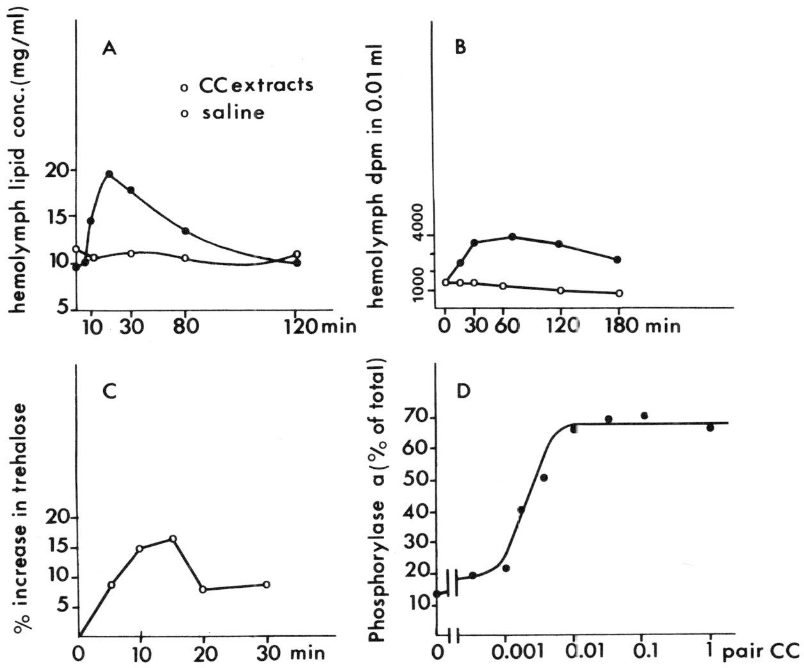

FIGURE 100. Demonstration of the existence of metabolic hormones in insects. (A) Effects of injection of corpora cardiaca extracts on the hemolymph lipid level in *Schistocerca.* (B) Effects of injection of corpora cardiaca extracts on lipid release in *Locusta.* (C) Hypertrehalosemic response to excitation in *Periplaneta.* (D) Dose-response curve for the stimulation of fat body glycogen phosphorylase activity by the hyperglycemic factor from the corpora cardiaca of *Manduca.* Corpora cardiaca extracts (●); controls (○). [Modified from Mayer and Candy, 1969 (A); Beenakkers, 1969 (B); Downer, 1979 (C); Ziegler, 1979 (D).]

glycerides, but also to induce the formation of diglycerides in the hemolymph from the fatty acids contained in the latter (Beenakkers, 1969a).

The diglycerides are transported by specific lipoproteins. Thorough and recent investigations have been devoted to the structural transformation of the lipoproteins of locusts, and it has been demonstrated that changes in the level of hemolymph diglycerides are accompanied by changes in the pattern of hemolymph carrier lipoproteins. In flying insects, the lipoproteins of resting locusts (O and A yellow) are replaced by a higher molecular weight lipoprotein (A^+) which displays higher diglyceride-binding capacities (Mwangi and Goldsworthy, 1977b). Several recent studies have shown that pure AKH causes the appearance of the lipoprotein A^+ and a decrease in A yellow lipoprotein (Wheeler and Goldsworthy, 1983a,b; see Goldsworthy, 1983) (Fig. 101).

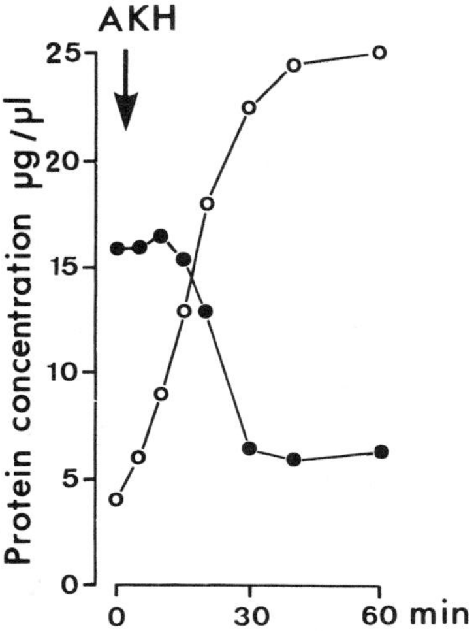

FIGURE 101. Effects of AKH on the concentration of A yellow (●) and A+ (○) lipoproteins in the hemolymph of *Locusta.* (Modified from Goldsworthy, 1983.)

Adipokinetic hormone also acts at muscle level, where it enhances lipid uptake (Robinson and Goldsworthy, 1976), and a membrane-bound lipase appears to be involved in this action.

The reactivity of insects to adipokinetic hormone may vary; hence in *Locusta,* the nymph reacts very slightly and the adult only displays typical reactions between days 5 and 35 (Mwangi and Goldsworthy, 1977a). These reduced responses are not due to the adipokinetic hormone deficiency in the CC because the CC extracts of locusts of different ages exhibit the same activity (Gäde and Beenakkers, 1977).

The chemical structure of adipokinetic hormone (AKH) was elucidated in 1976 (Stone *et al.*, 1976) (see Section 11.4), an achievement which initiated a series of fruitful investigations.

11.1.1.2. Hypertrehalosemic Hormones

The discovery in the CC of a hyperglycemic neurohormone was made in *Periplaneta* by Steele (1961, 1963). Injection of CC extracts into this insect markedly enhanced hemolymph trehalose, which rose by 50% in 30 min and reached a peak in 5 hr (150%). The extracts exhibited very strong activity, since maximum reactions were obtained with a dose corresponding to 1/10 pair of CC, and even 1/500 pair induced a clear reaction.

Simultaneously with this increase in hemolymph trehalose, hyperglycemic hormone (HGH) or hypertrehalosemic hormone or factor (HTH, HTF) causes a drop in fat body glycogen. It appears to have no effect on the glycogen of the digestive tract and muscles, but acts on that of the perineurium of the ventral nerve cord, in which it activates trehalose formation, at least in extreme cases (Hart and Steele, 1973). The main target for HTH-HGH-HTF neurohormones is the fat body, where they activate glycogen phosphorylase (see Steele, 1985).

The rises caused by CC extracts in the hemolymph trehalose concentration and fat body glycogen phosphorylase content were confirmed by various authors through *in vivo* and *in vitro* investigations performed on numerous insects, including cockroaches (Ralph, 1962a; Bowers and Friedman, 1963; Wiens and Gilbert, 1965; Natalizi and Frontali, 1966), *Carausius* (Dutrieu and Gourdoux, 1967a; Gäde, 1980), *Locusta* (Chalaye, 1969; Goldsworthy, 1969; Loughton and Orchard, 1981), *Tenebrio* (Dutrieu and Gourdoux, 1974), *Manduca* (Ziegler, 1979) (Fig. 100D), *Phormia* (Friedman, 1967), *Calliphora erythrocephala* (Normann and Duve, 1969; Vejbjerg and Normann, 1974), and the bug *Lohita grandis* (Islam and Roy, 1982).

The hypertrehalosemic activity of CC cannot be demonstrated at all times. Hence in the fly *Phormia,* CC extracts only displayed an effect *in vivo* in starved flies because in this insect, which is a high-carbohydrate consumer, trehalose can be formed directly from glucose and is only synthesized from glycogen if glucose is lacking. Since the hormone acts through glycogen, it is inactive in glucose-rich, fed insects (Friedman, 1967). Locusts also display irregular reactions. In different periods of their life, they are unresponsive to the hypertrehalosemic hormone of CC. While a clear response was observed in 6-day-old adult males, there was no reaction to injections of CC extracts in younger, mature, or unfed males (Goldsworthy, 1969). The ineffectiveness of hypertrehalosemic hormone in young and unfed adults is probably due to a low glycogen storage and, in the mature male, to saturation of the phosphorylase system, or to modifications in the phosphorylase content of the fat body. This content appears to increase from the onset of eclosion to the sixth day and then remains at a high level. The absence of a response to CC hypertrehalosemic hormone has also been observed in unfed *Periplaneta* (Ralph and McCarthy, 1964).

The irregular responses of the locust and its low sensitivity compared to that of the cockroach led investigators to use the latter to test the hypertrehalosemic activity of locust CC. In *Carausius* (Gäde, 1980) the CC contain a factor that stimulated lipid mobilization in *Locusta* and raised the carbohydrate level in *Periplaneta* but had no effect on *Carausius*. It might therefore be that in some cases the activity of a neurohormone from an organ such as the CC may be masked by the presence of an antagonistic factor, as already suggested for the neurohormones involved in diuresis regulation. In *Locusta* and *Carausius,* only neck-ligated insects respond to CC extract injection by increasing their hemolymph carbohydrates (Loughton and Orchard, 1981; Gäde and Lohr, 1982).

11.1.1.3. Identification of the Metabolic Neurohormones of the AKH Family

The purification of metabolic hormones has progressed considerably in the past few years, during which the primary structure of eight metabolic neurohormones was defined.

These important advances were made possible by the continuous progress achieved in developing increasingly sensitive techniques. For purification, various chromatographic procedures, HPLC, and reversed phase HPLC were used, while sequencing was performed by the Edman degradation method, fast-atom bombardment mass spectrometry, and gas phase amino acid sequencing (see Stone and Mordue, 1980; Mordue and Morgan, 1985; Schaffer, 1985).

11.1.1.3a. *Identification of Adipokinetic Hormone.* Stone *et al.* (1976) and Cheeseman *et al.* (1977) identified the structure of AKH by using the lipid mobilization capacity of the adult male *Locusta* as a bioassay. One AKH unit was defined as the quantity of hormone causing the release of 1 μg lipid per μl hemolymph in 1 hr.

The starting material for isolation of the hormone consisted of 3000 glandular lobes of CC of *Schistocerca* and *Locusta,* comprising a mass of 750 μg. The dissected material was placed in methanol and subjected to ultrasonication, centrifugation, and uptake by distilled water. This was followed by chromatography and elution with distilled water. The residue was dissolved in methanol and subjected to thin-layer chromatography on silica gel. The chromatogram was developed in isopropanol–water–acetic acid which separated out an ultraviolet absorbing area of R_f 0.67 containing adipokinetic hormone activity. The amino acid composition of the hormone was determined by acid hydrolysis followed by mass spectrometry and supplemented by enzymatic digestion and Edman's degradation.

The AKH proved to be a decapeptide with a molecular weight of 1158; it was devoid of amine and free carboxyl groups. Its formula, given by Stone, is:

$$\text{PCA-Leu-Asn-Phe-Thr-Pro-Asn-Trp-Gly-Thr-NH}_2$$

In subsequent investigations, Cheeseman *et al.* (1977) further clarified the formula of AKH, which they showed to be as follows:

$$\text{Glu-Leu-Asn-Phe-Thr-Pro-Asn-Trp-Gly-Thr-NH}_2$$

Adipokinetic hormone increases the hemolymph lipid concentration in a dose-dependent manner (Fig. 102), and 188 pmol of AKH is present in one pair of CC. The half-life of the AKH was estimated to be 19–24 min.

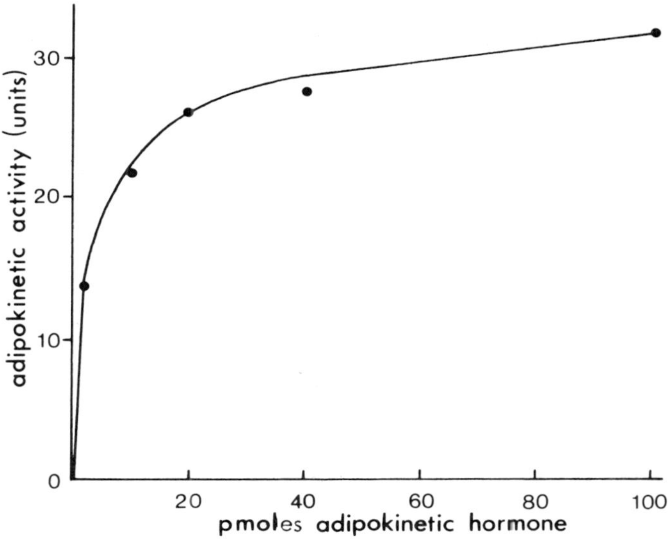

FIGURE 102. Dose-response curve for adipokinetic hormone which elevates hemolymph lipid levels in *Locusta*. A unit of adipokinetic activity is defined as the amount of hormone causing an increase of 1 μm lipid per μl hemolymph per hr. (Modified from Jones *et al.*, 1977.)

Adipokinetic hormone was synthesized in 1977 by Broomfield and Hardy. Its three-dimensional configuration was suggested to include a β bend formed by residues 5–8, the amino acids involved being Pro-Asp-Trp (Stone *et al.*, 1978).

Subsequently, in 1979, Carlsen *et al.* isolated another peptide from the CC of *Schistocerca* in addition to AKH. This peptide displayed adipokinetic activity in the locust, and its amino acid composition was determined. It was designated as AKHII and AKH became AKHI.

The identification of AKHII gave rise to investigations by several teams (Fig. 103A,B). Gäde and co-workers (1984) showed that the AKHII of *Locusta* was different from that of *Schistocerca,* although both are octapeptides. They were completely identified by Siegert *et al.* (1985) (Table XXXV) as having the following molecular structure:

AKHII *Locusta*	pGlu-Leu-Asn-Phe-Ser-Ala-Gly-Trp-NH_2
AKHII *Schistocerca*	pGlu-Leu-Asn-Phe-Ser-Thr-Gly-Trp-NH_2

Note that the only difference between AKHII L (*Locusta*) and AKHII S (*Schistocerca*) is at residue 6. Like AKHI, both AKHII are blocked at N and C

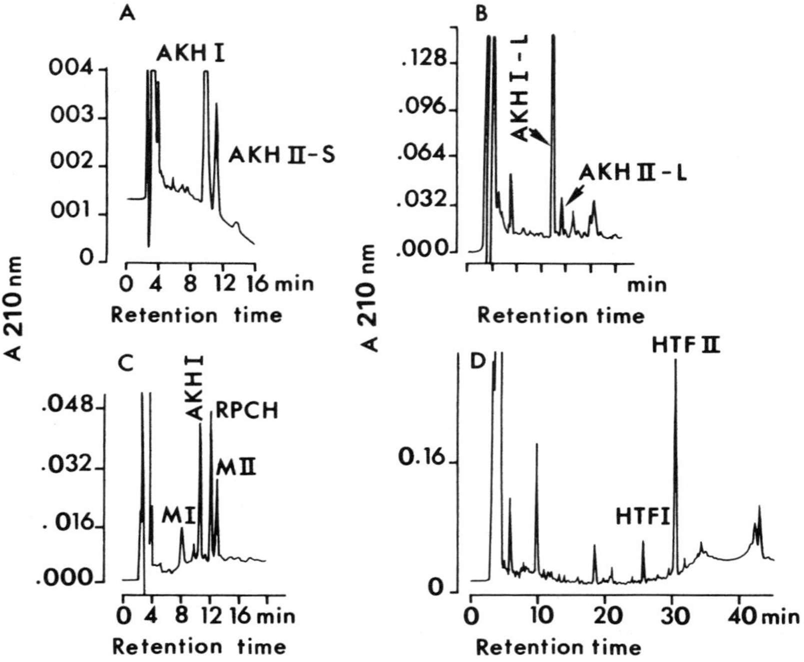

FIGURE 103. Separation of insect metabolic neurohormones by reverse phase HPLC. (A) AKHI and AKHII from the corpora cardiaca of *Schistocerca*. (B) AKHI and AKHII from the corpora cardiaca of *Locusta*. (C) Synthetic MI, MII, AKHI, and RPCH. (D) Hypertrehalosemic factors I and II from the corpora cardiaca of *Carausius*. [Modified from Siegert and Mordue, 1986a (A); Gäde *et al.*, 1884 (B); Gäde, 1986 (C); Gäde, 1985b (D).]

terminals, but they have different residues from those of AKHI in positions 5, 6, and 7 and are thus devoid of Pro 6.

11.1.1.3b. *Adipokinetic Hormones and Crustacean RPCH.* As mentioned in Chapter 9, many investigators have demonstrated the presence in insect CC of a substance causing pigment concentration in the red chromatophores of crustaceans. This substance (RPCH) was identified by Fernlund and Josefsson (1972), and was shown to be an octapeptide with a molecular structure resembling that of AKHI and II, thus explaining the action of insect CC on crustacean chromatophores. Residues 1, 2, 3, 4, and 8 are common to AKHI, AKHII L and S, and RPCH, and residues 5 and 7 are common to RPCH and both AKHII only. Conversely, residue 6 is proline in AKHI and RPCH only. The retention times of AKHI and RPCH differ slightly (see Fig. 103C).

AKHI	pGlu-Leu-Asn-Phe-Thr-Pro-Asn-Trp-Gly-
AKHIIL	Thr-NH_2
AKHII S	pGlu-Leu-Asn-Phe-Ser-Ala-Gly-Trp-NH_2
RPCH	pGlu-Leu-Asn-Phe-Ser-Thr-Gly-Trp-NH_2
	pGlu-Leu-Asn-Phe-Ser-Pro-Gly-Trp-NH_2

Both the insect hormones AKHI and AKHII induce pigment concentration in the red chromatophores of prawns and shrimps (Mordue and Stone, 1976, 1977; Carlsen *et al.*, 1979), and RPCH, the red pigment-concentrating hormone, provokes the release of fat body diglycerides in insects (Herman *et al.*, 1977; Mordue and Stone, 1976, 1977). However, insect hormones are more potent in insects than in crustaceans, and vice versa.

11.1.1.3c. *Cockroach Hypertrehalosemic Hormones.* The isolation and identification of the metabolic factors of cockroach CC began more than 20 years ago and rapidly showed that two factors were involved. Investigations by Brown (1965), Natalizi and Frontali (1966), and Natalizi *et al.* (1970) showed that the active factors are destroyed by proteolytic enzymes and are different from serotonin. Using heads of *Periplaneta* or *Apis* and CC of *Periplaneta,* two peaks exhibiting both hypertrehalosemic and cardioacceleratory activity were demonstrated. About the same time, Mordue and Goldsworthy (1969) separated four fractions from the CC of *Locusta* and *Schistocerca.* Two of them increased the hemolymph trehalose concentration in *Periplaneta,* but the third, located in the neurohemal lobe, was only slightly active; the fourth, originating in the glandular lobe, displayed far greater activity.

Cockroach CC were also shown to contain two active factors other than AKH; one of them was lipemic in the locust and the other, hyperglycemic in the cockroach (Holwerda *et al.* 1977a,b). Furthermore, the retention times of AKHI and II were shown by RPHPLC to be different from those of two hyperglycemic peaks (Gäde, 1984). This observation agrees with the results of Downer and Steele (1972) who, by purification of the hypertrehalosemic factor of *Periplaneta,* showed that it was devoid of adipokinetic activity. It also explains the differences between the reactions of the cockroach and locust to the same extract.

The molecular structure of two factors both trehalosemic and myotropic was finally obtained in 1984 by Baumann and Penzlin, O'Shea *et al.*, Scarborough *et al.*, and Witten *et al.* (Table XXXVI) (see Chapter 7), and confirmed by Gäde (1985a) and Siegert and Mordue (1986b). These factors were shown to be octapeptides as AKHII with both the N and C terminals blocked. Their retention times are shown in Figure 103C. They have been termed factor MI or CCI or neurohormone D, and factor MII or CCII, respectively. They exhibit strong homologies with AKHI, AKHII, and RPCH, and together with the three latter peptides constitute the AKH family. Factors MI and MII have the same residues

TABLE XXVI. Molecular Structures of Members of the AKH Family[a]

Name	Insect	Molecular structure	MW	Reference
AKHI	*Locusta*	pGlu-Leu -Asn -Phe-Thr -Pro -Asn -Trp-Gly-Thr-NH_2	1159	Stone *et al.*, 1976
AKHIIL	*Locusta*	pGlu-Leu -Asn -Phe-Ser -Ala -Gly -Trp-NH_2	940	Siegert *et al.*, 1985
AKHIIS	*Schistocerca*	pGlu-Leu -Asn -Phe-Ser -Thr -Gly -Trp-NH_2	910	
MI	*Periplaneta*	pGlu-Val -Asn -Phe-Ser -Pro -Asn -Trp-NH_2	973	O'Shea *et al.*, 1984; Scarborough *et al.*, 1984
MII		pGlu-Leu -Thr -Phe-Thr -Pro -Asn -Trp-NH_2	988	
HTH	*Nauphoeta, Blaberus*	pGlu-Val -Asn -Phe-Ser -Pro -Gly -Trp-Gly-Thr-NH_2		Gäde *et al.*, 1986; Hayes *et al.*, 1986
HTFII	*Carausius*	pGlu-Leu -Thr -Phe-Thr -Pro -Asn -Trp-Gly-Thr-NH_2		Gäde and Rinehart, 1987
AKHM	*Manduca, Heliothis*	pGlu-Leu -Thr -Phe-Thr -Ser -Ser -Trp-Gly-NH_2	1007	Ziegler *et al.*, 1985; Jaffé *et al.*, 1986
RPCH	Crustaceans	pGlu-Leu -Asn -Phe-Ser -Pro -Gly -Trp-NH_2	930	Fernlund and Josefsson, 1972

[a]MI = CC1; MII = CC2 = neurohormone D.

1, 4, and 8 as AKHI, II L, and II S, and the same proline in position 6 as AKHI. Other homologies occur between either MI or MII on the one hand and AKH peptides on the other. It is of interest to note that MII displays homologies with glucagon N-terminal sequences 1–12.

A single hypertrehalosemic factor was identified in two other cockroaches, *Nauphoeta* and *Blaberus,* using fast-atom bombardment mass spectrometry and Edman degradation method (Gäde and Rinehart, 1986; Hayes *et al.*, 1986). Like AKH, it is a decapeptide and resembles the *Periplaneta* factor MI, with which it shares its five first residues (Table XXXVI). It is also active *in vitro* (Hayes *et al.,* 1986).

11.1.1.3d. *Stick Insect and Lepidopteran Metabolic Peptides.* Extracts of the CC of *Carausius* increase the hemolymph lipid level in the locust and have properties resembling those of locust AKH. They are heat-stable and withstand incubation with trypsin, carboxypeptidase A, and leucine aminopeptidase, but not with thermolysin or α-chymotrypsin (Gäde, 1980). They have been shown to contain two hypertrehalosemic factors, HTFI and II, but only HTFII has been identified (Gäde, 1985b; Gäde and Rinehart, 1987). It is a decapeptide with blocked C and N terminals. It resembles both AKHI, from which it only differs by its Asn3, and MI, with which it has the first eight residues in common (Table XXXVI). It is therefore not surprising that HTFII is both adipokinetic and trehalosemic.

The identification of metabolic peptides has also been completed in two lepidopterans: *Manduca* (Ziegler *et al.*, 1985) and *Heliothis* (Jaffé *et al.*, 1986) (Table XXXVI). It is interesting to note that in *Manduca,* the CC extracts which increases fat body glycogen phosphorylase activity in the larva (Ziegler, 1979) cause lipid mobilization in the adult (Ziegler, 1984). It has not been possible to determine in preliminary separation experiments whether both of these processes are activated by one or by two factors (Ziegler and Gäde, 1984).

Investigations were also conducted in the grasshopper *Romalea microptera* (Spring and Gäde, 1987) and the cricket *Gryllus* (Gäde, 1987). Corpora cardiaca homogenates of both species induced hyperlipemia in locusts and *Gryllus,* and the CC of *Romalea* also triggered hypertrehalosemia in *Periplaneta.* In *Romalea,* however, the CC homogenates only induced lipid mobilization and had no effect on sugars. Two absorbance peaks were demonstrated with a peak I/peak II ratio of 6 : 1. In *Gryllus,* AKH activity was confined to a single absorbance peak which corresponded to a heat-stable octapeptide resembling AKHII S. Its amino acid composition was found to be:

Asx, Thr, Ser, Glx, Gly, Val, Phe, Trp

11.1.1.3e. *Conclusion.* Eight of the peptides regulating metabolism are now known, and several others are being identified. It is interesting to note that all these peptides closely resemble each other but are not identical except those of the two lepidopterans *Heliothis* and *Manduca* and the two cockroaches *Blaberus* and *Nauphoeta*. The metabolic peptides constituting the AKH family have several common features, particularly the presence of blocked terminals. In most cases, two distinct peptides are present in a single species.

11.1.1.4. Respective Roles of Members of the AKH Family

As shown above, the members of the AKH family, especially AKHI, not only affect lipid metabolism but also stimulate the production of diacylglycerol transport lipoproteins and intervene in flight muscle metabolism, carbohydrate mobilization, and myotropic activity.

11.1.1.4a. *Lipids*. Both AKHI and AKHII have been observed to raise the lipid titer of the hemolymph and to stimulate the production of the lipoproteins involved in transporting diacylglycerol. Both these peptides are indeed released during flight (Orchard and Lange, 1983a,b), but AKHII is less potent than AKHI as far as lipid and lipoprotein synthesis are concerned, although, surprisingly, AKHII is more potent than AKHI in increasing fat body cAMP (Goldsworthy *et al.*, 1986a) (Fig. 104).

The availability of synthetic AKHI and II has permitted quantification of the hormones present in the CC. In *Locusta* and *Schistocerca* CC, AKHI is clearly more abundant than AKHII (Siegert and Mordue, 1986a). Attempts to elucidate the respective roles of both hormones were made in which the presence of metabolic hormones was demonstrated in the larvae of locusts (Mwangi and Goldsworthy, 1977a) and of *Carausius* (Gäde, 1980). Smaller amounts of both AKHI and AKHII were found in larvae than in adults, and their proportions changed during development (Siegert and Mordue, 1986a). As regards their function in the larvae, all we know is that AKHI enhances both larval lipid and sugar levels (Van Marrewijk *et al.*, 1984) and that lipemia is induced by both AKHI and II (Pannabecker *et al.*, 1987).

11.1.1.4b. *Carbohydrates*. Carbohydrates are known to be affected by AKHI, at least in *Periplaneta* (Jones *et al,* 1977; Mordue and Stone, 1977). In *Locusta,* the fat body glycogen phosphorylase content was reported to increase after AKHI injection (Van Marrewijk *et al.*, 1983) and also after either AKHI or AKHII injection (Goldsworthy *et al.*, 1986a,b), but to a lesser degree with AKHII (Fig. 104). The effects of AKH on locust carbohydrate metabolism in fact seem to depend on the developmental stage of the insect. Adipokinetic hormone does not affect the trehalosemia of the adult but does act on the larva (Van Marrewijk *et al.*, 1984). In *Manduca,* CC extracts were observed to cause

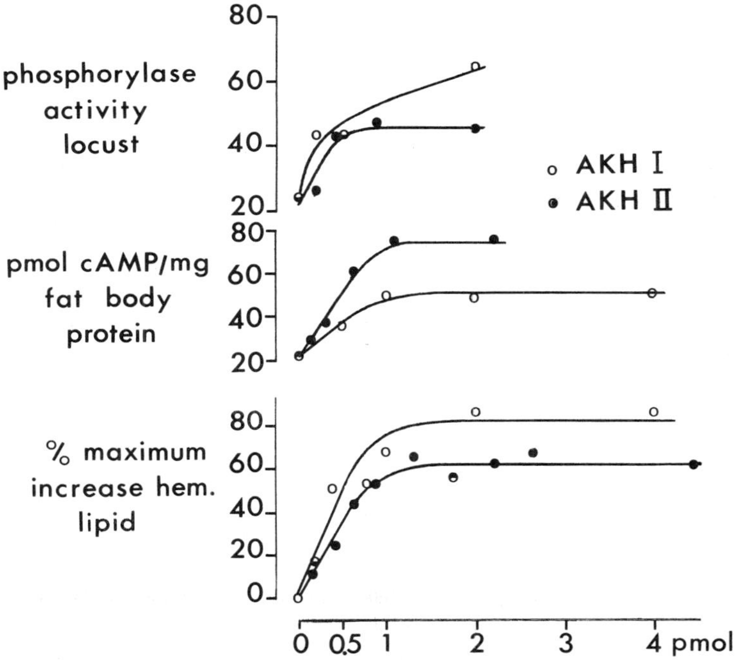

FIGURE 104. Comparative effects of natural AKHI and II on glycogen phosphorylase activity, fat body cAMP, and the hemolympth total lipid titer. hem, Hemolymph. (Modified from Goldsworthy and Wheeler, 1986.)

lipid mobilization in the adult, but only glycogen phosphorylase activation in the larva (Ziegler, 1984).

Loughton and Orchard (1981) proposed that AKHII was the hyperglycemic factor in *Locusta*. On the basis of chromatographic separation of the CC factors released after electrical stimulation of NCCII, these authors showed in neck-ligated locusts that a CC factor distinct from AKHI exerted hyperglycemic effects (Loughton and Orchard, 1981) thus acting as vertebrate glucagon on both lipids and carbohydrates. However, comparison of the timing of the release of AKHI and AKHII does not indicate that AKHII was released earlier, i.e., during the period of carbohydrate utilization for flight. Moreover, comparison of the effects of AKHI and AKHII on fat body glycogen phosphorylase clearly shows that AKHI is more potent than AKHII (Goldsworthy *et al.*, 1986a,b), thus suggesting that AKHII is not the hypertrehalosemic factor of the adult locust. The physiological significance of AKHII therefore remains to be established, as do the factors involved in stimulating carbohydrate release during the first minutes of flight. As indicated below, these factors might well be octopamine or a glucagon-like peptide.

11.1.1.4c. *Interspecific Activity.* Given the closely related molecular structures of the members of the AKH family, it was not surprising that they were found to exert interspecific effects in numerous studies, whose results are summarized below and in Table XXXVII. These results show that the hyperlipemic response of *Locusta* and the hyperglycemic response of *Periplaneta* are obviously stimulated by CC extracts of many species, including coleopterans and lepidopterans (Fig. 105). Conversely, homogenates from the CC of *Locusta* and *Periplaneta* stimulated these responses in *Tenebrio* and *Manduca*. In some cases, however, only weak reactions were obtained. For instance, there was no hyperlipemic response by *Periplaneta* to CC extracts which were active in locusts, and no hyperglycemic response by *Locusta* to extracts from *Periplaneta*. *Carausius*, on the other hand, only reacted to its own CC and even then the reaction was confined to neck-ligated insects; neither AKH nor RPCH was active in this insect. The present possibility to dispose of pure natural or synthetic hormones made it possible to determine the amount of hormone necessary to evoke 50% of the maximal lipemic response by locusts and glycemic response by *Periplaneta* (Gäde, 1986; Goldsworthy *et al.*, 1986b and Table XXXVIII). These determinations revealed virtually opposite orders of magnitude for the mobilization of lipids and carbohydrates, respectively, as follows:

Lipids	AKHI > AKHII L > HTFII *Carausius* > RPCH > MI-MII
Carbohydrates	HTFII *Carausius* > MII > MI > RPCH > AKHI

Although no clear correlation has been obtained between molecular structure and lipemic and glycemic activities, it might be that decapeptides are more active than octapeptides. Proline 6 is present in most members of the AKH family and has been suggested to be essential for binding to adipokinetic hormone receptors. However, this residue is replaced by an Ala residue in *Locusta* AKHII, which is nevertheless very potent.

11.1.1.4d. *Structure–Activity Relationships.* Stone and co-workers (1978) investigated structure–activity relationships with synthetic compounds structurally related to AKH. They showed that AKH activity requires blocked terminals and at least eight amino acid residues. Moreover, the L enantiomer of pyroglutamic acid cannot be replaced by D-pyroglutamic acid without great loss of activity. In addition, the dipeptide Gly_9-Thr_{10} appears to be of importance because the inversion of position also results in a great loss of activity.

TABLE XXXVII. Interspecific Activity of Corpora Cardiaca Extracts[a]

CC donor insect	Effector insect	L	G	Reference
Periplaneta	*Periplaneta*	0		Downer, 1972; Goldsworthy *et al.*, 1972b
	Locusta	+		
	Danaus	0		Dallmann and Herman, 1978
Periplaneta and other cockroaches	*Periplaneta*		+	Gäde, 1984; Gäde and Scheid, 1986
	Locusta	+		
	Carausius		(+)	Gäde, 1979, 1984
Carausius	*Carausius*		0	
	Head-ligated *Carausius*		+	
	Periplaneta		+	
	Locusta	+		
	Tenebrio		+	Dutrieu and Gourdoux, 1967
Locusts	Locusts	+	0	Goldsworthy *et al.*, 1972b
	Periplaneta	0	+	Chalaye, 1969; Goldsworthy *et al.*, 1972b
Schistocerca	*Tenebrio*	+		Goldsworthy *et al.*, 1972b
Locusta and *Gryllus*	*Periplaneta*		+	Gäde, 1984
	Locusta	+		
	Manduca	+	+	Ziegler and Gäde, 1984
Gryllus	*Gryllus*	+	+	Gäde and Scheid, 1986
	Periplaneta		+	Gäde, 1984
	Locusta	+		
Tenebrio	*Tenebrio*	+		Goldsworthy *et al.*, 1972b
	Danaus	+		Dallmann and Herman, 1978
	Locusta	(+)		Gäde, 1984
	Schistocerca	+		Goldsworthy *et al.*, 1972b
	Carausius		0	Gäde, 1984
Leptinotarsa	*Leptinotarsa*		+	Gäde and Scheid, 1986
	Periplaneta		+	
	Locusta	+		
Manduca	*Locusta*	+	+	Ziegler and Gäde, 1984
Danaus	*Danaus*	+		Dallmann and Herman, 1978
Phormia	*Periplaneta*		+	Friedman, 1967

[a]0, no effect; (+), slight effect; +, clear-cut effect; G, glycemia; L, lipemia.

11.1.2. Glycogenolytic and Glucosemic Factors

The results obtained in some insects are also suggestive of the existence of a glycogenolytic hormone. Thus, PI removal was found to cause glycogen accumulation in the fat body in *Calliphora* (Thomsen, 1952), *Aedes* (Van Handel

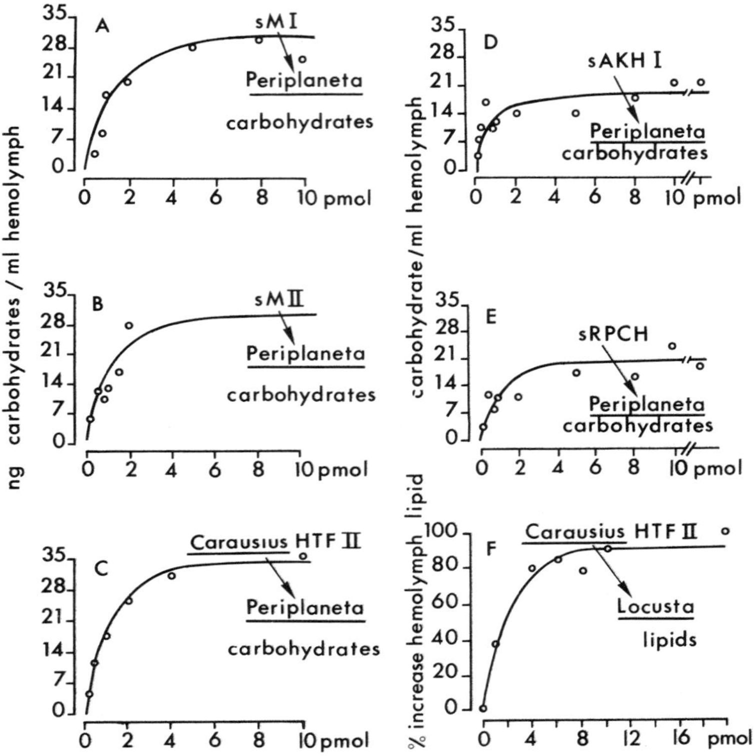

FIGURE 105. Effects of increasing doses of MI, MII, HTFII, AKHI, and RPCH on carbohydrate mobilization in *Periplaneta* (A–E) and lipid mobilization in *Locusta* (F); s, synthetic. (Modified from Gäde and Rinehart, 1987.)

and Lea, 1965), and *Locusta* (Goldsworthy, 1971), and the implantation of the PI into parsectomized *Aedes* restored a normal glycogen level in the fat body (Lea and Van Handel, 1970).

The factors involved in sugar regulation seem numerous. Apart from their action on hemolymph trehalose, the CC appear to act on the blood glucose concentration, but this action is controversial. Some investigators reported that the CC do not act on hemolymph glucose (*Periplaneta*, Steele, 1961; McCarthy and Ralph, 1962) but other authors (Bowers and Friedman, 1963) were able to cause a rise in hemolymph glucose by CC extracts in *Blaberus*. Lafon-Cazal and Roussel (1971) obtained the same effect on hemolymph trehalose and hemolymph glucose concentration with the glandular part of the CC of *Locusta* and

TABLE XXXVIII. Relative Potencies of Some Members of the AKH Family[a]

	ED_{50} values[b]	
	Lipid mobilization in *Locusta*	Increase in hemolymph sugars in *Periplaneta*
sAKHI	1	5.9
nAKHII	1,3	
nHTFII *Carausius*	2	0,9
n, sMHI	5	1,9
n, sMHII	5	1,8
sRPCH	4,6	4,5

[a]From the data of Gäde, 1986 and Goldsworthy *et al.*, 1986.
[b]ED_{50}, amount of hormone necessary to evoke 50% of the response; n, natural; s, synthetic.

Schistocerca, but opposite effects with the neurohemal part, which proved to be hypertrehalosemic and hypoglucosemic.

11.1.3. Hypolipemic and Hypoglycemic Factors

The occurrence of both hypolipemic and hypoglycemic factors has been demonstrated in certain insects.

The hypotrehalosemic factor was demonstrated in *Periplaneta* (Downer and Steele, 1972), and in the flies *Calliphora* and *Phormia,* in which removal of the PI and CC raised trehalosemia (Normann, 1975; Chen and Friedman, 1977a; Duve, 1978), which was restored by injection of brain homogenates (Anelli and Friedman, 1986). In other species such as *Manduca,* CC-CA extracts were also shown to induce hypotrehalosemia but in addition significantly reduced fat body glycogen, which may correspond to the action of a glycogenolytic factor.

In locusts and *Carausius,* the existence of a hypoglycemic factor was suggested by experiments in which normal and neck-ligated insects were injected with CC homogenates. Only the latter were able to react (Van Marrewijk *et al.*, 1980; Loughton and Orchard, 1981; Gäde, 1981).

11.1.4. Insulin- and Glucagon-like Peptides

Recent immunocytological and RIA investigations revealed that insulin- and glucagon-like peptides were present in the nervous system, CC, midgut, and hemolymph of several insect species (see Chapter 2), suggesting that these

TABLE XXXIX. Physiological Effects of Both Vertebrate and Insect Insulins on Carbohydrate Metabolism

Insect	Vertebrate insulin	Insect insulin-like fraction	Vertebrate glucagon	Insect glucagon-like fraction	Reference
Periplaneta	↓ Glycemia	↓ Glycemia			Barrett and Loughton, 1987
	0 Lipemia	0 Lipemia			
Locusta	↓ Trehalosemia	↓ Trehalosemia	↑ Trehalosemia		Moreau *et al.*, 1982
	↓ Glycemia	↓ Glycemia			Barrett and Loughton, 1987
Vespa	↓ Glucosemia				Ishay *et al.*, 1976
Apis	↓ Trehalosemia		↑ Trehalosemia		Bounias and Pacheco, 1979
		↓ Glycemia			Bounias *et al.*, 1986
Tenebrio		↓ Glucogenogenesis[a]			Gourdoux *et al.*, 1983
Manduca	0	↓ Glycemia[b]	0	↓ Fat body glycogene	Tager *et al.*, 1976
Calliphora		↓ Trehalosemia[c]			Duve and Thorpe, 1979
		↓ Glucosemia			

[a]Intestinal insulin-like fraction.
[b]CC-CA insulin-like fraction.
[c]Head insulin-like fraction.

peptides might be involved in carbohydrate and lipid regulation. Indeed, injection experiments with vertebrate molecules and insect insulin- and glucagon-like molecules, respectively, cause a decrease in hemolymph carbohydrate titer and an increase in trehalosemia (see Table XXXIX). In addition, the midgut insulin-like peptide of *Locusta* modifies glucose catabolism (Ben Khay *et al.*, 1987). As regards lipids, fewer results are available. Insulin was shown to block lipid release in incubated prelabeled fat body of *Hyalophora* and *Periplaneta* (Bhakthan and Gilbert, 1968).

Attempts to purify and sequence insect peptides reactive to vertebrate metabolic peptides (Tager *et al.*, 1976; Duve *et al.*, 1979) showed a definite similarity as regards their insulin β chain (see Loughton, 1987). However, more than one peak was revealed. It shall be noted that the 1–12 portion of glucagon displayed similarities to the MII hypertrehalosemic factor of the cockroach.

An interesting observation was made in the bee, in which somatostatin inhibited the hypoglycemic and hyperlipemic effects of insulin and the hypolipemic effects of glucagon (Bounias and Dubois, 1982; Bounias, 1986; Bounias *et al.*, 1986) (Table XXXX).

11.1.5. Octopamine and Other Biogenic Amines

11.1.5.1. Octopamine

As will be seen in Section 11.5.2, the stimulation of the release of AKH from the glandular lobe of the CC is triggered by octopamine which is transported by the NCCII. Octopamine also appears to have another direct humoral function in lipid metabolism since it is present in the hemolymph of locusts (David and Lafon-Cazal, 1979), where its level increases rapidly during flight, especially during the first 10 min (Goosey and Candy, 1980), i.e., before AKHI and II release (Fig. 106). In addition, octopamine was shown to exert hyper-

TABLE XL. Trehalose, Glucose, 1,2-, 1,3-Diglyceride, Small and Large Triglyceride Concentration in Honeybee Hemolymph in Response to a Bee Insulin-Immunoreactive Peptide (BIRP) or to BIRP and Somatostatin[a,b]

Injection	Hemolymph concentration after 3hr						
	Trehalose	Glucose	1,2-DG	1,3-DG	sTG	lTG	FFA
Saline	22.7	4.38	0.11	0.06	0.07	0.03	0.13
BIRP	7.54	2.47	0.05	0.02	0.05	0.08	0.06
BIRP + somatostalin	21.0	5.82	0.06	0.02	0.05	0.01	0.15

[a]Modified from Bounias *et al.*, 1986.
[b]DG, diplycerides; FFA, free fatty acids; lTG, sTG, large and small triglycerides.

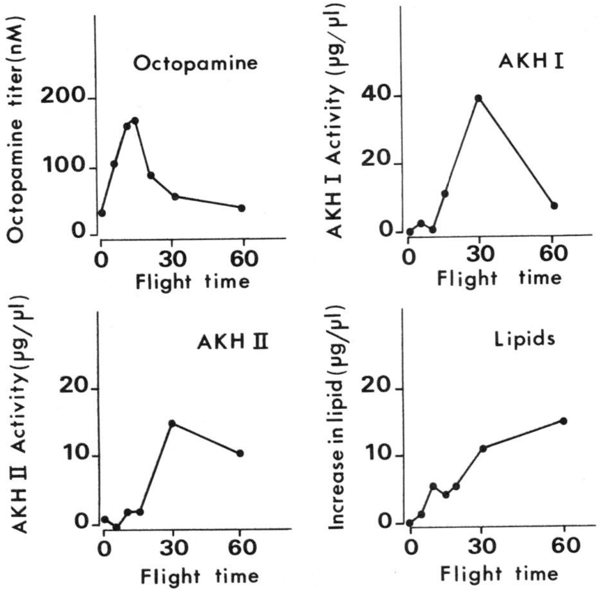

FIGURE 106. Hemolymph hormonal titer and lipid concentration during the first hour of flight in locusts. (Modified from Orchard, 1987.)

lipemic effects on the fat body both *in vivo* (Orchard *et al.*, 1981, 1983; Lafon-Cazal and Morandini, 1982) and *in vitro,* where its effects were shown to be dose-dependent (Van Heusden *et al.*, 1984). The effects of octopamine are blocked by α-adrenergic blocking agents but not by β-blocking agents (Orchard *et al.*, 1982). The mobilization of lipids for proline synthesis which takes place in *Glossina morsitans* is also triggered by octopamine (Pimley, 1985).

As regards carbohydrates, the role of octopamine is still being debated. In cockroaches it has been demonstrated that the stress caused by manipulation of the insect induces hypertrehalosemia, apparently due to the action of octopamine (Downer, 1979). Thus, determination of the hemolymph octopamine level after stress or during flight indicated that there is a very rapid increase in octopamine (Bailey *et al.*, 1983) which reaches a peak in less than 2 min of flight. Although octopamine was found to increase the glycogen phosphorylase content of the fat body within a minute in *Periplaneta,* this effect was not observed in *Locusta* (Van Marrewick *et al.*, 1983). In addition, neither octopamine nor other biogenic amines exerted hyperglycemic effects *in vitro* in *Blaberus* (Hayes and Keeley, 1985).

The half-life of octopamine has been shown to be very short (about 30 min)

(Goosey and Candy, 1982). Contact with the Malpighian tubules *in vitro* was found to accelerate its breakdown (Fig. 107).

11.1.5.2. Other Biogenic Amines

The effect of biogenic amines on lipid metabolism was first shown by Bhakthan and Gilbert (1968) and Gole and Downer (1979). Norepinephrine stimulates lipid release, but the lipids released in the cockroach, as in *Hyalophora,* are free fatty acids and not diglycerides, as normally occurs in the insect. Hence in these insects, epinephrine exerts the same effects on lipid release as in vertebrates, but these effects are abnormal for the insect. In *Danaus,* it appears to cause the release of tri- and diglycerides (Chang, 1974).

As regards carbohydrates, dopamine and tyramine, but not serotonin, were reported to have slight effects (Downer, 1979). When the effects of norepinephrine and dopamine were investigated in adult honeybees, injection of norepinephrine was observed to raise the glucose and trehalose blood concentrations while dopamine and dopa appeared to require conversion into an active form in order to exert such activity (Bounias, 1980).

11.1.6. Diapause Hormone

As explained in Chapter 5, the offspring of certain strains of *Bombyx* exhibit diapause. The existence of a diapause hormone was demonstrated, originating in the subesophageal ganglion (Hasegawa, 1957). A relationship exists between diapause and the carbohydrate content of the ovary; and ovaries which produce diapause eggs are far richer in glycogen. The diapause hormone exerts its action during vitellogenesis and acts on ovarian trehalase (Yamashita and Hasegawa,

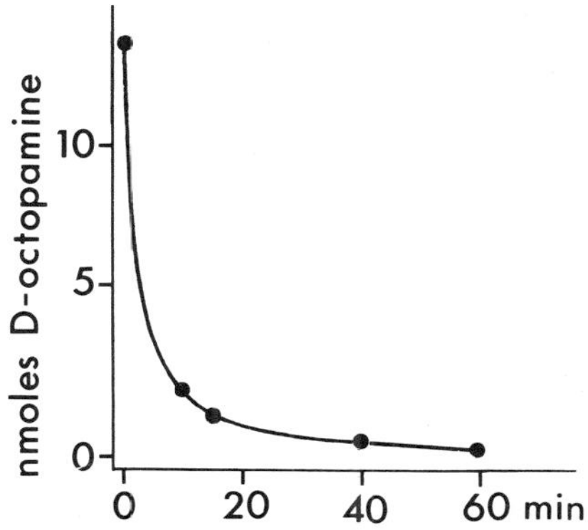

FIGURE 107. Breakdown of octopamine by incubation with Malpighian tubules. (Modified from Goosey and Candy, 1982.)

1967; Shimada and Yamashita, 1979). Removal of the subesophageal ganglion lowers its level in the ovary, but the other enzymes active in glycogen synthesis remain unchanged. Diapause hormone depresses fat body glycogen and hemolymph trehalose (Hasegawa and Yamashita, 1965). It also affects the lipids, which become more abundant in the ovary and less so in the fat body.

11.1.7. Juvenile Hormone

Allatectomy and JH application reveal the part played by the CA in both carbohydrate and lipid metabolism. Glycemia and trehalosemia are increased by allatectomy while fat body glycogen decreases after this operation. This was observed in several species, i.e., *Carausius* (L'Helias, 1955), *Periplaneta* (McCarthy and Ralph, 1962), flies (Thomsen, 1952), bugs (Janda and Sláma, 1965; Islam and Roy, 1982; Saha *et al.*, 1986), and *Schizodactylus* (Mandal *et al.*, 1984). As regards hemolymph glucose, the results are still contradictory. In *Periplaneta,* the CA seems to play an inhibitory role whereas in *Locusta* they are stimulatory (McCarthy and Ralph, 1962; Lafon-Cazal and Roussel, 1971). During vitellogenesis, the action of the CA on carbohydrate metabolism was also observed. Their removal inhibits the deposition of ovarian glycogen and modifies the trehalose level of the adult female but this effect is probably indirect and is due to the absence of ovarian development (Chen and Friedman, 1977b); however, CA has been shown to exert direct action on lipid metabolism *in vitro* (Gilbert, 1967; Hill and Izatt, 1974).

Lipid metabolism has often been reported to be influenced by JH. A decrease in the hemolymph lipid titer was demonstrated after allatectomy by Vroman and co-workers (1965) in *Periplaneta,* by Odhiambo (1966), Lafon-Cazal and Morandini (1982), and others in locusts, by El-Ibrashy and Boctor (1970) in *Spodoptera,* and by Saha *et al.* (1986) in *Chrysocoris.* In *Hyalophora,* a correlation was observed during metamorphosis between the fatty acid and JH titers (Stephen and Gilbert, 1970). However, no effect was observed after JH application in *Chilo* diapausing pupae (Yagi and Fukaya, 1974; Tsumuki and Kanehisa, 1981).

11.1.8. Ecdysone

In the silkworm, ligation of the pupa behind the PG arrests development and causes an increase in glycogen comparable to that of the diapausing insect. Prothoracic gland implantation causes the resumption of development and diminution of glycogen reserves (Kobayashi, 1957). Moreover, if ecdysone is injected into decerebrated diapausing pupae of *Samia* and *Bombyx,* incorporation of labeled glucose is very low in the fat body glycogen but high in the hemolymph trehalose, which increases considerably in quantity (Kobayashi and Kimura,

1967). The direct or indirect role of ecdysone in carbohydrate metabolism is under debate and could be better established by *in vitro* studies.

11.2. PROTEIN METABOLISM

Protein synthesis is of considerable importance to the physiology of the insect during its development, at metamorphosis, and in the adult stage during reproduction. Proteins are the main constituent of the tissues and their pattern changes during the developmental phases, metamorphosis, and vitellogenesis. For example, in the lepidopteran *Manduca,* certain proteins are synthesized during larval life but the production of larval proteins has to stop at larval–pupal commitment and to be replaced by the synthesis of new proteins involved in the secretion of pupal tissues, particularly cuticula. Throughout their life, insects synthesize the new proteins needed by the growing and developing tissues. In holometaboles, storage proteins which can be considered as amino acid stores for metamorphosis are present in large amounts in the hemolymph of the larvae. They are synthesized in the fat body.

Protein synthesis is of major importance during vitellogenesis because oocyte maturation requires the uptake of specific vitellogenins from the hemolymph. In addition, various secretory processes, some of them associated with reproduction, involve protein synthesis (see Chapter 6).

As far as the amino acids are concerned, proline is known to be used as a fuel for flight in certain insect species, and other amino acids such as tryptophan or tyrosine play an important part in the processes of sclerotization, melanization, and production of ommochrome pigment. In addition, certain amino acids are oxidized and converted to fatty acids and carbohydrates. After deamination, the residual carbon skeleton is incorporated into the glycolytic pathway.

11.2.1. Ecdysteroids, Juvenile Hormone, and Protein Metabolism

Ecdysteroids and JH are considered to influence protein synthesis processes, particularly those involved in growth and reproduction. During development, 20-HE induces growth and differentiation and stimulates RNA synthesis. In the last larval instar, the first ecdysteroid commitment peak arrests larval RNA and protein synthesis until the second peak triggers a new pattern for these syntheses (see Riddiford, 1985). In the adult female, ovarian ecdysone stimulates the production of vitellogenin by the fat body as shown in mosquitoes (Fallon *et al.*, 1974) and a number of other species (see Chapter 6).

Juvenile hormone plays an important role in reproduction because it triggers the main processes involved (see Chapter 6). In the last larval instar and pupa of

Bombyx, however, it is the disappearance of JH which causes biosynthesis of a high molecular weight protein (Izumi *et al.*, 1984).

11.2.2. Neurohormones and Protein Metabolism

11.2.2.1. Stimulatory Neurohormones

11.2.2.1a. *Vitellogenesis.* In addition to JH and ecdysone, the ns cells of the PI contribute to protein regulation in the adult female. They act in several ways, controlling JH and ecdysone release by the CA and the ovary, respectively (see Chapters 3 and 6), and regulating fat body activity. This regulation has been demonstrated in several insects, including locusts, other orthopterans, and coleopterans. In other insects the role of neurohormones in triggering vitellogenin synthesis is either controversial or believed to be nonexistent.

In *Locusta* (Hill, 1962), in which histophysiological data suggested that the ns cells of the brain affect protein regulation, removal of the PI and CC demonstrated that neurohormones influence hemolymph protein titer *in vivo* (Fig. 108) and increase fat body protein synthesis *in vitro* (Osborne *et al.*, 1968). Furthermore, incorporation *in vivo* of [^{14}C]glycine or [^{3}H]leucine into fat body proteins is reduced by cautery of the PI, as also occurs after removal of the CA (Hill, 1963). The presence of the latter organs is necessary but not sufficient for vitellogenin synthesis. In two other orthopterans, *Gomphocerus* and *Melanoplus,* both the PI and CA were shown to be necessary for protein synthesis (Loher, 1966; Elliott and Gillott, 1977), and in the coleopterans *Tenebrio* and *Leptinotarsa,* vitellogenin production is also dependent on the presence of the CA and PI (Mordue, 1965; De Loof and De Wilde, 1970).

The results for cockroaches are different. In several species of cockroach, the role of the CA in vitellogenin synthesis has been firmly established, but the

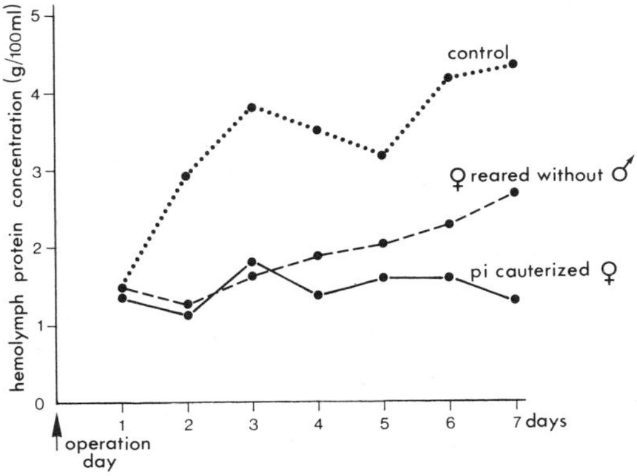

FIGURE 108. Hemolymph protein concentration in *Schistocerca* females following parsectomy. (Modified from Hill, 1962.)

data obtained for the PI are not really convincing as shown by Engelmann and Penney (1966), Lüscher (1968a), and Engelmann (1969). In *Leucophaea,* however, it was shown that both the PI and the CA play a part in regulating protein metabolism, but that each organ controls the synthesis of different proteins. Thus, experiments conducted on both normal fasting females and decapitated females implanted with brain, CC, or CA showed that all hemolymph proteins do not behave in the same way, since the brain and CC implants caused a sharp rise in certain proteins while CA implants induces a significant increase of another protein which is a vitellogenin (Scheurer, 1969a,b).

11.2.2.1b. *Tanning and Pigmentation.* As far as tanning and pigmentation are concerned, distinct neurohormones such as bursicon or MRCH appear to be involved in the production of specific amino acids (tyrosine, tryptophan) (see Chapter 9).

11.2.2.1c. *Flight.* As already mentioned, proline is used as a fuel for flight by certain insects like tsetse flies and Colorado beetles. In the tsetse fly, a factor from the CC, probably a member of the AKH family, is involved in proline synthesis, and one or two factors from the same organs stimulate hydrolysis of the triglycerides, which helps to convert alanine to proline (Pimley and Langley, 1982). Proline synthesis is also stimulated by octopamine (see Section 11.1.5). In the Colorado beetle, the titers of hemolymph carbohydrates and lipids did not change after a half-hour flight in the laboratory but the hemolymph proline titer dropped considerably (Mordue and De Kort, 1978) and the proline concentration in the flight muscles also decreased (Weeda *et al.,* 1979). In this insect, proline synthesis takes place in the fat body and is stimulated by a factor from the CC and by synthetic AKH (Weeda, 1981).

11.2.2.1d. *Feeding.* The relationship between feeding and ns product release was shown in *Locusta,* and the presence was demonstrated in the brain and CC of a peptide which stimulates protein synthesis both *in vivo* and *in vitro* (Carlisle *et al.,* 1987).

11.2.2.1e. *Urates.* Neurohormonal control of protein metabolism is confirmed by the study of urates, which are protein degradation products. Bodenstein (1953b) investigated the urate cells of the fat body in *Periplaneta* after removing the CC-CA complex or the CA alone; he observed that the urates persist in the latter case, whereas in the former they disappear within 2 weeks but reappear with CC implantation. The transaminases of the fat body, flight muscles, and Malpighian tubules decrease significantly in locusts after cautery of the PI and reappear if CC extracts are injected into operated insects. A decrease in uric acid excretion and both glutamate-pyruvate and glutamate-oxalacetate aminotransferase activity was also observed in *Periplaneta* after parsectomy. Injection of CC extracts was followed by a return to normal (Mordue and Goldsworthy, 1973).

11.2.2.2. Inhibitory Neurohormones

In addition to their stimulatory effects, neurohormones also inhibit protein synthesis. The experiments dealing with this inhibition mostly concern *Locusta,* in which CC glandular lobe extracts were shown to suppress labeled leucine incorporation by the fat body (Carlisle and Loughton, 1979). This inhibition appears to be general, since the amounts in the hemolymph of both the storage protein arylphorin and vitellogenins diminish after injection of CC extracts (Carlisle and Loughton, 1986).

The factor involved in the inhibitory action of the CC on protein synthesis appears to be AKHI, which acts both *in vivo* and *in vitro* in a dose-dependent manner and is effective at low doses (Carlisle and Loughton, 1986). Neither the extract of the nh lobe of the CC nor octopamine displays such effects. However, a brain factor distinct from AKHI also reduces protein synthesis in the locust (Moshitzky *et al.*, 1987a). Furthermore, in *Manduca,* an inhibitory factor from the head which appears to originate in the brain is also known to be present (Caglayan and Gilbert, 1987).

If the factor inhibiting protein synthesis in the locust is indeed AKHI, its location in the glandular lobe of the CC is not surprising. A factor that reduces the hemolymph protein titer in the young adult female was also found in the PO of *Carausius* (Raabe and Demarti-Lachaise, 1973). These nh organs, however, failed to react to AKH antibodies (Raabe, 1986a). The occurrence in the central nervous system of a factor exerting AKH-like effects but distinct from AKHI is probable.

11.3. PROTEASES AND AMYLASES

The protease level in the midgut of insects is not constant and rises during feeding periods. Thus in the bug *Pyrrhocoris* the intestinal protease content increases when the adult female feeds at the beginning of each oogenetic cycle; in the adult male, feeding is relatively rare, and the protease level is therefore low, as in starved insects (Hrubéšová and Sláma, 1967). In the cockroach *Leucophaea* (Engelmann, 1969) and in blood-feeders, the level in the midgut depends on the amount of proteins present in the digestive tract.

Different experiments gave grounds for suspecting that a neurohormone is involved in controlling the secretion of intestinal enzymes. Decapitation of *Tenebrio* 1 day before eclosion prevents the normal increase in proteases and injection of the blood of fed insects into starved insects raises the protease content of the midgut (Dadd, 1961). Head ligation also reduces the protease level in the tsetse fly *Glossina morsistans* (Langley, 1967). The PI was shown to cause a decrease in midgut protease content in *Calliphora* (Thomsen and Møller, 1959), *Tenebrio* (Mordue, 1967), *Aedes* (Briegel and Lea, 1979), and *Dysdercus* (Muraleedharan and Prabhu, 1979).

Removal of the CC–CA complex also causes a drop in the protease level in the adult female of *Pyrrhocoris* (Hrubéšová and Sláma 1967), and in the orthopteran *Schizodactylus* (Mandal *et al.*, 1981) both the brain and CA are necessary for protease synthesis. Implantations and injections performed on operated insects restored normal situations, as shown after implantation of CC in *Pyrrhocoris* (Hrubéšová and Sláma, 1967), CC-CA complex in *Calliphora* (Thomsen and Møller, 1963), after brain extract injection into neck-ligated *Glossina* (Langley, 1967), and after implantation of the ns cells of the PI in parsectomized *Dysdercus* (Muraleedharan and Prabhu, 1979). Midgut amylase activity was also shown to increase after brain implantation in the caterpillar *Hyblaea puera* (Muraleedharan and Prabhu, 1981).

In adult *Tenebrio* and in *Morimus* larvae, exposure of the insect to a constant high temperature reduces the enzyme content of the gut. Injection of neuroendocrine complex extracts restores a normal rate of proteolytic activity in both insects (Ivanović *et al.*, 1976; Janković-Hladni *et al.*, 1976). These extracts also act on the amylolytic activity of *Tenebrio* larvae and adults (Ivanović *et al.*, 1976; Janković-Hladni *et al.*, 1978). During the annual cycle, both amylases and proteases display compensatory reactions which are dependent on the phase of the annual cycle and which differ from one another (Ivanović *et al.*, 1982, 1987), suggesting that these enzymes might be regulated independently.

A neurohormone which originates in the PI and is released into the CC could therefore directly ensure regulation of the functioning of the endocrine cells of the midgut. Certain authors, however, consider that this regulation is accomplished by the food intake. A number of immunoreactive endocrine cells and axons recently discovered in the wall of the midgut (see Chapter 2) certainly play a regulatory role. It is not yet know, however, whether they respond directly to sensory inputs from the gut, or whether the release of their content is controlled by nervous or neurohormonal action exerted by the nervous system. The release of ns products after food intake observed in *Calliphora* (Thomsen, 1965) and *Melanoplus* (Dogra and Gillott, 1971) supported this hypothesis.

11.4. RESPIRATORY METABOLISM

During the life of the insect, its respiratory metabolism changes significantly with the changes in energy expenditure. Processes such as morphogenesis, vitellogenesis, locomotor activity, and metabolic processes require large energy inputs and accordingly lead to increases in respiratory metabolism; diapause and fasting, on the other hand, cause a decrease in respiration. It is therefore obvious that all the hormones and neurohormones which act on the above processes affect respiratory metabolism.

Ecdysone, which triggers RNA and protein synthesis, brings about changes in respiratory metabolism. The CA, which also take part in protein synthesis, stimulate the respiratory metabolism of larvae, and of the adult females, during the reproductive processes. However, these organs also seem to have a direct action on respiratory metabolism, as shown under various experimental conditions and in incubated fat body and isolated mitochondria (see Raabe, 1982).

The different neurohormones controlling metabolism, reproduction, and locomotor activity exert a definite indirect influence on respiratory metabolism, which declines sharply after removal of the CC in *Pyrrhocoris* (Sláma, 1964a,b), locusts (Lafon-Cazal and Roussel, 1971), and cockroaches (Keeley and Friedman, 1967), and after destruction of the PI in *Pyrrhocoris* (Sláma, 1964a,b) and *Oncopeltus* (Conradi-Larsen, 1970). In addition, changes in respiratory metabolism were obtained *in vitro* on pieces of fat body incubated with CC extracts in *Leucophaea* (Wiens and Gilbert, 1965; Lüscher, 1968a; Müller and Engelmann, 1968) and *Blaberus* (Keeley and Friedman, 1967) and with PI extracts in *Leucophaea* (Lüscher, 1968a). The brain and CC also act on homogenates of the fat body and muscles of *Periplaneta, Leucophaea,* and *Blaberus* (Ralph and Matta, 1965). It therefore appears that the active factors concerned originate, at least partly, in the PI. They may include hypertrehalosemic hormone from the PI and also AKH originating in the cells of the CC.

By what pathway do neurohormones act on respiratory metabolism? Examination of the respiration rate of fat body mitochondria showed a significant decline in *Blaberus* which had been deprived of CC for 30 days, apparently due to the loss of electron transport capacity of the mitochondria (Keeley, 1971). It was also shown in *Blaberus* (Hayes and Keeley, 1981) that cytochrome synthesis by fat body mitochondria, which occurs during the first days of imaginal life, is stimulated fivefold by CC extracts. The factor involved, termed the cytochromogenic factor, is a heat-labile protein. Its secretory pattern does not depend on food intake or photoperiod but was stimulated by JIII injection (Hayes and Keeley, 1983). It probably enables the fat body cells to meet increased biosynthetic requests.

11.5. ORIGIN AND RELEASE SITES OF METABOLIC HORMONES

11.5.1. Adipokinetic Hormone Origin Cells

The investigations of Goldsworthy *et al.* (1972a,b) clearly showed that in locusts, AKH is absent from the PI and very abundant in the glandular lobe of the CC even after sectioning of the NCCI. In *Carausius* and *Manduca,* the locations of the hyperglycemic factor(s) are the same as in locusts. Extracts of the CC are very potent but extracts of the brain and subesophageal, thoracic, and abdominal ganglia are devoid of AKH activity (Ziegler, 1979; Gäde, 1980).

Comparative ultrastructural studies of the glandular part of the CC of *Locusta* in insects at rest and after 5 and 60 min of flight showed that after prolonged flight, the secretory activity of the Golgi apparatus increases significantly together with cellular exocytotic activity (Rademakers and Beenakkers, 1977), thus confirming that AKH originates in the glandular cells of the CC. It was also observed that both AKHI and AKHII were produced in the same cell and even the same granule (Diederen *et al.*, 1987).

The presence of an adipokinetic activity in the nh lobe of the CC of locusts observed by several investigators (Goldsworthy, 1969; Carlsen *et al.*, 1979; Lafon-Cazal and Morandini, 1982; Moshitzky *et al.*, 1987a,b) is difficult to explain. According to the latter authors, it might be AKHI. The occurrence in the brain of molecules closely related to AKH is also suggested by immunocytological studies. Thus, in *Locusta*, antisera raised against AKH N terminal stained the glandular cells of the CC, and also a number of neurons (Schooneveld *et al.*, 1986, 1987a). In the ventral nerve cord, no AKH activity was found in *Carausius* (Gäde, 1980), but the thoracic ganglia of *Danaus* were shown to be potent in this respect (Dallmann and Herman, 1978).

11.5.2. Regulation of Adipokinetic Hormone Release

Lipids are only used by the locust for flight after an initial period of high-speed flight during which, as already stated, trehalose is used as fuel. The question therefore arises as to whether the hemolymph trehalose level is involved in AKH release. Trehalose injection during flight prevents AKH release (Houben and Beenakkers, 1974), causes reduction of the electrical activity of the CC (Bloemen *et al.*, 1987), and maintains a low AKH hemolymph titer (Cheeseman *et al.*, 1976). On the other hand, the hemolymph lipid concentration exerts a negative feedback on AKH release (Cheeseman and Goldsworthy, 1979).

In locusts, the release of AKH requires intact nerve connections between the brain and CC, even though it is synthesized in the latter (Goldsworthy *et al.*, 1972a). The pathway from the brain to the CC runs through the NCCII, whose sectioning prevents AKH release (Rademakers, 1977a; Goldsworthy *et al.*, 1972a,b) and whose electrical stimulation results in the release of a hyperlipemic factor into the incubation medium of the CC (Orchard and Loughton, 1981; Orchard and Lange, 1983a) (Fig. 109). The substance involved in stimulating the release of the content of the glandular lobe cells appears to be octopamine, which mimicks the effects of NCCII electrical stimulation and whose action is suppressed in the presence of α-adrenergic receptor antagonists (Orchard and Loughton, 1981). Octopamine was shown to intervene in the release of both AKHI and AKHII (Pannabecker and Orchard, 1986b).

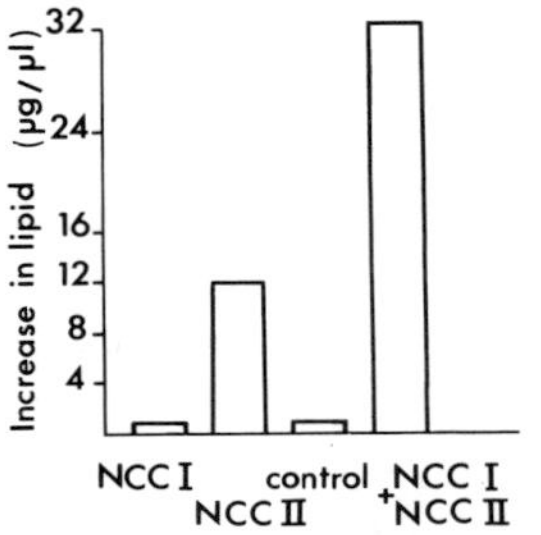

FIGURE 109. *In vitro* lipid mobilization induced by NCCI and NCCII stimulation in *Locusta*. Note that NCCI stimulation alone does not cause hyperlipemia but that stimulation of both NCCI and II potentiates the response. (Modified from Orchard and Loughton, 1981a; Orchard and Lange, 1983a.)

As regards the way in which the glandular cells of the CC are stimulated, a recent study showed that the electrical activity recorded after electrical stimulation of the NCCII does not originate in the cells themselves but in a local neuronal network (Bloemen *et al.,* 1986). The origin of the NCCII octopaminergic axons was identified in the lateral ns cell of the protocerebron of locusts (Rademakers, 1977b).

11.5.3. Source of Hypertrehalosemic Hormones

The place of origin of the hyperglycemic factors is not yet clear. Although in numerous species they are indisputably present in the CC (see Raabe, 1982), their presence in the brain does not seem certain.

In addition to CC glandular lobe extracts, nh lobe extracts were observed to display hypertrehalosemic activity (Mordue and Goldsworthy, 1969). Furthermore, experiments of sectioning of the NCCI and II which made it possible to obtain pure nh regenerates from the sectioned nerve extremity revealed the presence of hypertrehalosemic activity in the regenerates (Highnam and Goldsworthy, 1972), suggesting a cerebral origin of the hypertrehalosemic factors. Besides, the results of the removal of the glandular lobe of *Locusta* confirmed that the nh lobe contains a hypertrehalosemic-hypoglucosemic factor whereas the glandular lobe contains a hypertrehalosemic-hyperglucosemic factor (Cazal, 1971).

As far as the brain is concerned, it is known to be the source of the substances present in the nh lobe of the CC. However, brain extracts are mostly devoid of hypertrehalosemic activity. In *Locusta, Manduca,* and *Carausius,* CC extracts are very potent in enhancing this activity but brain extracts are ineffective (Highnam and Goldsworthy, 1972; Ziegler, 1979; Gäde, 1980). This might be due to the presence of an antagonistic inhibitory factor; in fact, the hypoglucosemic–hypolipemic factor that was found in the nh lobe of the CC (Cazal, 1971; Loughton, 1987) was also found in the lateral parts of the protocerebrum, and in flies PI removal enhanced trehalosemia, as shown in Section 11.1.3.

As indicated earlier, the different metabolic hormones have several common sequences and therefore display cross-reactivity. Consequently, it is difficult to establish which molecules are responsible for the effects observed with crude extracts from different organs, unless the extracts are diluted so much that their relative potencies can be estimated, or unless they are separated chromatographically and compared with known molecules.

As regards the ventral nerve cord, the investigations by Steele (1961), Gersch (1974) in *Periplaneta,* Gäde (1980) in *Carausius,* and Ziegler (1979) in *Manduca* failed to demonstrate any hypertrehalosemic hormone activity.

11.5.4. Control of Hypertrehalosemic Hormone Release

The release of the hypertrehalosemic factor from the CC is controlled by the brain, as seen by electrical stimulation of this organ in *Phormia* (Normann and Duve, 1969). In *Periplaneta,* brain stimulation is transmitted to the CC by the NCCI and involves adrenergic mechanisms (Gersch *et al.*, 1970); NCCII severance, on the contrary, causes a considerable decrease in hemolymph trehalose (Gersch, 1974), and after removal of the brain, subesophageal ganglion, or mesothoracic ganglion, the blood trehalose concentration remains at its lowest level (Gersch, 1976). Octopamine was shown to mimic the effects of neural stimulation (Downer *et al.*, 1984) and to elicit a dose-dependent increase in cAMP in incubated CC. The same effect is also brought about by dopamine and serotonin (Gole *et al.*, 1987). The hemolymph trehalose concentration appears to exert a feedback regulation on the fat body, at least in some insects. In *Hyalophora* (Murphy and Wyatt, 1965), the presence of large quantities of trehalose in the fat body incubation medium increases the incorporation of [^{14}C]glucose by the fat body glycogen to the detriment of trehalose, and in *Phormia,* the quantities of trehalose synthesized *in vitro* depend on the amount of trehalose in the incubation medium (Friedman, 1967). In locusts, however, these findings were not substantiated (Goldsworthy, 1969).

11.5.5. Origin of Octopamine, Hypolipemic, and Hypoglycemic Factors

In locusts, the lateral ns cells of the brain have been demonstrated to produce the octopamine which controls AKH release from the glandular cells of the CC. Whether the same cells also send processes to the neural lobe, which releases octopamine into the hemolymph, is not yet known.

The origin and release sites of the hypoglycemic and hypolipemic factors were also investigated in the locust. In the CC, they seem to include the nh lobe (Cazal, 1971; Loughton, 1987). The active factor, which appears to be an insulin-like peptide, is also found in the lateral part of the protocerebrum.

11.6. MODE OF ACTION OF METABOLIC HORMONES

The mode of action of metabolic hormones has been fairly well investigated (Fig. 110). Evidence is available to show that the diapause hormone of *Bombyx,* which raises the ovary glycogen level, acts through an ovarian trehalase (Yamashita and Hasegawa, 1967). The hypertrehalosemic hormone activates a fat body phosphorylase, giving rise to the formation of trehalose from the stored glycogen (Steele, 1963; Wiens and Gilbert, 1967). In addition, hypertrehalosemic hormone appears to diminish the activity of another enzyme, UDP glucose/glucose phosphate glucosyltransferase (Chen and Friedman, 1977b). Adipokinetic hormone brings about the formation of diglycerides from triglycerides and free fatty acids, and also acts on the muscle, where it enhances the penetration of diglycerides (Mayer and Candy, 1969; Robinson and Goldsworthy, 1976). It also performs a very different function, that of stimulating the formation of lipoproteins, which are diglyceride carriers (see Section 11.1.1.1).

Besides the effects just described, the metabolic hormones, like all peptidic hormones, use a second messenger, cAMP. The function of cAMP in the action of hypertrehalosemic and adipokinetic hormones has been demonstrated by measuring the cAMP level in fat body subjected to the action of these hormones, and by comparing the hemolymph trehalose and lipid concentrations induced by the hormones, by nucleotides, and by phosphodiesterase inhibitors.

Corpora cardiaca extracts elevate the fat body cAMP level in *Periplaneta*

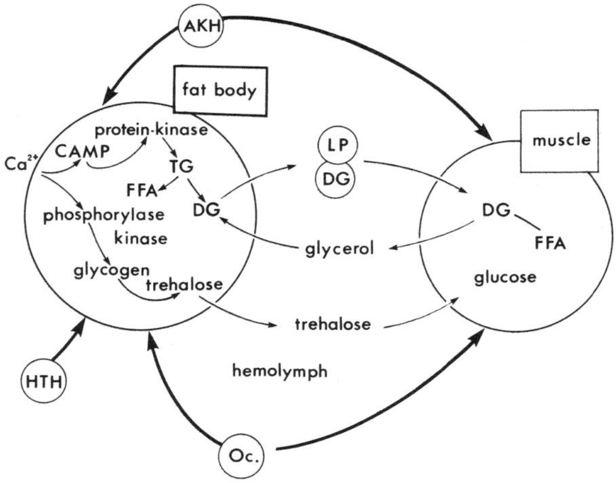

FIGURE 110. Schematic representation of the endocrine control of lipid and carbohydrate metabolism. HTH, Hypertrehalosemic hormone; DG, diacylglycerol; FFA, fatty acids; LP, lipoprotein; Oc, octopamine; TG, triglycerol. (Modified from Goldsworthy, 1983.)

(Hanaoka and Takahashi, 1977; Gäde and Beenakkers, 1977) and *Locusta* (Gäde and Holwerda, 1976; Gäde, 1979). An increase in adenylate cyclase is also observed in *Periplaneta* (Hanaoka and Takahashi, 1977). In the latter insect, the increase in cAMP caused by injection of CC extracts is related to their hypertrehalosemic effect, and in *Locusta,* to their adipokinetic effect. cAMP is elevated not only under experimental conditions but also under natural conditions, as in the locust after flight (Gäde and Holwerda, 1976; Orchard and Lange, 1983b). When cAMP is added to the fat body incubation medium, it brings about a hypertrehalosemic reaction in *Periplaneta,* thus stimulating the phosphorylase activity of the fat body (Steele, 1964; Hart and Steele, 1973). It also has an adipokinetic effect on the adult male of *Periplaneta* (Bhakthan and Gilbert, 1968) and on *Danaus* larvae (Chang, 1974), causing high fatty acid release in the former and triglyceride release in the latter.

Dibutyric cAMP mimics the hypertrehalosemic effect in *Periplaneta* (Hanaoka and Takahashi, 1977) and the adipokinetic effect in *Schistocerca* (Spencer and Candy, 1976) and *Locusta* (Gäde and Holwerda, 1976). This is also true of caffeine and theophylline in *Periplaneta* (Steele, 1964; Hanaoka and Takahashi, 1977) and of theophylline in *Schistocerca* (Spencer and Candy, 1976).

Using natural AKHI and AKHII, Goldsworthy and co-workers (1986a) showed that both hormones raised fat body cAMP, but that AKHII enhanced it more than AKHI, even though it was less effective than the latter in lipid mobilization and glycogen phosphorylase elevation. Whatever the reason for this, both hormones probably act through different receptors. The MI and MII factors were also reported to have various effects on fat body cAMP in *Periplaneta.* An increase was observed *in vivo* (Gäde, 1985c) but not *in vitro* (Orr *et al.,* 1985). The effect of synthetic AKH on cAMP induction was also demonstrated in dispersed locust fat body cells (Asher *et al.,* 1984). Adipokinetic hormone stimulates the synthesis and degradation of amino acids, probably by acting on fat body aminotransferases.

Octopamine has been demonstrated to exert direct metabolic action on the fat body in addition to its action on AKH release from the CC. Both effects were shown to involve cAMP (Orchard *et al.,* 1983a; Pannabecker and Orchard, 1986b). In addition, carbohydrate mobilization by octopamine was also accompanied by a rise in fat body cAMP (Gole and Downer, 1979) and even in *Glossina,* whose fuel for flight is the amino acid proline, octopamine raised the cAMP level (Pimley, 1985).

Octopamine receptors appear to be distinct from AKH receptors. They were suggested to belong to the octopamine-2 class (Pannabecker and Orchard, 1986a), and Lafon-Cazal and Morandini (1982) demonstrated that they are inhibited by phentolamine and propranolol, but not by haloperidol.

Study of the Ca^{2+} influx into fat body cells *in vitro* made it possible to observe that HTH, purified from CC extracts, triggers this process. Increased cell membrane permeability allows an extracellular Ca^{2+} influx which results in an increase in Ca^{2+} content of the cells, which in turn activates phosphorylase kinase (McClure and Steele, 1981).

11.7. BREAKDOWN OF METABOLIC HORMONES

As already mentioned, octopamine is broken down by the Malpighian tubules and its half-life is very short (30 min). The Malpighian tubules were also observed to breakdown AKHI (Siegert and Mordue, 1987). It appears that the three main breakdown products do not possess lipid-mobilizing activity.

11.8. CONCLUSION

Metabolism is the field in which the most progress has been made in identifying the primary structure of neurohormones. These neurohormones are octa- or decapeptides with blocked C and N terminals. They exhibit sequence homologies and are considered to constitute the AKH family, one member of which is known to exist in crustaceans. In insects, they perform important functions in lipid, carbohydrate, and protein regulation.

At present, the best known process is the lipid metabolism of locusts during flight. In this process, AKH is known to play a part in the release of diglycerides from the fat body, the synthesis of their carrier lipoproteins, and the incorporation of diglycerides at the muscle level.

On the whole, metabolic regulation appears to involve complicated processes. One surprising fact is the presence in several species of two factors influencing the same processes, as, for example, AKHI and II in locusts and MI and MII in *Periplaneta*. The significance of this duality is not yet understood.

The existence is also known of antagonistic factors involved in lipid, carbohydrate, and protein metabolism. One of them might be identical to an insulin-like peptide. Ecdysone and JH also have direct or indirect effects on metabolism.

In locust CC, the site of AKH synthesis has been clearly identified as being in the glandular lobe. The release of this hormone is triggered by octopamine, transported via the NCCII. The hypertrehalosemic factors appear to be secreted and released in the same manner in cockroaches and stick insects. However, the neural lobe and brain also appear to play a part which has not yet been precisely defined in the synthesis of octopamine, other metabolic factors, and perhaps some members of the AKH family.

12

Miscellaneous

Existing data regarding the functions of insect neurohormones indicate that some of them might participate in the control of certain processes such as locomotion, flight, stress endogenous nerve activity, and polymorphism.

12.1. LOCOMOTOR ACTIVITY

Locomotor activity in insects can be more or less intense and often follows circadian rhythms.

12.1.1. Circadian Rhythms

Circadian rhythms, which are often photoperiodically dependent, control several types of processes including various reproductive behaviors, locomotion, ecdysis, and changes in the activity level.

A number of experiments concerning the regulation of circadian rhythms were performed in *Periplaneta.* It was proposed that a humoral factor from the subesophageal ganglion was involved (Harker, 1956; 1960), but this was not confirmed in subsequent investigations (Brady, 1967a,b; Nishiitsutsuji-Uwo *et al.*, 1967). The results of investigations of other species did not agree either. In the crickets *Acheta* and *Teleogryllus,* removal of the PI made the insects arrhythmic (Cymborowski, 1970a; Sokolove and Loher, 1975) and parabioses caused the activity rhythm to reappear, while in insects deprived of the PI, implantation of a brain from a rhythmic donor failed to restore the normal situation, although it periodically reduced locomotor activity.

In the stick insect *Carausius,* which displays both circadian color change and circadian activity rhythm, it was shown that the mechanisms triggering these two processes are different, i.e., neuroendocrine control in the former (see Chapter 9) and nervous control in the latter. Removal of the CC-CA and PI had no effect on behavior. However, removal of the brain or severance of the circumesophageal connectives deleted the circadian rhythm and increased activity, showing that control of this rhythm is exerted by the brain. Removal of the subesophageal ganglion or cutting of the connectives behind this organ permanently immobilized the insects. These results (Eidmann, 1956; Dupont-Raabe, 1957), similar to those obtained by Brady (1967a,b), Nishiitsutsuji-Uwo *et al.* (1967), and Nishiitsutsuji-Uwo and Pittendrigh (1968) on cockroaches, suggest that the subesophageal ganglion stimulates activity while the brain inhibits it. The severance of the ventral nerve cord behind the subesophageal ganglion allowed circadian activity to continue in the anterior part of the body, while the posterior part remained motionless (Dupont-Raabe, 1957).

Besides experimental investigations, histological, histochemical, and ultrastructural studies were carried out in order to demonstrate putative circadian changes in the ns cell load. In crickets and cockroaches, the activity of the ns cells in the PI displayed a circadian rhythmic activity which is also found in the A-type ns cells of the ventral nerve cord of the cockroach (Cymborowski, 1970b; De Bessé, 1967). RNA and protein synthesis also follow a temporal rhythm, observed in studies of labeled cystein incorporation and acetylcholinesterase level which drops with the rise in nocturnal activity. Serotonin determinations in the brain and hemolymph also revealed cyclic modifications, but no correlation was found between the serotonin content of the hemolymph and the circadian rhythm. However, a close correlation was shown between this rhythm and the serotonin content of the central body, known to be an important connection center of the brain (Cymborowski, 1983).

In *Drosophila,* neuroendocrine control of the circadian rhythm was suggested for mutant flies whose locomotor rhythm altered. Brain implantation indeed changed the activity pattern of the implanted receiver (Handler and Konopka, 1979). Experiments on a cockroach identified the location of the circadian clock as being in the optic lobes (Sokolove and Loher, 1975) whose reimplantation in arrhythmic insects reestablished a circadian activity rhythm (Page, 1982).

12.1.2. Activity Levels

Although the rhythm and level of activity are linked, they are not necessarily governed by the same mechanisms. Thus, in *Locusta,* the postprandial rest, which is not part of the circadian rhythm, appears to be under neurohormonal control, since locomotor activity diminishes in fasting insects injected with the hemolymph of recently fed insects or with CC homogenates (Bernays, 1980). In *Carausius,* the action potentials recorded from *in vitro* preparations showed that the electrical activity of the peripheral ns cells varies according to the origin of the bathing hemolymph: thus, light phase hemolymph appears to contain an inhibitory factor which depresses the electrical activity and which is not found in dark phase hemolymph (Finlayson *et al.,* 1983).

12.2. FLIGHT

Experiments performed mostly on locusts suggest that neurohormones may play a role in flight determination. Removal of the ns cells of the PI causes the disappearance of nocturnal flight in noctuiid moths (Hinks, 1967), and in the blowfly *Calliphora* the CC are indispensable to the achievement of flight (Vejbjerg and Normann, 1974). In *Schistocerca,* the nh part of the CC and the PI are necessary for sustained flight to take place (Michel, 1973; Michel and Bernard, 1973). In *Locusta,* removal of the PI disturbs flight in young insects, but this effect is probably exerted through the CA, since injections of JHA reduces the effects caused by destruction of the PI (Goldsworthy *et al.*, 1977). Nevertheless, the glandular part of the CC is indispensable to flight because its AKH plays a major role in the release from the fat body of diglycerides that are used by the locust in prolonged flight; trehalose, whose release appears to depend on a neurohormone, is used in short flights (see Chapter 11).

Biogenic amines seem to be involved in flight regulation through metabolic and other pathways. In experiments on the lepidopterans *Noctua pronuba* and *Agrotis ipsilon,* serotonin injections enhanced the duration and amplitude of night flight, whereas other amines had no effect (Hinks, 1967).

Flight, which in many insects is involved in sexual behavior and often precedes ovogenesis and oviposition, is also dependent on the JH titer (Rankin and Riddiford, 1978 in *Oncopeltus*; see also Coats *et al.*, 1987 in a coleopteran).

Following flight cessation, degeneration processes often take place in the muscles, which depend on the JH titer as well. These processes were extensively studied in *Leptinotarsa* (De Kort, 1969), *Dysdercus* (Edwards, 1970), and *Acheta* (Chudakova and Gutmann, 1978) (see Pener, 1985).

12.3. ENDOGENOUS NERVE ACTIVITY—STRESS

Studies of the spontaneous nerve activity of the central nervous system suggest that neurohormones participate in this activity. In the cockroaches *Blaberus* and *Periplaneta,* CC extracts cause a drop in the endogenous nerve activity of the ventral nerve cord *in vitro* and a decrease in coordination and locomotor activity *in vivo* (Ozbas and Hodgson, 1958). They also trigger bursts of rhythmic nerve impulses in the phallic nerve of *Periplaneta* (Milburn *et al.*, 1960) and act in stress situations. Prolonged immobilization of the insects giving rise to a continuous reaction of struggling, or to a series of mechanical or electrical stimulations causing hyperactivity, induces the release into the blood of a paralyzing substance which can be transmitted from one insect to another by injection or parabiosis (Beament, 1958). Hodgson and Geldiay (1959) and Sternburg *et al.* (1959) showed that stresses such as electrical stimulation or successive turning over of the insects induce the disappearance of ns material from the CC, and also cancels the effect exerted by the extracts of these organs on the spontaneous nerve activity of the central nervous system. Thus, a neurohormone probably acts on behavior as well as on endogenous nerve activity. It undoubtedly originates in the brain because CC extracts lose their activity when disconnected.

The CC factor is thermostable, resistant to chymotrypsin, and distinct from epinephrine, norepinephrine, serotonin, and dopamine (Milburn and Roeder, 1962). According to Gersch and Richter (1963), Strejckova *et al.* (1964), and Birkenbeil (1971), the substance that stimulates nerve activity in *Periplaneta* is the neurohormone D1, which possesses many other physiological functions, including the ability to inhibit associative learning processes in *Periplaneta* (Weiss *et al.,* 1987).

It should also be noted that biogenic amines were shown to modify the electrical activity of the central nervous system, causing bursts of action potential in the ganglia of the ventral nerve cord (Twarog and Roeder, 1957; Hodgson and Wright, 1963), and that food deprivation and stress conditions cause an increase in the octopamine titer of the hemolymph in both cockroaches and locusts (Davenport and Evans, 1984a,b).

12.4. POLYMORPHISM

Polymorphism is widespread in insects and has many determining factors and aspects, including the morphology and pigmentation of the insect, its reproductive mode, and its behavioral characters. On the whole, polymorphism allows specialization which facilitates the dispersion of the species and above all the sharing of work in social insects (see Hardie and Lees, 1985).

12.4.1. Corpora Allata

The CA are known to play a major part in determining polymorphism and phase. In locusts, their role has been established in the triggering of green pigmentation (see Chapter 9), wing development, ovarian maturation rate, certain behavioral characteristics of the solitary phase (Joly and Joly, 1953; Staal, 1961), and the phase determinism of their offspring (Cassier and Papillon, 1968) (see Chapter 6).

In aphids, the CA also appear to be involved in morph determination, which in response to crowding and depending on the photoperiod gives rise to viviparous, apterous, parthenogenetic females or alate, oviparous sexuparae. Experiments with topically applied JH or JHA seemed to divert alate development toward the apterous condition, an effect similar to that obtained by long-day photoperiod treatment. Moreover, JH acts on the morph determination of the pedogenetic embryos developing in the abdomen of the treated larvae, which switch from oviparae to parthenogenetic viviparae (Lees, 1977; Hardie, 1987).

Caste determination in social insects depends on many factors including pheromones and nutritional factors. Juvenile hormone was shown to play an important part in caste regulation in *Polistes* (Röseler *et al.*, 1980), *Bombus* (Röseler, 1977), bees (Engels, 1978; Asencot and Lensky, 1984; Rembold, 1987), melipones (Hartfelder, 1987), ants (Passera and Suzzoni, 1979; Ono, 1982; Wheeler and Nijhout, 1983), and termites (Lüscher, 1958; Lebrun, 1967).

12.4.2. Neurohormones

The involvement of a neurohormone has been demonstrated in the aphid *Megoura viciae* by microcautery of different groups of brain ns cells in insects reared under various conditions (Steel and Lees, 1977; Hardie, 1987). The morph of the offspring can be distinguished in young larvae by the presence of growing eggs in oviparae and by chains of embryos located in the ovarioles in the virginoparae. The operated insects behave normally when reared under short-day conditions, but under long-day conditions oviparae appear in the offspring; the response to long days in insects previously exposed to short days, i.e., production of virginoparae instead of oviparae, is prevented by microcautery of brain median ns cells. These ns cells therefore appear to be the site of origin of a factor controlling virginoparae production in response to increased day length. Since they send one of their processes to the ventral nerve cord, their secretory products might be delivered directly to the abdomen and perhaps to the reproductive system itself. It is not yet known if the action of the brain in connection with this process is direct or if it is exerted through the CA.

In the lepidopterans *Polygonia c-aureum, Papilio xuthus,* and *Lycaena phlaeas,* which may develop in distinct spring or summer morphs, a hormonal role of the brain has been shown (Endo and Funatsu, 1985; Endo and Kamata, 1985).

In addition, it was observed in *Locusta* that gregarious locusts have less octopamine than solitary ones (David and Fuzeau-Braesch, 1979). These results, however, were not confirmed in *Schistocerca* (Morton and Evans, 1983).

Concluding Remarks

The foregoing survey of the data concerning the different types of neuroendocrine regulation of insects demonstrates the great progress achieved in various aspects of this field, especially identification of neurohormones. Immunocytochemistry and other new techniques for the detection of biogenic amines have enabled the latter to be precisely located in the nervous system and neurohemal organs, where the presence of vertebrate and invertebrate peptides was also demonstrated. Concepts regarding the structure of the neurosecretory–neurohemal systems have evolved, and the existence of numerous diversified sites of release of neurohormones has been shown. In addition, the neurons synthesizing several neurohormones have been identified. Finally, new data concerning the regulatory mechanisms involved in various processes, especially the functioning of endocrine glands, have demonstrated the complexity and diversity of these controls.

The first two insect neurohormones, proctolin and AKH, were identified in 1975 and 1976. By 1989 the molecular structures of more than 20 neurohormones were known. The metabolic–myotropic neurohormones which they include are octa- and decapeptides. Their molecular structures are similar and they belong for the most part to the AKH family. Other neurohormones, such as PTTH and MRCH, are large molecules whose structures have not been completely elucidated. It should be noted that the four insect neurohormones PTTH, MRCH, MII, and LSK display sequence homologies with vertebrate peptides: the insulin A chain, the carboxyl terminal of insulin-like growth factor II, glucagon, and gastrin/CCK, respectively. Comparatively, in crustaceans, progress has been made in studies of the molt-inhibiting hormone from the eyestalk, the vitellogenin-inhibiting hormone, and the hyperglycemic hormone, but the molecular structures of these large peptides are not known. On the other hand, two chromatophorotropic hormones, both of them octapeptides, have been identified. They are the red pigment concentrating hormone (Fernlund and Josefsson, 1972) and the distal retinal pigment hormone (Fernlund, 1976). In mollusks, on the other hand, two neurohormones have also been identified: the small cardioexcitatory peptide of a bivalve clam, the FMRFamide (Price and Greenberg, 1977), and the egg-laying hormone of *Aplysia,* a much larger peptide (Chiu *et al.*, 1979).

It is interesting that in many cases not one active peptide but two and sometimes three related peptides were separated, with slightly different retention times but similar activities.

In insects as in other invertebrates, numerous investigations revealed that neurons not always recognized as neurosecretory, and also the midgut, contained molecules immunoreactive to various vertebrate peptide antibodies. This opened the way to new researches, in particular to comparative biochemical studies of invertebrate and vertebrate peptides, and to experiments designed to reveal the physiological role played by vertebrate peptides in invertebrates.

Many interesting results were obtained. Insulin appears to control carbohydrate metabolism, and vasopressin seems to be implicated in the regulation of water metabolism. Further, immunoreactivity to FMRFamide, the cardioexcitatory peptide of mollusks, was identified in insect neurons and neurohemal organs, and this peptide was shown to have a neuromodulatory function in a skeletal muscle. The presence of a FMRFamide-like peptide was also revealed in most invertebrate and vertebrate phyla. Proctolin, the myotropic factor of the insect hindgut, was shown to be present in a hirudinean (Li and Calabrese, 1985) and in the central nervous system and neurohemal pericardial organs of crustaceans (Stangier *et al.*, 1986), and also to act on heart muscles. Adipokinetic hormones which regulate lipid metabolism in insects have been found in the nervous system and in neurohemal organs and areas of several invertebrates including two crustaceans, a myriapod, and a mollusk (Schooneveld *et al.*, 1987b).

With regard to the three peptides just mentioned, it is noteworthy that they act at one and the same time as neurohormones and neuromodulators since they are found both in blood and neurohemal organs, and in the neuropil of nerve ganglia. Their presence modifies the contractility of certain muscles as well as the bursting activity of certain neurons. From this point of view, immunocytochemistry has allowed great strides to be made in fundamental neuroendocrinology. It also allowed the demonstration that a single neuron can contain either two distinct peptides or a peptide and a biological amine such as proctolin and serotonin.

Immunocytological methods are very useful, too, for identifying the neurons involved in the synthesis of given peptides whose somata and pathways can be revealed. This is all the more interesting as antibodies can be raised not only against identified peptides but also against peptides which have only been purified. In insects, the neurons of origin of certain neurohormones, e.g., eclosion hormone, bursicon, proctolin, have been identified in this way. The molecules produced in identified neurons were also determined by RIA analysis of their content. The most striking fact to emerge from immunocytochemical investigations in insects is that all the new concepts established mostly in vertebrates are valid in insects, e.g., peptide ubiquity, colocation, and neuromodulation.

The acquisition of knowledge about biogenic amines has made significant parallel advances to those achieved in the field of immunoreactive peptides. New methods have established the qualitative and quantitative distribution of these amines and experimental studies have made it possible to understand their physiological significance as well. Thus, octopamine was shown to have important functions as a neurotransmitter/neuromodulator involved in the regulation of AKH release from the CC, and the discharge of brain ovulation hormone was also shown to be aminergically controlled. In addition, biogenic amines, especially octopamine whose cells of origin have been identified, have a neurohormonal role as they modulate the contractions of both visceral and skeletal muscles. Biogenic amines are present in the nervous system of invertebrates and vertebrates, but in the latter octopamine is less important than in insects. In crustaceans and mollusks, both octopamine and serotonin are present in the nervous system and act as neurohormones and neurotransmitters/neuromodulators (see Beltz and Kravitz, 1986). In crustaceans, serotonin appears to be involved in the release of MIH (Mattson and Spaziani, 1985) and red pigment-dispersing hormone, while the release of the two eyestalk chromatophorotropins red pigment concentrating hormone and melanin-dispersing hormone depends on dopamine and norepinephrine, respectively (Fingerman, 1985). Octopamine is released in the pericardial organs along with dopamine, serotonin, and the peptide proctolin, and also from thoracic neurohemal nerves. It acts on heart and skeletal muscles. Like serotonin, it exerts centrally controlled actions that modify motoneural outputs.

Besides techniques allowing detection of biogenic amines and immunoreactive peptides, other methods, such as whole-neuron staining by vital stain injection into the somata, have modified ideas about neurosecretory/neurohemal systems. These methods revealed the complexity of the axonal pathways and neuronal arborizations, and the diversity of the release sites, which were shown to be much more numerous than was previously believed. In addition to the neurohemal organs such as the CC and PO, numerous narrow but extensive neurohemal areas were described at the periphery of different nerves; in *Locusta,* a large network of fibers extending throughout the thoracic and abdominal ganglia was shown to originate from only two neurosecretory neurons of the subesophageal ganglion while in cockroaches four other neurons from the same ganglion were shown to give rise to a large serotonergic plexus. In crustaceans, too, new neurohemal areas were discovered at the periphery of thoracoabdominal segmental nerves (Beltz and Kravitz, 1986). However, like insects these invertebrates possess typical concentrated neurohemal organs, the sinus glands, as well as pericardial and postcommissural organs. Other invertebrates appear to lack neurohemal organs, and in several phyla, as in mollusks and annelids, neurohormones are released in diffuse extended areas which probably constitute primitive structures that have been preserved in evolved groups.

One may wonder if functional differences do not exist between the neurohormones released in specialized concentrated neurohemal organs and those discharged in diffuse areas. It is conceivable that the latter structures are better adapted to sudden nonprogrammed massive discharges which significantly increase the titer of the substances released in the internal medium. Systematic investigations of the distribution of insect neurohormones in neurohemal organs and areas have not been performed, but some results seem to indicate that metabolic hormones, diuretic vasopressin-like hormone, and the oviposition-stimulating hormones are released in neurohemal areas and sometimes also in neurohemal organs, while other neurohormones whose release is programmed, such as eclosion hormone and PTTH, are only discharged by the neurohemal organs.

Demonstration of neuronal pathways revealed important morphological features. Thus, a number of identified neurons which terminate in neurohemal organs or areas were shown in insects, as in other invertebrates, to send out extensive branches into the neuropil. Such pictures are indicative of a neurotransmitter/neuromodulator role exerted by the concerned peptide. Moreover, it has been shown in certain species that the firing pattern of particular neurons is modified by the presence of certain peptides such as eclosion hormone. The best known example of this is the action of the mollusk egg-laying hormone onto the RI5 neurons of the abdominal ganglion of *Aplysia* (Mayeri *et al.,* 1985).

The investigations performed during the last decade represent an important contribution to understanding the various types of hormonal and neurohormonal regulations, which were demonstrated to be both complex and diversified. The most significant advances concerned the regulation of synthesis and release of PTTH, the control of JH production, and the regulation of various aspects of reproduction and metabolism. Several other systems of control were also investigated and the number of neurohormones whose presence was demonstrated is fairly large. In general, the neurohormones constitute the most widespread hormonal means of communication in invertebrates, since endocrine glands are lacking in several phyla, although the presence of ecdysteroids has been often reported in arachnids, mollusks, and nematodes.

From different points of view (see Kleinholtz and Keller, 1979), crustaceans resemble insects as regards the mechanisms involved in regulating their growth, reproduction, and other processes. The crustacean Y organ, homologous to the prothoracic gland in insects, synthesizes the ecdysteroids and its activity is controlled by an inhibitory neurohormone from the eyestalk, the molt-inhibiting hormone, whose activity was recently demonstrated *in vitro*. Its release is in turn controlled by serotonin and by a feedback loop exerted by the hemolymph ecdysteroid titer (Mattson and Spaziani, 1985, 1986). The existence of another crustacean endocrine gland was recently revealed; this gland, the mandibular organ, synthesizes *in vitro* a compound closely related to juvenile hormone, the methylfarnesoate (Laufer *et al.*, 1987). The mandibular organ, which is apparently regulated by neurohormones from the eyestalk, might intervene in growth and reproduction, which would constitute another similarity between crustaceans and insects. Schematically speaking, the regulation of crustacean vitellogenesis (see Charniaux-Cotton and Payen, 1988) also displays several common features in crustaceans and insects, although the role of juvenoids has not been established in the latter. Vitellogenesis regulation involves a nonecdysteroid factor from the ovary, the vitellogenin-stimulating ovarian hormone, ecdysteroids produced in the Y organ, and one or more inhibitory neurohormones controlling the Y organ and the ovary.

In other invertebrates including annelids and mollusks, neurohormones have been shown to play a part in regulating various processes associated with reproduction (see Durchon, 1984; Joose, 1984). An ovulation and egg-laying neurohormone was identified in *Aplysia* and *Lymnea,* but vitellogenesis seems to be regulated in these invertebrates by a hormone produced in the endocrine glands (the dorsal bodies in pulmonates and the optic gland in cephalopods). Comparison with insects in this respect is difficult.

In insects, both diversity and complexity characterize the neuroendocrine regulations. In the case of the PG, brain PTTH plays the main stimulatory role but it also receives nervous messages from the central nervous system and is modulated negatively or positively by the hemolymph JH level according to the intermolt period; the PG is also influenced by the presence in the blood of a stimulatory factor from the fat body. The regulation of JH synthesis appears to be complicated as well since both inhibitory and stimulatory nervous and neurohormonal factors are involved. The regulation of vitellogenesis is both complex and diversified, and the regulatory mechanisms seem to have developed in diverging directions. While JH plays a major role in triggering vitellogenin synthesis and vitellogenesis in numerous insects, ecdysteroids from the ovary take over these functions in other species, in which JH is only involved during the previtellogenesis stage. Brain hormones are involved in varying degrees, either directly or indirectly, through the CA and ovaries. Other hormones that originate in the ventral nerve cord and brain control egg laying and parturition. In the tsetse fly, the latter process involves both stimulatory and inhibitory neurohormones and is also modulated by the ratio of ecdysone/20-HE in the hemolymph. Together with an extremely specialized mode of reproduction, these viviparous insects apparently developed particularly evolved regulations.

The large body of data obtained during the last few years has opened up many lines of research, most of them based on new information regarding the molecular structure of neurohormones. It therefore appears necessary to continue this fruitful type of investigation while at the same time applying the results to further research.

In addition to other methods, the recombinant nucleic acid techniques would be helpful in the sequencing and identification of neuropeptides. This was undertaken for mollusk egg-laying hormone and insect AKH (see Schaffer and Noyes, 1987).

Once a peptide has been identified, antibodies may be obtained which allow immunocytochemical recognition of both the neurons involved in its synthesis and their pathways and release sites, and also an RIA procedure from which the timing of synthesis and release can be ascertained by quantitative determination of the peptide in the hemolymph and tissues. The antibodies may be also used for immunoneutralization of the peptide, which is a sophisticated method for its functional elimination from the body.

Another promising field that has already given rise to a number of investigations is that of neurohormone receptor recognition and, more generally, the mechanism of neurohormonal action.

As far as the choice of the insect studied is concerned, it should be remembered that although most of the current studies deal with *Manduca,* cockroaches, or locusts, about 100 other insect species are also being investigated. This might favor the detection of new types of regulation but, on the other hand, excessive complexity of the data may hamper progress. A reduction of the number of species investigated could therefore be fruitful. The progress achieved in *Manduca* supports this conception. In this insect, the various physiological stages and hormonal titers are well known along with the way in which several of the main physiological processes are regulated. Moreover, the results obtained for *Manduca* illustrate the improvement achieved with *in vitro* methods. The latter, which are also used in several other species, will no doubt be widely applied in the future.

Finally, the investigations of insects can lead to practical applications in the struggle against pests. Once the neurohormones are identified, it is possible to synthesize agonists and antagonists, as well as antibodies able to inhibit the action of the peptides. Knowledge of the metabolism of neurohormones also makes it possible to act on the enzymes involved in their synthesis and breakdown.

Addendum

A.1. SYNTHESIS AND RELEASE SITES OF NEUROHORMONES

A.1.1. Neuroanatomical Studies

Using cobalt chloride or nickel chloride preparation, the origin cells of the CC, CA, aortic wall, antennal heart, and NCA2 nh areas were determined in *Rhodnius* (Chiang and Davey, 1988) and *Teleogryllus* (Moore and Loher, 1988).

The distribution of the cell bodies and the axonal processes of peripheral ns cells located along the thoracic median nerves were traced in the locust and shown to terminate in the PO (Myers and Evans, 1988).

A.1.2. Peripheral ns Ganglia

The occurrence of a series of peripheral ns ganglia located on the median nerve was demonstrated in *Stomoxys* in the vicinity of the accessory glands (Meola, 1988). The ultrastructural study revealed that they contain two ns cells that terminate in several sites: on the secretory cells of the accessory glands, on muscle fibers of the ejaculatory duct, and in nh areas. The study of such peripheral ganglia whose ns cells may intervene in both neurohemal and neuroeffectory regulations is quite new.

A.1.3. Biogenic Amines

Thanks to the recent appearance of antibodies raised against serotonin and octopamine, the distribution of these biogenic amines was thoroughly investigated and reinvestigated in the central nervous system of several insect species, i.e., locusts (serotonin and octopamine: Konings *et al.*, 1988a,b; Pflüger and Watson, 1988), *Tenebrio* (serotonin: Breidbach, 1987), *Calliphora* (serotonin: Cantera, 1988; Cantera and Carlberg, 1988), *Drosophila* (serotonin and catecholamine: Budnik and White, 1988; Vallés and White, 1988) and *Rhodnius* (serotonin:

Lange *et al.*, 1988a). The localization of cell bodies and axonal tracts as well as the organization of the neuropil has been described. Moreover, a few peripheral aminergic cells were discovered. In *Calliphora*, 10–12 serotonergic cell bodies were revealed in the ring gland, which send branches to the aorta, to the neurohemal part of the ring gland, and to the part corresponding to the PG. These neurons not only contain serotonin but also a G/CCK-like peptide (Cantera, 1988). Among the sensory neurons of *Schistocera*, which are mostly cholinergic, some serotonergic neurons were identified (Lutz and Tyrer, 1988).

The target organs of aminergic fibers were studied in several insect species. In *Locusta* the storage lobe of the CC revealed a dense network of serotonergic fibers that originate in 20 neurons located in the PI and reach the CC through the NCCI. The lateral ns cells of the brain and the NCCII, which innervate the glandular lobe of the CC, were not observed to be immunoreactive (Konings *et al.*, 1988a,b).

The presence of octopamine was demonstrated in the wall of the antennal heart of *Periplaneta*, which is known to contain a nh area. The octopaminergic tract originates in the DUM neurons of the subesophageal ganglion. It follows a complicated pathway through the brain, the NCCIII, and the CC (Pass *et al.*, 1988a,b).

A new serotonergic plexus covering peripheral nerves has been described in *Rhodnius* (Lange *et al.*, 1988a) from which serotonin is released in a calcium-dependent manner in response to high-potassium saline.

In the genital chamber of *Acheta* an efferent serotonergic network was shown, whose origin neurons were identified (Elekes and Hustert, 1988).

Among the varied organs innervated by serotonergic fibers, it has been shown in *Rhodnius* that the epidermis receives an extensive supply. The amount of serotonin demonstrated in the epidermis in both unfed and fed insects suggests a possible involvement of this amine in the plasticization of the abdominal cuticle that follows the blood meal. The amount of serotonin is more pronounced in unfed insects than just-fed ones (Orchard *et al.*, 1988).

Investigations were performed during postembryonic development of several species, i.e., *Calliphora* (Cantera and Carlberg, 1988), *Drosophila* (Budnik and White, 1988; Vallès and White, 1988) and *Tenebrio* (Breidbach, 1988). In *Drosophila* it was shown that the larval serotonin and catecholamine neurons are conserved throughout metamorphosis and that only a few new clusters appear at this time, in particular in the optic lobe. In the beetle, *Tenebrio,* it was observed that only thoracic serotonergic neurons display obvious changes during development. Determination of serotonin level in the central nervous system of *Calliphora* revealed the occurrence of variations during development. It is maximal at the onset of pupariation, drops at the start of adult development, and then increases again.

A.2. VERTEBRATE AND INVERTEBRATE NEUROPEPTIDES IN INSECTS

A.2.1. Vertebrate Peptides

Immunoreactive elements of G/CCK were demonstrated in the central nervous system and nh organs of *Locusta* and *Calliphora*. Their distribution was studied in the embryo, larva, and adult of *Locusta* (Tamarelle *et al.*, 1988) and during postembryonic development of *Calliphora* (Nässel *et al.*, 1988). Moreover, G/CCK-like secretory cells were described in the midgut of several species including *Aeshna, Locusta, Carausius,* and *Periplaneta* (Andriès and Beauvillain, 1988).

Met-enkephalin-like peptides were revealed in both the nervous system and the ovary (trophocytes and oocytes) of the blow fly (Schoofs *et al.*, 1988a).

The coexistence in the same neuron of two distinct immunological reactions was shown in the brain of *Locusta* (Tamarelle *et al.*, 1988) for G/CCK and met-enkephalin, in the thoracoabdominal nerve mass of *Calliphora* (Nässel *et al.*, 1988) for G/CCK and serotonin, and in the central nervous system of *Locusta* (Schoofs *et al.*, 1988b) for MCH and MSH.

A.2.2. Molluscan FMRFamide

The investigations concerning FMRFamide-like neurons or neurosecretory cells were continued in *Locusta* (Rémy *et al.*, 1988), *Schistocerca* (Walther and Schäfer, 1988), and *Aedes* (Brown and Lea, 1988). These neurons are widespread in the central nervous system and most often react with vertebrate PP. Positive reactions were also observed with antisera raised against neuropeptide Y, in *Locusta*.

A.2.3. Insect Peptides

Pars intercerebralis ns neurons and axonal tracts in the nh lobe of the CC of *Locusta* were shown to stain with a monoclonal antibody raised against a synthetic decapeptide corresponding to the N-terminal of the A-chain of bombyxin (Zachary *et al.*, 1988).

The presence of proctolinergic neurons and nerve fibers was investigated in *Drosophila* and *Rhodnius*. In *Drosophila,* not only the hindgut muscles but also segmental bodywall muscles receive a proctolinergic innervation (Anderson *et al.*, 1988). In *Rhodnius,* ovaries, oviducts, and hindgut were shown to contain substantial amounts of proctolin (Lange *et al.*, 1988b).

A.3. CONTROL OF PROTHORACIC GLAND ACTIVITY

A.3.1. Hemolymph Protein Factor—Juvenile Hormone

Studies concerning the mode of action of the hemolymph protein factor (HPF), which is known to increase ecdysteroid release by the PG *in vitro* in *Manduca,* reveal unexpected facts. Indeed, the ecdysteroids secreted by the incubated PG were shown to be mostly 2-dehydroecdysone and 3-dehydroxyecdysone, which are converted to ecdysone by the HPF (Warren *et al.*, 1988). In last instar larvae of *Manduca* the titer of the HPF was shown to be regulated by the JH titer (Watson *et al.*, 1988), which also acts on the competency of the PG; the drop in JH titer that takes place at the beginning of the instar triggers the competency of the PG (Watson and Bollenbacher, 1988).

A.3.2. Ecdysteroid Production Outside Prothoracic Gland

It is well known that in certain insects an ecdysteroid peak occurs in the absence of the PG and it was suggested that ecdysteroids might be present in an inactive undetectable form at certain periods. Experiments using labeled ecdysone injected into larvae of *Tenebrio* do not support this hypothesis. Indeed, no radioactive ecdysone–20-HE could be detected at the time of pupal peak and the polar products did not transform into active hormones (Delbecque *et al.*, 1988).

A.3.3. Brain–Ring Gland Interrelationships

In the blowfly, intact connections between brain and ring gland were shown to be necessary at the time of the commitment peak only. Pharmacologic agents assayed on brain–ring gland complexes revealed that GABA and acetylcholine both inhibit ecdysteroid release, the sensitivity of the PG being stage dependent (Käuser *et al.*, 1988).

A.4. CONTROL OF JUVENILE HORMONE ACTIVITY

A.4.1. Juvenile Hormone Diversity

Contrary to prior reports, JHI, JHII, JHIII, JH acid, JH diols, methylfarnesoate, and farnesol were found to be absent in the adult and eggs of *Oncopeltus,* suggesting that a new JH might occur in this insect (Baker *et al.*, 1988). In *Phormia,* it was shown that the major labeled products released in the incubation medium of the CC–CA were not JHI and JHIII but an unknown product more polar than JHIII (Liu *et al.*, 1988).

A.4.2. Azadirachtin

The effects of azadirachtin on JH and ecdysteroid hemolyphatic titer in the cold-induced supernumerary larvae of *Galleria* were investigated. Azadirachtin decreased the JH elevation due to cold but had no effect on brain allatotropic activity. It elevates the ecdysteroid peak, suggesting a possible effect on PTTH production (Malczewska *et al.*, 1988).

A.4.3. Regulation by Octopamine of CA Synthetic Activity

Numerous data have accumulated concerning both nervous and endocrine regulation of CA activity in insects. It is now shown that octopamine enhances JHIII release from isolated CA of locust. The mode of action of the biogenic amine is still uncertain; it may involve either peptidergic nerve endings or octopaminergic receptors of the secretory cells of the CA (Lafon-Cazal and Baehr, 1988).

The release of JH by the locust CA was demonstrated to be stimulated by Ca^{2+} as shown in experiments using a calcium channel blocker and calcium-free incubation medium. Cyclic AMP, which plays a part in JH synthesis in *Diploptera,* does not appear to be involved in Ca^{2+} stimulation (Dale and Tobe, 1988).

A.4.4. JH Binding Proteins and JH Esterases

In *Locusta,* a high-weight lipoprotein was isolated and characterized. It has a high affinity for JHIII, one molecule containing six hormone-specific binding sites (Koopmanschap and De Kort, 1988). The fluctuations of hemolymph JH binding proteins (JHB) and JH esterases (JHE) were investigated during the final instar of *Leucophaea* and their interrelationships studied. Juvenile hormone esterase activity was shown to be strongly increased by removal of JHB using precipitation with a polyclonal specific antibody. High JHB titers do not depend on JH presence in the hemolymph, at least in last instar larvae (Engelmann *et al.*, 1988). As far as JHE are concerned, the respective roles played by JH and head factors were further investigated in *Manduca* five-instar larvae (Venkatesh and Roe, 1988).

A.5. EMBRYONIC DIAPAUSE

Determination of ecdysteroid concentration in diapausing and nondiapausing eggs of *Bombyx* showed that it was higher in the former. As far as free ecdysteroids are concerned, their amount is reduced in early diapause, but an increase takes place in the late diapause (Coulon, 1988).

A.6. REPRODUCTION

A.6.1. Vitellogenesis Regulation in Mosquitoes

In order to confirm the role of JH in vitellogenin synthesis and fat body responsiveness in mosquitoes, the latter was incubated in varied experimental conditions; this *in vitro* study clearly showed that JH is a prerequisite for the development of the competence of the fat body to 20-HE (Ma *et al.*, 1988). In the anautogenous mosquito, *Culex pipiens,* the synthetic activity of the CA was shown to be dependent to the blood meal, which arrested JHIII production by the Ca (Readio *et al.*, 1988).

A.6.2. Vitellogenesis Regulation in Flies

The respective roles played by the ovary and the different elements of the neuroendocrine system of *Musca* in vitellogenin synthesis and vitellogenesis were further investigated. The comparison of the vitellogenin levels of the various experimental insects demonstrated that vitellogenin synthesis requires both JH and 20-HE (Adams and Filipi, 1988) and cannot be achieved with only one of these hormones. Characterization of hemolymph ecdysteroids suggested that ecdysone plays the major role; 20-HE probably does not originate in the ovary (Adams *et al.*, 1988).

A.6.3. Vitellogenesis Regulation in Firebrats

The role played by 20-HE in ovarian maturation was investigated in the firebrat *Thermobia domestica,* both *in vivo* and *in vitro*. Ecdysteroids were demonstrated to exert in this primitive insect a dose-dependent direct role in follicle formation and growth. They do not seem to intervene in further processes associated with the vitellogenesis onset (Bitsch and Bitsch, 1988).

A.6.4. Embryonic Ecdysteroids

Several studies were devoted to the characterization of the embryonic ecdysteroids and to their fluctuations during embryonic development (see Fournier and Radallah, 1988; Slinger and Isaac, 1988; Thompson *et al.*, 1988). It is generally believed that free ecdysteroids that peak before each embryonic cuticle secretion are released from ecdysteroid conjugates. The trigger for such releases that may be of major importance in the embryonic development appears to depend on the state of degradation of yolk proteins (Bownes *et al.*, 1988).

A.6.5. A New Function for Oostatic Hormone

The relationships existing between the neuroendocrine complex, the ovary, and digestion were investigated in the mosquito *Aedes aegypti*. Purified oostatic hormone that inhibits egg development was shown to display another role, that of preventing the activity of proteolytic enzymes of the gut. The effect of oostatic hormone on the midgut secretory cells appears to be a direct one, mediated neither by the cephalic endocrine complex nor by the ovary (Borovsky, 1988).

A.6.6. Spermatogenesis

The role played in spermatogenesis regulation by ecdysteroids from the testis is known in *Heliothis,* where an ecdysiotropin has been shown to be released from the brain. The triggering of ecdysteroid release by the testis was also demonstrated in *Lymantria* in which brain extracts of both insects stimulate ecdysteroid synthesis (Loeb *et al.*, 1988). 20-Hydroxyecdysone was also shown to intervene in early spermatogenesis of *Manduca,* in which it induces meiotic metaphase (Friedländer and Reynolds, 1988). In *Spodoptera* it is involved in the stimulation of DNA synthesis (Sridevi *et al.*, 1988).

The regulation of apyrene spermatogenesis is not yet clearly understood; investigations made in *Ostrinia* showed that both ecdysone and 20-HE stimulate the appearance of this process in the incubated young larval testis (Gelman *et al.*, 1988a). A comparative study of hemolymph and testis ecdysteroid levels was carried out in this species during larval–pupal–pharate adult development (Gelman *et al.*, 1988b).

A.7. MUSCLE ACTIVITY

A.7.1. Hyperneural Muscle

The complexity of visceral muscle regulation was illustrated in a study of the hyperneural muscle of *Periplaneta,* a thin muscle that lacks myogenic activity. Both glutamate and proctolin induce dose-dependent contractions while octopamine and GABA play an inhibitory role (Moss and Miller, 1988).

A.7.2. Modulatory Role of Proctolin

Immunocytochemistry has revealed in several insect species the presence of proctolin-like arborizations in the ganglionic neuropil. The increase in duration and frequency of bursts recorded in the ventral nerve cord of *Periplaneta* in the presence of proctolin confirms that this neuropeptide could exert a neuromodulatory function (Fitch and Djamgoz, 1988).

A.8. PIGMENT SYNTHESIS AND BREAKDOWN

Studies in hormonal control of cuticular melanization in *Spodoptera* confirmed the inhibitory role of JH in the larval melanization and showed that the subesophageal ganglion and the brain do not play the same role; while the former appears to secrete the melanization and reddish coloration hormone, the brain appears to display a regulatory role (Morita *et al.*, 1988).

A.9. OSMOREGULATION

A.9.1. Identification of Active Factors

Further studies of the antidiuretic factors from the nh and glandular lobe of the CC of *Locusta* were performed. Two factors were separated and characterized. Using antibodies to neoparsin, it is shown that one of these factors, the factor from the nh lobe, was removed by these antibodies (Herault *et al.*, 1988; Fournier and Girardie, 1988).

A.9.2. Biogenic Amines

The chief biogenic amines were assayed on the Malpighian tubules of *Aedes*. Only serotonin was observed to modify both fluid secretion and transepithelial voltage (Veenstra, 1988).

A.10. METABOLISM

A.10.1. AKH and Lipid Synthesis

Lipid synthesis in incubated fat body of *Schistocerca* was shown to be inhibited by both CC extracts and the adipokinetic hormones AKHI and AKHII, a new effect of the hormones known to stimulate the release of diacylglycerol (Gokuldas *et al.*, 1988).

A.10.2. New Members of the AKH Family

A second metabolic peptide from the CC of *Heliothis* moth was recently identified, which causes elevation of hemolymph trehalose and lipid levels. It is a decapeptide that belongs to the AKH family (Jaffé *et al.*, 1988a). In horseflies, two neuropeptides were isolated from the CC from which the amino acid composition was determined (Jaffé *et al.*, 1988b). Two hyperglycemic myotropic factors were found in the CC of *Tenebrio*, which stimulate carbohydrate release in larvae and adults and accelerate or inhibit heart beat (Rosinski and Gäde, 1988).

A.10.3. Other Metabolic Factors

Hypertrehalosemic and hyperlipemic effects were demonstrated to be induced in locusts by the neoparsins isolated from the A-type ns cells of the PI. Although these effects recall the effects triggered by AKH, they are less potent, more rapid, and of shorter duration. In addition, they are not followed by a decrease in total glycogen content (Moreau *et al.*, 1988).

A.10.4. Proteolytic Enzymes

Blood digestion in mosquitoes is of fundamental importance for vitellogenesis. It has been shown recently that the oostatic hormone from the ovary inhibits the synthesis by the midgut of proteolytic enzyme (Borovsky, 1988).

A.10.5. Mode of Action of Metabolic Hormones

The involvement of cyclic nucleotides in mediating the effects of AKH and HTH was often reported. A recent *in vitro* study using *Periplaneta* fat body exposed to synthetic MI and MII neurohormones indicated that these factors do not promote cAMP synthesis (Steele *et al.*, 1988). The roles played by Ca^{2+} ions and calmodulin in these processes were analyzed (Pallen and Steele, 1988).

A.11. MISCELLANEOUS

In the lepidopteran *Polygonia c-aureum*, some characteristics of the factor producing summer-morphs were established and its amounts in the brain of summer and autumn insects compared. Summer-morph-producing hormone activity was detected in the brain extracts of *Papilio, Lycaena,* and *Bombyx* (Endo *et al.*, 1988).

References

Abboud, Y. M., Charnley, A. K., Reynolds, S. E., and Williams-Wynn, C. A., 1983, Bursicon in the mealworm, *Tenebrio molitor* L. and its role in the control of postecdysial tanning, *J. Insect Physiol.* **29:**947–951.

Abou-Halawa, S., and Slama, K., 1986, A neurohemal organ in the prothorax of Lepidoptera, *Acta Ent. Bohemoslav.* **83:**241–252.

Abu-Hakima, R., 1981, Vitellogenin synthesis induced in locust fat body by juvenile hormone analog in vitro, *Experientia* **37:**1309–1311.

Abu-Hakima, R., and Davey, K. G., 1975, Two actions of juvenile hormone one the follicle cells of *Rhodnius prolixus* Stal, *Can. J. Zool.* **53:**1187–1188.

Abu-Hakima, R., and Davey, K. G., 1977, The action of juvenile hormone on follicle cells of *Rhodnius prolixus in vitro:* The effect of colchicine and cytochalasin B, *Gen. Comp. Endocrinol.* **32:**360–370.

Abu-Hakima, R., and Davey, K. G., 1979, A possible relationship between ouabain-sensitive (Na^+-K^+) dependent ATPase and the effect of juvenile hormone on the follicle cells of *Rhodnius prolixus, Insect Biochem.* **9:**195–198.

Adams, C. M. W., and Sloper, J. C., 1956, The hypothalamic elaboration of posterior pituitary principles in man, the rat and the dog. Histochemical evidence derived from a performic acid-alcian blue reaction for cystine, *J. Endocr.* **13:**221–228.

Adams, M. E., and O'Shea, M., 1983, Peptide cotransmitter at a neuromuscular junction, *Science,* **221:**286–289.

Adams, T. S., 1974, The role of juvenile hormone in housefly ovarian follicle morphogenesis, *J. Insect Physiol.* **20:**263–276.

Adams, T. S., 1981, Activation of successive ovarian gonotrophic cycles by the corpus allatum in the house fly, *Musca domestica* (Diptera: Muscidae), *Int. J. Invert. Reprod.* **3:**41–48.

Adams, T. S., and Filipi, P. A., 1988, Interaction between juvenile hormone, 20-hydroxyecdysone, the corpus cardiacum–allatum complex, and the ovaries in regulating vitellogenin levels in the housefly, *Musca domestica, J. Insect Physiol.* **34:**11–19.

Adams, T. S., and Hintz, A. M., 1969, Relationship of age, ovarian development, and the corpus allatum to mating in the house-fly, *Musca domestica, J. Insect Physiol.* **15:**201–215.

Adams, T. S., Hintz, A. M., and Pomonis, J. G., 1968, Oöstatic hormone production in houseflies *Musca domestica,* with developing ovaries, *J. Insect Physiol.* **14:**983–993.

Adams, T. S., Grugel, S., Ittycheria, P. I., Olstad, G., and Caldwell, J. M., 1975, Interactions of the ring gland, ovaries and juvenile hormone with brain neurosecretory cells in *Musca domestica, J. Insect Physiol.* **21:**1027–1044.

Adams, T. S., Hagedorn, H. H., and Wheelock, G. D., 1985, Hemolymph ecdysteroid in the housefly, *Musca domestica,* during oögenesis and its relationship with vitellogenin levels, *J. Insect Physiol.* **31:**91–97.

Adams, T. S., Kelly, T. J., and Woods, C. W., 1988, Relationship between the corpus cardiacum–allatum complex and ovaries with the haemolymph ecdysteroid profile in the housefly, *Musca domestica, J. Insect Physiol.* **34:**1105–1109.

Adiyodi, K. G., 1974, Extracerebral cephalic neuroendocrine complex of the blattids, *Periplaneta americana* (L.) and *Neostylopyga rhombifolia* (Stoll): An in situ study, *J. Morphol.* **144:**469–483.

Agricola, H., Eckert, M., Ude, J., Birkenbeil, H., and Penzlin, H., 1985, The distribution of a proctolin-like immunoreactive material in the terminal ganglion of the cockroach, *Periplaneta americana* L, *Cell Tissue Res.* **239:**203–209.

Agui, N., 1975, Activation of prothoracic glands by brains in vitro, *J. Insect Physiol.* **21:**903–913.

Agui, N., and Hiruma, K., 1977a, Ecdysone as a feedback regulator for the neurosecretory brain cells in *Mamestra brassicae, J. Insect Physiol.* **23:**1393–1396.

Agui, N., and Hiruma, K., 1977b, In vitro activation of neurosecretory brain cells in *Mamestra brassicae* by β-ecdysone, *Gen. Comp. Endocrinol.* **33:**467–472.

Agui, N., Kimura, Y., and Fukaya, M., 1972, Action of the prothoracic gland on the insect integument *in vitro, Appl. Entomol. Zool.* **7:**71–78.

Agui, N., Granger, N. A., Gilbert, L. I., and Bollenbacher, W. E., 1979, Cellular localization of the insect prothoracicotropic hormone: In vitro assay of a single neurosecretory cell, *Proc. Natl. Acad. Sci. USA* **76:**5694–5698.

Agui, N., Bollenbacher, W. E., Granger, N. A. and Gilbert, L. I., 1980, Corpus allatum is release site for insect prothoracicotropic hormone, *Nature* **285:**669–670.

Agui, N., Bollenbacher, W. E., and Gilbert, L. I., 1983, In vitro analysis of prothoracicotropic hormone specificity and prothoracic gland sensitivity in Lepidoptera, *Experientia* **39:**984–988.

Agui, N., Izumi, S., and Tomino, S., 1985, The role of ecdysteroids and juvenoids in vitellogenin levels and follicle development in the housefly, *Musca domestica, Appl. Entomol. Zool.* **20:** 179–188.

Aizono, Y., Matsuo, N., Yoshida, Y., Funatsu, G., Funatsu, M., and Kobayashi, M., 1986, In vitro action of prothoracicotropic hormone on RNA-synthesis in the prothoracic gland of the silkworm, *Bombyx mori, J. Insect Physiol.* **32:**711–717.

Alexander, N. J., 1970, A regulatory mechanism of ecdysone release in *Galleria mellonella, J. Insect Physiol.* **16:**271–276.

Alexandrowicz, J. S., 1926, The innervation of the heart of the cockroach (*Periplaneta orientalis*), *J. Comp. Neurol.* **41:**291–309.

Alrubeai, H. F., and Gorell, T. A., 1982, Electrophoretic analysis of testicular protein components in developing *Tenebrio molitor, Insect Biochem.* **12:**171–175.

Altmann, G., 1956, Hormonale Steuerung der Wasserhausshaltes der Honigbiene, *Zool. Anz.* **19:** 107–112.

Ampleford, E. J., 1985, Ecdysteroids influence the circadian system timing ecdysis in the insect *Rhodnius prolixus* (Hemiptera), *J. Comp. Physiol. A* **157:**699–704.

Ampleford, E. J., and Steel, C. G. H., 1982, Circadian control of ecdysis in *Rhodnius prolixus* (Hemiptera), *J. Comp. Physiol.* **147:**281–286.

Anderson, M. S., Halpern, M. E., and Keshishian, H., 1988, Identification of the neuropeptide transmitter proctolin in *Drosophila* larvae: Characterization of muscle fiber-specific neuromuscular endings, *J. Neurosci.* **8:**242–255.

Andrewartha, H. G., Miethke, P. M., and Wells, A., 1974, Induction of diapause in the pupa of *Phalaenoides glycinae* by a hormone from the subesophageal ganglion, *J. Insect Physiol.* **20:** 679–701.

Andriès, J. C., and Beauvillain, J. C., 1988, Ultrastructural study of cholecystokinin-like immunoreactivity in endocrine cells of the insect midgut, *Cell Tissue Res.* **254:**75–81.

Andriès, J. C., and Tramu, G., 1985a, Distribution patterns of mammalian-like peptide immunoreactive cells in the midgut of *Aeshna cyanea* (Insecta, Odonata), *Experientia* **41:**500–503.

Andriès, J. C., and Tramu, G., 1985b, Ultrastructural and immunohistochemical study of endocrine cells in the midgut of the cockroach *Blaberus craniifer* (Insecta, Dictyoptera), *Cell Tissue Res.* **240:**323–332.

Andriès, J. C., Belemtougri, G., and Tramu, G., 1984, Immunohistochemical identification of growth hormone-releasing factor-like material in the nervous system of an insect, *Aeshna cyanea* (Odonata), *Neuropeptides* **4:**519–528.

Anelli, C. M., and Friedman, S., 1986, Presence and regulation of trehalose in the body of adult *Phormia regina, Arch. Insect Biochem Physiol.* **3:**381–395.

Anstee, J. H., Bell, D. M., and Hyde, D., 1980, Some factors affecting Malpighian tubule fluid secretion and transepithelial potential in *Locusta migratoria* L, *Experientia* **36:**198–199.

Anwyl, R., and Finlayson, L. H., 1973, The ultrastructure of neurons with both a motor and a neurosecretory function in the insect, *Rhodnius prolixus, Z. Zellforsch.* **146:**367–374.

Arpagaus, M., 1987, Vertebrate insulin induces diapause termination in *Pieris brassicae* pupae, *Roux's Arch. Develop. Biol.* **196:**527–530.

Arvy, L., and Gabe, M., 1953, Données histologiques sur la neurosécrétion chez les insectes Paléoptères, *Z. Zellforsch.* **38:**591–610.

Arvy, L., and Gabe, M., 1962, Histochemistry of the neurosecretory product of the pars intercerebralis of Pterygote insects, *3rd International Symposium on Neurosecretion* (H. Heller and R. B. Clark, eds.). Academic Press, London, pp. 331–344.

Asencot, M., and Lensky, Y., 1984, Juvenile hormone induction of "queenliness" on female honey bee (*Apis mellifera L.*) larvae reared on worker jelly and on stored royal jelly, *Comp. Biochem. Physiol.* **78B:**109–117.

Asher, C., Moshitzky, P., Ramchandran, J., and Applebaum, S. W., 1984, The effects of synthetic locust adipokinetic hormone on dispersed locust fat body cell preparations; cAMP induction, lipid mobilization and inhibition of protein synthesis, *Gen. Comp. Endocrinol.* **55:**167–173.

Aston, R. J., 1975, The role of adenosine 3' : 5'-cyclic monophosphate in relation to the diuretic hormone of *Rhodnius prolixus, J. Insect Physiol.* **21:**1873–1878.

Aston, R. J., 1979, Studies on the diuretic hormone of *Rhodnius prolixus*. Some observations on the purification and nature of the hormone and the dynamic of its release in vitro, *Insect Biochem.* **9:** 163–176.

Aston, R. J., and White, A. E., 1974, Isolation and purification of the diuretic hormone from *Rhodnius prolixus, J. Insect Physiol.* **20:**1673–1682.

Atzler, M., 1931, Untersuchungen über den morphologischen und physiologischen Farbenwechsel von *Dixippus (Carausius) morosus, Z. Vgl Physiol.* **13:**509–533.

Aucoin, R. R., Rankin, S. M., Stay, B., and Tobe, S. S., 1987, Calcium and cyclic AMP involvement in the regulation of juvenile hormone biosynthesis in *Diploptera punctata, Insect Biochem.* **17:**965–969.

Awasthi, V. B., 1975, Neurosecretory system of the earwig *Labidura riparia* and the role of the aorta as a neurohemal organ, *J. Insect Physiol.* **21:**1713–1720.

Awasthi, V. B., 1977, Ultrastructure of *corpus cardiacum* of the house cricket. *Gryllodes sigillatus* (Walk) (Orthoptera: Gryllidae), *Zool. Jahrb. Abt. All. Anat.* **98:**476–490.

Awasthi, V. B., 1980, Neurosecretion in the milkweed bug, *Lygaeis Pandurus* scop. (Heteroptera: *Lygaeidae*), *J. Hirnforsh.* **21:**169–181.

Awasthi, V. B., and Singh, U. V., 1981, An analytical study of the neurosecretory cells of the brain in *Philosamia ricini* (Hutt.) (Lepidoptera: Saturniidae) during post-embryonic development, *Experientia* **37:**372–375.

Awiti, L. R. S., and Hidaka, T., 1982, Neuroendocrine mechanism involved in pupal colour dimorphism in swallowtail *Papilio xuthus* L., *Insect Sci. Appl.* **3:**181–191.

Awiti, L. R. S., Odhiambo, T. R., and Ogoma, N. T., 1985, An electron microscopic study of the aortic wall and corpus cardiacum in relation to their neurohemal function in the tsetse, *Glossina morsitans morsitans* Westwood, *Insect Sci. Appl.* **6:**489–501.

Baehr, J. C., 1968, Etude histologique de la neurosécrétion du cerveau et du ganglion sous-oesophagien de *Rhodnius prolixux* Stal (Hémiptère), *C.R. Acad. Sci. Ser. D* **267:**2364–2367.

Baehr, J. C., 1973, Contrôle neuroendocrine du fonctionnement du *corpus allatum* chez *Rhodnius prolixus, J. Insect Physiol.* **19:**1041–1055.

Baehr, J. C., 1976, Étude du contrôle neuro-endocrine du fonctionnement du *corpus allatum* chez les larves du quatrième stade de *Rhodnius prolixus, J. Insect Physiol.* **22:**73–82.

Baehr, J. C., Pradelles, P., Lebreux, C., Cassier, P., and Dray, F., 1976, A simple and sensitive radiommunoassay of insect juvenile hormone using an iodinated tracer, *FEBS Lett.* **69:**123–128.

Baehr, J. C., Cassier, P., Caussanel, C., and Porcheron, P., 1982, Activity of corpora allata, endocrine balance and reproduction in female *Labidura riparia* (Dermaptera), *Cell Tissue Res.* **225:**267–282.

Baehr, J. C., Casas, J., Messeguer, A., Pradelles, P., and Grassi, J., 1987, Enzyme immunoassays of JHIII and JHIII diol using acetylcholinesterase as tracer: An alternative to radioimmunoassay, *Insect Biochem.* **17:**929–932.

Bahjou, A., Moreau, R., Bounias, M., Gourdoux, L., and Dutrieu, J., 1986, Interaction moleculaire in vitro d'une substance apparentée à l'insuline avec l'activité trehalasique du corps gras du coleoptere *Tenebrio molitor L, C.R. Acad. Sci. Ser. III* **303:**655–658.

Bailey, B. A., Martin, R. J., and Downer, R. G. H., 1983, Hemolymph octopamine levels during and following flight in the American cockroach, *Periplaneta americana L., Can. J. Zool.* **62:** 19–22.

Baker, F. C., Tsai, L. W., Reuter, C. C., and Schooley, D. A., 1987, In vivo fluctuation of JH, JH acid and ecdysteroid titer, and JH esterase activity, during development of fifth stadium *Manduca sexta, Insect Biochem.* **17:**989–996.

Baker, F. C., Tsai, L. W., Reuter, C. C., and Schooley, D. A., 1988, The absence of significant levels of the known juvenile hormones and related compounds in the milkweed bug, *Oncopeltus fasciatus, Insect Biochem.* **18:**453–462.

Banner, S. E., Osborne, R. H., and Cattell, K. J., 1987, The pharmacology of the isolated foregut of the locust *Schistocerca gregaria.* I. The effect of a range of putative neurotransmitters, *Comp. Biochem. Physiol.* **88C:**131–138.

Barde, S. P., 1981, Neurosecretory cells in the optic lobe of two aquatic coleopteran species, *Dineutes indicus* Aube and *Cybister rugulosus* Redt, *Hydrobiologia.* **76:**97–101.

Bargmann, W., 1949, Uber die Neurosekretorische Verknüpfung von Hypothalamus und Hypophyse, *Z. Zellforsch.* **34:**610–634.

Bargmann, W., 1966, Neurosecretion, *Int. Rev. Cytol.* **19:**183–201.

Barker, J. F., and Davey, K. G., 1981, Neuroendocrine regulation of protein accumulation by the transparent accessory reproductive gland of male *Rhodnius prolixus, Int. J. Invert. Reprod.* **3:** 291–296.

Barker, J. F., and Davey, K. G., 1983, A polypeptide from the brain and corpus cardiacum of male *Rhodnius prolixus* which stimulates in vitro protein synthesis in the transparent accessory reproductive gland, *Insect Biochem.* **13:**7–10.

Barker, J. F., and Herman, W. S., 1973, On the neuroendocrine control of ovarian development in the monarch butterfly, *J. Exp. Zool.* **183:**1–10.

Barrett, B. A., and Loughton, B. G., 1987, A hypoglycemic factor from the corpus cardiacum of the American cockroach *Periplaneta americana, Gen. Comp. Endocrinol.* **66:**79–84.

Barth, R. H., 1962, The endocrine control of mating behavior in the cockroach *Byrsotria fumigata* (Guérin), *Gen. comp. Endocrinol.* **2:**53–69.

Barth, R. H., 1965, Insect mating behavior: Endocrine control of a chemical communication system, *Science* **149:**882–883.

Barth, R. H., Lester, L. J., Sroka, P., Kessler, T., and Hearn, R., 1975, Juvenile hormone promotes dominance behavior and ovarian development in social wasps (*Polistes annularis*), *Experientia* **31:**691–692.

Baudry, N., 1968, Etude histologique de la neurosécrétion dans la chaîne nerveuse ventrale de *Rhodnius prolixus* Stal (Hémiptère), *C.R. Acad. Sci. Ser. D.* **267:**2356–2359.

Baudry, N., 1969, Recherches histophysiologiques sur la neurosécrétion dans la chaîne nerveuse ventrale de *Rhodnius prolixus* Stal (Hémiptère) au cours de la vie imaginale, *C.R. Acad. Sci. Ser. D* **268:**147–150.

Baudry-Partiaoglou, N., 1978, Anatomie et histologie des organes neurohémaux de quelques Hémiptères, *Int. J. Insect Morph. Embryol.* **7:**1–31.

Baudry-Partiaoglou, N., 1983a, Ultrastructure of perisympathetic organs in insects, in *Neurohemal Organs in Arthropods* (A. P. Gupta, ed.), Charles C Thomas, pp. 513–551.

Baudry-Partiaoglou, N., 1983b, Diversity of neurohormonal release sites in insects: Coexistence of two different (α and β) types of neurohemal structure in the perisympathetic organs, *Gen. Comp. Endocrinol.* **62:**254–267.

Baudry-Partiaoglou, N., 1986, Diversity of neurohormonal release sites in insects: Coexistence of two different (α and β) types of neurohemal structure in the perisympathetic organs, *Gen. Comp. Endocrinol.* **62:**254–267.

Baudry-Partiaoglou, N., 1987, Diversity of neurohormonal release sites in insects: Neurohemal areas associated with peripheral neurosecretory cells in *Periplaneta americana* L. (Dictyoptera: Blattidae) and *Locusta migratoria* R. & F. (Orthoptera: Locustidae), *Int. J. Insect Morph. Embryol.* **16:**295–307.

Baudry-Partiaoglou, N., and Grillot, J. P., 1975, Ultrastructure des organes périsympathiques de la glossine, *Glossina fuscipes quanzenis* (Muscidae), *C.R. Acad. Sci. Ser. D* **281:**921–924.

Baumann, E., and Gersch, M., 1982, Purification and identification of neurohormone D, a heart accelerating peptide from the corpora cardiaca of the cockroach *Periplaneta americana, Insect Biochem.* **12:**7–14.

Baumann, E., and Penzlin, H., 1984, Sequence analysis of neurohormone D, a neuropeptide of an insect, *Periplaneta americana, Biomed. Biochim. Acta* **43:** K13–K16.

Baumann, E., and Penzlin, H., 1987, Inactivation of neurohormone D by Malpighian tubules in an insect, *Periplaneta americana, J. Comp. Physiol.* **157:**511–517.

Beament, J. W. L., 1958, A paralysing agent in the blood of cockroaches, *J. Insect Physiol.* **2:**199–214.

Bean, D. W., and Beck, S. D., 1983, Hemolymph ecdysteroid titres in diapause and non diapause larvae of the European corn borer, *Ostrinia nubilalis, J. Insect Physiol.* **29:**687–693.

Beattie, T. M., 1971, Vital staining of neurosecretory material with acridine orange in the insect, *Periplaneta americana, Experientia* **27:**110–111.

Beaulaton, J., 1968a, Etude ultrastructurale et cytochimique des glandes prothoraciques de vers à soie aux quatrième et cinquième âges larvaires. II. Les cellules interstitielles et les fibres nerveuses, *J. Ultrastruc. Res.* **22:**499–515.

Beaulaton, J., 1968b, Modifications ultrastructurales des cellules sécrétrices de la glande prothoracique de vers à soie au cours des deux derniers âges larvaires. II. Le glycogène, ses relations avec le chondriome et le réticulum endoplasmique, *J. Microsc.* **7:**673–692.

Beaulaton, J., 1980, L'hémocytopoïèse chez les Lépidoptères. Etude des effets de ligatures et de la 20-hydroxyecdysone sur l'évolution de l'hémocytogramme en période prénymphale chez *Antheraea pernyi* Guér, *C.R. Acad. Sci. Paris. Ser. D* **290:**727–730.

Beckemeyer, E. F., and Lea, A. O., 1980, Induction of follicle separation in the mosquito by physiological amounts of ecdysterone, *Science* **209:**819–821.

Beenakkers, A. M. T., 1965, Transport of fatty acids in *Locusta migratoria* during sustained flight, *J. Insect Physiol.* **11:**879–888.

Beenakkers, A. M. T., 1969a, Carbohydrate and fat as a fuel for insect flight. A comparative study, *J. Insect Physiol.* **15:**353–361.

Beenakkers, A. M. T., 1969b, The influence of corpus allatum and corpus cardiacum on lipid metabolism in *Locusta migratoria, Gen. Comp. Endocrinol.* **13:**492.

Beenakkers, A. M. T., Van ber Horst, D. J., and Van Marrewijk, W. J. A., 1985, Insect lipids and lipoproteins, and their role in physiological processes, *Prog. Lipid Res.* **24:**19–67.

Bell, W. J., 1969, Dual role of juvenile hormone in the control of yolk formation in *Periplaneta americana, J. Insect Physiol.* **15:**1279–1290.

Bell, W. J., 1971, Starvation-induced oocyte resorption and yolk protein salvage in *Periplaneta americana, J. Insect Physiol.* **17(6):**1099–1111.

Bell, W. J., 1973, Factors controlling initiation of vitellogenesis in a primitively social bee, *Lasioglossum zephyrum* (Hymenoptera: Halictidae), *Insectes Soc.* **20:**253–260.

Bell, W. J., and Barth, R. H., 1970, Quantitative effects of juvenile hormone on reproduction in the cockroach *Byrsotria fumigata, J. Insect Physiol.* **16:**2303–2313.

Bell, W. J., and Barth, R. H., 1971, Initiation of yolk deposition by juvenile hormone, *Nature* **230:** 220–221.

Bell, W. J., and Bohm, M. K., 1975, Oosorption in insects, *Biol. Rev.* **50:**373–396.

Beltz, B., and Kravitz, E. A., 1986, Aminergic and peptidergic neuromodulation in crustacea, *J. Exp. Biol.* **124:**115–141.

Benedeczky, I., Mala, J., and Sehnal, F., 1980, Ultrastructural study on the innervation of prothoracic glands in *Galleria mellonella, Gen. Comp. Endocrinol.* **41:**400–407.

Benedeczky, I., Kiss, J. Z., and Somogyi, P., 1982, Light and electron localization of substance P-like immunoreactivity in the cerebral ganglion of locust with a monoclonal antibody, *Histochemistry* **75:**123–131.

Ben Khay, A., Gourdoux, L., Moreau, R., and Dutrieu, J., 1987, Effects of intestinal insulin-like peptide on glucose catabolism in male adult *Locusta migratoria, Arch. Insect Biochem. Physiol.* **4:**233–239.

Benoit, J., and Assenmacher, I., 1953, Rapport entre la stimulation sexuelle préhypophysaire et la neurosécrétion chez l'oiseau, *Arch. Anat. Micr.* **42:**334–386.

Bergot, B. J., Jamieson, G. C., Ratcliff, M. A., and Schooley, D. A., 1980, JH zero: New naturally occurring insect juvenile hormone from developing embryos of the tobacco hornworm, *Science* **210:**336–338.

Bergot, B. J., Schooley, D. A., and De Kort, C. A. D., 1981, Identification of JH III as the principal juvenile hormone in *Locusta migratoria, Experientia* **37:**909–910.

Berlind, A., 1981, Mobilization of a peptide neurohormone for release during a physiological secretion cycle, *Gen. Comp. Endocrinol.* **44:**444–453.

Berlind, A., and Maddrell, S. H. P., 1979, Changes in hormone activity of single neurosecretory cell bodies during a physiological secretion cycle, *Brain Res.* **161:**459–467.

Bernays, E. A., 1980, The post-prandial rest in *Locusta migratoria* nymphs and its hormonal regulation, *J. Insect Physiol.* **26:**119–123.

Berridge, M. J., 1966, The physiology of excretion in the cotton stainer *Dysdercus fasciatus* Signoret. IV. Hormonal control of excretion, *J. Exp. Biol.* **44:**553–566.

Berthold, G., 1980, Microtubules in the epidermal cells of *Caausius morosus* (Br.). Their pattern and relation to pigment migration, *J. Insect Physiol.* **26:**421–425.

Berthold, G., 1982, Histochemical localization of Mg^{2+} activated ATPases on microtubule bundles in epidermal cells of *Carausius morosus:* Possible significance to pigment migration, *J. Insect Physiol.* **28:**137–141.

Berthold, G., and Bückmann, D., 1976, Pterin und Ommochromgehalt in Integument grüner und schwartzbrauner Imagines von *Carausius morosus* Br, *J. Comp. Physiol. B* **111:**33–37.

Beydon, P. and Lafont, R., 1983, Feedback inhibition of ecdysone production by 20-hydroxyecdysone in *Pieris brassicae* pupae, *J. Insect Physiol.* **29:**529–533.

Beydon, P., Mauchamp, B., L'Honore, J., Bouthier, A., and Lafont, R., 1980, The epidermal cell cycle during last larval instar of *Pieris brassicae* (Lepidoptera). II. Endocrine control of xanthommatin content, *J. Comp. Physiol.* **136:**21–29.

Bhakthan, N. M. G., and Gilbert, L. I., 1968, Effects of some vertebrate hormones on lipid mobilization in the insect fat body, *Gen. Comp. Endocrinol.* **11:**186–197.

Bhaskaran, G., Jones, G. and Jones, D., 1980, Neuroendocrine regulation of corpus allatum activity in *Manduca sexta:* Sequential neurohormonal and nervous inhibition in the last instar larva, *Proc. Natl. Acad. Sci. USA* **77:**4407–4411.

Bidmon, H. J., 1986, Ultrastrukturelle Veränderungen der Prothoraxdrüsen unbehandelter und mit Azadirachtin behandelter, *Entomol. Gen.* **12:**1–17.

Bilinski, S. M., Hage, W. J., and Bluemink, J. G., 1985, Gap junctions between the follicle cells and the oocyte during oogenesis in an insect, *Tribolium destructor*/Coleoptera, *Roux's Arch. Dev. Biol.* **194:**247–256.

Birkenbeil, H., 1971, Untersuchungen zur Wirkung von Neurohormon D und von anderen aktiven Substanzen auf das Zentralnervensystem der Schabe *Periplaneta americana* (L.), *Zool. Jahrb. Abt. Allg. Zool. Physiol. Tiere* **75:**493–525.

Birkenbeil, H., 1983, Ultrastructural and immunocytochemical investigation of ecdysteroid secretion by the prothoracic gland of the moth *Galleria mellonella, Cell Tiss. Res.* **229:**433–441.

Birkenbeil, H., and Agricola, H., 1980, Die Innervierung der Prothorakaldrüse von *Periplaneta americana* L., *Zool. Anz.* **204:**331–336.

Birnbaum, M. J., Kelly, T. J., Woods, C. W., and Imberski, R. B., 1984, Hormonal regulation of ovarian ecdysteroid production in the autogenous mosquito, *Aedes atropalpus, Gen. Comp. Endocrinol* **56:**9–18.

Bishop, C. A., and O'Shea, M., 1982, Neuropeptide proctolin (H-Arg-Try-Leu-Pro-Thr-OH): Immunocytochemical mapping of neurons in the central nervous system of the cockroach, *J. Comp. Neurol.* **207:**223–238.

Bishop, C. A., and O'Shea, M., 1983, Serotonin immunoreactive neurons in the central nervous system of an insect (*Periplaneta americana*), *J. Neurobiol.* **14:**251–269.

Bishop, C. A., O'Shea, M., and Miller, R. J., 1981, Neuropeptide proctolin (H-Arg-Tyr-Leu-Pro-Thr-OH): Immunological detection and neuronal localization in insect central nervous system, *Proc. Natl. Acad. Sci. USA* **78:**5899–5902.

Bishop, C. A., Wine, J. J., and O'Shea, M., 1984, Neuropeptide proctolin in postural motoneurons of the crayfish, *J. Neurosci.* **4:**2001–2009.

Bitsch, C., and Bitsch, J., 1988, 20-Hydroxyecdysone and ovarian maturation in the firebrat *Thermobia domestica* (Thysanura: Lepismatidae), *Arch. Insect Biochem. Physiol.* **7:**281–293.

Bitsch, J., 1986, Ultrastructure of the cephalic neurohemal organs and of the corpora cardiaca in *Thermobia domestica* (Packard) (Thysanurolepismatidae) and *Trigoniophthalmus alternatus* (Silv.) (Microcoryphia = Machilidae), *Int. J. Insect Morphol. Embryol.* **15:**35–47.

Blaine, W. D., and Dixon, S. E., 1976, Testicular development in the cockroach, *Periplaneta americana* (L.), *Proc. Ent. Soc. Ont.* **106:**60–65.

Blazsek, I., Balazs, A., Novak, V. J. A., and Mala, J., 1975, Ultrastructural study of the prothoracic glands of *Galleria mellonella* L. in the penultimate, last larval and pupal stages, *Cell Tissue Res.* **158:**269.

Bloch, B., Thomsen, E., and Thomsen, M., 1966, The neurosecretory system of the adult *Calliphora erythrocephala.* III. Electron microscopy of the medial neurosecretory cells of the brain and some adjacent cells, *Z. Zellforsch.* **70:**185–208.

Bloemen, R. E. B., and De Vlieger, T. A., 1985, Spontaneous electrical activity recorded extracellularly from the glandular lobe of the corpus cardiacum of *Locusta migratoria, Comp. Biochem. Physiol.* **82A:**627–633.

Bloemen, R. E. B., Beenakkers, A. M. T., and De Vlieger, T. A., 1986, Influences of the NCC axons on electrical activity in the glandular lobe of the corpus cardiacum of *Locusta migratoria, Comp. Biochem. Physiol.* **84A:**319–324.

Bloemen, R. E. B., De Vlieger, T. A., and Beenakkers, A. M. T., 1987, The inhibitory effect of trehalose on electrical activity in the glandular lobe of in vitro corpus cardiacum of *Locusta migratoria, Comp. Biochem. Physiol.* **87A:**193–195.

Bodenstein, D., 1938, Untersuchungen zum Metamorphose-problem. III. über die Entwicklung der Ovarien im thoraxlosen Puppenabdomen, *Biol. Zentr.* **58:**329–332.

Bodenstein, D., 1946, Developmental relations between genital ducts and gonads in *Drosophila, Biol. Bull. (Woods Hole, Mass.)* **91:**288–294.

Bodenstein, D., 1953a, Studies on the humoral mechanisms in growth and metamorphosis of the cockroach, *Periplaneta americana.* II. The function of the prothoracic gland and the corpus cardiacum, *J. Exp. Zool.* **123:**413–433.

Bodenstein, D., 1953b, Studies of the humoral mechanism in growth and metamorphosis of the cockroach *Periplaneta americana.* III. Humoral effects of the metabolism, *J. Exp. Zool.* **124:** 105–115.

Bodenstein, D., and Sprague, I. B., 1959, The developmental capacities of the accessory sex glands in *Periplaneta americana, J. Exp. Zool.* **142:**177–202.

Bodnaryk, R. P., 1977, Stages of diapause development in the pupa of *Mamestra configurata* based on β-ecdysone sensitivity index, *J. Insect Physiol.* **23:**537–542.

Bodnaryk, R. P., 1978, Levels of brain cyclic AMP and cyclic GMP during the initiation of adult development in the Bertha armyworm, *Mamestra configurata* Wlk, *Insect Biochem.* **8:**383–387.

Bodnaryk, R., 1979, Characterization of an octopamine-sensitive adenylate cyclase from insect brain (*Mamestra configurata* Wlk.), *Can. J. Biochem.* **57:**226–232.

Bodnaryk, R. P., 1980, Changes in brain octopamine levels during metamorphosis of the moth, *Mamestra configurata* Wlk, *Insect Biochem.* **10:**169–173.

Bodnaryk, R. P., 1982, Biogenic amine-sensitive adenylate cyclases in insects, *Insect Biochem.* **12:** 1–6.

Bodnaryk, R. P., 1986a, Ecdysteroid hormone levels in the pupa of the Bertha armyworm, *Mamestra configurata* Wlk, during spermatogenesis, *J. Insect Physiol.* **32:**931–935.

Bodnaryk, R. P., 1986b, Feedback inhibition of ecdysone production by 20-hydroxyecdysone during pupal-adult metamorphosis of *Mamestra configurata* Wlk, *Arch. Insect Biochem. Physiol.* **3:**53–60.

Bodnaryk, R. P., 1987, Dual control of pupal diapause by cyclic nucleotides in the Bertha armyworm, *Mamestra configurata* WLK, *J. Insect Physiol.* **33:**33–37.

Boer, H. H., Schot, L. P. C., Veenstra, J. A., and Reichelt, D., 1980, Immunocytochemical identification of neural elements in the central nervous system of a snail, some insects, a fish, and a mammal with an antiserum to the molluscan cardio-excitatory tetrapeptide FMRFamide, *Cell Tissue Res.* **213:**21–27.

Bogus, M. I., and Cymborowski, B., 1984, Induction of supernumerary moults in *Galleria mellonella:* Evidence for an allatotropic function of the brain, *J. Insect Physiol.* **30:**557–561.

Bogus, M. I., Wisniewski, J. R., and Cymborowski, B., 1986, Effects of injury to the neuroendocrine system of last-instar larvae of *Galleria mellonella, J. Insect Physiol.* **32:**1011–1017.

Bohm, M. K., 1972, Effects of environment and juvenile hormone on ovaries of the wasp, *Polistes metricus, J. Insect. Physiol.* **18:**1875–1883.

Bollenbacher, W. E., and Gilbert, L. I., 1981, Neuroendocrine control of postembryonic development in insects. The prothoracicotropic hormone, in *Neurosecretion: Molecules, Cells, Systems, VIII International Symposium on Neurosecretion,* (D. S. Farmer and K. Lederis, eds.), Vol. 31, Plenum Press, New York, pp. 361–370.

Bollenbacher, W. E., and Granger, N. A., 1985, Endocrinology of the prothoracicotropic hormone, in *Comprehensive Insect Physiology, Biochemistry and Pharmacology,* Vol. 7, Pergamon Press, New York, pp. 109–151.

Bollenbacher, W. E., Vedeckis, W. V., Gilbert, L. I., and O'Connor, J. D., 1975, Ecdysone titers and prothoracic gland activity during the larval-pupal development of *Manduca sexta, Dev. Biol.* **44:**46–53.

Bollenbacher, W. E., Agui, N., Granger, N. A., and Gilbert, L. I., 1979, *In vitro* activation of insect prothoracic glands by the prothoracicotropic hormone, *Proc. Natl. Acad. Sci. USA* **76:**5148–5152.

Bollenbacher, W. E., Smith, S. L., Goodman, W., and Gilbert, L. I., 1981, Ecdysteroid titer during larval-pupal-adult development of the tobacco hornworm, *Manduca sexta, Gen. Comp. Endocrinol.* **44:**302–306.

Bollenbacher, W. E., O'Brien, M. A., Katahira, E. J., and Gilbert, L. I., 1983, A kinetic analysis of the action of the insect prothoracicotropic hormone, *Mol. Cell. Endocrinol.* **32:**27–46.

Bollenbacher, W. E., Katahira, E. J., O'Brien, M., and Gilbert, L. I., 1984, Insect prothoracicotropic hormone: Evidence for two molecular forms, *Science* **224:**1243–1245.

Borg, T. K., and Bell, R. A., 1977, Ultrastructure of the neurosecretory cells in the brain of diapausing pupae of the tobacco hornworm, *Manduca sexta* (L), *Tissue Cell* **9:**567–574.

Borg, T. K., Bell, R. A., and Picard, D. J., 1973, Ultrastructure of neurosecretory cells in the frontal ganglion of the tobacco hornworm, *Manduca sexta* (L.), *Tissue Cell* **5:**259–267.

Borovsky, D., 1982, Release of egg development neurosecretory hormone in *Aedes aegypti* and *Aedes taeniorhynchus* induced by an ovarian factor, *J. Insect Physiol.* **28:**311–316.

Borovsky, D., 1985a, Isolation and characterization of highly purified mosquito oostatic hormone, *Arch. Insect Biochem Physiol.* **2:**333–349.

Borovsky, D., 1985b, The role of the male accessory gland fluid in stimulating vitellogenesis in *Aedes taeniorhynchus, Arch. Insect Biochem Physiol.* **2:**405–413.

Borovsky, D., 1988, Oostatic hormone inhibits biosynthesis of midgut proteolytic enzymes and egg development in mosquitoes, *Arch. Insect Biochem. Physiol.* **7:**187–210.

Borovsky, D., and Thomas, B. R., 1985, Purification and partial characterization of mosquito egg development neurosecretory hormone: Evidence for gonadotropic and steroidogenic effects, *Arch. Insect Biochem. Physiol.* **2:**265–281.

Borovsky, D., Whisenton, L. R., Thomas, B. R., and Fuchs, M. S., 1986, Biosynthesis and distribution of ecdysone and 20-ecdysone in *Aedes aegypti, Arch. Insect Biochem. Physiol.* **3:** 19–30.

Borst, D. W., and O'Connor, J. D., 1974, Arthropod molting hormone radioimmune assay, *Science* **178:**418–419.

Bouhniol, J. J., 1938, Recherches expérimentales sur le déterminisme de la métamorphose chez les Lépidoptères, *Bull. Biol. Fr. Belg.* (Suppl.) **24:**1–199.

Bouletreau-Merle, J., 1973, Réceptivité sexuelle et vitellogénèse chez les femelles de *Drosophila melanogaster:* Effets d'une application d'hormone juvénile et de deux analogues hormonaux, *C.R. Acad. Sci. Ser D* **277:**2045–2048.

Bounias, M., 1980, Cinétiques d'action de la noradrenaline, de la dopa et de la dopamine sur la glycémie de l'Abeille in vivo, *Arch. Int. Biochem. Physiol.* **88:**93–98.

Bounias, M., 1986, Modification de la lipémie de l'abeille sous l'action de l'AMP cyclique et du glucagon injectés in vivo, seuls ou en présence de somatostatine, *C.R. Soc. Biol.* **180:**77–88.

Bounias, M., and Dubois, M. P., 1982, Inhibition par la somastostatine de l'action hypoglycémiante de l'insuline chez l'abeille in vivo et localisation cytochimique d'un antigène apparenté à la somatostatine dans l'épithélium digestif ainsi que les tubes de Malphighi de l'abeille, *C.R. Acad. Sci. Ser. III* **294:**1029–1034.

Bounias, M., and Pacheco, H., 1979, Sensibilité de la glycémie de l'abeille à l'action de l'insuline et du glucagon injectés *in vivo, C.R. Acad. Sci. Ser. D* **289:**201–204.

Bounias, M., Moreau, R., and Gourdoux, L., 1985, Une inter-relation entre glucose et acides gras, induite dans l'hémolymphe d'abeilles par l'injection d'insuline endospecifique et supprimée par la somatostatine de vertébrés, *C.R. Acad. Sci. Ser III* **300:**623–626.

Bounias, M., Moreau, R., and Gourdoux, L., 1986, Effect of honeybee insulin-immunoreactive peptide on hemolymph lipid and carbohydrate. Interaction of vertebrate somatostatin, *Insect Biochem.* **16:**721–731.

Bourême, D., Tamarelle, M., and Girardie, J., 1987, Production and characterization of antibodies to neuroparsins A and B isolated from the corpora cardiaca of the locust, *Gen. Comp. Endocrinol.* **67:**169–177.

Bouthier, A., and Lafont, R., 1976, Biosynthèse et excrétion de substances à rôle pigmentaire lors de la mue nymphale des Lépidoptères, *Ann. Endocrinol.* **37:**537–538.

Bowen, M. F., Bollenbacher, W. E., and Gilbert, L. I., 1984, *In vitro* studies on the role of the brain and prothoracic glands in the pupal diapause of *Manduca sexta, J. Exp. Biol.* **108:**9–24.

Bowen, M. F., Irish, R., Whisenton, L. R., Granger, N. A., Gilbert, L. I., and Bollenbacher, W. E., 1985, Endocrine events during pre-diapause and non-diapause larval-pupal development of the tobacco hornworm, *Manduca sexta, J. Insect Physiol.* **31(2):**83–90.

Bowers, B., and Johnson, B., 1966, An electron microscope study of the corpora cardiaca and secretory neurons in the aphid, *Myzus persicae* (Sulz.), *Gen. Comp. Endocrinol.* **6:**213–230.

Bowers, W. S., and Blickenstaff, C. C., 1966, Hormonal termination of diapause in the alfalfa weevil, *Science* **154:**1673–1674.

Bowers, W. S., and Friedman, S., 1963, Mobilization of fat body glycogen by an extract of corpus cardiacum, *Nature* **198:**685.

Bowers, W. S., Thompson, M. J., and Uebel, E. C., 1965, Juvenile and gonadotrophic hormone activity of 10, 11-epoxyfarnesenic acid methyl ester, *Life Sci.* **23:**2323–2331.

Bowers, W. S., Ohta, T., Cleere, J. S., and Marsella, P. A., 1976, Discovery of insect anti-juvenile hormones in plants. Plants yield a potential fourth-generation insecticide, *Science* **193:**542–547.

Bownes, M., 1982, The role of 20-hydroxyecdysone in yolk-polypeptide synthesis by male and female fat bodies of *Drosophila melanogaster, J. Insect Physiol.* **28:**317–328.

Bownes, M., Duebendorfer, A., and Smith, T., 1984, Ecdysteroids in adult males and females of *Drosophila melanogaster, J. Insect Physiol.* **10:**823–830.

Bownes, M., Shirras, A., Blair, M., Collins, J., and Coulson, A., 1988, Evidence that insect embryogenesis is regulated by ecdysteroids released from yolk proteins, *Proc. Natl. Acad. Sci. USA* **85:**1554–1557.

Bradfield, J. Y., and Denlinger, D. L., 1980, Diapause development in the tobacco hornworm: A role for ecdysone or juvenile hormone ?, *Gen. Comp. Endocrinol.* **41:**101–107.

Brady, J., 1967a, Control of the cirdadian rhythm of activity in the cockroach. I. The role of the corpora cardiaca, brain and stress, *J. Exp. Biol.* **47:**153–163.

Brady, J., 1967b, Control of the circadian rhythm of activity in the cockroach. II. Role of the sub-oesophageal ganglion and ventral nerve cord, *J. Exp. Biol.* **47:**165–178.

Brady, J., and Maddrell, S. H. P., 1967, Neurohemal organs in the medial nervous system of insects, *Z. Zellforsch. Mikrost. Anat.* **76:**389–404.

Bräuer, R., Gersch, M., Böhm, G. A., and Baumann, E., 1977, *In vitro*-Stimulierung der Synthese von α-Ecdyson in Prothoracaldrüsen von *Periplaneta americana* durch den Aktivations-faktor I, *Zool. Jahrb. Abt. Allg. Zool. Physiol. Tiere* **81:**1–12.

Bräunig, P., 1987, The satellite nervous system—an extensive neurohemal network in the locust head, *J. Comp. Physiol. A* **160:**69–77.

Breidbach, O., 1987, Constancy and variation of the serotonin-like immunoreactive neurons in the metamorphosing ventral nerve cord of the meal beetle, *Tenebrio molitor* L. (Coleoptera: Tenebrionidae), *Int. J. Insect Morphol. Embryol.* **16:**17–26.

Breitfuss, A., Dürnberger, H., and Pohlhammer, K., 1980, Das Neurohämalorgan der paramedianen neurosekretorischen Zellen des Suboesophagealganglions bei *Teleogryllus commodus* Walker, *Mikroskopie* **36:**16–22.

Brennan, M. D., Weiner, A. J., Goralski, T. J., and Mahowald, A. P., 1982, The follicle cells are a major site of vitellogenin synthesis in *Drosophila melanogaster*, *Dev. Biol.* **89:**225–236.

Brian, M. V., 1974, Caste differentiation in *Myrmica rubra:* The role of hormones, *J. Insect Physiol.* **20:**1351–1365.

Bridges, J. R., 1982, Effects of juvenile hormone on pheromone synthesis in *Dendroctonus frontalis*, *Environ. Entomol.* **11:**417–420.

Briegel, H., and Lea, A. O., 1979, Influence of the endocrine system on tryptic activity in female *Aedes aegypti*, *J. Insect Physiol.* **25:**227–230.

Briers, T., and De Loof, A., 1980, The molting hormone activity in *Sarcophaga bullata* in relation to metamorphosis and reproduction, *Int. J. Invert. Reprod.* **2:**363–372.

Briers, T., and De Loof, A., 1981, Moulting hormone activity in the adult Colorado potato beetle, *Leptinotarsa decemlineata* Say in relation to reproduction and diapause, *Int. J. Invert. Reprod.* **3:**145–155.

Briers, T., Peferoen, M., and De Loof, A., 1982a, Ecdysteroids and adult diapause in the Colorado potato beetle, *Leptinotarsa decemlineata, Physiol. Entomol.* **7:**379–386.

Briers, T., Stoppie, P., and De Loof, A., 1982b, *In vivo* uptake and metabolism of [^{3}H]ecdysone in the vitellogenic follicles of *Sarcophaga bullata* (Diptera) and its localisation by autoradiography, *Int. J. Invert. Reprod.* **4:**343–357.

Briers, T., De Clerck, D., Van Beek, E., and De Loof, A., 1983, Metabolism of injected [^{3}H]ecdysone in male *Sarcophaga bullata* (Diptera), *Gen. Comp. Endocrinol.* **52:**379–387.

Brookes, V. J., 1949, Protein synthesis in the fat body of *Leucophaea maderae* during vitellogenesis, *J. Insect Physiol.* **22:**1649–1657.

Brookes, V. J., 1969, The maturation of the oocytes in the isolated abdomen of *Leucophaea maderae, J. Insect Physiol.* **15:**621–631.

Broomfield, C. E., and Hardy, P. M., 1977, The synthesis of locust adipokinetic hormone, *Tretrahedron Lett.* **25:**2201–2204.

Brousse-Gaury, P., 1967, Généralisation, à divers insectes, de l'innervation deutocérébrale des corpora cardiaca, et rôle neurosécrétoire des nervi corporis cardiaci IV, *C.R. Acad. Sci. Ser. D* **265:**2043–2046.

Brousse-Gaury, P., 1971a, Influence de stimuli externes sur le comportement neuro-endocrinien de blattes. I. Les organes sensoriels céphaliques, point de départ de réflexes neuroendocriniens, *Ann. Sci. Nat.* **13:**181–332.

Brousse-Gaury, P., 1971b, Influence de stimuli externes sur le comportement neuro-endocrinien de blattes. II. Histophysiologie des voies réflexes neuro-endocriniennes, *Ann. Sci. Nat.* **13:**333–450.

Brown, B. E., 1965, Pharmacologically active constituents of the cockroach corpus cardiacum: Resolution and some characteristics, *Gen. Comp. Endocrinol.* **5:**387–401.

Brown, B. E., 1967, Neuromuscular transmitter substance in insect visceral muscle, *Science* **155:** 595–597.

Brown, B. E., 1975, Proctolin: A peptide transmitter candidate in insects, *Life Sci.* **17:**1241–1252.

Brown, B. E., 1977, Occurrence of proctolin in six orders of insects, *J. Insect Physiol.* **23:**861–864.

Brown, B. E., and Starratt, A. N., 1975, Isolation of proctolin a myotropic peptide, from *Periplaneta americana, J. Insect Physiol.* **21:**1879–1881.

Brown, F. A., and Meglitsch, A., 1940, Comparison of the chromatophorotropic activity of insect corpora cardiaca with that of crustacean sinus glands, *Biol. Bull. (Woods Hole, Mass.).* **79:**409–418.

Brown, M. R., and Lea, A. O., 1988, FMRFamide- and adipokinetic hormone-like immunoreactivity in the nervous system of the mosquito, *Aedes aegypti, J. Comp. Neurol.* **270:**606–614.

Brown, M. R., Crim, J. W., and Lea, A. O., 1986, FMRFamide- and pancreatic polypeptide-like immunoreactivity of endocrine cells in the midgut of a mosquito, *Tissue Cell* **18:**419–428.

Broza, M., and Pener, M. P., 1969, Hormonal control of the reproductive diapause in the grasshopper, *Oedipoda miniata, Experientia* **25:**414–415.

Brüning, E., and Lanzrein, B., 1987, Function of juvenile hormone III in embryonic development of the cockroach, *Nauphoeta cinerea, Int. J. Invert. Reprod.* **12:**29–44.

Bückmann, D., 1956a, Die Umfärbung dr Raupen von *Cerura vinula* unter verschiedenen experimentellen Bedingungen, *Naturwissenschaften* **43:**44.

Bückmann, D., 1956b, Experimentelle Auslösung der Umfärbung bei *Cerura*-Raupen durch Prothorakaldrüsenhormon, *Naturwissenschaften* **44:**45.

Bückmann, D., 1959, Die Auslösung der Umfärbung durch das Häutungshormon bei *Cerura vinula* L. (Lepidoptera, Notodontidae), *J. Insect Physiol.* **3:**159–189.

Bückmann, D., 1960, Die Determination der Puppenfarbung bei *Vanessa urticae* L., *Naturwissenschaften* **47:**610–611.

Bückmann, D., 1971, Melanisierungsverlauf und Melanisierungshemmung bei der Kohlweisslingspuppe *Pieris brassicae* L., *Wilhelm Roux Arch. Entw. Mech. Org.* **166:**236–253.

Bückmann, D., 1974a, Die Wirkung von Ecdyson und von Tryptophanüberschuss bei der Ommochrombildung von *Cerura vinula* L. vor der Verpuppung, *Zool. Jahrb. Abt. Allg. Zool. Physiol. Tiere* **78:**257–278.

Bückmann, D., 1974b, Die hormonale Steuerung der Pigmentierung und des morphologischen Farbwechsels bei den Insekten, *Forts. Zool.* **22:**1–22.

Bückmann, D., 1985, Color change in insects, in *Pigment Cell 1985* (J. Bagnara, S. N. Klaus, E. Paul, and M. Schartl, eds.), University of Tokyo Press.

Bückmann, D., and Dustmann, J. H., 1962, Biochemische Untersuchungen über den morphologischen Farbwechsel von *Carausius morosus, Naturwissenschaften* **16:**1–3.

Bückmann, D., and Maisch, A., 1987, Extraction and partial purification of the pupal melanization reducing factor (PMRF) from *Inachis io* (Lepidoptera), *Insect. Biochem.* **17:**841–844.

Bückmann, D., Willig, A., and Linzen, B., 1966, Veränderungen der Hämolymphe vor der Verpuppung von *Cerura vinula* L: der Gehalt an Eiweiss, Aminosäuren, Ommochrom-Vorstufen und Ommochromen, *Z. Naturforsch.* **21b:**1184–1195.

Budnik, V., and White, K., 1988, Catecholamine-containing neurons in *Drosophila melanogaster:* Distribution and development, *J. Comp. Neurol.* **268:**400–413.

Burgess, L., 1973, Axon pathways of the intermediate neurosecretory cells in *Culex tarsalis* Coquillett (Diptera: Culicidae) *Can. J. Zool.* **51:**379–382.

Burov, V. N., Reutskaya, O. E., and Sazonov, A. P., 1972, Activation of diapausing bugs of the species *Eurygaster integriceps* Put. by means of analogues of juvenile hormone, *Dokl. Akad. Nauk SSSR, Washington* **204:**295–297.

Butenandt, A., and Karlson, P., 1954, Über der Isolierung eines Metamorphose-Hormons der Insekten in kristellisierten Form, *Naturforschung* **96:**389–391.

Buys, C. M., and Gibbs, D., 1981, The anatomy of neurons projecting to the corpus cardiacum from the larval brain of the tobacco hornworm, *Manduca sexta* (L.), *Cell Tissue Res.* **215:**505–513.

Caglayan, S. H., and Gilbert, L. I., 1987, In vitro synthesis, release and uptake of storage proteins by the fat body of *Manduca sexta:* Putative hormonal control, *Comp. Biochem. Physiol.* **87:** 989–997.

Calvez, B., Hirn, M., and De Reggi, M., 1976, Ecdysone changes in the hemolymph of two silkworms (*Bombyx mori* and *Philosamia cynthia*) during larval and pupal development, *Febs Letters* **71:**1367–1370.

Cameron, M. L., 1953, Secretion of an orthodiphenol in the corpus cardiacum of the insect, *Nature* **172:**349–350.

Camier, M., Girardie, J., Remy, C., Girardie, A., and Cohen, P., 1980, Identification of immunoreactive neurophysin-like proteins in the central nervous system of an insect, *Locusta migratoria, Biochem. Biophys. Res. Commun.* **93:**792–796.

Candy, D. J., 1974, The control of muscle trehalase activity during locust flight, *Biochem. Soc. Trans.* **2:**1107–1109.

Cantacuzène, A. M., 1967a, Histologie des glandes annexes mâles de *Schistocerca gregaria* F. (Orthoptères). Effet de l'allatectomie sur leur structure et leur activité, *C.R. Acad. Sci. Ser. D* **264:**93–96.

Cantacuzène, A. M., 1967b, Effets comparés de l'allatectomie sur l'activité des glandes annexes mâles et le comportement sexuel de 2 Acridiens: *Schistocerca gregaria* et *Locusta migratoria* (souches migratorioides et "Kazalinsk"), *C.R. Acad. Sci. Ser. D* **265:**224–227.

Cantacuzène, A. M., and Seureau, C., 1970, Altérations de la spermiogénèse de *Locusta migratoria migratorioides* après implantation de corps allates et après irradiation, *Z. Zellforsch.* **103:**351–364.

Cantera, R., 1988, Serotonin and gastrin/cholecystokinin-like immunoreactive neurons in the larval retrocerebral complex of the blowfly *Calliphora erythrocephala, Cell Tissue Res.* **253:**425–433.

Cantera, R., and Carlberg, M., 1988, Serotonin levels in the central nervous system of the blowfly *Calliphora erythrocephala* during postembryonic development, *Insect Biochem.* **18:**507–509.

Cantera, R., and Nässel, D. R., 1987, Postembryonic development of serotonin-immunoreactive neurons in the central nervous system of the blowfly. II. The thoracico-abdominal ganglia, *Cell Tissue Res.* **250:**449–459.

Carlisle, D. B., and Knowles, F. G. W., 1953, Neurohaemal organs in crustaceans, *Nature* **172:** 404–405.

Carlisle, D., Dupont-Raabe, M., and Knowles, F. G. W., 1955, Recherches préliminaires relatives à la séparation et à la comparaison des substances chromactives des Crustacés et des Insectes, *C.R. Acad. Sci.* **240:**665–667.

Carlisle, J., and Loughton, B., 1979, Adipokinetic hormone inhibits protein synthesis in Locusta, *Nature* **282:**420–421.

Carlisle, J., and Loughton, B. G., 1986, The inhibition of protein synthesis in *Locusta migratoria* by adipokinetic hormone, *J. Insect Physiol.* **32:**573–578.

Carlisle, J., Loughton, B., and Ampleford, E., 1987, Feeding causes the appearance of a factor in the hemolymph that stimulates protein synthesis, *J. Insect Physiol.* **33:**493–499.

Carlsen, J., Herman, W. S., Christensen, M., and Josefsson, L., 1979, Characterization of a second peptide with adipokinetic and red pigment-concentrating activity from the locust corpora cardiaca, *Insect Biochem.* **9:**497–501.

Carlsson, A., Falck, B., and Hillarps, N. A., 1962, Cellular localisation of brain monoamines, *Acta Physiol. Scand.* **56:**1–27.

Carroll, L. S., Carrow, G. M., and Calabrese, R. L., 1986, Localization and release of FMRFamide-like immunoreactivity in the cerebral neuroendocrine system of *Manduca sexta, J. Exp., Biol.* **126:**1–14.

Carrow, G. M., Calabrese, R. L., and Williams, C. M., 1981, Spontaneous and evoked release of prothoracicotropin from multiple neurohemal organs of the tobacco hornworm, *Proc. Natl. Acad. Sci. USA* **78:**5866–5870.

Carrow, G. M., Calabrese, R. L., and Williams, C. M., 1984, Architecture and physiology of insect cerebral neurosecretory cells, *J. Neurosci.* **4:**1034–1044.

Cassagneau, P., and Juberthie, C., 1983, Neurohemal organs in apterygota, in *Neurohemal Organs of Arthropods* (A. P. Gupta, ed.), Charles C Thomas, pp. 281–318.

Cassier, P., 1966, L'activité des corps allates et la reproduction du criquet migrateur africain *Locusta migratoria migratorioides* R. et F., *Bull. Soc. Zool. F.* **91:**133–148.

Cassier, P., 1969, The corpora allata of insects, *Int. Rev. Cytol.* **57:**1–73.

Cassier, P., and Fain-Maurel, M. A., 1970a, Contribution à l'étude infrastructurale du système neurosécréteur rétrocérébral chez *Locusta migratoria migratorioides* (R. et F.). I. Les corpora cardiaca, *Z. Zellforsch.* **111:**471–482.

Cassier, P., and Fain-Maurel, M. A., 1970b, Contribution à l'étude infrastructurale du système neurosécréteur rétrocérébral chez *Locusta migratoria migratorioides* (R. et F.). II. Le transit des neurosécrétions, *Z. Zellforsch.* **111:**483–492.

Cassier, P., and Fain-Maurel, M. A., 1970c, Contrôle plurifactoriel de l'évolution post-imaginale des glandes ventrales chez *Locusta migratoria* L. Données expérimentales et infrastructurales, *J. Insect Physiol.* **16:**301–318.

Cassier, P., and Papillon, M., 1968, Effets des implantations de corps allates sur la reproduction des femelles groupées de *Schistocerca gregaria* (Forsk.) et sur le polymorphisme de leur descendance, *C.R. Acad. Sci. Ser. D* **266:**1048–1051.

Cazal, M., 1967, Activité cardiaque de *Locusta migratoria in vitro* et influence des corpora cardiaca. Mise au point expérimentale et premiers résultats, *C.R. Acad. Sci. Ser. D* **264:**842–845.

Cazal, M., 1969, Action d'extraits de corpora cardiaca sur le péristaltisme intestinal de *Locusta migratoria, Arch. Zool. Exp. Gén.* **110:**83–89.

Cazal, M., 1971, Action des corpora cardiaca sur la tréhalosémie et la glycémie de *Locusta migratoria* L., *C.R. Acad. Sci. Ser. D* **272:**2596–2599.

Cazal, M., and Girardie, A., 1968, Control humoral de l'équilibre hydrique chez *Locusta, J. Insect. Physiol.* **14:**655–668.

Cazal, M., Joly, L., and Porte, A., 1971, Etude ultrastructurale des corpora cardiaca et de quelques formations annexes chez *Locusta migratoria* L., *Z. Zellforsch.* **114:**61–72.

Chadwick, L. E., 1956, Removal of prothoracic glands from the nymphal cockroach, *J. Exp. Zool.* **131:**291–306.

Chalaye, D., 1967, Neurosécrétion au niveau de la chaîne nerveuse ventrale de *Locusta migratoria,* R et F (Orthoptère Acridien), *Bull. Soc. Zool. Fr.* **92:**87–108.

Chalaye, D., 1969, La tréhalosémie et son contrôle neuroendocrine chez le Criquet migrateur, *Locusta migratoria migratorioides.* 2. Rôle des corpora cardiaca et des organes périsympathiques, *C.R. Acad. Sci. Ser. D* **268:**3111–3114.

Chaminade, M., and Laverdure, A. M., 1980, La vitellogénèse endogène chez *Tenebrio Molitor* (Coléoptère), *Bull. Soc. Zool. Fr.* **105:**437–444.

Chang, F., 1974, The effect of oxygen on tanning in ligated Tachinid larvae of *Lespesia archippivora* (Diptera: Tachinidae), *Ann. Entomol. Soc. Am.* **67:**427–432.

Chang, H. C., and Bradley, J. T., 1983, Vitellogenin synthesis and secretion in ovariectomized crickets, *Comp. Biochem. Physiol.* **75B:**733–737.

Chan-Palay, V., Johnsson, G., and Palay, S. L., 1978, On the coexistence of serotonin and substance P in neurons of the rats central nervous system, *Proc. Natl. Acad. Sci. USA* **75:**1582–1586.

Chanussot, B., and Pentreath, V. W., 1973, Etude des monoamines du système nerveux stomatogastrique extracéphalique de *Periplaneta americana, C.R. Soc. Biol.* **167:**1405–1409.

Chanussot, B., Dando, J., Moulins, M., and Laverack, M. S., 1969, Mise en évidence d'une amine biogène dans le système nerveux stomatogastrique des Insectes : Etude histochimique et ultrastructurale, *C.R. Acad. Sci. Ser. D* **268:**2101–2104.

Charlet, M., 1974, Mise en évidence d'un centre neurosécréteur protocérébral intervenant dans l'équilibre hydrique de la larve d'*Aeshna cyanea* Müll. (Insecte, Odonate), *C.R. Acad. Sci. Ser. D* **279:**835–838.

Charlet, M., and Schaller, F., 1975, Effets de la destruction de la pars intercerebralis sur la mue chez la larve d'*Aeshna cyanea* Müll. (Insecte, Odonate), *C.R. Acad. Sci. Ser. D* **281:**831–834.

Charlet, M., and Schaller, F., 1976, Blocage de l'exuviation chez la larve d'*Aeshna cyanea* (Insecte

Odonate) après électrocoagulation d'un centre neurosécréteur du protocérébron antérieur, *C.R. Acad. Sci. Ser. D* **283:**1539–1541.

Charlet, M., Schaller, F., and Joly, P., 1974, Données sur les phénomènes de neurosécrétion chez les Odonates, *Zool. Jahrb. Abt. Allg. Zool. Physiol. Tiere* **78:**279–288.

Charlet, M., Goltzene, F., and Hoffmann, J. A., 1979, Experimental evidence for a neuroendocrine control of ecdysone biosynthesis in adult female of *Locusta migratoria, J. Insect Physiol.* **25:** 463–466.

Charniaux-Cotton, H., and Payen, G., 1988, Crustacean reproduction, in *Endocrinology of Selected Invertebrate Types* (H. Laufer and G. H. Downer, eds.), Alan R. Liss, New York, pp. 279–303.

Chatani, F., and Ohnishi, E., 1976, Effect of ecdysone on the ovarian development of *Bombyx* silkworm, *Dev. Growth Diff.* **18:**481–484.

Chaudhury, M. F. B., and Dhadialla, T. S., 1976, Evidence of hormonal control of ovulation in tsetse flies, *Nature* **260:**243–244.

Chaudhury, M. F. B., Dhadialla, T. S., and Kunyiha, R. W., 1981, Evidence of neuroendocrine relationships between mating and ovulation in the tsetse fly, *Glossina morsitans morsitans, Insect Sci. Appl.* **1:**161–166.

Cheeseman, P., and Goldsworthy, G. J., 1979, The release of adipokinetic hormone during flight and starvation, *Gen. Comp. Endocrinol.* **37:**35–43.

Cheeseman, P., Jutsum, A. R., and Goldsworthy, G. J., 1976, Quantitive studies on the release of locust adipokinetic hormone, *Physiol. Entomol.* **1:**115–121.

Cheeseman, P., Goldsworthy, G. J., and Mordue, W., 1977, Studies on the purification of locust adipokinetic hormone, *Life Sci.* **21:**231–236.

Chen, A. C., and Friedman, S., 1977a, Hormonal regulation of trehalose metabolism in the blowfly, *Phormia regina:* Interaction between hypertrehalosemic and hypotrehalosemic hormones, *J. Insect Physiol.* **23:**1223–1232.

Chen, A. C., and Friedman, S., 1977b, Hormonal regulation of trehalose metabolism in the blowfly *Phormia regina* Meig.: Effects of cardiacectomy and allatectomy at the subcellular level, *Comp. Biochem. Physiol.* **58B:**339–344.

Chen, H. J., Fugo, H., Nakajima, M., Nagasawa, H., and Suzuki, A., 1987, Neurohormones in developing embryos of the silkworm, *Bombyx mori:* The presence and characteristics of prothoracicotropic hormone B, *J. Insect Physiol.* **33:**407–411.

Chen, T. T., Couble, P., De Lucca, F. L., and Wyatt, G. R., 1976, Juvenile hormone control of vitellogenin synthesis in *Locusta migratoria,* in *The Juvenile Hormones* (L. I. Gilbert, ed.), Plenum Press, New York, pp. 505–529.

Chernysh, S. I., 1983, Mammalian adrenocoticotropic hormone possesses properties of insect prothoracotropic hormone, *Translated from Dokl. Akad. Nauk. SSSR.* **271:**247–249.

Chervin, D., and Baudry-Partiaoglou, N., 1984, Aires neurohémales distales associées à des cellules neurosecrétrices périphériques chez la blatte *Periplaneta americana, C.R. Acad. Sci. Ser III* **299:**227–230.

Chiang, R. G., and Davey, K. G., 1988, Morphology of neurosecretory cells delineated with cobalt applied extracellularly to the cephalic aorta of the insect *Rhodnius prolixus, J. Morph.* **195:**17–29.

Chippendale, G. M., 1983, Larval and pupal diapause, in *Endocrinology of Insects* (R. G. H. Downer and H. Laufer, eds.), Alan R. Liss, New York, pp. 348–356.

Chippendale, G. M., and Yin, C. M., 1973, Endocrine activity retained in diapause insect larvae, *Nature* **246:**511–513.

Chippendale, G. M., and Yin, C. M., 1979, Larval diapause of the European corn borer, *Ostrinia nubilalis:* Further experiments examining its hormonal control, *J. Insect Physiol.* **25:**53–58.

Chiu, A. Y., Hunkapiller, M. W., Heller, E., Stuart, D. K., Hood, L. E., and Strumwasser, F., 1979, Purification and primary structure of the neuropeptide egg laying hormone of *Aplysía californica, Proc. Natl. Acad. Sci. USA* **76:**6656–6660.

Choudhuri, D. K., Ghosh, B., and Roy, S., 1984, Histochemical studies of the A_1 and A_2 neurosecretory cells of the adult, *Schizodactylus monstrosus* D. (Orthoptera, Schizodactylidae), *J. Anim. Morphol. Physiol.* **31:**119–126.

Chudakova, I., and Gutmann, E., 1978, Developmental changes of succinate dehydrogenase, ATPase and acid phosphatase activity in flight muscles of the normal and allatectomized adult cricket, *Acheta domestica* (Orthoptera). *Zool. Jahrb. Abt. Allg. Zool. Physiol. Tiere* **82:**1–15.

Chudakova, I., Masslenikova, V., and Luchnikova, E., 1982, Effect of 20-hydroxyecdysone on *Acheta domestica* L. (Orthoptera) and *Drosophila melanogaster* Meig (Diptera) reproduction, *Zool. Jahrb. Abt. Allg. Zool. Physiol. Tiere* **86:**45–52.

Ciancio, M. J., Watson, R. D., and Bollenbacher, W. E., 1986, Competency of *Manduca sexta* prothoracic glands to synthesize ecdysone during development, *Mol. Cell. Endocrinol.* **44:**171–178.

Claret, J., 1966, Recherche du centre photorécepteur lors de l'induction de la diapause chez *Pieris brassicae* L., *C.R. Acad. Sci. Ser. D* **262:**1464–1465.

Claret, J., Porcheron, P., and Dray, F., 1978, La teneur en ecdysones circulantes au cours du dernier stade larvaire de l'hyménoptère endoparasite *Pimpla instigator,* et l'entrée en diapause, *C.R. Acad. Sci. Ser. D* **286:**639–644.

Clements, A. N., Potter, S. A., and Scales, M. D. C., 1985, The cardiacal neurosecretory system and associated organs of an adult mosquito, *Aedes aegypti, J. Insect Physiol.* **31:**821–830.

Cloutier, E. J., Beck, S. D., McLeod, D. G. R., and Silhacek, D. L., 1962, Neural transplants and insect diapause, *Nature* **22:**1222–1224.

Coats, S. A., Mutchmor, J. A., and Toffefson, J. J., 1987, Regulation of migratory flight by juvenile hormone mimic and inhibitor in the Western corn rootworm (Coleoptera: Chrysomelidae), *Ann. Ent. Soc. Am.* **80:**697–708.

Colhoun, E. H., 1963, Synthesis of 5-hydroxytryptamine in the American cockroach, *Experientia* **19:**9–10.

Collins, C., and Miller, T., 1977, Studies on the action of biogenic amines on cockroach heart, *J. Exp. Biol.* **67:**1–15.

Compton, D. A., and Mills, R. R., 1982, Bursicon-induced adenyl cyclase activity in the hemocytes of the American cockroach, *J. Insect Physiol.* **28:**221–227.

Conradi-Larsen, E. M., 1970, Influence of neurosecretion and nutrition on 0^2-consumption in the bug, *Oncopeltus fasciatus, J. Insect Physiol.* **16:**471–482.

Cook, B. J., 1981, The action of proctolin and 5-hydroxytryptamine on the oviduct of the horsefly, *Tabanus proximis, Int. J. Invert. Reprod.* **3:**209–212.

Cook, B. J., and Holman, G. M., 1975a, The neural control of muscular activity in the hindgut of the cockroach *Leucophaea maderae:* Prospects of its chemical mediation, *Comp. Biochem. Physiol.* **50C:**137–146.

Cook, B. J., and Holman, G. M., 1975b, Sites of action of neurohormone that controls hindgut muscle activity in the cockroach *Leucophaea maderae, J. Insect Physiol.* **21:**1187–1192.

Cook, B. J., and Holman, G. M., 1978, Comparative pharmacological properties of muscle function in the foregut and the hindgut of the cockroach *Leucophaea maderae, Comp. Biochem. Physiol.* **61C:**291–295.

Cook, B. J., and Holman, G. M., 1979, The action of proctolin and L-glutamic acid on the visceral muscles of the hindgut of the cockroach *Leucophaea maderae, Comp. Biochem. Physiol.* **64C:** 21–28.

Cook, B. J., and Holman, 1985, Peptides and kinins, in *Comprehensive Insect Physiology, Bio-*

chemistry and Pharmacology (G. A. Kerkut and L. I. Gilbert, eds.), Vol. 11, Pergamon Press, pp. 531–596.

Cook, B. J., and Meola, S., 1978, The oviduct musculature of the horsefly, *Tabanus sulcifrons,* and its response to 5-hydroxytryptamine and proctolin, *Physiol. Entomol.* **3:**273–280.

Cook, B. J., and Meola, S., 1983, Heart structure and beat in the stable fly, *Stomoxys calcitrans, Physiol. Entomol.* **8:**139–149.

Cook, B. J., and Milligan, J. V., 1972, Electrophysiology and histology of the medial neurosecretory cells in the adult male cockroaches, *Periplaneta americana, J. Insect Physiol.* **18:**1197–1214.

Cook, B. J., Eraker, J., and Anderson, G. R., 1969, The effect of various biogenic amines on the activity of the foregut of the cockroach, *Blaberus giganteus, J. Insect Physiol.* **15:**445–455.

Cook, B. J., Holman, G. M., and Marks, E. P., 1975, Calcium and cyclic AMP as possible mediators of neurohormone action in the hindgut of the cockroach, *Leucophaea maderae, J. Insect Physiol.* **21:**1807–1814.

Cook, B. J., Holman, G. M., and Meola, S., 1984, The oviduct musculature of the cockroach *Leucophaea maderae* and its response to various neurotransmitters and hormones, *Arch. Insect Biochem. Physiol.* **1:**167–178.

Copenhaver, P. F., and Truman, J. W., 1982, The role of eclosion hormone in the larval ecdyses of *Manduca sexta, J. Insect Physiol.* **28:**695–701.

Copenhaver, P. F., and Truman, J. W., 1986a, Identification of the cerebral neurosecretory cells that contain eclosion hormone in the moth *Manduca sexta, J. Neurosci.* **6:**1738–1747.

Copenhaver, P. F., and Truman, J. W., 1986b, Metamorphosis of the cerebral neuroendocrine system in the moth *Manduca sexta, J. Comp. Neurol.* **249:**186–204.

Copenhaver, P. F., and Truman, J. W., 1986c, Control of neurosecretion in the moth *Manduca sexta:* Physiological regulation of the eclosion hormone cells, *J. Comp. Physiol.* **158A:**445–455.

Cottrell, C. B., 1962a, The imaginal ecdysis of blowflies. The control of cuticular hardening and darkening, *J. Exp. Biol.* **39:**395–411.

Cottrell, C. B., 1962b, The imaginal ecdysis of blowflies. Detection of the blood-borne darkening factor and determination of some of its properties, *J. Exp. Biol.* **39:**413–430.

Cottrell, C. B., 1962c, The imaginal ecdysis of blowflies. Evidence for a change in the mechanical properties of the cuticle at expansion, *J. Exp. Biol.* **39:**449–458.

Couillaud, F., 1986, Influence of sexual organs on corpora allata biosynthetic activity in *Locusta migratoria, Physiol. Entomol.* **11:**397–403.

Couillaud, F., and Girardie, A., 1984, Ovariectomie et biosynthèse de l'hormone juvénile C_{16} (JH-3) chez le criquet migrateur, *C.R. Acad. Sci. Ser. III* **298:**157–162.

Couillaud, F., and Girardie, A., 1985, Control of C-16 juvenile hormone biosynthesis in active corpora allata of female African locusts, *Int. J. Invert. Reprod. Dev.* **8:**303–315.

Couillaud, F., Girardie, J., Tobe, S. S., and Girardie, A., 1984, Activity of disconnected corpora allata in *Locusta migratoria:* Juvenile hormone biosynthesis in vitro and physiological effects in vivo, *J. Insect Physiol.* **30:**551–556.

Couillaud, F., Mauchamp, B., and Girardie, A., 1987, Biological, radiochemical and physicochemical evidence for the low activity of disconnected corpora allata in locust, *J. Insect Physiol.* **33:**223–228.

Coulon, M., 1988, Comparative changes of ecdysteroid content in *Bombyx mori* eggs in diapausing and non-diapausing development, *Comp. Biochem. Physiol.* **89A:**503–509.

Cupo, A., and Proux, J., 1983, Biochemical characterization of a vasopressin-like neuropeptide in *Locusta migratoria.* Evidence of high molecular weight protein encoding vasopressin sequence, *Neuropeptides* **3:**309–318.

Curtis, A. T., Hori, M., Green, J. M., Wolfgang, W. J., Hiruma, K., and Riddiford, L. M., 1984, Ecdysteroid regulation of the onset of cuticular melanization in allatectomized and black mutant *Manduca sexta* larvae, *J. Insect Physiol.* **30:**597–606.

Cymborowski, B., 1970a, Investigations on the neurohormonal factors controlling circadian rhythm of locomotor activity in the house cricket (*Acheta domesticus* L.). I. The role of the brain and subesophageal ganglion, *Zool. Poloniae* **20:**103–125.

Cymborowski, B., 1970b, Investigations on the neurohormonal factors controlling circadian rhythm of locomotor activity in the house cricket (*Acheta domesticus*) L. II. Daily histochemical changes in the neurosecretory cells of the pars intercerebralis and subesophageal ganglion, *Zool. Poloniae* **20:**127–149.

Cymborowski, B., 1983, Daily changes in synthesis and accumulation of neurosecretion in the brain of the house crickets, *J. Interdiscipl. Cycle Res.* **14:**111–116.

Cymborowski, B., and Stolarz, G., 1979, The role of juvenile hormone during larval-pupal transformation of *Spodoptera littoralis:* Switchover in the sensitivity of the prothoracic gland to juvenile hormone, *J. Insect Physiol.* **25:**939–942.

Cymborowski, B., and Zimowska, G., 1984, Switchover in the sensitivity of the prothoracic glands to juvenile hormone in the cotton leafworm, *Spodoptera littoralis, J. Insect Physiol.* **30:**911–918.

Dadd, R. M., 1961, Evidence for humoral regulation of digestive secretion in the beetle *Tenebrio molitor, J. Exp. Biol.* **38:**259–266.

Dale, J. F., and Tobe, S. S., 1988, The effect of a calcium ionophore, a calcium channel blocker and calcium-free medium on juvenile hormone release *in vitro* from corpora allata of *Locusta migratoria, J. Insect Physiol.* **34:**451–456.

Dallmann, S. H., and Herman, W. S., 1978, Hormonal regulation of hemolymph lipid concentration in the monarch butterfly, *Danaus plexippus, Gen. Comp. Endocrinol.* **36:**142–150.

Davenport, A. P., and Evans, P. D., 1984a, Changes in hemolymph octopamine levels associated with food deprivation in the locust, *Schistocerca gregaria, Physiol. Entomol.* **9:**269–274.

Davenport, A. P., and Evans, P. D., 1984b, Stress-induced changes in the octopamine levels of insect haemolymph, *Insect Biochem.* **14:**135–143.

Davenport, A. P., and Evans, P. D., 1986, Sex-related differences in the concentration of met-enkephalin-like immunoreactivity in the nervous system of an insect, *Schistocerca gregaria,* revealed by radioimmunoassay, *Brain Res.* **383:**319–322.

Davenport, A. P., and Wright, D. J., 1986, Octopamine distribution in the larvae and adults of the species of moth, *Spodoptera littoralis* and *Manduca sexta, J. Insect Physiol.* **32:**987–993.

Davey, K. G., 1961, Substance controlling the rate of beating of the heart of *Periplaneta, Nature* **192:**284.

Davey, K. G., 1962a, The release by feeding of a pharmacologically active factor from the corpus cardiacum of *Periplaneta americana, J. Insect Physiol.* **8:**205–208.

Davey, K. G., 1962b, The mode of action of the corpus cardiacum on the hindgut of *Periplaneta americana, J. Exp. Biol.* **39:**319–324.

Davey, K. G., 1963, The release by enforced activity of the cardiac accelerator from the corpus cardiacum of *Periplaneta americana, J. Insect Physiol.* **9:**375–381.

Davey, K. G., 1965, Copulation and egg production in *Rhodnius prolixus:* The role of the spermathecae, *J. Exp. Biol.* **42:**373–378.

Davey, K. G., 1984, Neuroendocrine controls on insect reproduction, in *Insect Neurochemistry and Neurophysiology* (A. B. Bořkovec, ed.), Plenum Press, New York, pp. 93–113.

Davey, K. G., and Huebner, E., 1974, The response of the follicle cells of *Rhodnius prolixus* to juvenile hormone and antigonadotropin in vitro, *Can. J. Zool.* **52:**1407–1412.

Davey, K. G., and Kriger, F. L., 1985, Variations during the gonotrophic cycle in the titer of the myotropic ovulation hormone and the response of the ovarian muscles in *Rhodnius prolixus, Gen. Comp. Endocrinol.* **58:**452–457.

Davey, K. G., and Kuster, J. E., 1981, The source of an antigonadotropin in the female of *Rhodnius prolixus* Stal, *Can. J. Zool.* **59:**761–764.

David, J. C., and Coulon, J. F., 1985, Octopamine in invertebrates and vertebrates. A review, *Progr. Neurobiol.* **24:**141–185.

David, J. C., and Fuzeau-Braesch, S., 1979, Amines biogènes et AMP cyclique chez *Locusta migratoria:* Recherche de récepteurs à octopamine, *C.R. Acad. Sci. Ser. D* **288:**1207–1210.

David, J. C., and Lafon-Cazal, M., 1979, Octopamine distribution in the *Locusta migratoria* nervous and non-nervous systems, *Comp. Biochem. Physiol.* **64C:**161–164.

Davis, N. T., 1985, Serotonin-immunoreactive visceral nerves and neurohemal system in the cockroach *Periplaneta americana* (L.), *Cell Tissue Res.* **240:**593–600.

Davis, N. T., 1987, Neurosecretory neurons and their projections to the serotonin neurohemal system of the cockroach *Periplaneta americana* (L.), and identification of mandibular and maxillary motor neurons associated with this system, *J. Comp. Neurol.* **259:**604–621.

Dawson, A. B., 1953, Evidence for the termination of neurosecretory fibres in the pars intermedia of the hypophysis of the frog, *Rana pipiens, Anat. Rec.* **115:**63–69.

Day, M. F., 1943, The function of the corpus allatum in Muscoid Diptera, *Biol. Bull. (Woods Hole, Mass.)* **84:**127–140.

Deb, D. C., and Chakravorty, S., 1981, Juvenoid-induced effects on the growth and differentiation of testis in the rice moth, *Corcyra cephalonica, J. Insect Physiol.* **27:**397–402.

Deb, D. C., and Chakravorty, S., 1982, Effect of precocene II, applied independently or subsequent to hydropene treatment, on the morphogenesis of female reproductive organs of *Corcyra cephalonica, J. Insect Physiol.* **28:**703–712.

De Bessé, N., 1967, Neurosécrétion dans la chaîne nerveuse ventrale de deux blattes, *Leucophaea maderae* (F.) et *Periplaneta americana* (L.), *Bull. Soc. Zool. Fr.* **92:**73–86.

De Bessé, N., 1978, Mise en évidence des cellules neurosécrétrices protocérébrales impliquées dans la régulation de l'équilibre hydrique et de la sécrétion salivaire chez les adultes de *Leucophaea maderae, C.R. Acad. Sci. Ser. D* **286:**1695–1698.

De Bessé, N., and Cazal, M., 1968, Action des extraits d'organes périsympathiques et de corpora cardiaca sur la diurèse de quelques insectes, *C.R. Acad. Sci. Ser. D* **266:**615–618.

De Clerck, D., Eechaute, W., Leusen, I., Diederik, H., and De Loof, A., 1983, Identification of testosterone and progesterone in hemolymph of larvae of the fleshfly *Sarcophaga bullata, Gen. Comp. Endocrinol.* **52:**368–378.

De Kort, C. A. D., 1969, Hormones and the structure and function of insect flight-muscles, *J. Endocrinol.* **43:**158.

De Kort, C. A. D., 1981, Hormonal and metabolic regulation of adult diapause in the Colorado beetle, *Leptinotarsa decemlineata* (Coleoptera : Chrysomelidae), *Entomol. Gen.* **7:**261–271.

De Kort, C. A. D., and Granger, N. A., 1981, Regulation of the juvenile hormone titer, *Annu. Rev. Entomol.* **21:**1–28.

De Kort, C. A. D., Wieten, M., and Kramer, S. J., 1979, The occurrence of juvenile hormone specific esterases in insects. A comparative study, *Proc. Konink. Nederl. Akad. Wetenschap.* **82:**325–331.

De Kort, C. A. D., Koopmanschap, A. B., and Ermens, A. A. M., 1984, A new class of juvenile hormone binding proteins in insect haemolymph, *Insect Biochem.* **14:**619–623.

Delachambre, J., 1976, Contrôle du métabolisme de la tyrosine par l'hormone de tannage (bursicon) chez *Tenebrio molitor* L. (Insecta, Col.) étude *in vitro* de l'activité tyrosine hydroxylasique, *Coll. Int. CNRS* **251:**493.

Delachambre, J., Delbecque, J. P., Provansal, A., De Reggi, M. L., and Cailla, H., 1979a, Induction of epidermal cyclic AMP by bursicon in mealworm, *Tenebrio molitor, Experientia* **35:**701–702.

Delachambre, J., Delbecque, J. P., Provansal, A., Grillot, J. P., De Reggi, M. L., and Cailla, H. L., 1979b, Total and epidermal cyclic AMP levels related to the variations of ecdysteroids and bursicon during the metamorphosis of the mealworm *Tenebrio molitor* L., *Insect Biochem.* **9:** 95–99.

Delbecque, J. P., and Slama, K., 1980, Ecdysteroid titres during autonomous metamorphosis in a dermestid beetle, *Z. Naturforsch.* **35C:**1066–1080.

Delbecque, J. P. A., Delachambre, J., Hirn, M., and De Reggi, M. L., 1978, Abdominal production of ecdysterone and pupal-adult development in *Tenebrio molitor* (Insecta, Coleoptera), *Gen. Comp. Endocrinol.* **35:**436–444.

Delbecque, J. P., Connat, J. L., and Lafont, R., 1988, Polar and apolar metabolites of ecdysteroids during the metamorphosis of *Tenebrio molitor, J. Insect Physiol.* **34:**619–624.

Delepine, Y., 1965, Recherches sur la neurosécrétion dans l'ensemble du systéme nerveux central d'un lépidoptère, Galleria Mellonella (*), *Bull. Soc. Zool.* **90:**525–540.

Della-Cioppa, G., and Engelmann, F., 1980, Juvenile hormone-stimulated proliferation of endoplasmic reticulum in fat body cells of a vitellogenic insect, *Leucophaea maderae* (Blattaria), *Biochem. Biophys. Res. Commun.* **93:**825–832.

Della-Cioppa, G., and Engelmann, F., 1984, Juvenile hormone regulation of phospholipid synthesis in the endoplasmic reticulum in vitellogenic fat body cells from *Leucophaea maderae, Insect Biochem.* **14:**27–36.

Della-Cioppa, G., and Engelmann, F., 1987, The vitellogenin of *Leucophaea maderae.* Synthesis as a large phosphorylated precursor, *Insect Biochem.* **17:**401–415.

De Loof, A., and De Wilde, J., 1970, Hormonal control of synthesis of vitellogenic female protein in the Colorado beetle, *Leptinotarsa decemlineata, J. Insect Physiol.* **16:**1455–1466.

De Loof, A., Van Loon, J., and Vanderroost, C., 1979, Influence of ecdysterone, precocene and compounds with juvenile hormone activity on induction, termination and maintenance of diapause in the parasitoid wasp, *Nasonia vitripennis, Physiol. Entomol.* **4:**319–328.

De Loof, A., Briers, T., and Huybrechts, R., 1984, Presence and function of ecdysteroids in adult insects, *Comp. Biochem. Physiol.* **79B:**505–509.

Delphin, F., 1965, The histology and possible functions of neurosecretory cells in the ventral ganglia of *Schistocerca gregaria* Forskål (Orthoptera: Acrididae), *Trans. R. Entomol. Soc. London* **117:** 167.

Dempster, A., and McFarlane, J. E., 1985, In vitro maintenance and sperm development in the testis of the house cricket *(Acheta domestica), Int. J. Invert. Reprod. Dev.* **8:**293–296.

Denlinger, D. L., 1975, Insect hormones as tsetse abortifacients, *Nature* **253:**347–348.

Denlinger, D. L., 1976, Preventing insect diapause with hormones and cholera toxin, *Life Sci.* **19:** 1485–1490.

Denlinger, D. L., 1985, Hormonal control of diapause, in *Comprehensive Insect Physiology Biochemistry and Pharmacology* (G. A. Kerkut and L. I. Gilbert, eds.), Vol. 11, Pergamon Press, New York, pp. 354–412.

Denlinger, D. L., Chaudhury, M. F. B., and Dhadialla, T. S., 1978, Cyclic AMP is a likely mediator of ovulation in the tsetse fly, *Experientia* **34:**1296–1297.

Denlinger, D. L., Saini, R. K., and Chaudhury, M. F. B., 1983, Parturition in the tsetse fly *Glossina morsitans:* Pattern of activity, sound production and evidence for control by the mother's brain, *J. Insect Physiol.* **29:**715–721.

Denlinger, D. L., Shukle, M., and Faustini, D. L., 1984, Juvenile hormone involvement in pupal diapause of the flesh fly *Sarcophaga crassipalpis:* Regulation of infradian cycles of 0_2 consumption, *J. Exp. Biol.* **109:**191–199.

De Robertis, E., 1965, Histophysiologie des synapes et neurosécrétions, in *Monographies de Physiologie causale,* Vol. 4, Gauthier-Villars, Paris, pp. 1–252.

De Wilde, J., 1966, Endocrine aspects of nutrition and excretion in the Colorado beetle, *J. Endocrinol.* **37**:XI.

De Wilde, J., and De Boer, J. A., 1969, Humoral and nervous pathways in photoperiodic induction of diapause in *Leptinotarsa decemlineata, J. Insect Physiol.* **15**:661–675.

De Wilde, J., Staal, G. B., De Kort, C. A. D., De Loof, A., and Baard, G., 1968, Juvenile hormone titer in the haemolymph as a function of photoperiodic treatment in the adult Colorado beetle (*Leptinotarsa decemlineata* Say), *Proc. R. Neth. Acad. Sci.* **71**:321–326.

Dhadialla, T. S., and Wyatt, G. R., 1983, Juvenile hormone-dependent vitellogenin synthesis in *Locusta migratoria* fat body: Inducibility related to sex and stage, *Dev. Biol.* **96**:436–444.

Diederen, J. H. B., Maas, H. A., Pel, H. J., Schooneveld, H., Jansen, W. F., and Vullings, H. G. B., 1987, Co-localization of the adipokinetic hormones I and II in the same glandular cells and in the same secretory granules of corpus cardiacum of *Locusta migratoria* and *Schistocerca gregaria.* An immuno-electronmicroscopic study, *Cell Tissue Res.* **249**:379–389.

Dietz, A., Hermann, H. R., and Blum, M. S., 1979, The role of exogenous JH-1, JH-3 and anti-JH (Precocene II) on queen induction of 3, 5-day old worker honey bee larvae, *J. Insect Physiol.* **25**:503–512.

Dillwith, J. W., Lenz, C. Z., and Chippendale, G. M., 1985, An immunochemical study of the diapause-associated protein of the southwestern corn borer, *Diatraea grandiosella, Insect Biochem.* **15**:711–722.

Dixon, S. E., and Blaine, W. D., 1973, Hormonal control of male accessory gland development in the cockroach, *Periplaneta americana* (L.), *Proc. Entomol. Soc. Ont.* **103**:97–103.

Dockray, G. J., Duve, H., and Thorpe, A., 1981a, Immunochemical characterization of gastrin/cholecystokinin-like peptides in the brain of the blowfly, *Calliphora vomitoria, Gen. Comp. Endocrinol.* **45**:491–496.

Dockray, G. J., Vaillant, C., Williams, R. G., Gayton, R. J., and Osborne, N. N., 1981b, Vertebrate brain-gut peptides related to FMRFamide and met-enkephalin Arg^{6}Phe^7, *Peptides* **2**:25–30.

Doerr-Schott, J., Joly, L., and Dubois, M. P., 1978, Sur l'existence dans la *pars intercerebralis* d'un insect (*Locusta migratoria* R. et F.) de cellules neurosécrétrices fixant un antiserum anti-somatostatine, *C.R. Acad. Sc. Ser. D* **286**:93–95.

Dogra, G. S., and Gillott, C., 1971, Neurosecretory activity and protease synthesis in relation to feeding in *Melanoplus sanguinipes* (Fab.), *J. Exp. Zool.* **177**:41–49.

Dogra, G. S., and Tandan, B. K., 1964, Adaptation of certain histological techniques for in situ demonstration of the neuro-endocrine system of insects and other animals, *Q. J. Microsc. Sci.* **105**:445–466.

Dogra, G. S., and Tandan, B. K., 1965, Ontogenic fate of the neurosecretory cells in the larval brain of *Sarcophaga ruficornis* (Fabricius, 1794) (Diptera: Cyclorrhapha), *Experientia* **21**:216–218.

Dores, R. M., Dallmann, S. H., and Herman, W. S., 1979, The regulation of post-eclosion and post-feeding diuresis in the monarch butterfly, *Danaus plexippus, J. Insect Physiol.* **25**:895–901.

Dorn, A., 1976, Neurosecretion in the frontal ganglion of the stick insect: Electron microscopic study and extirpation experiments, in *Neurosecretion and Neuroendocrine Activity,* 7th International Symposium on Neurosecretion, Leningrad (W. Bargmann, A. Oksche, A. Polenav and B. Scharrez), Springer-Verlag, Berlin, p. 74.

Dorn, A., and Sehn, E., 1987, Ecdysteroid-dependent larval-adult oviduct transformation in the milkweed bug *Oncopeltus fasciatus* requires absence of juvenile hormone, *Experientia* **43**: 1115–1117.

Dorn, A., Rademacher, J. M., and Sehn, E., 1986, Effects of azadirachtin on the moulting cycle, endocrine system and ovaries in last-instar larvae of the milkweed bug, *Oncopeltus fasciatus, J. Insect Physiol.* **32:**231–238.

Dorn, A., Gilbert, L. I., and Bollenbacher, W. E., 1987, Prothoracicotropic hormone activity in the embryonic brain of the tobacco hornworm, *Manduca sexta, J. Comp. Physiol.* **157B:**279–283.

Douglas, W. W., and Poisner, A. M., 1964, Stimulus-secretion coupling in a neurosecretory organ: The role of calcium in the release of vasopressin from the neurohypophysis, *J. Physiol.* **172:**1–18.

Downer, R. G. H., 1972, Interspecificity of lipid-regulating factors from insect corpus cardiacum, *Can. J. Zool.* **50:**63–65.

Downer, R. G. H., 1979, Induction of hypertrehalosemia by excitation in *Periplaneta americana, J. Insect Physiol.* **25:**59–63.

Downer, R. G. H., and Parker, G. H., 1979, Glycogen utilization during flight in the American cockroach, *Periplaneta americana* L., *Comp. Biochem. Physiol.* **64A:**29–32.

Downer, R. G. H., and Steele, J. E., 1972, Hormonal stimulation of lipid transport in the American cockroach, *Periplaneta americana, Gen. Comp. Endocrinol.* **19:**259–265.

Downer, R. G. H., Orr, G. L., Gole, J. W. D., and Orchard, I., 1984, The role of octopamine and cyclic AMP in regulating hormone release from corpora cardiaca of the American cockroach, *J. Insect Physiol.* **30:**457–462.

Dumbar, S., and Huddart, H., 1982, Calcium movements in insect visceral muscle, *Comp. Biochem. Physiol.* **71A:**425–437.

Dumser, J. B., 1980, In vitro effects of ecdysterone on the spermatogonial cell cycle in *Locusta, Int. J. Invert. Reprod.* **2:**165–174.

Dumser, J. B., and Davey, K. G., 1974, Endocrinological and other factors influencing testis development in *Rhodnius prolixus, Can. J. Zool.* **52:**1011–1022.

Dumser, J. B., and Davey, K. G., 1975, The *Rhodnius* testis : hormonal effects on germ cell division, *Can. J. Zool.* **53:**1682–1689.

Dunbar, S. J., and Huddart, H., 1981, Calcium movements in insect visceral muscle, *Comp. Biochem. Physiol.* **71A:**425–437.

Dupont-Raabe, M., 1949, Les chromatophores de la larve de Corèthre, *Arch. Zool. Exp. Gén.* **86:** 32–39.

Dupont-Raabe, M., 1951, Etude expérimentale de l'adaptation chromatique chez le phasme, *Carausius morosus,* Br, *C.R. Acad. Sci.* **232:**886–888.

Dupont-Raabe, M., 1952a, Substances chromactives de crustacés et d'insectes. Activité réciproque, répartition, différences qualitatives, *Arch. Zool. Exp. Gén.* **89:**102–112.

Dupont-Raabe, M., 1952b, Contribution à l'étude du rôle endocrine du cerveau et notamment de la *pars intercerebralis* chez les Phasmides, *Arch. Zool. Exp. Gén.* **89:**128–138.

Dupont-Raabe, M., 1954, Répartition des activités chromatiques dans le ganglion susoesophagien des Phasmides: Mise en évidence d'une région sécrétoire dans la partie deuto et tritocérébrale, *C.R. Acad. Sci.* **238:**950–951.

Dupont-Raabe, M., 1957, Les mécanismes de l'adaptation chromatique chez les insectes, *Arch. Zool. Exp. Gén.* **94:**61–292.

Durchon, M., 1984, Peptidic hormones in annelids, in *Biosynthesis, Metabolism and Mode of Action of Invertebrate Hormones* (J. Hoffmann and M. Porchet, eds.), Springer-Verlag, Berlin.

Dürnberger, H., Pohlhammer, K., and Weinbormair, G., 1978, The paramedial neurosecretory cells of the subesophageal ganglion of the cricket, *Teleogryllus commodus* (Walk.). I. Effect of ovariectomy on stainability and incorporation of cysteine, *Cell Tissue Res.* **187:**489–494.

Düstmann, J. H., 1964, Die Redoxpigmente von *Carausius morosus* und ihre Bedeutung für den morphologischen Farbwechsel, *Z. Vgl. Physiol.* **49:**28–57.

Dutrieu, J., and Gourdoux, L., 1967, Le contrôle neuro-endocrinien de la tréhalosémie de *Carausius morosus, C.R. Acad. Sci. Ser D* **265:**1067–1070.

Dutrieu, J., and Gourdoux, L., 1974, Rôle des corps cardiaques et des corps allates dans le métabolisme du glycogène et du tréhalose au cours du développement du coléoptère *Tenebrio molitor, Ann. Endocrinol.* **35:**375–382.

Duve, H., 1978, The presence of a hypoglucemic and hypotrehalocemic hormone in the neurosecretory system of the blowfly *Calliphora erythrocephala, Gen. Comp. Endocrinol.* **36:**102–110.

Duve, H., and Thorpe, A., 1979, Immunofluorescent localization of insulin-like material in the median neurosecretory cells of the blowfly, *Calliphora vomitoria* (Diptera), *Cell Tissue Res.* **200:**187–191.

Duve, H., and Thorpe, A., 1980, Localisation of pancreatic polypeptide (PP)-like immunoreactive material in neurones of the brain of the blowfly, *Calliphora erythrocephala* (Diptera), *Cell Tissue Res.* **210:**101–109.

Duve, H., and Thorpe, A., 1981, Gastrin/cholecystokinin (CCK)-like immunoreactive neurones in the brain of the blowfly, *Calliphora erythrocephala* (Diptera), *Gen. Comp. Endocrinol.* **43:** 381–391.

Duve, H., and Thorpe, A., 1982, The distribution of pancreatic polypeptide in the nervous system and gut of the blowfly, *Calliphora vomitoria* (Diptera), *Cell Tissue Res.* **277:**67–77.

Duve, H., and Thorpe, A., 1983, Immunocytochemical identification of α-endorphin-like material in neurones of the brain and corpus cardiacum of the blowfly, *Calliphora vomitoria* (Diptera), *Cell Tissue Res.* **233:**415–426.

Duve, H., and Thorpe, A., 1984, Immunocytochemical mapping of gastrin/CCK-like peptides in the neuroendocrine system of the blowfly *Calliphora vomitoria* (Diptera), *Cell Tissue Res.* **237:** 309–320.

Duve, H., and Thorpe, A., 1988, Mapping of enkephalin-related peptides in the nervous system of the blowfly, *Calliphora vomitoria,* and their co-localization with cholecystokinin (CCK)- and pancreatic polypeptide (PP)-like peptides, *Cell Tissue Res.* **251:**399–415.

Duve, H., Thorpe, A., and Lazarus, N. R., 1979, Isolation of material displaying insulin-like immunological and biological activity from the brain of the blowfly *Calliphora vomitoria, Biochem. J.* **184:**221–227.

Duve, H., Thorpe, A., Lazarus, N. R., and Lowry, P. J., 1982, A neuropeptide of the blowfly *Calliphora vomitoria* with an amino acid composition homologous with vertebrate pancreatic polypeptide, *Biochem. J.* **201:**429–432.

Du Vigneaud, V., Ressler, V., Swan, C., Roberts, J., Katsoyaninis, P., and Gordon, J., 1953, The synthesis of an octapeptide amine with the hormonal activity of oxytocin, *J. Am. Chem. Soc.* **75:**4879–4880.

Dymond, G. R., and Evans, P. D., 1979, Biogenic amines in the nervous system of the cockroach, *Periplaneta americana:* Association of octopamine with mushroom bodies and dorsal unpaired median (Dum) neurones, *Insect Biochem.* **9:**535–545.

Eckert, M., Agricola, H., and Penzlin, H., 1981, Immunocytochemical identification of proctolinlike immunoreactivity in the terminal ganglion and hindgut of the cockroach *Periplaneta americana* (L.), *Cell Tissue Res.* **217:**633–645.

Edwards, J., 1970, Endocrine control of flight muscle histolysis in *Dysdercus intermedius, J. Insect Physiol.* **16:**2027–2032.

Edwards, J. P., Bergot, B. J., and Staal, G. B., 1983, Effects of three compounds with antijuvenile hormone activity and a juvenile hormone analogue on endogenous juvenile hormone levels in the tobacco hornworm, *Manduca sexta, J. Insect Physiol.* **29:**83–89.

Edwards, J. S., 1966, Neural control of metamorphosis in *Galleria mellonella* (Lepidoptera), *J. Insect Physiol.* **12:**1423–1433.

Eidmann, H., 1956, Uber rhythmische Erscheinungen bei der Stabheuschrecke *Carausius morosus* Br., *Z. Vgl. Physiol.* **38:**370–390.

Ejezie, G. C., and Davey, K. G., 1976, Some effects of allatectomy in the female tsetse, *Glossina austeni, J. Insect Physiol.* **22:**1743–1749.

Elekes, K., and Hustert, R., 1988, The efferent innervation of the genital chamber by an identified serotonergic neuron in the female cricket *Acheta domestica, Cell. Tissue Res.* **252:**449–457.

Elekes, K., Hustert, R., and Geffard, M., 1987, Serotonin-immunoreactive and dopamine-immunoreactive neurones in the terminal ganglion of the cricket, *Acheta domestica:* Light- and electron-microscopic immunocytochemistry, *Cell Tissue Res.* **250:**167–180.

El-Ibrashy, T., and Boctor, I. Z., 1970, Effect of allatectomy upon lipid metabolism of the female moth of *Spodoptera littoralis* Biosd., *Z. Vgl. Physiol.* **68:**111–116.

Elliott, R. H., and Gillott, C., 1977, Changes in the protein concentration and volume of the haemolymph in relation to yolk deposition, ovariectomy, allatectomy, and cautery of the median neurosecretory cells in *Melanoplus sanguinipes, Can. J. Zool.* **55:**97–103.

Ellis, P. E., and Carlisle, D. B., 1961, The prothoracic gland and colour change in locusts, *Nature* **190:**368–369.

El-Salhy, M., Abou-el-Ela, R., Falkmer, S., Grimelius, L., and Wilander, E., 1980, Immunohistochemical evidence of gastro-entero-pancreatic neurohormonal peptides of vertebrate type in the nervous system of the larva of a dipteran insect, the hoverfly *Eristalis aenus, Regulatory Peptides* **1:**187–204.

El-Salhy, M., Falkmer, S., Kramer, K. J., and Speirs, R. D., 1983, Immunohistochemical investigations of neuropeptides in the brain, corpora cardiaca, and corpora allata of an adult lepidopteran insect, *Manduca sexta* (L), *Cell Tissue Res.* **232:**295–317.

El-Salhy, M., Falkmer, S., Kramer, K. J., and Speirs, R. D., 1984, Immunocytochemical evidence for the occurrence of insulin in the frontal ganglion of a lepidopteran insect, the tobacco hornworm moth, *Manduca sexta* L., *Gen. Comp. Endocrinol.* **54:**85–88.

Emson, P. C., 1979, Peptides as neurotransmitter candidates in the mammalian CNS, *Progr. Neurobiol.* **13:**61–116.

Enami, E., 1954, The functional significance of beta-neurosecretory cells in *Sesarma, Pubbl. Staz. Zool. Napoli, Suppl.* **24:**70–71.

Enders, E., 1955, Die hormonal Steuerung rhythmischer Bewegungen von Insekten-Ovidukten, *Verh. Dtsch. Zool. Ges.* **19:**113–116.

Endo, K., and Funatsu, S., 1985, Hormonal control of seasonal-morph determination in the swallowtail butterfly, *Papilio xuthus* L. (Lepidoptera: Papilionidae), *J. Insect Physiol.* **31:**669–674.

Endo, K., and Kamata, Y., 1985, Hormonal control of seasonal-morph determination in the small copper butterfly, *Lycaena phlaeas daimio* Seitz., *J. Insect Physiol.* **31:**701–706.

Endo, Y., Iwanaga, T., Fujita, T., and Nishiitsutsuji-Uwo, J., 1982, Localization of pancreatic polypeptide (PP)-like immunoreactivity in the central and visceral nervous systems of the cockroach *Periplaneta, Cell Tissue Res.* **277:**1–9.

Endo, K., Masaki, T., and Kumagai, K., 1988, Neuroendocrine regulation of the development of seasonal morphs in the Asian comma butterfly, *Polygonia c-aureum* L.: Difference in activity of summer-morph-producing hormone from brain-extracts of the long-day and short-day pupae, *Zool. Sci.* **5:**145–152.

Engelmann, F., 1957, Die Steuerung der Ovarfunktion bei der ovoviviparen Schabe *Leucophaea maderae, J. Insect Physiol.* **1:**257–278.

Engelmann, F., 1959, The control of reproduction in *Diploptera punctata* (Blattaria), *Biol. Bull. (Woods Hole, Mass.)* **116:**406–419.

Engelmann, F., 1969, Food-stimulated synthesis of intestinal proteolytic enzymes in the cockroach *Leucophaea maderae, J. Insect Physiol.* **15:**217–235.

Engelmann, F., 1979, Insect vitellogenin: Identification, biosynthesis and role in vitellogenesis, *Adv. Insect Physiol.* **14:**49–108.

Engelmann, F., 1984, Regulation of vitellogenesis in insects: The pleiotropic role of juvenile hormones, in *Biosynthesis, Metabolism and Mode of Action of Invertebrate Hormones* (J. A. Hoffmann and M. Porchet, eds.), Springer-Verlag, Berlin, pp. 444–453.

Engelmann, F., and Barth, R. H., 1968, Endocrine controle of female receptivity in *Leucophaea maderae* (Blattaria), *Ann. Ent. Soc. Am.* **61:**503–505.

Engelmann, F., and Ladduwahetty, A. M., 1974, The female specific protein, an extraovarian protein essential for egg maturation in insects, *Zool. Jahrb. Abt Allg. Zool. Physiol. Tiere* **78:** 289–300.

Engelmann, F., and Lüscher, M., 1957, Die hemmende Wirkung des Gehirn auf die corpora allata bei *Leucophaea maderae* (Orthoptera), *Verh. Dtsch. Zool. Ges.* **1956:**215–220.

Engelmann, F., and Penney, D., 1966, Studies on the endocrine control of metabolism in *Leucophaea maderae* (Blattaria). 1. The hemolymph proteins during egg maturation, *Gen. Comp. Endocrinol.* **7:**314–325.

Engelmann, F., Hill, L., and Wilkens, J. L., 1971, Juvenile hormone control of female specific protein synthesis in *Leucophaea maderae, Schistocerca vaga,* and *Sarcophaga bullata, J. Insect Physiol.* **17:**2179–2191.

Engelmann, F., Mala, J., and Tobe, S. S., 1987, Cytosolic and nuclear receptors for juvenile hormone in fat bodies of *Leucophaea maderae, Insect Biochem.* **17:**1045–1052.

Engelmann, F., Mala, J., and Borst, D., 1988, Juvenile hormone-binding protein and juvenile hormone esterase activity in hemolymph from last-instar nymphs of *Leucophaea maderae, Arch. Insect Biochem. Physiol.* **8:**11–23.

Engels, W., 1978, Der Einfluss verschiedener Juvenilhormone und Dosierungen auf den Vitellogenin-Stoffwechsel eierlegender Bienenköniginnen (Apis mellifica), *Mitt. Dtsch. Ges. Allg. Angew. Ent.* **1:**308–312.

Eusebio, E. J., and Moody, W. J., 1986, Calcium-dependent action potentials in the prothoracic gland of *Manduca sexta, J. Exp. Biol.* **126:**531–536.

Evans, P. D., 1978, Octopamine distribution in the insect nervous system, *J. Neurochem.* **1978:** 1009–1013.

Evans, P. D., 1981, Multiple receptor types for octopamine in the locust, *J. Physiol. (London)* **318:** 99–122.

Evans, P. D., 1984a, Studies on the mode of action of octopamine, 5-hydroxytryptamine and proctolin on a myogenic rhythm in the locust, *J. Exp. Biol.* **110:**231–251.

Evans, P. D., 1984b, The role of cyclic nucleotides and calcium in the mediation of the modulatory effects of octopamine on locust skeletal muscle, *J. Physiol. (London)* **348:**325–340.

Evans, P. D., 1985, Octopamine, in *Comprehensive Insect Physiology, Biochemistry and Pharmacology* (G. A. Kerkut and L. I. Gilbert, eds.), Vol. 8, Pergamon Press, New York, pp. 491–550.

Evans, P. D., and Myers, C. M., 1986, The modulatory actions of FMRFamide and related peptides on locust skeletal muscle, *J. Exp. Biol.* **126:**403–422.

Evans, P. D., and O'Shea, M., 1978, The identification of an octopaminergic neurone and the modulation of a myogenic rhythm in the locust, *J. Exp. Biol.* **73:**235–260.

Ewen, A. B., 1962, An improved aldehyde fuchsin staining technique for neurosecretory products in insects, *Trans. Am. Microsc. Soc.* **81:**94–96.

Ewen, A. B., 1966, A possible endocrine mechanism for inducing diapause in the eggs of *Adelphocoris lineolatus* (Goeze) (Hemiptera: Miridae), *Experientia* **22:**470.

Fain-Maurel, M. A., and Cassier, P., 1968, Etude infrastructurale des glandes de mue de *Locusta migratoria migratorioides* (R. et F.). I. Evolution cyclique au cours des stades larvaires, *Arch. Zool. Exp. Gen.* **109:**445–476.

Fain-Maurel, M. A., and Cassier, P., 1969, Etude infrastructurale des glandes de mue de *Locusta migratoria migratorioides* (R. et F.). II. Analyse morphologique des étapes de la dégénérescence chez les imagos grégaires, *Arch. Zool. Exp. Gen.* **110:**91–124.

Fain, M. J., and Riddiford, L. M., 1975, Juvenile hormone titers in the hemolymph during late larval development of the tobacco hornworm, *Manduca sexta* (L), *Biol. Bull. (Woods Hole, Mass.)* **149:**506–521.

Fain, M. J., and Riddiford, L. M., 1976, Reassessment of the critical periods for prothoracicotropic hormone and juvenile hormone secretion in the larval molt of the tobacco hornworm *Manduca sexta, Gen. Comp. Endocrinol.* **30:**131–141.

Falck, B., Hillarp, N. A., Thieme, G., and Torp, A., 1962, Fluorescence of catecholamines and related compounds condensed with formaldehyde, *J. Histochem. Cytochem.* **10:**348–354.

Fallon, A. M., Hagedorn, H. H., Wyatt, G. R., and Laufer, H., 1974, Activation of vitellogenin synthesis in the mosquito *Aedes aegypti* by ecdysone, *J. Insect Physiol.* **20:**1815–1823.

Farmer, J., Maddrell, S. H. P., and Spring, J. H., 1981, Absorption of fluid by the midgut of *Rhodnius, J. Exp. Biol.* **94:**301–316.

Ferenz, H. J., 1977, Two-step photoperiodic and hormonal control of reproduction in the female beetle, *Pterostichus nigrita, J. Insect Physiol.* **23:**671–676.

Ferenz, H. J., and Diehl, I., 1983, Stimulation of juvenile hormone biosynthesis in vitro by locust allatotropin, *Z. Naturforsch.* **386:**856–858.

Fernlund, P., 1976, Structure of a light adapting hormone from the shrimp *Pandulus borealis, Biochim. Biophys. Acta.* **439:**17–25.

Fernlund, P., and Josefsson, L., 1972, Crustacean color-change hormone: Amino acid sequence and chemical synthesis, *Science* **177:**173–175.

Feyereisen, R., 1985, Radiochemical assay for juvenile hormone III biosynthesis in vitro, *Meth. Enzymol.* **111:**442–453.

Fifield, S. M., and Finlayson, L. H., 1978, Peripheral neurons and peripheral neurosecretion in the stick insect, *Carausius morosus, Proc. R. Soc. London Ser. B* **200:**63–85.

Fingerman, M., 1985, The physiology and pharmacology of crustacean chromatophores, *Am. Zool.* **25:**233–252.

Finlayson, L. H., and Osborne, M. P., 1968, Peripheral neurosecretory cells in the stick insect (*Carausius morosus*) and the blowfly larva (*Phormia terrae-novae*), *J. Insect Physiol.* **14:**1793–1801.

Finlayson, L. H., and Osborne, M. P., 1970, Electrical activity of neurohaemal tissue in the stick insect, *Carausius morosus, J. Insect Physiol.* **16:**791–800.

Finlayson, L. H., Osborne, M. P., and Anwyl, R., 1976, Effects of acetylcholine, physostigmine, and hemicholinium-3 on spontaneous electrical activity of neurosecretory nerves in *Carausius* and *Rhodnius, J. Insect Physiol.* **22:**1321–1326.

Finlayson, L. H., Orchard, I., and Day, N., 1983, A factor in the haemolymph of the stick insect that depresses the activity of extraganglionic neurosecretory cells, *J. Comp. Physiol.* **152:**283–288.

Fitch, J., and Djamgoz, M. B. A., 1988, Proctolin potentiates synaptic transmission in the central nervous system of an insect, *Comp. Biochem. Physiol.* **89C:**109–112.

Flanagan, T. R., 1984, Cytological evidence for serotonin-containing fibers in an abdominal neurohemal organ in a hemipteran, *Brain Res.* **306:**235–242.

Flanagan, T. R., 1986, Serotonin-containing, catecholamine-containing and dopamine-sequestering neurones in the ventral nerve cord of the hemipteran *Rhodnius prolixus, J. Insect Physiol.* **32:** 17–26.

Flanagan, T. R., and Berlind, A., 1984, Serotonin modulation of release of sequestered [^{3}H]-serotonin from nerve terminals in an insect neurohemal organ *in vitro, Brain Res.* **306:**243–250.

Flanagan, T. R., and Hagedorn, H. H., 1977, Vitellogenin synthesis in the mosquito: The role of juvenile hormone in the development of responsiveness to ecdysone, *Physiol. Entomol.* **2:**173–178.

Fogal, W., and Fraenkel, G., 1969, The role of bursicon in melanization and endocuticle formation in the adult fleshfly, *Sarcophaga bullata, J. Insect Physiol.* **15:**1235–1247.

Ford, R., Jackson, D. M., Tetrault, L., Torres, J. C., Assanah, P., Harper, J., Leung, M. K., and Stefano, G. B., 1986, A behavioral role for enkephalins in regulating locomotor activity in the insect *Leucophaea maderae:* Evidence for high affinity kappa-like opioid binding sites, *Comp. Biochem. Physiol.* **85C:**61–66.

Foster, W. A., 1967, Hormone-mediated nutritional control of sexual behavior in male dung flies, *Science* **158:**1596–1597.

Foster, W. A., 1974, Surgical inhibition of ovulation and gestation in the tsetse fly *Glossina austeni* Newst, (Dipt., *Glossinidae*), *Bull. Ent. Res.* **63:**483–493.

Fournier, B., and Girardie, J., 1988, A new function for the locust neuroparsins: Stimulation of water reabsorption, *J. Insect Physiol.* **34:**309–313.

Fournier, B., and Radallah, D., 1988, Ecdysteroids in *Carausius* eggs during embryonic development, *Arch. Insect. Biochem. Physiol.* **7:**211–224.

Fraenkel, G., 1935, A hormone causing pupation in the blowfly, *Calliphora erythrocephala, Proc. R. Soc. London Ser. B* **118:**1–12.

Fraenkel, G., and Hollowell, M., 1979, Actions of the juvenile hormone, 20-hydroxyecdysone, and the oöstatic hormone during oögenesis in the flies *Phormia regina* and *Sarcophaga bullata, J. Insect Physiol.* **25:**305–310.

Fraenkel, G., and Hsiao, C., 1962, Hormonal and nervous control of tanning in the fly, *Science* **138:** 27–29.

Fraenkel, G., and Hsiao, C., 1965, Bursicon, a hormone which mediates tanning of the cuticle in the adult fly and other insects, *J. Insect Physiol.* **11:**513–640.

Fraenkel, G., and Hsiao, C., 1968, Morphological and endocrinological aspects of pupal diapause in a fleshfly *Sarcophaga argyrostoma, J. Insect Physiol.* **14:**707–717.

Fraenkel, G., and Su, J., 1984, Hormonal control of eclosion of flies from the puparium, *Proc. Natl. Acad. Sci.* **81:**1457–1459.

Fraenkel, G., Hsiao, C., and Seligman, M., 1966, Properties of bursicon: An insect protein hormone that controls cuticular tanning, *Science* **151:**91–93.

Fraenkel, G., Su, J., and Zdarek, J., 1984, Neuromuscular and hormonal control of post-eclosion processes in flies, *Arch. Insect Biochem. Physiol.* **1:**345–366.

Fraser, A., 1958, Humoral control of metamorphosis and diapause in the larvae of certain *Calliphoridae* (Diptera: Cyclorrhapha), *Proc. R. Soc. Edinburgh Sect. B* **67:**127–140.

Fraser, A., and Pipa, R., 1977, Corpus allatum regulation during the metamorphosis of *Periplaneta americana:* Axon pathways, *J. Insect Physiol.* **23:**975–984.

Freeman, M. A., 1966, The effect of drugs on the alimentary canal of the African migratory locust *Locusta migratoria, Comp. Biochem. Physiol.* **17:**755–764.

Freon, G., 1964, Contribution à l'étude de la neurosécrétion dans la chaîne nerveuse ventrale du criquet migrateur, *Locusta migratoria* (L.), *Bull. Soc. Zool. Fr.* **89:**819–830.

Friedel, T., 1974, Endocrine control of vitellogenesis in the harlequin bug, *Dindymus versicolor, J. Insect Physiol.* **20:**717–733.

Friedel, T., and Gillott, C., 1976, Male accessory gland substance of *Melanoplus sanguinipes:* An oviposition stimulant under the control of the corpus allatum, *J. Insect Physiol.* **22:**489–495.

Friedel, T., Loughton, B. G., and Andrew, R. D., 1980a, A neurosecretory protein from *Locusta migratoria, Gen. Comp. Endocrinol.* **41:**487–498.

Friedel, T., Feyereisen, R., Mundall, E. C., and Tobe, S. S., 1980b, The allatostatic effect of 20-hydroxyecdysone on the adult viviparous cockroach, *Diploptera punctata, J. Insect Physiol.* **26:** 665–670.

Friedel, T., Orchard, I., and Loughton, B. G., 1981, Release of neurosecretory protein from insect neurohaemal tissue following electrical stimulation, *Brain Res.* **208:**451–455.

Friedländer, M., 1982, Juvenile hormone and regulation of dichotomous spermatogenesis during the larval diapause of the codling moth, *J. Insect Physiol.* **28:**1009–1012.

Friedländer, M., and Benz, G., 1982, Control of spermatogenesis resumption in post-diapause larvae of the codling moth, *J. Insect Physiol.* **28**:349–355.

Friedländer, M., and Reynolds, S. E., 1988, Meiotic metaphases are induced by 20-hydroxyecdysone during spermatogenesis of the tobacco hornworm, *Manduca sexta, J. Insect Physiol.* **34**:1013–1019.

Friedman, S., 1967, The control of trehalose synthesis in the blowfly, *Phormia regina* Meig., *J. Insect Physiol.* **13**:397–405.

Frontali, N., 1968, Histochemical localization of catecholamines in the brain of normal and drug-treated cockroaches, *J. Insect Physiol.* **14**:881–886.

Frontali, N., and Häggendal, J., 1969, Noradrenaline and dopamine content in the brain of the cockroach *Periplaneta americana, Brain Res.* **14**:540–542.

Fuchs, M. S., Sundland, B. R., and Kang, S. H., 1980, In vivo induction of ovarian development in *Aedes atropalpus* by a head extract from *Aedes aegypti, Int. J. Invert. Reprod.* **2**:121–129.

Fugo, H., Iwata, Y., and Nakajima, M., 1984, Eclosion hormone activity in haemolymph of eclosing silkmoth, *Bombyx mori, J. Insect Physiol.* **30**:471–475.

Fugo, H., Saito, H., Nagasawa, H., and Suzuki, A., 1985, Eclosion hormone activity in developing embryos of the silkworm, *Bombyx mori, J. Insect Physiol.* **31**:293–298.

Fugo, H., Chen, J. H., Nakajima, M., Nagasawa, H., and Suzuki, A., 1987, Neurohormones in developing embryos of the silkworm, *Bombyx mori:* The presence and characteristics of prothoracicotropic hormone-S, *J. Insect Physiol.* **33**:243–248.

Fujishita, M., and Ishizaki, H., 1981, Circadian clock and prothoracicotropic hormone secretion in relation to the larval-larval ecdysis rhythm of the saturniid *Samia cynthia ricini, J. Insect Physiol.* **27**:121–128.

Fujita, T., Yui, R., Iwanaga, T., Nishiitsutsuji-Uwo, J., Endo, Y., and Yanaihara, N., 1981, Evolutionary aspects of "brain-gut peptides": An immunohistochemical study, *Peptides* **2**:123–131.

Fukaya, M., and Mitsuhashi, J., 1957, The hormonal control of larval diapause in the rice stem borer, *Chilo suppressalis.* Some factors in the head maintaining larval diapause, *Jpn. J. Appl. Entomol. Zool.* **1**:145–154.

Fukuda, S., 1944, The hormonal mechanism of larval molting and metamorphosis in the silkworm, *J. Fac. Sci. Imp. Univ. Tokyo, Sect. IV* **6**:477–532.

Fukuda, S., 1951, The production of the diapause eggs by transplanting the suboesophageal ganglion in the silkworm, *Proc. Jpn. Acad.* **27**:672–677.

Fukuda, S., 1952, Function of the pupal brain and suboesophageal ganglion in the production of non-diapause and diapause eggs in the silkworm, *Annot. Zool. Jpn.* **25**:149–155.

Fukuda, S., 1953, Alteration of voltinism in the silkworm following transection of pupal oesophageal connectives, *Proc. Jpn. Acad.* **29**:389–391.

Fukuda, S., 1963, Déterminisme hormonal de la diapause chez le ver à soie, *Bull. Soc. Zool. Fr.* **88**:151–179.

Fukuda, S., and Takeuchi, S., 1967, Diapause factor-producing cells in the suboesophageal ganglion of the silkworm, *Bombyx mori* L., *Proc. Jpn. Acad.* **43**:51–56.

Fukushima, T., and Yagi, S., 1975, Hormonal effect on cultivated insect tissues, III. Effects of a β-ecdysone or prothoracic glands on spermiogenesis in two noctuid insects in vitro, *Appl. Entomol. Zool.* **10**:220–225.

Füller, H. B., 1960, Morphologische und experimentelle Untersuchungen über die neurosekretorischen Verhältnisse im Zentralnervensystem von Blattiden und Culiciden, *Zool. Jahrb. Abt. Allg. Zool. Physiol. Tiere* **69**:223–250.

Furtado, A. F., 1971, Recherches sur le contrôle endocrine cérébral de la vitellogenèse et de la parturition chez une punaise vivipare, *Stilbocoris natalensis* (Hétéroptères, Lygéidés), *C.R. Acad. Sci. Ser. D* **272**:2468–2471.

Furtado, A. F., 1979, The hormonal control of mitosis and meiosis during oögenesis in a blood-sucking bug *Panstrongylus megistus, J. Insect Physiol.* **25:**561–570.

Furtado, A. F., Porcheron, P., and Dray, F., 1976, Evolution du taux des ecdysones au cours des deux dernières intermues de *Panstrongylus megistus* (Heteroptera: Reduviidae) *C.R. Acad. Sci. Ser. D* **283:**1077–1080.

Fuxe, K., 1963, Cellular localization of monoamines in the median eminence and infundibular stem of some mammals, *Acta Physiol. Scand.* **58:**383–384.

Fuzeau-Braesch, S., 1968, L'effet de groupe chez le grillon: Niveaux et relativité de ses manifestations, *Coll. Internat. CNRS* **173:**115–125.

Fuzeau-Braesch, S., Nicolas, G., Baehr, J. C., and Porcheron, P., 1982, A study of hormonal levels of the locust *Locusta migratoria cinerascens* artificially changed to the solitary state by a chronic CO_2 treatment of one minute per day, *Comp. Biochem. Physiol.* **71A:**53–58.

Gabe, M., 1953, Sur quelques applications de la coloration par la fuchsineparaldéhyde, *Bull. Microsc. Appl.* **21:**153–162.

Gabe, M., 1954, La neurosécrétion chez les invertébrés, *Ann. Biol.* **30:**5–62.

Gäde, G., 1979, Studies on the influence of synthetic adipokinetic hormone and some analogs on cyclic AMP levels in different arthropods systems, *Gen. Comp. Endocrinol.* **37:**122–130.

Gäde, G., 1980, Further characteristics of adipokinetic and hyperglycaemic factor(s) of stick insects, *J. Insect Physiol.* **26:**351–360.

Gäde, G., 1981, Activation of fat body glycogen phosphorylase in *Locusta migratoria* by corpus cardiacum extract and synthetic adipokinetic hormone, *J. Insect Physiol.* **27:**155–161.

Gäde, G., 1984, Adipokinetic and hyperglycaemic factors of different insect species : separation with high performance liquid chromatography, *J. Insect Physiol.* **30:**729–736.

Gäde, G., 1985a, Amino acid composition of cockroach hypertrehalosaemic hormones, *Z. Naturforsch.* **40c:**42–46.

Gäde, G., 1985b, Isolation of the hypertrehalosaemic factors I and II from the corpus cardiacum of the Indian stick insect, *Carausius morosus,* by reversed-phase high-performance liquid chromatography, and amino acid composition of factor II, *Biol. Chem. Hoppe-Seyler* **366:**195–199.

Gäde, G., 1985c, Mode of action of the hypertrehalosaemic peptides from the American cockroach, *Z. Naturforsch.* **40c:**670–676.

Gäde, G., 1986, Relative hypertrehalosaemic activities of naturally occurring neuropeptides from the AKH/RPCH family, *Z. Naturforsch.* **41c:**315–320.

Gäde, G., 1987, Isolation and amino acid composition of the adipokinetic neuropeptide from the corpus cardiacum of the cricket, *Gryllus bimaculatus, Physiol. Entomol.* **12:**309–316.

Gäde, G., and Beenakkers, A. M. T., 1977, Adipokinetic hormone-induced lipid mobilization and cyclic AMP accumulation in the fat body of *Locusta migratoria* during development, *Gen. Comp. Endocrinol.* **32:**481–487.

Gäde, G., and Holwerda, D. A., 1976, Involvement of adenosine 3′–5′ cyclic monophosphate in lipid mobilization in *Locusta migratoria, Insect Biochem.* **6:**535–540.

Gäde, G., and Lohr, P., 1982, Restricted specificity of a hyperglycaemic factor from the corpus cardiacum of the stick insect *Carausius morosus, J. Insect Physiol.* **28:**805–811.

Gäde, G., and Rinehart, K. L., 1986, Amino-acid sequence of a hypertrehalosaemic neuropeptide from the corpus cardiacum of the cockroach, *Nauphoeta cinerea, Biochem. Biophys. Res. Commun.* **141:**774–781.

Gäde, G., and Rinehart, K. L., 1987, Primary structure of the hypertrehalosaemic factor II from the corpus cardiacum of the Indian stick insect, *Carausius morosus,* determined by fast atom bombardment mass spectrometry, *Biol. Chem. Hoppe-Seyler* **368:**67–75.

Gäde, G., and Scheid, M., 1986, A comparative study on the isolation of adipokinetic and hypertrehalosaemic factors from insect corpora cardiaca, *Physiol. Entomol.* **11:**145–157.

Gäde, G., Goldsworthy, G. J., Kegel, G., and Keller, R., 1984, Single step purification of locust adipokinetic hormones I and II by reversed-phase high-performance liquid chromatography and amino-acid composition of the hormone II, *Hoppe-Seyler's Z. Physiol. Chem.* **365:**393–398.

Gäde, G., Goldsworthy, G. J., Schaffer, M. H., Carter Cook, J., and Rinehart, K. L., 1986, Sequence analyses of adipokinetic hormones II from corpora cardiaca of *Schistocerca nitans, Schistocerca gregaria* and *Locusta migratoria* by fast atom bombardment mass spectrometry, *Biochem. Biophys. Res. Commun.* **134:**723–730.

Gadot, M., and Applebaum, S. W., 1985, Rapid in vitro activation of corpora allata by extracted locust brain allatotropic factor, *Arch. Insect Biochem. Physiol.* **2:**117–129.

Gadot, M., Rafaeli, A., and Applebaum, S. W., 1987, Partial purification and characterization of locust allatotropin 1, *Arch. Insect Biochem. Physiol.* **4:**213–223.

Gallois, D., 1979, Etude chronologique de l'organogenèse de l'appareil génital mâle chez *Locusta migratoria migratoroides* R. et F. (*Orthoptera Acridoidea*), *Bull. Biol. Fr. Belg.* **113:**31–51.

Ganagarajah, M., and Saleuddin, A. S. M., 1970, A simple manoeuver to prevent loss of sections for performic acid-Victoria blue technique and a comparison of various neurosecretory stains, *Can. J. Zool.* **48:**1457–1458.

Garcia, E. S., Furtado, A. F., and De Azambuja, P., 1987, Effect of allatectomy on ecdysteroid dependent development of *Rhodnius prolixus* larvae, *J. Insect Physiol.* **33:**729–732.

Gavin, J. A., and Williamson, J. H., 1976, Juvenile hormone-induced vitellogenesis in *Apterous,* a non-vitellogenic mutant in *Drosophila melanogaster, J. Insect Physiol.* **22:**1737–1742.

Gee, J. D., 1975, The control of diuresis in the tsetse fly *Glossina austeni:* A preliminary investigation of the diuretic hormone, *J. Exp. Biol.* **63:**391–401.

Gee, J. D., Whitehead, D. L., and Koolman, J., 1977, Steroids stimulate secretion by insect Malpighian tubules, *Nature* **269:**238–239.

Gelbic, I., and Metwally, M. M., 1981, Changes in the development of male germinal cells in *Spodoptera littoralis* caused by the effects of juvenoids (Lepidoptera, Noctuidae), *Acta Entomol. Bohemoslov.* **78:**10–17.

Geldiay, S., 1959, Neurosecretory cells in ganglia of the roach, *Blaberus craniifer, Biol. Bull. (Woods Hole, Mass.)* **117:**267–274.

Geldiay, S., 1967, Hormonal control of adult reproductive diapause in the Egyptian grasshopper, *Anacridium aegyptium,* L., *J. Endocrinol.* **37:**63–71.

Geldiay, S., and Edwards, J. S., 1973, The protocerebral neurosecretory system and associated cerebral neurohemal area of *Acheta domesticus, Z. Zellforsch. Mikrosk. Anat.* **145:**1–22.

Geldiay, S., and Karaçali, S., 1980, The neurosecretory system of the adult *Melanogryllus desertus* Pall. (Orthoptera, Gryllidae). II. Cerebral neurohaemal area, *Cell Tissue Res.* **211:**235–240.

Geldiay, S., and Karaçali, S., 1983, Rhabdome-bearing cells within the corpus cardiacum of *Melanogryllus desertus* (Orthoptera, Gryllidae), *Gen. Comp. Endocrinol.* **51:**239–245.

Gelman, D. B., and Woods, C. W., 1983, Haemolymph ecdysteroid titers of diapause- and non-diapause-bound fifth instars and pupae of the European corn borer, *Ostrinia nubilalis* (Hubner), *Comp. Biochem. Physiol.* **76A:**367–375.

Gelman, D. B., Woods, C. W., and Borkovec, A. B., 1988a, Effects of ecdysone and 20-hydroxyecdysone on apyrene spermiogenesis in the European corn borer, *Ostrinia nubilalis, J. Insect Physiol.* **34:**733–738.

Gelman, D. B., Woods, C. W., and Borkovec, A. B., 1988b, Ecdysteroid profiles for hemolymph and testes from larvae, pupae, and pharate adults of the European corn borer *Ostrinia nubilalis* Hubner, *Arch. of Insect Biochem. and Physiol.* **7:**267–279.

Gemmill, R. M., Hamblin, M., Glaser, R. L., Racioppi, J. V., Mary, J. L., White, B. N., Calvo, J. M., Wolfner, M. F., and Hagedorn, H. H., 1986, Isolation of mosquito vitellogenin genes and induction of expression by 20-hydroxyecdysone, *Insect Biochem.* **16:**761–774.

Gersch, M., 1955, Untersuchungen über Auslösung und Steuerung der Darmbewegungen bei der Larve von *Chaoborus* (Corethra), *Biol. Zentralbl.* **74:**603–628.

Gersch, M., 1956, Untersuchungen zur Frage der Hormonalen Beeinflüssung der Melanophoren bei der *Corethra*-Larve, *Z. Vgl. Physiol.* **39:**190–208.
Gersch, M., 1958, Neurohormonale Beeinflussung der Herztätigkaeit bei der larve von *Corethra, J. Insect Physiol.* **2:**281–297.
Gersch, M., 1962, The activation hormone of the metamorphosis of insects, *Gen. Comp. Endocrinol.* **1:**322–329.
Gersch, M., 1967, Experimental examinations on the hormonal control of the water balance and excretion of the larva of *Corethra (Chaoborus), Gen. Comp. Endocrinol.* **9:**453.
Gersch, M., 1969, Experimentelle Untersuchungen zur endocrinen Regulation des Wasserhaushaltes der Larve von *Corethra (Chaoborus), Zool. Jahrb. Abt. Allg. Zool. Physiol. Tiere* **75:**1–16.
Gersch, M., 1972, Experimentelle Untersuchungen zum Freisetzungsmechanismus von Neurohormonen nach elektrischer Reizung der Corpora cardiaca von *Periplaneta americana in vitro, J. Insect Physiol.* **18:**2425–2439.
Gersch, M., 1974, Selektive Freisetzung des hyperglykämischen Faktors aus den Corpora cardiaca von *Periplaneta americana* in vivo, *Experientia* **30:**767.
Gersch, M., 1976, Krontrolle der circadianen Rhythmik des Trehalosegehaltes der Haemolymphe von *Periplaneta americana, Zool. Jahrb. Abt. Allg. Zool. Physiol. Tiere* **80:**1–15.
Gersch, M., 1977, Häutungen von Insekten ohne Häutungsdrüse: Befunde mit Larven von *Periplaneta americana, Experientia* **33:**228–229.
Gersch, M., and Birkenbeil, H., 1980, C-AMP als möglischer Mediator der neurohormonalen Beeinflussung des Membranpotentials der Prothoracaldrüsenzellen von *Galleria mellonella, Zool. Jahrb. Abt. Allg. Zool. Physiol. Tiere* **84:**58–66.
Gersch, M., and Bräuer, R., 1974, In vitro-Stimulation der Prothoracaldrüsen von Insekten als Testsystem (Prothoracaldrüsentest), *J. Insect Physiol.* **20:**735–741.
Gersch, M., and Mothes, G., 1956, Neurohormonalen Wirkungsantagonismus bein Farbwechsel von *Dixippus morosus, Naturwissenschaften* **43:**542–543.
Gersch, M., and Richter, K., 1963, Auslösung von Nervenimpulsen durch ein Neurohormon bei *Periplaneta americana, Zool. Jahrb. Abt. Allg. Zool. Physiol.* **70:**301–308.
Gersch, M., and Sturzebecher, J., 1968, Weitere Untersuchungen zur Kennzeichnung des Aktivationshormons der Insektenhautung, *J. Insect Physiol.* **14:**87–96.
Gersch, M., and Sturzebecher, J., 1970, Experimentelle Stimulierung der Zellulären Aktivität der Prothorakaldrüsen von *Periplaneta americana* durch den Aktivationsfaktor, *J. Insect Physiol.* **16:**1813–1826.
Gersch, M., Unger, H., and Fischer, F., 1957, Die Isolierung eines Neurohormons aus den Nervensystem von *Periplaneta americana* und einige biologische Test verfahren, *Wiss. Z. Friedrich-Schiller Univ. Jena.* **6:**125–129.
Gersch, M., Fischer, F., Unger, H., and Koch, H., 1960, Die Isolierung Neurohormonalen Faktoren aus dem Nervensystem der Küchenschabe *Periplaneta americana, Z. Naturforsch. Teil C.* **15:**319–323.
Gersch, M., Fischer, F., Unger, H., and Kabitza, W., 1961, Vorkommen von Serotonin im Nervensystem von *Periplaneta americana* L. (Insecta), *Z. Naturforsch.* **15:**351–352.
Gersch, M., Richter, K., Böhm, G. A., and Sturzebecher, J., 1970, Selektive Ausschüttung von Neurohormonen nach elektrischer Reizung der Corpora cardiaca von *Periplaneta americana* in vitro, *J. Insect Physiol.* **16:**1991–2014.
Gersch, M., Bräuer, R., and Birkenbeil, H., 1973, Experimentelle Untersuchungen zur Wirkungsmechanismus der beiden entwicklungsphysiologisch aktiven Fraktionen des "Gehirnhormons" der Insekten (Aktivationsfaktor I und II) auf die Prothoracaldrüse, *Experientia* **29:**425–427.
Gersch, M., Hentschel, E., and Ude, J., 1974, Aminerge Substanzen im lateralen Herznerven und

im stomatogastrischen Nervensystem der Schabe *Blaberus craniifer* Burm., *Zool. Jahrb. Abt. Allg. Zool. Physiol. Tiere* **78:**1–15.

Gersch, M., Birkenbeil, H., and Ude, J., 1975, Ultrastructure of the prothoracic gland cells of the last instar of *Galleria mellonella* in relation to the state of development, *Cell Tissue Res.* **160:** 389–397.

Gersch, M., Birkenbeil, H., and Klare, G., 1976, Das Membranpotential der Prothorakaldrüsenzellen von *Galleria mellonella, Zool. Jahrb. Abt. Allg. Zool. Physiol. Tiere* **80:** 16–23.

Gibbs, D., 1976, The initiation of adult development in *Sarcophaga argyrostoma* by β-ecdysone, *J. Insect Physiol.* **22:**1195–1200.

Gibbs, D., and Riddiford, L. M., 1977, Prothoracicotropic hormone in *Manduca sexta:* Localization by a larval assay, *J. Exp. Biol.* **66:**255–266.

Giebultowicz, J. M., and Denlinger, D. L., 1986, Role of the brain and ring gland in regulation of pupal diapause in the flesh fly *Sarcophaga crassipalpis, J. Insect Physiol.* **32:**161–166.

Giebultowicz, J. M., Cymborowski, B., and Delbecque, J. P., 1984, Environmental control of larval behaviour and its consequences for ecdysteroid content and pupation in *Ephestia Kuehniella, Physiol. Entomol.* **9:**409–416.

Giebultowicz, J. M., Loeb, M. J., and Borkovec, A. B., 1987, In vitro spermatogenesis in lepidopteran larvae: Role of the testis sheath, *Int. J. Invert. Reprod.* **11:**211–224.

Giersberg, H., 1928, Über den morphologischen und physiologischen Farbwechsel der Stabheuschrecke *Dixippus (Carausius morosus), Z. Vgl. Physiol.* **7:**657–695.

Gilbert, L. I., 1962, Maintenance of the prothoracic gland by the juvenile hormone in insects, *Nature* **193:**1205–1207.

Gilbert, L. I., 1967, Changes in lipid content during the reproductive cycle of *Leucophaea maderae* and effects of the juvenile hormone on lipid metabolism in vitro, *Comp. Biochem. Physiol.* **21:** 237–257.

Gilbert, L. I., and Schneiderman, H. A., 1959, Prothoracic gland stimulation by juvenile hormone extracts of insects, *Nature* **184:**171–173.

Gilbert, L. I., and Schneiderman, H. A., 1960, The development of a bioassay for the juvenile hormone in insects, *Trans. Am. Microsc. Soc.* **79:**38–67.

Gillott, C., and Friedel, T., 1976, Development of accessory reproductive glands and its control by the corpus allatum in adult male *Melanoplus sanguinipes, J. Insect Physiol.* **22:**365–372.

Gillott, C., and Langley, P. A., 1981, The control of receptivity and ovulation in the tsetse fly, *Glossina morsitans, Physiol. Entomol.* **6:**269–281.

Gillott, C., and Venkatesh, K., 1985, Development of secretory ability in the spermatheca of the migratory grasshopper, *Melanoplus sanguinipes, J. Insect Physiol.* **31:**647–652.

Girardie, A., 1966a, Contribution à l'étude du contrôle du métabolisme de l'eau chez *Gryllus bimaculatus* (Orthoptère). Fonction diurétique de la pars intercerebralis, *C.R. Acad. Sci.* **262:** 1361–1364.

Girardie, A., 1966b, Contrôle de l'activité génitale chez *Locusta migratoria.* Mise en évidence d'un facteur gonadotrope et d'un facteur allatotrope dans la pars intercerebralis, *Bull. Soc. Zool. Fr.* **91:**423–439.

Girardie, A., 1967, Contrôle neuro-hormonal de la métamorphose et de la pigmentation chez *Locusta migratoria cinerascens* (Orthoptère), *Bull. Biol. Fr. Belg.* **101:**79–114.

Girardie, A., 1970, Mise en évidence, dans le protocérébron de *Locusta migratoria migratorioides* et de *Schistocerca gregaria,* de nouvelles cellules neurosécrétrices contrôlant le métabolisme hydrique, *C.R. Acad. Sci. Ser. D* **271:**504–507.

Girardie, A., 1974, Recherches sur le rôle physiologique des cellules neurosécrétrices latérales du protocérébron de *Locusta migratoria migratorioides* (Insecte Orthoptère), *Zool. Jahrb. Abt. Allg. Zool. Physiol. Tiere* **78:**310–326.

Girardie, A., and Cazal, M., 1965, Role de la pars intercerebralis et des corpora cardica sur la melanization dez *Locusta migratoria* (L.), *C. R. Acad. Sci.* **261:**4325–4328.
Girardie, A., and De Reggi, M., 1978, Moulting and ecdysone release in response to electrical stimulation of protocerebral neurosecretory cells in *Locusta migratoria, J. Insect Physiol.* **24:** 797–802.
Girardie, A., and Girardie, J., 1966, Mise en évidence d'une activité neurosécrétrice des cellules C de la pars intercerebralis de *Locusta migratoria* L. par étude comparative histologique et ultrastructurale, *C.R. Acad. Sci. Ser. D* **263:**1119–1122.
Girardie, A., and Girardie, J., 1967, Etude histologique, histochimique et ultrastructurale de la pars intercerebralis chez *Locusta migratoria* L. (Orthoptère), *Z. Zellforsch. Mikrosk. Anat.* **78:**54–75.
Girardie, A., and Granier, S., 1973, Systeme endocrine et physiologie de la drapaude imaginale chez le criquet egyptien *Anacridium aegyptium, J. Insect Physiol.* **19:**2341–2358.
Girardie, A., and Lafon-Cazal, M., 1972, Contrôle endocrine des contractions de l'oviducte isolé de *Locusta migratoria migratorioides* (R. et F.), *C.R. Acad. Sci. Ser. D* **274:**2208–2210.
Girardie, A., and Vogel, A., 1966, Etude du contrôle neurohumoral de l'activité sexuelle mâle de *Locusta migratoria* (L.), *C.R. Acad. Sci. Ser. D* **263:**543–546.
Girardie, A., Moulins, M., and Girardie, J., 1974, Rupture de la diapause ovarienne d'*Anacridium aegyptium* par stimulation électrique des cellules neurosécrétrices médianes de la pars intercerebralis, *J. Insect Physiol.* **20:**2261–2275.
Girardie, J., 1975, Recherche en microscopie photonique et électronique des éléments neurosécréteurs tritocérébraux de *Locusta migratoria* (Insecte Orthoptère), *Arch. Anat. Microsc. Morphol. Exp.* **64:**223–246.
Girardie, J., and Girardie, A., 1977, Libération provoquée in vitro du produit de neurosécrétion des cellules protocérébrales médianes chez le criquet migrateur, *J. Physiol. (Paris)* **73:**707–721.
Girardie, J., and Rossi, C., 1978, Preuves histologiques et ultrastructurales de 2 catégories de cellules neurosécrétrices protocérébrales médianes de type A chez le criquet migrateur, *C.R. Acad. Sci. Ser. D* **286:**97–100.
Girardie, J., Faddoul, A., and Girardie, A., 1985, Characterization of three neurosecretory proteins from the A median neurosecretory cells of *Locusta migratoria* by coupled chromatographic, electrophoretic and isoelectrofocusing methods, *Insect Biochem.* **15:**85–92.
Girardie, J., Bourême, D., and Girardie, A., 1987a, Production sites of the three neurosecretory proteins characterized in the corpora cardiaca of the migratory locust, *Insect Biochem.* **17:**29–36.
Girardie, J., Bourême, D., Couillaud, F., Tamarelle, M., and Girardie, A., 1987b, Anti-juvenile effect of neuroparsin A, a neuroprotein isolated from locust corpora cardiaca, *Insect Biochem.* **17:**977–983.
Glass, H., Emmerich, H., and Spindler, K. D., 1978, Immunohistochemical localization of ecdysteroids in the follicular epithelium of locust oocytes, *Cell Tissue Res.* **194:**237–244.
Glitho, I., Delbecque, J. P., and Delachambre, J., 1979, Prothoracic gland involution related to moulting hormone levels during the metamorphosis of *Tenebrio molitor* L., *J. Insect Physiol.* **25:**187–191.
Gokuldas, M., Hunt, P. A., and Candy, D. J., 1988, The inhibition of lipid synthesis *in vitro* in the locust, *Schistocerca gregaria,* by factors from the corpora cardiaca, *Physiol. Entomol.* **13:**43–48.
Goldbard, G. A., Sauer, J. R., and Mills, R. R., 1970, Hormonal control of excretion in the American cockroach. II. Preliminary purification of a diuretic and antidiuretic hormone, *Comp. Gen. Pharmacol.* **1:**82–86.
Goldsworthy, G. J., 1969, Hyperglycaemic factors from the corpus cardiacum of *Locusta migratoria, J. Insect Physiol.* **15:**2131–2140.

Goldsworthy, G. J., 1971, The effects of removal of the cerebral neurosecretory cells on hemolymph and tissue carbohydrate in *Locusta migratoria* migration, *J. Endocrinol.* **50:**237–240.

Goldsworthy, G. J., 1983, The endocrine control of flight metabolism in locusts, *Adv. Insect Physiol.* **17:**149–204.

Goldsworthy, G. J., and Wheeler, C. H., 1986, The endocrine control of flight metabolism in locusts, in *Insect Flight: Dispersal and Migration* (Danthanarayana, ed.), Springer-Verlag, Berlin.

Goldsworthy, G. J., Mordue, W., and Guthkelch, J., 1972a, Studies on insect adipokinetic hormones, *Gen. Comp. Endocrinol.* **18:**545–551.

Goldsworthy, G. J., Johnson, R. A., and Mordue, W., 1972b, In vivo studies on the release of hormones from the corpora cardiaca of locusts, *J. Comp. Physiol.* **79:**85–96.

Goldsworthy, G. J., Lee, S. S., and Jutsum, A. R., 1977, Cerebral neurosecretory cells and flight in the locust, *J. Insect Physiol.* **23:**717–721.

Goldsworthy, G. J., Mallison, K., and Wheeler, C. H., 1986a, The relative potencies of two known locust adipokinetic hormones, *J. Insect Physiol.* **32:**95–101.

Goldsworthy, G. J., Mallison, K., Wheeler, C. H., and Gäde, G., 1986b, Relative adipokinetic activities of members of the adipokinetic hormone/red pigment concentrating hormone family, *J. Insect Physiol.* **32:**433–438.

Gole, J. W. D., and Downer, R. G. H., 1979, Elevation of adenosine 3′, 5′-monophosphate by octopamine in fat body of the American cockroach, *Periplaneta americana* L., *Comp. Biochem. Physiol.* **64C:**223–226.

Gole, J. W. D., Orr, G. L., and Downer, R. G. H., 1987, Pharmacology of octopamine-, dopamine-, and 5-hydroxytryptamine-stimulated cyclic AMP accumulation in the corpus cardiacum of the American cockroach, *Periplaneta americana* L., *Arch. Insect Biochem. Physiol.* **5:**119–128.

Goltzené, F., Lagueux, M., Charlet, M., and Hoffmann, J. A., 1978, The follicle cell epithelium of maturing ovaries of *Locusta migratoria:* A new biosynthetic tissue for ecdysone, *Hoppe-Seyler's Z. Physiol. Chem.* **359:**1427–1434.

Gomori, G., 1941, Observations with differential stains on human islets of Langerhans, *Am. J. Pathol.* **17:**395–406.

Gomori, G., 1950, Aldehyde-fuchsin. A new stain for elastic tissue, *Am. J. Clin. Pathol.* **20:**665–668.

Goodman, W. G., 1985, Relative hemolymph juvenile hormone binding capacity during larval, pupal and adult development of *Manduca sexta, Insect Biochem.* **15:**557–564.

Goodman, W. G., and Chang, E. S., 1985, Juvenile hormone cellular and hemolymph binding proteins, in *Comprehensive Insect Physiology, Biochemistry, and Pharmacology* (G. A. Kerkut and L. I. Gilbert, eds.), Pergamon Press, New York, pp. 491–510.

Goodman, W. G., and Gilbert, L. I., 1978, The hemolymph titer of juvenile hormone binding protein and binding sites during the fourth larval instar of *Manduca sexta, Gen. Comp. Endocrinol.* **35:**27–34.

Goosey, M. W., and Candy, D. J., 1980, The D-octopamine content of the haemolymph of the locust, *Schistocerca americana gregaria* and its elevation during flight, *Insect Biochem.* **10:** 393–397.

Goosey, M. W., and Candy, D. J., 1982, The release and removal of octopamine by tissues of the locust *Schistocerca americana gregaria, Insect Biochem.* **12:**681–685.

Gourdoux, L., Lequellec, Y., Moreau, R., and Dutrieu, J., 1983, Gluconeogenesis from some amino acids and its endocrine modification in *Tenebrio molitor* L. (Coleoptera), *Comp. Biochem. Physiol.* **74B:**273–276.

Granger, N. A., 1978, Innervation of the prothoracic glands in *Galleria mellonella* larvae (Lepidoptera: Pyralidae), *Int. J. Insect Morph. Embryol.* **7:**315–324.

Granger, N. A., and Janzen, W. P., 1987, Inhibition of *Manduca sexta* corpora allata in vitro by a cerebral allatostatic neuropeptide, *Mol. Cell. Endocrinol.* **49:**237–248.

Granger, N. A., and Sehnal, F., 1974, Regulation of larval corpora allata in *Galleria mellonella, Nature* **251:**415–417.

Granger, N. A., Bollenbacher, W. E., Vince, R., Gilbert, L. I., Baehr, J. C., and Dray, F., 1979, In vitro biosynthesis of juvenile hormone by the larval corpora allata of *Manduca sexta:* Quantification by radio-immunoassay, *Mol. Cell Endocrinol.* **16:**1–17.

Granger, N. A., Niemiec, S. M., Gilbert, L. I., and Bollenbacher, W. E., 1982, Juvenile hormone synthesis in vitro by larval and pupal corpora allata of *Manduca sexta, Mol. Cell. Endocrinol.* **28:**587–604.

Granger, N. A., Mitchell, L. J., Janzen, W. P., and Bollenbacher, W. E., 1984, Activation of *Manduca sexta* corpora allata in vitro by a cerebral neuropeptide, *Mol. Cell. Endocrinol.* **37:** 349–358.

Granger, N. A., Whisenton, L. R., Janzen, W. P., and Bollenbacher, W. E., 1987, Interendocrine control by 20-hydroxyecdysone of the corpora allata of *Manduca sexta, Insect Biochem.* **17:** 949–953.

Gray, R., Meola, R., and Holman, G. M., 1987, Fat body stimulation of ecdysone synthesis in *Heliothis zea, J. Insect Physiol.* **33:**325–331.

Greenplate, J. T., Glaser, R. L., and Hagedorn, H. H., 1985, The role of factors from the head in the regulation of egg development in the mosquito *Aedes aegypti, J. Insect Physiol.* **31:**323–329.

Gregg, P. C., Roberts, B., and Wentworth, S. L., 1987, Levels of ecdysteroids in diapause and non-diapause eggs of the Australian plague locust, *Chortoicetes terminifera* (Walker), *J. Insect Physiol.* **33:**237–242.

Griffiths, A. C., and Finlayson, L. H., 1982, Extra-ganglionic neurosecretory and non-neurosecretory neurons in the larva of the moth, *Agrotis segetum* Schiff (Lepidoptera: Noctuidae), *Int. J. Insect Morphol. Embryol.* **11:**341–349.

Grillot, J. P., 1968, Description d'organes neurohémaux métamériques associés à la chaîne nerveuse ventrale chez deux coléoptères: *Chrysocarabus auronitens* Fabr. (*Carabidae*) et *Oryctes rhinoceros* L., *C.R. Acad. Sci. Ser. D* **267:**772–775.

Grillot, J. P., 1976a, Les organes périsympathiques des insectes. Essai sur leur évolution, données structurales et physiologiques chez les coléoptères et les diptères, *Ann. Sci. Nat. Zool. Biol. Anim.* **18:**311–365.

Grillot, J. P., 1976b, Types morphologiques et phylogénie des organes périsympathiques des insectes, *Bull. Biol. Fr. Belg.* **110:**143–201.

Grillot, J. P., 1983, Morphology and evolution of perisympathetic organs, in *Neurohemal Organs of Arthropods* (A. P. Gupta, ed.), Charles C Thomas, 1983, pp. 481–512.

Grillot, J. P., and Raabe, M., 1973, Recherches préliminaires sur les organes périsympathiques des diptères, *C.R. Acad. Sci. Ser. D* **277:**425–428.

Grillot, J. P., Delachambre, J., and Provansal, A., 1971, Les organes périsympathiques de *Tenebrio molitor* L. (Coléoptère). Description. Rôle possible dans la décharge d'un facteur de tannage, *C. R. Acad. Sci. Ser. D* **273:**2599–2602.

Grillot, J. P., Delachambre, J., and Provansal, A., 1976, Rôle des organes périsympathiques et dynamique de la sécrétion de la bursicon chez *Tenebrio molitor, J. Insect Physiol.* **22:**763–780.

Grillot, J. P., Baudry-Partiaoglou, N., and Provansal-Baudez, A., 1987, Organes périsympathiques et évolution chez les insectes, *Ann. Soc. Ent. Fr.* **23:**333–349.

Grillou, H., 1978, Effet de l'injection de corps allates sur la réceptivité sexuelle de *Blaberus craniifer* Burm. (Blattaria) allatectomisées au dernier stade larvaire, *C.R. Acad. Sci. Ser. D* **286:**689–692.

Gros, C., Lafon-Cazal, M., and Dray, F., 1978, Présence de substances immunoréactivement apparentées aux enképhalines chez un insecte, *Locusta migratoria, C.R. Acad. Sci. Ser. D* **287:** 647–650.

Gruetzmacher, M. C., Gilbert, L. I., Granger, N. A., Goodman, W., and Bollenbacher, W. E., 1984a, The effect of juvenile hormone on prothoracic gland function during the larval-pupal development of *Manduca sexta:* An in situ and in vitro analysis, *J. Insect Physiol.* **30:**331–340.

Gruetzmacher, M. C., Gilbert, L. I., and Bollenbacher, W. E., 1984b, Indirect stimulation of the prothoracic glands of *Manduca sexta* by juvenile hormone: Evidence for a fat body stimulatory factor, *J. Insect Physiol.* **30:**771–778.

Guilvard, E., De Reggi, M., and Rioux, J. A., 1984, Changes in ecdysteroid and juvenile hormone titers correlated to the initiation of vitellogenesis in two *Aedes* species, *Gen. Comp. Endocrinol.* **53:**218–223.

Gündel, J., Müller, A., Müller, E., and Gersch, M., 1982a, Zur Charakterisierung der Ionenverhältnisse an den Prothorakaldrüsenzellen von *Galleria mellonella* L., *Zool. Jahrb. Abt. Allg. Zool. Physiol. Tiere* **86:**34–44.

Gündel, J., Müller, A., Müller, E., and Gersch, M., 1982b, Anderungen der Kaliuminnenkonzentrationen als Ursache für die entwicklungsabhängigen Anderungen der Ruhepotentiale von Prothorakaldrüsenzellen von *Galleria mellonella, Zool. Jahrb. Abt. Allg. Zool. Physiol. Tiere* **86:** 183–192.

Gündel, M., and Penzlin, H., 1980, Identification of neuronal pathways between the stomatogastric nervous system and the retrocerebral complex of the cockroach *Periplaneta americana* (L.), *Cell Tissue Res.* **208:**283–297.

Gwadz, R. W., 1972, Neuro-hormonal regulation of sexual receptivity in female *Aedes aegypti, J. Insect Physiol.* **18:**259–266.

Gwadz, R. W., and Spielman, A., 1973, Corpus allatum control of ovarian development in *Aedes aegypti, J. Insect Physiol.* **19:**1441–1448.

Hadorn, E., and Frizzi, G., 1949, Experimentelle Untersuchungen zur Melanophorenreaktion von *Corethra, Rev. Suisse Zool.* **56:**5–29.

Hagedorn, H. H., 1985, The role of ecdysteroids in reproduction, in *Comprehensive Insect Physiology Biochemistry and Pharmacology* (G. A. Kerkut and L. I. Gilbert, eds.), Vol. 8, Pergamon Press, New York, pp. 205–262.

Hagedorn, H. H., and Fallon, A. M., 1973, Ovarian control of vitellogenin synthesis by the fat body in *Aedes aegypti, Nature* **244:**103–105.

Hagedorn, H. H., and Kunkel, J. G., 1979, Vitellogenin and vitellin in insects, *Ann. Rev. Entomol.* **24:**475–505.

Hagedorn, H. H., O'Connor, J. D., Fuchs, M. S., Sage, B., Schlaeger, D. A., and Bohm, M. K., 1975, The ovary as a source of α-ecdysone in an adult mosquito, *Proc. Natl. Acad. Sci. USA* **72:**3255–3259.

Hagedorn, H. H., Turner, S., Hagedorn, E. A., Pontecorvo, D., Greenbaum, P., Pfeiffer, D., Wheelock, G., and Flanagan, T. R., 1977, Postemergence growth of the ovarian follicles of *Aedes aegypti, J. Insect Physiol.* **23:**203–206.

Hagedorn, H. H., Shapiro, J. P., and Hanaoka, K., 1979, Ovarian ecdysone secretion is controlled by a brain hormone in an adult mosquito, *Nature* **282:**92–94.

Halmi, N. S., 1952, Differentiation of two types of basophils in the adenohypophysis of the rat and mouse, *Stain Technol.* **27:**61–64.

Hanaoka, K., and Hagedorn, H. H., 1980, Brain hormone control of ecdysone secretion by the ovary in a mosquito, in *Progress in Ecdysone Research* (J. A. Hoffman, ed.), Elsevier North-Holland, Amsterdam, pp. 467–480.

Hanaoka, K., and Takahashi, S. Y., 1977, Adenylate cyclase system and the hyperglycemic factor in the cockroach, *Periplaneta americana, Insect Biochem.* **7:**95–99.

Handler, A. M., and Konopka, R. J., 1979, Transplantation of a circadian pacemaker in *Drosophila, Nature* **279:**236–238.

Hanrahan, J. W., and Phillips, J. E., 1985, Further observations on the regulation of KCL absorption across locust rectum, *J. Exp. Biol.* **116:**153–167.

Hansen, B. L., Hansen, G. N., and Scharrer, B., 1982, Immunoreactive material resembling vertebrate neuropeptides in the corpus cardiacum and corpus allatum of the insect *Leucophaea maderae, Cell Tissue Res.* **225:**319–329.

Hansen, G. N., Hansen, B. L., and Scharrer, B., 1987, Gastrin/CCK-like immunoreactivity in the corpus cardiacum-corpus allatum complex of the cockroach *Leucophaea maderae, Cell Tissue Res.* **248:**595–598.

Hanström, B., 1938, Zwei Probleme betreffs der hormonalen Lokalisation im Insektenkopf, *Lunds Univ. Arsskrift N.F.* **239:**1–17.

Hanström, B., 1940, Die chromatophoraktivierende Substanz des Insektenkopfes, *Kungl. Fysiogr. Sällsk. Handl. Lund. N.F.* **51:**1–20.

Hanström, B., 1953, Neurosecretory pathways in the head of crustaceans, insects and vertebrates, *Nature* **171:**72–75.

Hardie, J., 1987, The corpus allatum, neurosecretion and photoperiodically controlled polymorphism in an aphid, *J. Insect Physiol.* **33:**201–205.

Hardie, J., and Lees, A. D., 1985, Endocrine control of polymorphism and polyphenism, in *Comprehensive Insect Physiology, Biochemistry and Pharmacology* (G. A. Kerkut and L. I. Gilbert, eds.), Vol. 13, Pergamon Press, New York, pp. 441–490.

Harker, J. E., 1956, Factors controlling the diurnal rhythm of activity of *Periplaneta americana* L., *J. Exp. Biol.* **33:**224–234.

Harker, J. E., 1960, Endocrine and nervous factors in insect circadian rhythms, *Cold Spring Harbor Symp. Quant. Biol.* **25:**279–287.

Hart, D. E., and Steele, H. E., 1973, The glycogenolytic effect of the corpus cardiacum on the cockroach nerve cord, *J. Insect Physiol.* **19:**927–939.

Hartfelder, K. H., 1987, Rates of juvenile hormone synthesis control caste differentiation in the stingless bee *Scaptotrigona postica depilis, Roux's Arch. Develop. Biol.* **196:**522–526.

Hartmann, R., 1971, Der einfluss endokriner Faktoren auf die männlichen akzessorischen Drüsen und die Ovarien bei der Keulenheuschrecke *Gomphocerus rufus* L. (Orthoptera, Acrididae), *Z. Vgl. Physiol.* **74:**190–216.

Hartmann, R., and Loher, W., 1974, Control of sexual behaviour pattern 'secondary defence' in the female grasshopper, *Chorthippus curtipennis, J. Insect Physiol.* **20:**1713–1728.

Hartmann, R., Jendrsczok, C., and Peter, M. G., 1987, The occurrence of a juvenile hormone binding protein and in vitro synthesis of juvenile hormone by the serosa of *Locusta migratoria* embryos, *Roux's Arch. Develop. Biol.* **196:**347–355.

Hasegawa, K., 1951, Studies on the voltinism in the silkworm, *Bombyx mori* L., with special reference to the organs concerning determination of voltinism, *Proc. Jpn. Acad.* **27:**667.

Hasegawa, K., 1952, Studies on the voltinism in the silkworm, *Bombyx mori* L., with special reference to the organs concerning determination of voltinism, *J. Fac. Agric., Tottori Univ.* **1:** 83–124.

Hasegawa, K., 1957, The diapause hormone of the silkworm, *Bombyx mori, Nature* **179:**1300–1301.

Hasegawa, K., 1964, Studies on the mode of action of the diapause hormone in the silkworm, *Bombyx mori* L., II. Content of diapause hormone in the suboesophageal ganglion, *J. Exp. Biol.* **41:**855–863.

Hasegawa, K., and Shimuzu, I., 1987, *In vivo* and *in vitro* photoperiodic induction of diapause using isolated brain-suboesophageal ganglion complexes of the silkworm, *Bombyx mori, J. Insect Physiol.* **33:**959–966.

Hasegawa, K., and Yamashita, O., 1965, Studies on the mode of action of the diapause hormone in the silkworm, *Bombyx mori* L., VI. The target organ of the diapause hormone, *J. Exp. Biol.* **43:** 271–277.

Hasegawa, K., Isobe, M., and Goto, T., 1972, Highly purified diapause hormone from the silkworm, *Naturwissenschaften* **8:**364–365.

Hasegawa, K., Isobe, M., Kubota, I., and Goto, T., 1974, Activity of two species of the diapause hormone in the silkworm, *Bombyx mori, Zool. Jahrb. Abt. Allg. Zool. Physiol. Tiere* **78:**327–332.

Hashiguchi, T., 1960, On the relations between pupal color and ligature of a black pupa strain in the silkworm, *Bombyx mori, Jpn. J. Genet.* **35:**269–270.

Hashiguchi, T., 1964, Source of the hormone-like factor controlling the manifestation of the black pupa in the silkworm, *Bombyx mori* L., *Mem. Fac. Agric. Kagoshima Univ.* **5:**33–38.

Hashiguchi, T., Yoshitake, N., and Takahashi, N., 1965, Hormone determining the black pupal colour in the silkworm, *Bombyx mori* L., *Nature* **206:**215.

Hayes, T. K., and Keeley, L. L., 1981, Cytochromogenic factor: A newly discovered neuroendocrine agent stimulating mitochondrial cytochrome synthesis in the insect fat body, *Gen. Comp. Endocrinol.* **45:**115–124.

Hayes, T. K., and Keeley, L. L., 1983, Patterns and control of neurosecretion for the cytochromogenic hormone in *Blaberus discoidalis* cockroaches, *J. Exp. Biol.* **107:**49–58.

Hayes, T. K., and Keeley, L. L., 1985, Properties of an in vitro bioassay for hypertrehalosemic hormone of *Blaberus discoidalis* cockroaches, *Gen. Comp. Endocrinol.* **57:**246–256.

Hayes, T. K., Keeley, L. L., and Knight, D. W., 1986, Insect hypertrehalosemic hormone: Isolation and primary structure from *Blaberus discoidalis* cockroaches, *Biochem. Biophys. Res. Commun.* **140:**674–678.

Heidenhain, M., 1916, Über die Mallorysche Bindegewebsfärbung mit Karmin und Azokarmin als Vorfarben, *Z. Wiss. Mikrosk* **32:**361–372.

Heller, H., and Lederis, K., 1962, The distribution of vasopressin and oxytocin in hypothalamic nuclei, in *Neurosecretion,* 3d Symposium on Neurosecretion (H. Heller and R. B. Clark, eds.), Academic Press, London, pp. 227–239.

Henrich, V. C., Pak, M. D., and Gilbert, L. I., 1987, Neural factors that stimulate ecdysteroid synthesis by the larval ring gland of *Drosophila melanogaster, J. Comp. Physiol.* **157:** 543–549.

Hentschel, E., 1972, Ovulation und aminerges neurosekretorisches System bei *Periplaneta americana* (L.) (Blattoidea, Insecta), *Zool. Jahrb. Abt. Allg. Zool. Physiol. Tiere* **76:**356–367.

Herault, J. P. E., and Proux, J. P., 1987, Cyclic AMP: The second messenger of an antidiuretic hormone from the glandular lobes of the migratory locust corpora cardiaca, *J. Insect Physiol.* **33:**487–491.

Herault, J. P. E., Girardie, J., and Proux, J. P., 1985, Separation and characteristics of antidiuretic factors from the corpora cardiaca of the migratory locust, *Int. J. Invert. Reprod. Dev.* **8:**325–335.

Herault, J. P. E., Girardie, J., and Proux, J., 1988, Further characterization of the antidiuretic factor from the glandular part of the corpora cardiaca of the migratory locust, *Int. J. Invert. Reprod. Dev.* **13:**183–192.

Herman, W. S., 1967, The ecdysial glands of arthropods, *Int. Rev. Cytol.* **22:**209–347.

Herman, W. S., 1975, Juvenile hormone stimulation of tubular and accessory glands in male monarch butterflies, *Comp. Biochem. Physiol.* **51A:**507–510.

Herman, W. S., and Barker, J. F., 1976, Ecdysterone antagonism, mimicry, and synergism of juvenile hormone action on the monarch butterfly reproductive tract, *J. Insect Physiol.* **22:**643–648.

Herman, W. S., and Gilbert, L. I., 1966, The neuroendocrine system of *Hyalophora cecropia* (L.) (Lepidoptera: Saturniidae). 1. The anatomy and histology of the ecdysial glands, *Gen. Comp. Endocrinol.* **7:**275–291.

Hertel, W., 1971, Untersuchungen zur neurohormonalen Steuerung des Herzens der Amerikanischen Schabe *Periplaneta americana* (L.), *Zool. Jahrb. Abt. Allg. Zool. Physiol. Tiere* **76:**152–184.

Hertel, W., 1975, Untersuchungen zur Beeinflussung des Herzschlages von *Periplaneta americana* L. unter den Bedingungen der Kurzzeitkultur von Herzsegment-präparaten in vitro, *Zool. Jahrb. Abt. Allg. Zool. Physiol. Tiere* **79:**70–76.

Hertel, W., and Penzlin, H., 1986, Electrophysiological studies of the effect of the neuropeptide proctolin on the hyperneural muscle of *Periplaneta americana* (L.), *J. Insect Physiol.* **32:**239–248.

Hertel, W., Pass, G., and Penzlin, H., 1985, Electrophysiological investigation of the antennal heart of *Periplaneta americana* and its reactions to proctolin, *J. Insect Physiol.* **31:**563–572.

Hidaka, T., 1956, Recherches sur le déterminisme hormonal de la coloration pupale chez les lépidoptères. I. Les effets de la ligature, de l'ablation des ganglions et de l'incision des nerfs chez prépupes et larves âgées de quelques papilionides, *Annot. Zool. Jpn.* **29:**69–74.

Hidaka, T., 1961, Recherches sur le mécanisme endocrine de l'adaptation chromatique morphologique chez les nymphes de *Papilio xuthus* L., *J. Fac. Sci. Tokyo Univ.* **9:**223–261.

Hidaka, T., and Ohtaki, T., 1963, Effet de l'hormone juvénile et du farnésol sur la coloration tégumentaire de la nymphe de *Pieris rapae crucivora* Boisd., *C. R. Soc. Biol.* **167:**928–930.

Highnam, K. C., 1958, Activity of the brain/corpora cardiaca system during pupal/diapause "Break" in *Mimas tiliae* (Lepidoptera), *Q. J. Microsc. Sci.* **99:**73–88.

Highnam, K. C., 1961a, The histology of the neurosecretory system of the adult female desert locust, *Schistocerca gregaria, Q. J. Microsc. Sci.* **102:**27–38.

Highnam, K. C., 1961b, Induced changes in the amounts of material in the neurosecretory system of the desert locust, *Nature* **191:**199–200.

Highnam, K. C., 1962, Neurosecretory control of ovarian development in *Schistocerca gregaria, Q. J. Microsc. Sci.* **103:**57–72.

Highnam, K. C., 1976, L'activité neurosécrétrice chez les acridiens migrateurs, *Acrida* **5:**3–24.

Highnam, K. C., and Goldsworthy, G. J., 1972, Regenerated corpora cardiaca and hyperglycemic factor in *Locusta migratoria, Gen. Comp. Endocrinol.* **18:**83–88.

Highnam, K. C., and Mordue, A. J., 1970, Estimates of neurosecretory activity by an autoradiographic method in adult female *Schistocerca grejaria* (Forsk.), *Gen. Comp. Endocrinol.* **15:**31–38.

Highnam, K. C., and West, M. W., 1971, The neuropilar neurosecretory reservoir of *Locusta migratoria migratorioides* R. and F., *Gen. Comp. Endocrinol.* **16:**574–585.

Highnam, K. C., Lusis, O., and Hill, L., 1963, The role of the corpora allata during oöcyte growth in the desert locust, *Schistocerca gregaria* Forsk., *J. Insect Physiol.* **9:**587–596.

Highnam, K. C., Hill, L., and Gingell, D. J., 1965, Neurosecretion and water balance in the male desert locust *(Schistocerca gregaria), J. Zool.* **147:**201–215.

Highnam, K. C., Hill, L., and Mordue, W., 1966, The endocrine system and oöcyte growth in *Schistocerca* in relation to starvation and frontal ganglionectomy, *J. Insect Physiol.* **12:**977–994.

Hill, L., 1962, Neurosecretory control of haemolymph protein concentration during ovarian development in the desert locust, *J. Insect Physiol.* **8:**609–620.

Hill, L., 1963, Endocrine control of protein synthesis in female desert locusts *(Schistocerca gregaria), J. Endocrinol.* **26:**17–18.

Hill, L., and Izatt, M. E. G., 1974, The relationships between corpora allata and fat body and haemolymph lipids in the adult female desert locust, *J. Insect Physiol.* **20:**2143–2156.

Hinks, C. F., 1967, Relationship between serotonin and the circadian rhythm in some nocturnal moths, *Nature* **214:**386–387.

Hintze-Podufal, C., 1970, The innervation of the prothoracic glands of *Cerura vinula* L. (Lepidoptera), *Experientia* **26:**1269–1271.

Hiripi, L., and S.-Rozsa, K., 1973, Fluorimetric determination of 5-hydroxytryptamine and catecholamines in the central nervous system and heart of *Locusta migratoria migratorioides, J. Insect Physiol.* **19:**1481–1485.

Hiripi, L., S.-Rozsa, K., and Miller, T. A., 1979, The effect of proctolin on the adenylate and guanylate cyclases in the *Locusta* brain at various developmental stages, *Experientia* **35:**1287–1288.

Hiruma, K., 1986, Regulation of prothoracicotropic hormone release by juvenile hormone in the penultimate and last instar larvae of *Mamestra brassicae, Gen. Comp. Endocrinol.* **63:**201–211.

Hiruma, K., and Agui, N., 1977, Relationship between histological changes and functions of the neurosecretory cells in the brain of the cabbage armyworm, *Mamestra brassicae* L., *Appl. Entomol. Zool.* **12:**42–49.

Hiruma, K., and Agui, N., 1982, Larval-pupal transformation of the prothoracic glands of *Mamestra brassicae, J. Insect Physiol.* **28:**89–95.

Hiruma, K., and Riddiford, L. M., 1984, Regulation of melanization of tobacco hornworm larval cuticle in vitro, *J. Exp. Zool.* **230:**393–403.

Hiruma, K., and Riddiford, L. M., 1985, Hormonal regulation of dopa decarboxylase during a larval molt, *Dev. Biol.* **110:**509–513.

Hiruma, K., and Riddiford, L. M., 1986, Inhibition of dopa decarboxylase synthesis by 20-hydroxyecdysone during the last larval moult of *Manduca sexta, Insect Biochem.* **16:**225–231.

Hiruma, K., Shimada, H., and Yagi, S., 1978a, Activation of the prothoracic gland by juvenile hormone and prothoracicotropic hormone in *Mamestra brassicae, J. Insect Physiol.* **24:**215–220.

Hiruma, K., Yagi, S., and Agui, N., 1978b, Action of juvenile hormone on the cerebral neurosecretory cells of *Mamestra brassicae* in vivo and in vitro, *Appl. Entomol. Zool.* **13:**149–157.

Hiruma, K., Matsumoto, S., Isogai, A., and Suzuki, A., 1984, Control of ommochrome synthesis by both juvenile hormone and melanization hormone in the cabbage armyworm, *Mamestra brassicae, J. Comp. Physiol. B* **154:**13–21.

Hiruma, K., Riddiford, L. M., Hopkins, T. L., and Morgan, T. D., 1985, Roles of dopa decarboxylase and phenoloxidase in the melanization of the tobacco hornworm and their control by 20-hydroxyecdysone, *J. Comp. Physiol. B* **155:**659–669.

Hodek, I., Ruzicka, Z., and Sehnal, F., 1973, Termination of diapause by juvenoids in two species of ladybirds, *Experientia* **29:**1146–1147.

Hodgetts, R. B., Sage, B., and O'Connor, J. D., 1977, Ecdysone titers during postembryonic development of *Drosophila melanogaster, Dev. Biol.* **60:**310–317.

Hodgson, E. S., and Geldiay, S., 1959, Experimentally induced release of neurosecretory materials from roach corpora cardiaca, *Biol. Bull. (Woods Hole, Mass.)* **117:**275–283.

Hodgson, E. S., and Wright, A. M., 1963, Action of epinephrine and related compounds upon the insect nervous system, *Gen. Comp. Endocrinol.* **3:**519–525.

Hodkova, M., 1976, Nervous inhibition of corpora allata by photoperiod in *Pyrrhocoris apterus, Nature* **263:**521–523.

Hodkova, M., 1977, Function of the neuroendocrine complex in diapausing *Pyrrhocoris apterus* females, *J. Insect Physiol.* **23:**23–28.

Hodkova, M., 1979, Does the insect brain count larval instars? *Experientia* **35:**135–136.

Hoffmann, J., 1980, Ecdysone et reproduction chez les femelles adultes d'insectes, *Reprod. Nutr. Dev.* **20:**443–456.

Hoffmann, J. A., 1986, Introduction. Ten years of ecdysone workshops. Retrospect and perspectives, in *Ecdysone: From Metabolism to Regulation of Gene Expression,* VIIth Ecdysone Workshop, pp. 1–9, *Insect Biochem.* **16,** special issue.

Hoffmann, J. A., and Lagueux, M., 1985, Endocrinological aspects of embryogenesis, in *Comprehensive Insect Physiology, Biochemistry and Pharmacology* (G. A. Kerkut and L. I. Gilbert, eds.), Vol. 1, Pergamon Press, New York, pp. 435–460.

Hoffmann, K. H., and Behrens, W., 1982, Free ecdysteroids in adult male crickets, *Gryllus bimaculatus, Physiol. Entomol.* **7:**269–279.

Hökfelt, T., Elfvin, L. G., Elde, R., Schultzberg, M., Goldstein, M., and Luft, R., 1977, Occurrence of somatostatin-like immunoreactivity in some peripheral sympathetic noradrenergic neurons, *Proc. Natl. Acad. Sci. USA* **74:**3587–3591.

Hökfelt, T., Millhorn, D., Seroogy, K., Tsuruo, Y., Ceccatelli, S., Lindh, B., Meister, B., Melander, T., Schalling, M., Bartfai, T., and Terenius, L., 1987, Coexistence of peptides with classical neurotransmitters, *Experientia* **43:**768–776.

Holets, V. R., Hökfelt, T., Ude, J., Eckert, M., and Hansen, S., 1984, Coexistence of proctolin with TRH and 5-HT in the rat CNS, *Soc. Neurosci.* **10:**692.

Hollander, A. L., and Yin, C. M., 1985, Lack of humoral control in calling and pheromone release by brain, corpora cardiaca, corpora allata and ovaries of the female gypsy moth, *Lymantria dispar* (L.), *J. Insect Physiol.* **31:**159–163.

Holman, G. M., and Cook, B. J., 1970, Pharmacological properties of excitatory neuromuscular transmission in the hindgut of the cockroach, *Leucophaea maderae, J. Insect Physiol.* **16:**1891–1908.

Holman, G. M., and Cook, B. J., 1972, Isolation, partial purification and characterization of a peptide which stimulates the hindgut of the cockroach, *Leucophaea maderae* (Fabr.), *Biol. Bull. (Woods Hole, Mass.)* **142:**446–460.

Holman, G. M., and Cook, B. J., 1979, Evidence for proctolin and a second myotropic peptide in the cockroach, *Leucophaea maderae,* determined by bioassay and HPLC analysis, *Insect Biochem.* **9:**149–154.

Holman, G. M., and Cook, B. J., 1983, Isolation and partial characterization of a second myotropic peptide from the hindgut of the cockroach, *Leucophaea maderae, Comp. Biochem. Physiol.* **76C:**39–43.

Holman, G. M., Cook, B. J., and Wagner, R. M., 1984, Isolation and partial characterization of five myotropic peptides present in head extracts of the cockroach *Leucophaea maderae, Comp. Biochem. Physiol.* **77C:**1–5.

Holman, G. M., Cook, B. J., and Nachman, R. J., 1986a, Isolation, primary structure and synthesis of two neuropeptides from *Leucophaea maderae:* Members of a new family of cephalomyotropins, *Comp. Biochem. Physiol.* **84C:**205–211.

Holman, G. M., Cook, B. J., and Nachman, R. J., 1986b, Primary structure and synthesis of two additional neuropeptides from *Leucophaea maderae:* Members of a new family of cephalomyotropins, *Comp. Biochem. Physiol.* **84C:**271–276.

Holman, G. M., Cook, B. J., and Nachman, R. J., 1986c, Primary structure and synthesis of a blocked myotropic neuropeptide isolated from the cockroach, *Leucophaea maderae, Comp. Biochem. Physiol.* **85C:**219–224.

Holman, G. M., Cook, B. J., and Nachman, R. J., 1986d, Isolation, primary structure and synthesis of leucomyosuppressin, an insect neuropeptide that inhibits spontaneous contractions of the cockroach hindgut, *Comp. Biochem. Physiol.* **85C:**329–333.

Holman, G. M., Cook, B. J., and Nachman, R. J., 1987a, Isolation, primary structure and synthesis of leucokinins V and VI: Myotropic peptides of *Leucophaea maderae, Comp. Biochem. Physiol.* **88C:**27–30.

Holman, G. M., Cook, B. J., and Nachman, R. J., 1987b, Isolation, primary structure and synthesis of leucokinins VII and VIII: The final members of this new family of cephalomyotropic peptides isolated from head extracts of *Leucophaea maderae, Comp. Biochem. Physiol.* **88C:**31–34.

Holwerda, D. A., Van Doorn, J., and Beenakkers, A. M. T., 1977a, Characterization of the adipokinetic and hyperglycaemic substances from the locust corpus cadiacum, *Insect Biochem.* **7:**151–157.

Holwerda, D. A., Weeda, E., and Van Doorn, J., 1977b, Separation of the hyperglycemic and adipokinetic factors from the cockroach corpus cardiacum, *Insect Biochem.* **7:**477–481.

Hori, M., and Riddiford, L. M., 1982, Regulation of ommochrome biosynthesis in the tobacco hornworm, *Manduca sexta,* by juvenile hormone, *J. Comp. Physiol.* **147:**1–9.

Hori, M., Hiruma, K., and Riddiford, L. M., 1984, Cuticular melanization in the tobacco hornworm larva, *Insect Biochem.* **14:**267–274.

Houben, N. M. D., and Beenakkers, A. M. T., 1974, The influence of haemolymph carbohydrate concentration on the release of adipokinetic hormone during locust flight, *J. Endocrinol.* **64:**66.

Hoyle, G., and Barker, D. L., 1975, Synthesis of octopamine by insect dorsal median unpaired neurons, *J. Exp. Zool.* **193:**433–439.

Hoyle, G., Dagan, D., Moberly, B., and Colquhoun, W., 1974, Dorsal unpaired median insect neurons make neurosecretory endings on skeletal muscle, *J. Exp. Zool.* **187:**159–165.

Hrubéšová, H., and Sláma, K., 1967, The effect of hormones on the intestinal proteinase activity of adult *Pyrrhocoris apterus* L. (Hemiptera), *Acta Entomol. Bohemoslav.* **64:**175–183.

Hsiao, C., and Hsiao, T. H., 1969, Insect hormones: Their effect on diapause and development of *Hymenoptera, Life Sci.* **8:**767–774.

Hsiao, T. H., Hsiao, C., and De Wilde, J., 1975, Moulting hormone production in the isolated larval abdomen of the Colorado beetle, *Nature* **255:**727–728.

Huebner, E., and Davey, K. G., 1973, An antigonadotropin from the ovaries of the insect *Rhodnius prolixus* Stal., *Can. J. Zool.* **51:**113–120.

Huebner, E., and Injeyan, H. S., 1980, Patency of the follicular epithelium in *Rhodnius prolixus:* A re-examination of the hormone response and technique refinement, *Can. J. Zool.* **58:**1617–1625.

Huet, C., 1971, Différenciation en culture in vitro des ébauches présomptives de l'appareil génital femelle de *Tenebrio molitor* L., en présence d'ecdysone, *C.R. Acad. Sci. Ser. D* **272:**1896–1899.

Hughes, J., 1975, Isolation of an endogenous compound from the brain with pharmacological properties similar to morphine, *Brain Res.* **88:**295–305.

Hughes, P. R., and Renwick, J. A. A., 1977a, Neural and hormonal control of pheromone biosynthesis in the bark beetle, *Ips paraconfusus, Physiol. Entomol.* **2:**117–123.

Hughes, P. R., and Renwick, J. A. A., 1977b, Hormonal and host factors stimulating pheromone synthesis in female western pine beetles *Dentrotonus brevicomis, Physiol. Entomol.* **2:**289–292.

Hugues, L., 1977, High molecular weight forms of diuretic hormone from *Rhodnius prolixus, Biochem. Soc. Trans.* **5:**1060–1063.

Hugues, L., 1979, Further investigation of the isolation of diuretic hormone from *Rhodnius prolixus, Insect Biochem.* **9:**247–255.

Huignard, J., 1964, Recherches histophysiologiques sur le contrôle hormonal de l'ovogénèse chez *Gryllus domesticus* L., *C.R. Acad. Sci.* **259:**1557–1560.

Huignard, J., 1968, Influence de la durée de la copulation sur la fécondité des femelles chez *Acanthoscelides obtectus* Say (Insecte Coléoptère), *C.R. Acad. Sci.* **266:**1311–1313.

Huignard, J., Quesneau-Thierry, A., and Barbier, M., 1977, Isolement, action biologique et évolution des substances paragoniales contenues dans le spermatophore d'*Acanthoscelides obtectus* (Coléoptère), *J. Insect Physiol.* **23:**351–357.

Hustert, R., and Topel, U., 1986, Location and major postembryonic changes of identified 5-HT-immunoreactive neurones in the terminal ganglion of a cricket *(Acheta domesticus), Cell Tissue Res.* **245:**615–621.

Huybrechts, R., and De Loof, A., 1977, Induction of vitellogenin synthesis in male *Sarcophaga bullata* by ecdysterone, *J. Insect Physiol.* **23:**1359–1362.

Huybrechts, R., and De Loof, A., 1981, Effect of ecdysterone on vitellogenin synthesis in male *Sarcophaga bullata, Int. J. Invert. Reprod.* **3:**157–168.

Ichikawa, M., and Ishizaki, H., 1961, Brain hormone of the silkworm, *Bombyx mori, Nature* **191:**933–934.

Ichikawa, M., and Ishizaki, H., 1963, Protein nature of the brain hormone of insects, *Nature* **198:**308–309.

Ichikawa, M., and Nishiitsutsuji-Uwo, J., 1959, Studies on the role of the corpus allatum in the erisilkworm, *Philosamia cynthia ricini, Biol. Bull. (Woods Hole, Mass.)* **116:**88–94.

Ichimasa, Y., 1976, Sterol accumulation in developing ovaries of the silkworm *Bombyx mori* in relation to the diapause hormone, *J. Insect Physiol.* **22:**1071–1074.

Ichimasa, Y., and Hasegawa, K., 1973, Studies on the mode of action of the diapause hormone with special reference to lipid metabolism in the silkworm, *Bombyx mori* L. I. Effect of the hormone on lipid in pupal ovaries and matured eggs, *J. Seric. Sci. Tokyo* **42:**379–392.

Ikémoto, H., 1983, Further studies on the role of juvenile hormone in the density-related color variation in larvae of *Cephonodes hylas* L. (Lepidoptera: Sphingidae), *Appl. Ent. Zool.* **18:**438–439.

Ilenchuk, T. T., and Davey, K. G., 1982, Some properties of Na^+-K^+–ATPase in the follicle cells of *Rhodnius prolixus, Insect Biochem.* **12:**675–679.

Ilenchuk, T. T., and Davey, K. G., 1985, The binding of juvenile hormone to membranes of follicle cells in the insect *Rhodnius prolixus, Can. J. Biochem. Cell Biol.* **63:**102–105.

Ilenchuk, T. T., and Davey, K. G., 1987, Effects of various compounds on Na/K–ATPase activity, JH I binding capacity and patency response in follicles of *Rhodnius prolixus, Insect Biochem.* **17:**1085–1088.

Imboden, H., Lanzrein, B., Delbecque, J. P., and Lüscher, M., 1978, Ecdysteroids and juvenile hormone during embryogenesis in the ovoviviparous cockroach *Nauphoeta cinerea, Gen. Comp. Endocrinol.* **36:**628–635.

Irving, S. N., and Miller, T. A., 1980, Octopamine and proctolin mimic spontaneous membrane depolarisations in *Lucilia* larvae, *Experientia* **36:**566–568.

Ishay, J., Gitter, S., Galun, R., Doron, M., and Laron, Z., 1976, The presence of insulin in and some effects of exogenous insulin on hymenoptera tissues and body fluids, *Comp. Biochem. Physiol.* **54:**203–206.

Ishizaki, H., 1969, Changes in titer of the brain hormone during development of the silkworm, *Bombyx mori, Dev. Growth Differ.* **11:**1–7.

Ishizaki, H., and Ichikawa, M., 1967, Purification of the brain hormone of the silkworm *Bombyx mori, Biol. Bull. (Woods Hole, Mass.)* **133:**355–368.

Ishizaki, H., Suzuki, A., Isogai, A., Nagasawa, H., and Tamura, S., 1977, Enzymatic and chemical inactivation of partially purified prothoracicotropic (brain) hormone of the silkworm, *Bombyx mori, J. Insect Physiol.* **23:**1219–1222.

Ishizaki, H., Suzuki, A., Moriya, I., Mizoguchi, A., Fujishita, M., O'Oka, H., Kataoka, H., and Suzuki, A., 1983a, Prothoracicotropic hormone bioassay: Pupal-adult *Bombyx* assay, *Dev. Growth Differ.* **25:**585–592.

Ishizaki, H., Mizoguchi, A., Fujishita, M., Suzuki, A., Moriya, I., O Oka, H., Kataoka, H., Isogai, A., Nagasawa, H., Tamura, S., and Suzuki, A., 1983b, Species specificity of the insect prothoracicotropic hormone (PTTH): The presence of *Bombyx*- and *Samia*-specific PTTHs in the brain of *Bombyx mori, Dev. Growth Differ.* **25:**593–600.

Islam, A., and Roy, S., 1982, Effects of epinephrine, JH and corpora cardiaca extracts on the level of various sugars in *Lohita grandis* Gray (Hemiptera: Pyrrhocoridae), *Proc. Indian Natl. Sci. Acad. B.* **48:**31–34.

Ismail, S., Porcheron, P., and Fuzeau-Braesch, S., 1979, Taux des ecdystéroides chez des adultes de *Locusta migratoria cinerascens* solitaires, grégaires et traités par le gaz carbonique, *C. R. Soc. Biol.* **173:**553–557.

Ismail, S., Baehr, J. C., Porcheron, P., and Fuzeau-Braesch, S., 1984, The role of juvenile hormones in the larval diapause of the cricket *Gryllus campestris* (Orthoptera), *Comp. Biochem. Physiol.* **77A:**383–387.

Isobe, M., Hasegawa, K., and Goto, T., 1973, Isolation of the diapause hormone from the silkworm, *Bombyx mori, J. Insect Physiol.* **19:**1221–1239.

Isobe, M., Hasegawa, K., and Goto, T., 1975, Further characterization of the silkworm diapause hormone, *J. Insect Physiol.* **21:**1917–1920.

Itagaki, H., and Conner, W. E., 1986, Physiological control of pheromone release behaviour in *Manduca sexta* (L.), *J. Insect Physiol.* **32:**657–664.

Ittycheriah, P. I., and Marks, E. P., 1971, Performic acid-resorcin fuchsin: A technique for the in situ demonstration of neurosecretory material in insects, *Ann. Entomol. Soc. Am.* **64:**762–765.

Ittycheriah, P. I., and Nayar, K. K., 1968, Ovarian response to corpus allatum in *Iphita limbata* Stal., *Curr. Sci.* **36:**608–609.

Ivanović, J. P., Janković-Hladni, M. I., Stanić, V., Milanović, M. P., and Nenadović, V., 1976, The midgut of coleopteran larvae, the possible target organ for the action of neurohormones, in: *7th International Symposium on Neurosecretion, Leningrad,* Springer-Verlag, Berlin, Heidelberg, New York, p. 72.

Ivanović, J. P., Janković-Hladni, M. I., Milanović, M. P., Stanić, V., and Bozidarac, M., 1982, Possible role of neurohormones in the process of acclimatization and acclimation in *Morimus funereus* (Insecta). II. Changes in the haemolymph free amino-acid content and digestive enzyme activity, *Comp. Biochem. Physiol.* **71B:**695–701.

Ivanović, J. P., Janković-Hladni, M. I., Spasić, V., and Frusić, M., 1987, Compensatory reactions at the level of digestive enzymes in relation to acclimatization in *Morimus asper funereus* larvae, *Comp. Biochem. Physiol.* **86A:**217–224.

Iwanaga, T., Fujita, T., Nishiitsutsuji-Uwo, J., and Endo, Y., 1981, Immunohistochemical demonstration of PP-, somatostatin-, enteroglucagon- and VIP-like immunoreactivities in the cockroach midgut, *Biomed. Res.* **2:**202–207.

Iwanaga, T., Fujita, T., Takeda, N., Endo, Y., and Lederis, K., 1986, Urotensin I-like immunoreactivity in the midgut endocrine cells of the insects *Gryllus bimaculatus* and *Periplaneta americana, Cell Tissue Res.* **244:**565–568.

Izumi, S., Kiguchi, K., and Tomino, S., 1984, Hormonal regulation of biosynthesis of major plasma proteins in *Bombyx mori, Zool. Sci.* **1:**223–228.

Jaffé, H., Raina, A. K., Riley, C. T., Fraser, B. A., Holman, G. M., Wagner, R. M., Ridgway, R. L., and Hayes, D. K., 1986, Isolation and primary structure of a peptide from the corpora cardiaca of *Heliothis zea* with adipokinetic activity, *Biochem. Biophys. Res. Comm.* **135:**622–628.

Jaffé, H., Raina, A. K., Riley, C. T., Fraser, B. A., Bird, T. G., Tseng, C. M., Zhang, Y. S., and Hayes, D. K., 1988a, Isolation and primary structure of a neuropeptide hormone from *Heliothis zea* with hypertrehalosemic and adipokinetic activities, *Biochem. Biophys. Res. Commun.* **155:** 344–350.

Jaffé, H., Raina, A. K., Fraser, B. A., Keim, P., Ranga Rao, K., Zhang, Y. S., Lancaster, J. L., and Hayes, D. K., 1988b, Isolation of two neuropeptides in the AKH/RPCH-family from horseflies (*Diptera*), *Biochem. Biophys. Res. Comm.* **151:**656–663.

Janda, V., and Sláma, K., 1965, Über den Einfluss von Hormonen auf den Glykogen-Fett und Stickstoffmetabolismus bei Adulten *Pyrrhocoris apterus* L., *Zool. Jahrb. Abt. Allg. Zool. Physiol. Tiere* **71:**345–358.

Janković-Hladni, M. I., Stanić, V., and Ivanović, J. P., 1976, The effect of temperature and crowding on midgut amylolytic activity in *Tenebrio molitor* adults, *J. Insect Physiol.* **22:**851–853.

Janković-Hladni, M. I., Ivanovic, J. P., Stanic, V., and Milanovic, M. P., 1978, Possible role of hormones in the control of midgut amylolytic activity during adult development of *Tenebrio molitor, J. Insect Physiol.* **24:**61–63.

Jaros, P. P., and Gäde, G., 1982, Evidence for a crustacean hyperglycemic hormone-like molecule in the nervous system of the stick insect, *Carausius morosus, Cell Tissue Res.* **227:**555–562.

Johansson, A. S., 1958, Relation of nutrition to endocrine-reproductive functions in the milkweed bug *Oncopeltus fasciatus* (Dallas) (Heteroptera: *Lygaeidae), Nytt. Mag. Zool.* **7:**1–132.

Johansson, O., Olsson, A., and Wieslander, L., 1986, The distribution of putative neurotransmitters in the nervous system of the dipteran *Chironomus tentans* insect larva: An immunohistochemical study using antisera to 5-hydroxytryptamine, tyrosin hydroxylase, methionin-enkephalin, proctolin and bombesin, *Neurochem. Int.* **8:**311–326.

Johnson, B., 1963, A histological study of neurosecretion in aphids, *J. Insect Physiol.* **9:**727–739.

Johnson, B., 1966, Fine structure of the lateral cardiac nerves of the cockroach *Periplaneta americana* L., *J. Insect Physiol.* **12:**645–653.

Johnson, G. D., Stay, B., and Rankin, S. M., 1985, Ultrastructure of corpora allata of known activity during the vitellogenic cycle in the cockroach *Diploptera punctata, Cell Tissue Res.* **239:**317–327.

Joly, L., 1971, Action de l'implantation de glandes ventrales sur l'activité génitale des imagos de *Locusta migratoria* L., *C.R. Acad. Sci. Ser. D* **272:**2326–2329.

Joly, L., and Joly, P., 1975, Interaction entre l'hormone juvénile et l'ecdysone chez *Locusta migratoria* L., 3rd Coll. Physiologie des Insectes, Strasbourg, 2–3 mai 1975.

Joly, L., Porte, A., and Girardie, A., 1967, Caractères ultrastructuraux des corpora allata actifs et inactifs chez *Locusta migratoria, C.R. Acad. Sci.* **265:**1633–1635.

Joly, L., Joly, P., and Lagueux, M., 1977, Sur l'existence d'une seconde fonction physiologique de la glande prothoracique des insectes, *C.R. Acad. Sci.* **285:**543–546.

Joly, L., Goltzene, F., and Porte, A., 1978, L'action exercée par l'appareil endocrine sur les phases tardives de l'ovogenèse chez *Locusta migratoria, J. Insect Physiol.* **24:**187–193.

Joly, P., 1945, La fonction ovarienne et son contrôle humoral chez les Dytiscides, *Arch. Zool. Exp. Gen.* **84:**49–164.

Joly, P., 1951, Déterminisme endocrine de la pigmentation chez *Locusta migratoria* L., *C.R. Soc. Biol.* **145:**1362–1364.

Joly, P., 1952, Déterminisme de la pigmentation chez *Acrida turrita* L. (Insecte Orthoptéroïde), *C.R. Acad. Sci.* **235:**1054–1056.

Joly, P., and Joly, L., 1953, Résultats de greffes de corpora allata chez *Locusta migratoria* L., *Ann. Sci. Nat. Zool.* **11:**331–345.

Joly, P., Joly, L., and Halbwachs, M., 1956, Contrôle humoral du développement chez *Locusta migratoria, Ann. Sci. Nat. Zool.* **11:**257–261.

Jones, D., Jones, G., Wing, K. D., Rudnicka, M., and Hammock, B. D., 1982, Juvenile hormone esterases of Lepidoptera. I. Activity in the hemolymph during the last larval instar of 11 species, *J. Comp. Physiol. B* **148:**1–10.

Jones, G., Wing, K. D., Jones, D., and Hammock, B. D., 1981, The source and action of head factors regulating juvenile hormone esterase in larvae of the cabbage looper, *Trichoplusia ni, J. Insect Physiol.* **27:**85–91.

Jones, G., Click, A., Reck-Malleczewen, V., Loeb, M. J., Brandt, E. P., and Woods, C. W., 1986, Haemolymph ecdysteroid titre in larvae of the cabbage looper *Trichoplusia ni,* and its modification by juvenile hormone, *J. Insect Physiol.* **32:**561–566.

Jones, J., Stone, J. V., and Mordue, W., 1977, The hyperglycaemic activity of locust adipokinetic hormone, *Physiol. Entomol.* **2:**185–187.

Jones, T., 1978, The blood/germ cell barrier in male *Schistocerca gregaria:* The time of its establishment and factors affecting its formation, *J. Cell Sci.* **31:**145–153.

Joose, J., 1984, Recent progress in endocrinology of molluscs, in *Biosynthesis, Metabolism and Mode of Action of Invertebrate Hormones* (J. Hoffmann and M. Porchet, eds.), Springer-Verlag, Berlin, pp. 5–9.

Jowett, T., and Postlethwait, J. H., 1980, The regulation of yolk polypeptide synthesis in *Drosophila* ovaries and fat body by 20-hydroxyecdysone and a juvenile hormone analog, *Dev. Biol.* **80:** 225–234.

Judy, K. J., 1972, Diapause termination and metamorphosis in brainless tobacco hornworms (Lepidoptera), *Life Sci.* **11:**605–611.

Judy, K. J., Schooley, D. A., Dunham, L. L., Hall, M. S., Bergot, B. J., and Siddall, J. B., 1973, Isolation, structure and absolute configuration of a new natural insect juvenile hormone from *Manduca sexta, Proc. Natl. Acad. Sci. USA* **70:**1509–1513.

Jungreis, A. M., Ruhoy, M., and Cooper, P. D., 1982, Why don't tobacco hornworms (*Manduca sexta*) become dehydrated during larval-pupal and pupal-adult development, *J. Exp. Zool.* **222:** 265–276.

Jutsum, A. R., and Goldsworthy, G. J., 1975, Some problems of homeostasis of haemolymph metabolites in the locust, *J. Insect Physiol.* **21:**1899–1908.

Jutsum, A. R., Agarwal, H. C., and Goldsworthy, G. J., 1975, Starvation and haemolymph lipids in *Locusta migratoria migratorioides* (R and F)., *Acrida* **4:**47–56.

Kaiser, H., 1979, Nachweis von Neurosekretspeichern bei der Eintagsfliege *Ephemera danica* Müll. (Ephemeroptera: Ephemeridae), *Eur. J. Cell. Biol.* **19:**145–148.

Kambysellis, M. P., and Heed, W. B., 1974, Juvenile hormone induces ovarian development in diapausing cave-dwelling *Drosophila* species, *J. Insect Physiol.* **20:**1779–1786.

Kambysellis, M. P., and Williams, C. M., 1971a, In vitro development of insect tissues, I. A macromolecular factor prerequisite for silkworm spermatogenesis, *Biol. Bull. (Woods Hole, Mass.)* **141:**527–540.

Kambysellis, M. P., and Williams, C. M., 1971b, In vitro development of insect tissues, II. The role of ecdysone in the spermatogenesis of silkworms, *Biol. Bull. (Woods Hole, Mass.)* **141:**541–552.

Kambysellis, M. P., and Williams, C. M., 1972, Spermatogenesis in cultured testes of the *Cynthia* silkworm: Effects of ecdysone and of prothoracic glands, *Science* **175:**769–770.

Kamm, J., and Swenson, K. G., 1972, Termination of diapause in *Draeculacephalus crassicornis* with synthetic juvenile hormone, *J. Econ. Ent.* **65:** 364–367.

Kannan, S., and Prabhu, V. K. K., 1985, Probable role of brain-neurosecretory cells in water balance in *Oryctes rhinoceros* (Coleoptera: Scarabaeidae), *Entomon* **14:**313–315.

Kappler, C., Goltzene, F., Lagueux, M., Hetru, C., and Hoffmann, J. A., 1986, Role of the follicle cells and the oöcytes in ecdysone biosynthesis and esterification in vitellogenic females of *Locusta migratoria, Int. J. Invert. Reprod.* **9:**17–34.

Karlinsky, A., 1963, Effets de l'ablation des corpora allata imaginaux sur le développement ovarien de *Pieris brassicae* L. (Lépidoptère), *C.R. Acad. Sci.* **256:**4101–4103.

Karlinsky, A., 1967, Reprise de la vitellogenèse après implantation de corpora allata chez *Pieris brassicae* L. (Lépidoptère), *C.R. Acad. Sci. Ser. D.* **264:**1735–1738.

Kataoka, H., Nagasawa, H., Isogai, A., Tamura, S., Mizoguchi, A., Fujiwara, Y., Suzuki, C., Ishizaki, H., and Suzuki, A., 1987, Isolation and partial characterization of a prothoracicotropic hormone of the silkworm, *Bombyx mori, Agric. Biol. Chem.* **51:**1067–1076.

Kater, S. B., 1968, Cardioaccelerator release in *Periplaneta americana* (L.), *Science* **160:**765–767.

Käuser, G., Brandtner, H. M., Bidmon, H. J., and Koolman, J., 1988, Ecdysone synthesis and release by the brain–ring gland complex of blowfly larvae, *J. Insect Physiol.* **34:** 563–569.

Keeley, L. L., 1971, Endocrine effects on the biochemical properties of fat body mitochondria from the cockroach, *Blaberus discoidalis, J. Insect Physiol.* **17:**1501.

Keeley, L. L., and Friedman, S., 1967, Corpus cardiacus as a metabolic regulator in *Blaberus discoidalis* Serville (Blattidae). I. Long-term effects of cardiatectomy on whole body and tissue respiration and on tropic metabolism, *Gen. Comp. Endocrinol.* **8:**129–134.

Kelly, T. J., and Davenport, R., 1976, Juvenile hormone induced ovarian uptake of a female-specific blood protein in *Oncopeltus fasciatus, J. Insect Physiol.* **22:**1381–1393.

Kelly, T. J., Woods, C. W., Redfern, R. E., and Borkovec, A. B., 1981a, Makisterone A: The molting hormone of larval *Oncopeltus*? *J. Exp. Zool.* **218:**127–132.

Kelly, T. J., Fuchs, M. S., and Kang, S. H., 1981b, Induction of ovarian development in autogenous *Aedes atropalpus* by juvenile hormone and 20-hydroxyecdysone, *Int. J. Invert. Reprod.* **3:**101–112.

Kelly, T. J., Whisenton, L. R., Katahira, E. J., Fuchs, M. S., Borkovec, A. B., and Bollenbacher, W. E., 1986a, Inter-species cross-reactivity of the prothoracicotropic hormone of *Manduca sexta* and egg-development neurosecretory hormone of *Aedes aegypti, J. Insect Physiol.* **32:** 757–762.

Kelly, T. J., Masler, E. P., Schwartz, M. B., and Haught, S. B., 1986b, Inhibitory effects of oöstatic hormone on ovarian maturation and ecdysteroid production in Diptera, *Insect Biochem.* **16:**273–279.

Keshishian, H., and O'Shea, M., 1985, The distribution of a peptide neurotransmitter in the postembryonic grasshopper central nervous system, *J. Neurosci.* **5:**992–1004.

Key, K. H. L., 1957, Kentromorphic phases in three species of Phasmatodea, *Aust. J. Zool.* **5:**247–284.

Khan, M. A., and Buma, P., 1985, Neural control of the corpus allatum in the Colorado potato beetle, *Leptinotarsa decemlineata:* An electron microscope study utilizing the *in vitro* tannic acid ringer incubation method, *J. Insect Physiol.* **31:**639–645.

Khan, M. A., Koopmanschap, A. B., and De Kort, C. A. D., 1982, The effects of juvenile hormone, 20-hydroxyecdysone and precocene II on activity of corpora allata and the mode of negative-feedback regulation of these glands in the adult Colorado potato beetle, *J. Insect Physiol.* **28:**995–1001.

Khan, M. A., Koopmanschap, A. B., and De Kort, C. A. D., 1983, The relative importance of nervous and humoral pathways for control of corpus allatum activity in the adult Colorado potato beetle, *Leptinotarsa decemlineata* (Say), *Gen. Comp. Endocrinol.* **52:**214–221.

Khan, M. A., Romberg-Privee, H. M., and Schooneveld, H., 1984, Innervation of the corpus allatum in the Colorado potato beetle as revealed by retrograde diffusion with horse-radish peroxidase, *Gen. Comp. Endocrinol.* **5:**66–73.

Khan, M. A., Romberg-Privee, H. M., and Koopmanschap, A. B., 1986, Location of allatostatic centers in the pars lateralis regions of the brain of the Colorado potato beetle, *Experientia* **42:**836–838.

Khan, T. R., Singh, S. B., Singh, R. K., and Singh, T. K., 1978, Neurosecretory control of corpora allata activity in cockroach, *Periplaneta americana* L., *Experientia* **34:**49–51.

Khurad, A. M., and Thakare, V. K., 1983, Changes in the neuroendocrine system during post-embryonic development in the buffalo-fly, *Lyperosia exigua* (De Meijere) (Diptera: Muscidae), *J. Hirnforsch.* **23:**31–38.

Kiguchi, K., 1972, Hormonal control of the coloration of larval body and the pigmentation of larval markings in *Bombyx mori* (1) Endocrine organs affecting the coloration of larval body and the pigmentation of markings, *J. Seric. Sci. Tokyo* **41:**407–412.

Kiguchi, K., and Agui, N., 1981, Ecdysteroid levels and developmental events during larval moulting in the silkworm, *Bombyx mori, J. Insect Physiol.* **27:**805–812.

Kind, T. V., 1969, The role of neurosecretory cells of the suboesophageal ganglion in determination of embryonic diapause in *Orgyia antiqua* L., *Dokl. Akad. Nauk. SSSR* **187:**226–229.

Kind, T. V., 1976, Neurosecretory cells of the brain of *Mamestra brassicae* (Lepidoptera, Noctuidae) during metamorphosis and diapause development, in 7th *International Symposium on Neurosecretion,* Leningrad, Springer-Verlag, p. 81.

Kind, T. V., 1978, Dynamics of activity of the brain neurosecretory cells in *Pieris brassicae* under the photoperiodic control of development and diapause of pupae, *Zool. Zh.* **57:**1668–1680.

Kind, T. V., and Vaghina, N. F., 1976, The neurosecretory system of the brain in diapausing and developing pupae of *Acronycta rumicis* L. (Lepidoptera, *Noctuidae*), *Rev. Entomol. URSS* **55:** 286–299.

Kingan, T., and Titmus, M., 1983, Radioimmunologic detection of proctolin in arthropods, *Comp. Biochem. Physiol.* **74C:**75–78.

Kirby, P., Beck, R., and Clarke, K. U., 1984, The stomatogastric nervous system of the house cricket *Acheta domesticus* L. I. The anatomy of the system and the innervation of the gut, *J. Morphol.* **180:**81–103.

Kirchner, E., 1962, Untersuchungen über neurohormonale Faktoren bei *Melolontha vulgaris, Zool. Jahrb. Abt. Allg. Zool. Physiol. Tiere* **136:**43–62.

Kiss, T., Varanka, I., and Benedeczky, I., 1984, Neuromuscular transmission in the visceral muscle of locust oviduct, *Neuroscience* **12:**309–322.

Kleinhotz, L. H., and Keller, R., 1979, Endocrine regulation in crustacea, in *Hormone and Evolution* (E. J. W. Barrington, ed.), Vol. 1, Academic Press, New York, pp. 160–230.

Klemm, N., 1968, Monoaminhaltige Strukturen im Zentralnervensystem der Trichoptera (Insecta), *Z. Zellforsch. Mikrosk. Anat.* **92:**487–502.

Klemm, N., 1971, Monoaminhaltige Zellelemente im stomatogastrischen Nervensystem und in den Corpora cardiaca von *Schistocerca gregaria* Forsk., (Insecta, Orthoptera), *Z. Naturforsch.* **26b:** 1085–1086.

Klemm, N., 1972, Monoamine-containing fibres in fore gut and salivary gland of the desert locust, *Schistocerca gregaria* Forsk. (Orthoptera, Acrididae), *Comp. Biochem. Physiol.* **43A:** 207–211.

Klemm, N., 1983, Monoamine-containing neurons and their projections in the brain (supraoesophageal ganglion) of cockroaches, *Cell Tissue Res.* **229:**379–402.

Klemm, N., 1968, Monoaminhaltige Strukturen in Zentralneroensystem der Trichoptera (Insecta), *Z. Zellforsch. Mikrosk. Anat.* **92:**487–502.

Klemm, N., 1976, Histochemistry of putative transmitter substances in the insect brain, *Prog. Neurobiol.* **7:**99–169.

Klemm, N., and Axelsson, S., 1973, Determination of dopamine, noradrenaline, 5-hydroxytryptamine in the cerebral ganglion of the desert locust, *Schistocerca gregaria* Forsk. (Insecta, Orthoptera), *Brain Res.* **57:**289–298.

Klemm, N., and Falck, B., 1978, Monoamines in the pars intercerebralis-corpus cardiacum complex of locusts, *Gen. Comp. Endocrinol.* **34:**180–192.

Klemm, N., Hustert, R., Cantera, R., and Nässel, D. R., 1986, Neurons reactive to antibodies against serotonin in the stomatogastric nervous system and in the alimentary canal of locust and crickets (Orthoptera, Insecta), *Neurosci.* **17:**247–261.

Klowden, M. J., Davis, E. E., and Bowen, M. F., 1987, Role of the fat body in the regulation of host-seeking behaviour in the mosquito, *Aedes aegypti, J. Insect Physiol.* **33:**643–646.

Knowles, F. G. W., and Bern, H. A., 1966, The function of neurosecretion in endocrine regulation, *Nature* **210:**271–272.

Knowles, F. G. W., and Vollrath, L., 1966a, Neurosecretory innervation of the pituitary of the eels *Anguilla* and *Conger*. I. The structure and ultrastructure of the neuro-intermediate lobe under normal and experimental conditions, *Phil. Trans. B.* **250:**311–327.

Knowles, F. G. W., and Vollrath, L., 1966b, Neurosecretory innervation of the pituitary of the eels *Anguilla* and *Conger*. II. The structure and innervation of the pars distalis at different stages of the life-cycle, *Phil. Trans. B.* **250:**329–342.

Knowles, F. G. W., Carlisle, D. B., and Dupont-Raabe, M., 1955, Studies on pigment activating substances in animals. I. The separation by paper electrophoresis of chromactivating substances in arthropods, *J. Mar. Biol. Ass. UK* **34:**611–635.

Kobayashi, M., 1957, Studies on the neurosecretion in the silkworm, *Bombyx mori* L., *Bull. Ser. Exp. Stat.* **15:**181–273.

Kobayashi, M., and Kimura, S., 1967, Action of ecdysone on the conversion of 14C-glucose in dauer pupa of the silkworm, *Bombyx mori, J. Insect Physiol.* **13:**545–552.

Kobayashi, M., and Kirimura, J., 1958, The "brain" hormone in the silkworm, *Bombyx mori, Nature* **181:**1217–1218.

Kobayashi, M., and Muneoka, Y., 1986, Structural requirements for FMRFamide–like activity on the heart of the prosobranch *Rapana thomasiana, Comp. Biochem. Physiol.* **84C:**349–352.

Kobayashi, M., and Yamashita, Y., 1959, A function of corpus allatum in neurosecretory system in the silkworm, *Bombyx mori, J. Seric. Sci. Jpn.* **28:**335–339.

Kobayashi, M., Fukaya, M., and Mitsuhashi, J., 1960, Imaginal differentiation of "Dauer-pupae" in the silkworm, *Bombyx mori, J. Seric. Sci. Tokyo* **24:**337–340.

Kobayashi, M., Kirimura, J., and Saito, M., 1962, The "brain" hormone in an insect, *Bombyx mori* L., *Mushi* **36:**85–86.

Kobayashi, M., Ishitoya, Y., and Yamazaki, M., 1968, Action of proteinic brain hormone to the prothoracic gland in an insect, *Bombyx mori* L. (Lepidoptera: Bombycidae), *Appl. Entomol. Zool.* **3:**150–152.

Koch, P. B., and Bückmann, D., 1984, Vergleichende Untersuchung der Farbmuster und der Farbanpassung von Nymphalidenpuppen (Lepidoptera), *Zool. Beitr. N F* **28:**369–401.

Koch, P. B., and Bückmann, D., 1987, Hormonal control of seasonal morphs by the timing of ecdysteroid release in *Araschnia levana* L. (Nymphalidae: Lepidoptera), *J. Insect Physiol.* **33:** 823–829.

Koeppe, J., and Ofengand, J., 1976, Juvenile hormone-induced biosynthesis of vitellogenin in *Leucophaea maderae, Arch. Biochem. Biophys.* **173:**100–113.

Köller, G., 1948, Rhythmische Bewegung und hormonale Steuerung bei den Malpighischen Gefässen der Insekten, *Biol. Zentralbl.* **67:**201–211.

Köller, G., 1954, Zur Frage der hormonalen Steuerung bei rhythmischen Eingeweidebewegungen von Insekten, *Verh. Dtsch. Zool. Ges.* **27:**417–422.

König, R., and Lanzrein, B., 1985, Binding of vitellogenin to specific receptors in oöcyte membrane preparations of the ovoviviparous cockroach *Nauphoeta cinerea. Insect Biochem.* **15:**735–747.

Konings, P. N. M., Vullings, H. G. B., Siebinga, R., Diederen, J. H. B., and Jansen, W. F., 1988a, Serotonin-immunoreactive neurones in the brain of *Locusta migratoria* innervating the corpus cardiacum, *Cell Tissue Res.* **254:**147–153.

Konings, P. N. M., Vullings, H. G. B., Geffard, M., Buijs, R. M., Diederen, J. H. B., and Jansen, W. F., 1988b, Immunocytochemical demonstration of octopamine-immunoreactive cells in the nervous system of *Locusta migratoria* and *Schistocerca gregaria, Cell Tissue Res.* **251:**371–379.

Kono, Y., 1973, Light and electron microscopic studies on the neurosecretory control of diapause incidence in *Pieris rapae crucivora, J. Insect Physiol.* **19:**255–272.

Kono, Y., 1977, Ultrastructural changes of neurosecretory cells in the pars intercerebralis during diapause development in *Pieris rapae, J. Insect Physiol.* **23:**1461–1473.

Kono, Y., Kobayashi, M., and Claret, J., 1983, A putative photoreceptor-organelle in insect brain glial cell, *Appl. Entomol. Zool.* **18:**116–121.

Konopinska, D., Sobotka, W., Lesicki, A., Rosinski, G., and Sujak, P., 1986, Synthesis of proctolin analogues and their cardioexcitatory effect on cockroach, *Periplaneta americana* L., and yellow mealworm, *Tenebrio molitor* L., *Int. J. Peptide Protein Res.* **27:**597–603.

Koolman, J., Scheller, K., and Bodenstein, D., 1979, Ecdysteroids in the adult male blowfly, *Calliphora vicina, Experientia* **35:**134–135.

Koontz, M., and Edwards, J. S., 1980, The projection of neuroendocrine fibers (NCC I and II) in the brains of three orthopteroid insects, *J. Morphol.* **165:**285–299.

Koopmanschap, A. B., and De Kort, C. A. D., 1988, Isolation and characterization of a high molecular weight JH-III transport protein in the hemolymph of *Locusta migratoria, Arch. Insect Biochem. Physiol.* **7:**105–118.

Kopec, S., 1922, Physiological self differentiation on the wing germ grafted on caterpillars of the opposite sex, *J. Exp. Zool.* **36:**469–479.

Kramer, K. J., 1978, Regulation of the activity of JH specific esterases in the Colorado potato beetle, *Leptinotarsa decemlineata, J. Insect Physiol.* **24:**743–747.

Kramer, K. J., 1985, Vertebrate hormones in insects, in *Comprehensive Insect Physiology, Biochemistry and Pharmacology* (G. A. Kerkut and L. I. Gilbert, eds.), Vol. 7, Pergamon Press, New York, pp. 511–536.

Kramer, K. J., and Hopkins, Th. L., 1987, Tyrosine metabolism for insect cuticle tanning, *Arch. Insect Biochem. Physiol.* **6:**279–301.

Kramer, K. J., Tager, H. S., Childs, C. N., and Speirs, R. D., 1977a, Insulin-like hypoglycemic and immunological activities in honeybee royal jelly, *J. Insect Physiol.* **23:**293–295.

Kramer, K. J., Speirs, R. D., and Childs, C. N., 1977b, Immunochemical evidence for a gastrin-like peptide in insect neuroendocrine system, *Gen. Comp. Endocrinol.* **32:**423–426.

Kramer, K. J., Tager, H. S., and Childs, C. N., 1980, Insulin-like and glucagon-like peptides in insect hemolymph, *Insect Biochem.* **10:**179–182.

Kramer, K. J., Childs, C. N., Spiers, R. D., and Jacobs, R. M., 1982, Purification of insulin-like peptides from insect haemolymph and royal jelly, *Insect Biochem.* **12:**91–98.

Krauthamer, V., 1985a, Electrophysiology of identified neurosecretory and non-neurosecretory cells in the cockroach pars intercerebralis, *J. Exp. Zool.* **234:**207–219.

Krauthamer, V., 1985b, Morphology of identified neurosecretory and non-neurosecretory cells in the cockroach pars intercerebralis, *J. Exp. Zool.* **234:**221–230.

Kriger, F. L., and Davey, K. G., 1982, Ovarian motility in mated *Rhodnius prolixus* requires an intact cerebral neurosecretory system, *Gen. Comp. Endocrinol.* **48:**130–134.

Kriger, F. L., and Davey, K. G., 1983, Ovulation in *Rhodnius prolixus* Stal is induced by an extract of neurosecretory cells, *Can. J. Zool.* **61:**684–686.

Kriger, F. L., and Davey, K. G., 1984, Identified neurosecretory cells in the brain of female *Rhodnius prolixus* contain a myotropic peptide, *Can. J. Zool.* **62:**1720–1723.

Krishnakumaran, A., and Schneiderman, H. A., 1965, Prothoracotropic activity of compounds that mimic juvenile hormone, *J. Insect Physiol.* **11:**1517–1532.

Krishnakumaran, A., Oberlander, H., and Schneiderman, H. A., 1965, Rates of DNA and RNA synthesis in various tissues during a larval moult cycle of *Samia cynthia ricini* (Lepidoptera), *Nature* **205:**1131–1133.

Krolak, J. M., Clore, J. N., Petrovitch, E., and Mills, R. R., 1977, Vitellogenesis by the American cockroach: Haemolymph and follicle protein patterns during vitellogenin synthesis, *J. Insect Physiol.* **23:**381–385.

Kubben, F. J. G. M., Van Assche, C. L. M. V. J., and Bosman, F. T., 1986, FMRFamide immunoreactivity in the mammalian gastroenteropancreatic neuroendocrine system, *Histochemistry* **84:**439–444.

Kubota, I., Isobe, M., Goto, T., and Hasegawa, K., 1976, Molecular size of the diapause hormone of the silkworm *Bombyx mori, Z. Naturforsch.* **31c:**132–134.

Kudryavtseva, A. M., 1986, Morphology and functional activity of the neurosecretory cells of the synganglion and the perisympathetic organs of larvae of *Calliphora vicina* R.-D. and *C. uralensis* Vill. (Diptera, Calliphoridae) during active and diapausal development, *Entomol. Rev.* 35–43.

Kukukawa, S., and Tobe, S. S., 1986, Juvenile hormone biosynthesis in female larvae of *Diploptera punctata* and the effect of allatectomy on hemolymph ecdysteroid titer, *J. Insect Physiol.* **32:**981–986.

Kuster, J. E., and Davey, K. G., 1981, Fine structure of the abdominal neurosecretory organs of *Rhodnius prolixus* Stal., *Can. J. Zool.* **59:**765–770.

Kuster, J. E., and Davey, K. G., 1986, Mode of action of cerebral neurosecretory cells on the function of the spermatheca in *Rhodnius prolixus, Int. J. Invert. Reprod. Develop.* **10:**59–69.

Lafon-Cazal, M., 1976, Radioautographic detection of monoamines in a neuroendocrine gland (corpora cardiaca) of locust (Insects), *J. Microsc. Biol. Cell.* **27:**257–260.

Lafon-Cazal, M., and Arluison, M., 1976, Localization of monoamines in the corpora cardiaca and the hypocerebral ganglion of locusts, *Cell Tissue Res.* **172:**517–527.

Lafon-Cazal, M., and Baehr, J. C., 1988, Octopaminergic control of corpora allata activity in an insect, *Experientia.* **44:**895–896

Lafon-Cazal, M., and Morandini, M., 1982, Contrôle neuroendocrine de la lipémie chez le criquet, *Locusta migratoria, J. Physiol. (Paris)* **78:**566–573.

Lafon-Cazal, M., and Roussel, J. P., 1971, Modifications métaboliques après ablation simultanée des corpora allata et des corpora cardiaca chez *Schistocerca gregaria* et *Locusta migratoria, C. R. Acad. Sci. Ser. D* **273:**2613–2615.

Lafon-Cazal, M., Calas, A., and Bosc, S., 1973, Capture et rétention de monoamines tritiées dans les corpora cardiaca de *Locusta migratoria* L., étude in vitro par radioautographie à haute résolution, *J. Microsc.* **17:**223–226.

Lafon-Cazal, M., Gallois, D., Lehouelleur, J., and Bockaert, J., 1987, Stimulatory effects of male accessory-gland extracts on the myogenicity and the adenylate cyclase activity of the oviduct of *Locusta migratoria, J. Insect Physiol.* **33:**909–915.

Lafont, R., Mauchamp, B., Pennetier, J. L., and De Reggi, M. L., 1977, Endocrine significance of critical periods during insect development: Analysis of ligation experiments with *Pieris brassicae* last instar larvae, *Experientia* **33:**1662–1663.

Lagueux, M., Hirn, M., and Hoffmann, J. A., 1977, Ecdysone during ovarian development in *Locusta migratoria, J. Insect Physiol.* **23:**109–119.

Lagueux, M., Harry, P., and Hoffmann, J. A., 1981, Ecdysteroids are bound to vitellin in newly laid eggs of *Locusta, Mol. Cell. Endocrinol.* **24:**325–338.

Lagueux, M., Hoffmann, J. A., Goltzené, F., Kappler, C., Tsoupras, G., Hetru, C., and Luu, B., 1984, Ecdysteroids in ovaries and embryos of *Locusta migratoria,* in *Biosynthesis, Metabolism and Mode of Action of Invertebrate Hormones* (J. A. Hoffmann and M. Porchet, eds.), Springer-Verlag, Berlin, pp. 168–180.

Lamy, M., 1984, Vitellogenesis, vitellogenin and vitellin in the males of insects: A review, *Int. J. Invert. Reprod.* **7:**311–321.

Lamy, M., and Karlinsky, A., 1974, Vitellogenèse protéique en milieu mâle chez la piéride du chou, *Pieris brassicae* L. (Lépidoptères), *C.R. Acad. Sci. Ser. D* **278:**91–94.

Lamy, M., Julien-Laferrière, N., and Lavenseau, L., 1975, Vitellogenèse protéique en milieu mâle chez le ver à soie *Bombyx mori* L. (Lépidoptères): Contrôle hormonal, *C.R. Acad. Sci. Ser. D* **280:**1393–1395.

Lamy, M., Julien-Laferrière, N., and Allégret, P., 1978, Vitellogenèse protéique en milieu mâle chez le lépidoptère pyralide *Galleria mellonella* L., *C.R. Acad. Sci. Ser. D* **286:**883–885.

Landureau, J. C., and Szöllösi, A., 1974, Démonstration, par la méthode de culture in vitro, du rôle des hémocytes dans la spermatogenèse d'un insecte, *C.R. Acad. Sci.* **278:**3359–3362.

Langbein, V. L., 1981, Preliminary purification and physiological characterization of a heart-inhibiting factor from the thoracic ganglia of *Periplaneta americana* (L.), *Zool. Jahrb. Abt. Allg. Zool. Physiol. Tiere* **85:**278–303.

Lange, A. B., and Loughton, B. G., 1985, An oviposition-stimulating factor in the male accessory reproductive gland of the locust, *Locusta migratoria, Gen. Comp. Endocrinol.* **57:**208–215.

Lange, A. B., and Orchard, I., 1984, Dorsal unpaired median neurons, and ventral bilaterally paired neurons, project to a visceral muscle in an insect, *J. Neurobiol.* **15:**441–453.

Lange, A. B., and Orchard, I., 1986, Identified octopaminergic neurons modulate contractions of locust visceral muscle via adenosine 3′, 5′-monophosphate (cyclic AMP), *Brain Res.* **363:**340–349.

Lange, A. B., Orchard, I., and Loughton, B. G., 1984, Neural inhibition of egg-laying in the locust, *Locusta migratoria, J. Insect Physiol.* **30:**271–278.

Lange, A. B., Orchard, I., and Adams, M. E., 1986, Peptidergic innervation of insect reproductive tissue: The association of proctolin with oviduct visceral musculature, *J. Comp. Neurol.* **254:** 279–286.

Lange, A. B., Orchard, I., and Lam, W., 1987, Mode of action of proctolin on locust visceral muscle, *Arch. Insect Biochem. Physiol.* **5:**285–295.

Lange, A. B., Orchard, I., and Lloyd, R. J., 1988a, Immunohistochemical and electrochemical detection of serotonin in the nervous system of the blood-feeding bug, *Rhodnius prolixus, Arch. Insect Biochem. Physiol.* **8:**187–201.

Lange, A. B., Orchard, I., and Barrett, F. M., 1988b, The presence and distribution of proctolin in the blood-feeding bug, *Rhodnius prolixus, J. Insect. Physiol.* **34:**379–386.

Langley, P. A., 1967, Experimental evidence for a hormonal control of digestion in the tsetse fly, *Glossina morsitans* Westwood: A study of the larva, pupa, and teneral adult fly, *J. Insect Physiol.* **13:**1921–1931.

Langley, P. A., and Pimley, R. W., 1986, A role for juvenile hormone and the effects of so-called anti–juvenile hormones in *Glossina morsitans, J. Insect Physiol.* **32:**727–734.

Lanot, R., Thiebold, J., Lagueux, M., Goltzene, F., and Hoffmann, J. A., 1987, Involvement of ecdysone in the control of meiotic reinitiation in oöcytes of *Locusta migratoria* (Insecta, Orthoptera), *Dev. Biol.* **121:**174–181.

Lanzrein, B., 1975, Programming, induction, or prevention of the breakdown of the prothoracic gland in the cockroach, *Nauphoeta cinerea, J. Insect Physiol.* **21:**367–389.

Lanzrein, B., Whilhelm, R., and Buschor, J., 1981, On the regulation of the corpora allata activity in adult females of the ovoviviparous cockroach *Nauphoeta cinerea,* in *Juvenile Hormone Biochemistry* (G. E. Pratt and G. T. Books, eds.), Elsevier North-Holland, Amsterdam, pp. 147–160.

La Pointe, M. C., Koeppe, J. K., and Nair, K. K., 1985, Follicle cell polyploidy in *Leucophaea maderae:* Regulation by juvenile hormone, *J. Insect Physiol.* **31:**187–193.

Laufer, H., Landau, M., Homola, E., and Borst, D. W., 1987 Methyl farnesoata: Its site of synthesis and regulation of secretion, *Insect Biochem.* **17:**1129–1131.

Lauverjat, S., 1969, Effects de l'allatectomie sur le développement imaginal des voies génitales femelles de *Locusta migratoria migratorioides* (Insecte Orthoptère), *C.R. Acad. Sci. Ser. D* **268:**2109–2112.

Lauverjat, S., and Girardie, A., 1976, Différenciation postimaginale accélérée du corps gras et des voies génitales femelles chez *Locusta migratoria* après stimulation électrique des cellules neurosécrétrices "C" de la pars intercerebralis, in *Coll. Intern. C.N.R.S. n° 251, Actualités sur les Hormones d'Invertébrés* (Ed. C.N.R.S.), Villeneuve d'Ascq, pp. 307–315.

Lavenseau, L., Gadenne, C., and Trabelsi, M., 1984, Immunofluorescent localization of a substance immunologically related to insulin in the protocerebral neurosecretory cells of the European corn borer, *Cell Tissue Res.* **238:**207–208.

Lavenseau, L., Gadenne, C., and Trabelsi, M., 1985, Histological and ultrastructural study of the protocerebral neurosecretory cells of the European corn borer, *Ostrinia nubilalis* (Hübner) (Lepidoptera: Pyralidae), *Int. J. Insect Morphl. Embryol.* **14:**51–62.

Lavenseau, L., Gadenne, C., Hilal, A., and Peypelut, L., 1986, The endocrine control of diapause in insects, in *Adv. Invert. Reprod. 4* (M. Porchet, J. C. Andries and A. Dhainaut, eds.), Elsevier, Amsterdam, pp. 69–77.

Laverdure, A. M., 1971, Etude des conditions hormonales nécessaires à l'évolution de l'ovaire chez la nymphe de *Tenebrio molitor* (Coléoptère), *Gen. Comp. Endocrinol.* **17:**467–478.

Laverdure, A. M., 1972, L'évolution de l'ovaire chez la femelle adulte de *Tenebrio molitor.* La prévitellogenèse, *J. Insect Physiol.* **18:**1477–1491.

Lazarovici, P., and Pener, M. P., 1978, The relations of the pars intercerebralis, corpora allata, and juvenile hormone to oöcyte development and oviposition in the African migratory locust, *Gen. Comp. Endocrinol.* **35:**375–386.

Lazarovici, P., Shapira, D., Pisarev, V., and Shaaya, E., 1984, Ecdysteroid level and the change in sensitivity to a juvenile hormone analogue in *Ephestia cautella* larvae (Lepidoptera, Phycitidae), *Arch. Insect Biochem. Physiol.* **1:**409–415.

Lea, A. O., 1967, The medial neurosecretory cells and egg maturation in mosquitoes, *J. Insect Physiol.* **13:**419–429.

Lea, A. O., 1968, Mating without insemination in virgin *Aedes aegypti,* J. *Insect Physiol.* **14:**305–308.

Lea, A. O., 1969, Egg maturation in mosquitoes not regulated by the corpora allata, *J. Insect Physiol.* **15:**537–541.

Lea, A. O., 1972, Regulation of egg maturation in the mosquito by the neurosecretory system: The role of the corpus cardiacum, *Gen. Comp. Endocrinol.* **3:**602–608.

Lea, A. O., 1975, The control of reproduction by a blood meal: The mosquito as a model for vector endocrinology, *Acta Tropica* **32:**112–115.

Lea, A. O., and Van Handel, E., 1970, Suppression of glycogen synthesis in the mosquito by a hormone from the medial neurosecretory cells, *J. Insect Physiol.* **16:**319–321.

Lea, A. O., and Van Handel, E., 1982, A neurosecretory hormone-releasing factor from ovaries of mosquitoes fed blood, *J. Insect Physiol.* **28:**503–508.

Leahy, M. G., and Craig, G. B., 1965, Accessory gland substance as a stimulant for oviposition in *Aedes aegypti* and *Aedes Albopictus, Mosq. News* **25:**448–452.

Lebrun, D., 1967, La détermination des castes du termite à cou jaune (*Calotermes flavicollis* Fabr.), *Bull. Biol. Fr. Belg.* **101:**139–217.

Lees, A. D., 1977, Action of juvenile hormone mimics on the regulation of larval-adult and alary polymorphism in aphids, *Nature* **267:**46–48.

Lender, T., and Duveau-Hagege, J., 1963, La Culture organotypique de gonades de *Galleria mellonella* (L.), *Ann. Epiphyties* **14:**81–89.

Lensky, Y., Baehr, J. C., and Porcheron, P., 1978, Dosages radio-immunologiques des ecdysones et des hormones juvéniles au cours du développement post-embryonnaire chez les ouvrières et les reines d'abeille (*Apis mellifica* L. var. *Ligustica*), *C.R. Acad. Sci. Ser. D* **287:**821–824.

Leopold, R. A., 1976, The role of male accessory glands in insect reproduction, *Annu. Rev. Entomol.* **21:**199–221.

LeRoith, D., Lesniak, M. A., and Roth, J., 1981, Insulin in insects and annelids, *Diabetes* **30:**70–76.

Lessman, C. A., Rollins, L., and Herman, W. S., 1982, Effects of juvenile hormones I, II and III on reproductive tract growth in male and female monarch butterflies, *Comp. Biochem. Physiol.* **71A:**141–144.

Leviatan, R., and Friedländer, M., 1979, The eupyrene-apyrene dichotomous spermatogenesis of lepidoptera. I. The relationship with postembryonic development and the role of the decline in juvenile hormone titer toward pupation, *Dev. Biol.* **68:**515–524.

L'Helias, C., 1955, Recherches préliminaires sur les hormones du cerveau et du complexe post-cérébral des phasmes *Carausius morosus* et *Cuniculina cuniculina, C.R. Acad. Sci.* **240:**1141–1143.

L'Helias, C., Ebels, I., and Salemink, C., 1984, Association complexe de cellules neurosécrétrices avec les cellules réticulées dans les ocelles de *Pieris brassicae, C.R. Acad. Sci. Ser. III* **299:** 161–166.

Lhonoré, J., Bouthier, A., Beydon, P., Pennetier, J. L., and Lafont, R., 1980, The epidermal cell cycle during the last larval instar of *Pieris brassicae* (Lepidoptera), I. Cell ultrastructure and variations in pigment content, *J. Comp. Physiol.* **136:**11–20.

Li, C., and Calabrese, R. L., 1985, Evidence for proctolin-like substances in the central nervous system of the leech *Hirudo medicinalis, J. Comp. Neurol.* **232:**414–424.

Linzen, B., and Bückmann, D., 1961, Biochemische und histologische Untersuchungen zur Umfärbung der Raupe von *Cerura vinula* L., *Z. Naturforsch.* **16b:**6–18.

Liu, M. A., Jones, G. L., Stoffolano, J. G., and Yin, C. M., 1988, Conditions for estimation of corpus allatum activity in the blowfly, *Phormia regina, in vitro, Physiol. Entomol.* **13:**69–79.

Liu, T. P., and Davey, K. G., 1974, Partial characterization of a proposed antigonadotropin from the ovaries of the insect *Rhodnius prolixus* Stal., *Gen. Comp. Endocrinol.* **24:**405–408.

Lococo, D. J., and Tobe, S. S., 1984, Neuroanatomy of the retrocerebral complex, in particular the pars intercerebralis and partes laterales in the cockroach *Diploptera punctata* Eschscholtz (Dictyoptera: Blaberidae). *Int. J. Insect Morph. Embryol.* **13:**65–76.

Lococo, D. J., Thompson, C. S., and Tobe, S. S., 1986, Intercellular communication in an insect endocrine gland, *J. Exp. Biol.* **121:**407–419.

Loeb, M. J., Woods, C. W., Brandt, E. P., and Borkovec, A. B., 1982, Larval testes of the tobacco budworm: A new source of insect ecdysteroids, *Science* **218:**896–898.

Loeb, M. J., Brandt, E. P., and Birnbaum, M. J., 1984, Ecdysteroid production by testes of the tobacco budworm, *Heliothis virescens,* from last larval instar to adult, *J. Insect Physiol.* **30:** 375–381.

Loeb, M. J., Brandt, E. P., and Masler, E. P., 1985, Modulation of the rate of spermatogenesis by the central nervous system of the tobacco budworm, *Heliothis virescens, Int. J. Invert. Reprod.* **8:**39–49.

Loeb, M. J., Brandt, E. P., and Woods, C. W., 1986, Effects of exogenous ecdysteroid titer on endogenous ecdysteroid production *in vitro* by testes of the tobacco budworm, *Heliothis virescens, J. Exp. Zool.* **240:**75–82.

Loeb, M. J., Brandt, E. P., Woods, C. W., and Borkovec, A. B., 1987, An ecdysiotropic factor from brains of *Heliothis virescens* induces testes to produce immunodetectable ecdysteroid in vitro, *J. Exp. Zool.* **243:**275–282.

Loeb, M. J., Brandt, E. P., Woods, C. W., and Bell, R. A., 1988, Secretion of ecdysteroid by sheaths of testes of the gypsy moth, *Lymantria dispar,* and its regulation by testis ecdysiotropin, *J. Exp. Zool.* **248:**94–100.

Loher, W., 1961, The chemical acceleration of the maturation process and its hormonal control in the male of the desert locust, *Proc. R. Soc. Ser. B* **153:**380–397.

Loher, W., 1965, Hormonale Kontrolle der Oocytenentwicklung bei der Heuschrecke *Gomphocerus rufus* L. (Acridinae), *Zool. Jahrb. Abt. Allg. Zool. Physiol. Tiere* **71:**677–684.

Loher, W., 1966, Die Steuerung sescueller Verhaltens weisen und der Oocytenentwicklung bei *Gomphocerus rufus* L., *Z. Vergl. Physiol.* **53:**277–316.

Loher, W., 1974, Circadian control of spermatophore formation in the cricket *Teleogryllus commodus* Walker, *J. Insect Physiol* **20:**1155–1172.

Loher, W., Ganjian, I., Kubo, I., Stanley-Samuelson, D., and Tobe, S. S., 1981, Prostaglandins: Their role in egg-laying of the cricket *Teleogryllus commodus, Proc. Natl. Acad. Sci. USA* **78:** 7835–7838.

Loher, W., Schooley, D. A., and Baker, F. C., 1987, Influence of the ovaries on JH titer in *Teleogryllus commodus, Insect Biochem.* **17:**1099–1102.

Longley, A. J., and Longley, R. D., 1986, Serotonin immunoreactivity in the nervous system of the dragonfly nymph, *J. Neurobiol.* **17:**329–338.

Loughton, B. G., 1987, Studies on locust hypolipaemic hormone, *J. Insect Physiol.* **33:**569–573.

Loughton, B. G., and Orchard, I., 1981, The nature of the hyperglycaemic factor from the glandular lobe of the corpus cardiacum of *Locusta migratoria, J. Insect Physiol.* **27:**383–385.

Lu, Y. H., and Hagedorn, H. H., 1986, Egg development in the mosquito *Anopheles albimanus, Int. J. Invert. Reprod. Dev.* **9:**79–94.

Lüscher, M., 1958, Experimentelle Erzeugung von Soldaten bei der Termite *Kalotermes flavicollis* (Fabr.), *Naturwissenschaften* **3:**69–70.

Lüscher, M., 1968a, Hormonal control of respiration and protein synthesis in the fat body of the cockroach *Nauphoeta cinerea* during oöcyte growth, *J. Insect Physiol.* **14:**499–511.

Lüscher, M., 1968b, Oöcyte protection—A function of a corpus cardiacum hormone in the cockroach, *Nauphoeta cinerea, J. Insect Physiol.* **14:**685–688.

Lutz, E. M., and Tyrer, N. M., 1988, Immunohistochemical localization of serotonin and choline acetyltransferase in sensory neurones of the locust, *J. Comp. Neurol.* **267:**335–342.

Ma, M., Zhang, J. Z., Gong, H., and Gwadz, R., 1988, Permissive action of juvenile hormone on vitellogenin production by the mosquito, *Aedes aegypti, J. Insect Physiol.* **34:**593–596.

Maddrell, S. H. P., 1963, Excretion in the blood-sucking bug, *Rhodnius prolixus* Stal., I. The control of diuresis, *J. Exp. Biol.* **40:**247–256.

Maddrell, S. H. P., 1964a, Excretion in the blood-sucking bug, *Rhodnius prolixus* Stal. II. The normal course of diuresis and the effect of temperature, *J. Exp. Biol.* **41:**163–176.

Maddrell, S. H. P., 1964b, The excretion in the blood-sucking bug, *Rhodnius prolixus* Stal. III. The control of the release of the diuretic hormone, *J. Exp. Biol.* **41:**459–472.

Maddrell, S. H. P., 1966, The site of release of the diuretic hormone in *Rhodnius:* A new neurohaemal system in insects, *J. Exp. Biol.* **45:**499–508.

Maddrell, S. H. P., 1972, The functioning of insect Malpighian tubules, in *Role of Membranes in Secretory Processes* (L. Bolis, R. D. Keynes, and W. W. Wilbrant, eds.), Elsener North-Holland, Amsterdam, pp. 338–356.

Maddrell, S. H. P., 1977, Hormonal action in the control of fluid and salt transporting epithelia, in *Water Relations in Membrane Transport in Plants and Animals*, Academic Press, New York, pp. 303–313.

Maddrell, S. H. P., and Gee, J. D., 1974, Potassium-induced release of the diuretic hormones of *Rhodnius prolixus* and *Glossina austeni:* Ca dependence, time course and localization of neurohaemal areas, *J. Exp. Biol.* **61:**155–171.

Maddrell, S. H. P., and Klunsuwan, S., 1973, Fluid secretion by in vitro preparations of the Malpighian tubules of the desert locust *Schistocerca gregaria, J. Insect Physiol.* **19:**1369–1376.

Maddrell, S. H. P., and Phillips, J. E., 1976, Regulation of absorption in insect excretory systems, in *Perspectives in Experimental Biology* (P. Davies, ed.), Vol. 1, Zoology, Pergamon Press, New York, pp. 179–185.

Maddrell, S. H. P., and Phillips, J. E., 1978, Induction of sulphate transport and hormonal control of fluid secretion by Malpighian tubules of larvae of the mosquito *Aedes taeniorhynchus, J. Exp. Biol.* **72:**181–202.

Maddrell, S. H. P., Pilcher, D. E. M., and Gardiner, B. O. C., 1969, Stimulatory effect of 5-hydroxytryptamine (serotonin) on secretion by Malipighian tubules of insects, *Nature* **222:**784–785.

Maddrell, S. H. P., Pilcher, D. E. M., and Gardiner, B. O. C., 1971, Pharmacology of the Malpighian tubules of *Rhodnius* and *Carausius:* The structure-activity relationship of tryptamine analogues and the role of cyclic AMP, *J. Exp. Biol.* **54:**779–804.

Maisch, A., and Bückmann, D., 1987, The control of cuticular melanin and lutein incorporation in the morphological colour adaptation of a nymphalid pupa, *Inachis io* L., *J. Insect Physiol.* **33:** 393–403.

Malá, J., Granger, N. A., and Sehnal, F., 1977, Control of prothoracic gland activity in larvae of *Galleria mellonella, J. Insect Physiol.* **23:**309–316.

Malczewska, M., Gelman, D. B., and Cymborowski, B., 1988, Effect of azadirachtin on development, juvenile hormone and ecdysteroid titres in chilled *Galleria mellonella* larvae, *J. Insect Physiol.* **34:**725–732.

Mancini, G., and Frontali, N., 1970, On the ultrastructural localization of catecholamines in the Beta lobes (corpora pedunculata) of *Periplaneta* americana, *Z. Zellforsch. Mikrosk. Anat.* **103:**341–350.

Mandal, S., Roy, S., and Choudhuri, D. K., 1981, Effect of allatectomy and brain cauterization on the activity and distribution of three digestive enzymes in the gut lumen to *Schizodactylus monstrosus* Drury., *Curr. Sci.* **50:**871–872.

Mandal, S., Saha, L., and Choudhuri, D. K., 1984, Role of corpora allata, brain and juvenile hormone analogue on the biochemical changes of haemolymph composition of female *Schizodactylus monstrosus* Drury. (Schizodactylidae: Orthoptera), *Zool. Jahrb. Abt. Allg. Zool. Physiol. Tiere* **88:** 269–287.

Manning, A., 1966, Corpus allatum and sexual receptivity in female *Drosophila melanogaster, Nature* **211:**1321–1322.

Marcaillou, C., and Szöllösi, A., 1975, Variations de perméabilité du follicle testiculaire chez le criquet *Locusta migratoria migratorioides* (Orthoptère) au cours du dernier stade larvaire, *C.R. Acad. Sci. Ser. D* **281:**2001–2004.

Marks, E. P., and Holman, G. M., 1974, Release from the brain and acquisition by the corpus cardiacum of a neurohormone in vitro, *J. Insect Physiol.* **20:**2087–2093.

Marks, E. P., and Reinecke, J. P., 1965, Regenerating tissue from the cockroach leg: Nutrient media for maintenance in vitro, *J. Kansas Entomol. Soc.* **38:**179–182.

Marks, E. P., Ittycheriah, P. I., and Leloup, A. M., 1972, The effect of β-ecdysone on insect neurosecretion in vitro, *J. Insect Physiol.* **18:**847–850.

Maslennikova, V. A., 1970, Hormonal regulation of diapause in *Pieris brassicae* L., *Dokl. Akad. Nauk SSSR* **192:**412–414.

Maslennikova, V. A., 1973, Hormonal control of insect diapause, in *Hormonal Control of Growth and Differentiation* (J. La Bue and A. S. Gordon, eds.), Academic Press, New York.

Maslennikova, V. A., 1977, Different sensitivity of tissues of diapausing and diapauseless larvae of *Calliphora vicina* R.-D. to β-ecdysone, *Dokl. Adad. Nauk SSSR (Biological Sciences)* **234:** 229–232.

Masler, E. P., Fuchs, M. S., Sage, B., and O'Connor, J. D., 1980, Endocrine regulation of ovarian development in the autogenous mosquito, *Aedes atropalpus, Gen. Comp. Endocrinol.* **41:**250–259.

Masler, E. P., Hagedorn, H. H., Petzel, D. H., and Borkovec, A. B., 1983, Partial purification of egg development neurosecretory hormone with reverse-phase liquid chromatographic techniques, *Life Sci.* **33:**1925–1931.

Mason, C. A., 1973, New features of the brain-retrocerebral neuroendocrine complex of the locust *Schistocerca vaga* (Scudder), *Z. Zellforsch. Mikrosk. Anat.* **141:**19–32.

Matsumoto, S., Suzuki, A., Yamamoto, H., Ogura, N., and Ikemoto, H., 1979, Cyclic nucleotides and hormonal control of cuticular melanization in the armyworm larva, *Leucania* (Lepidoptera: Noctuidae), *Appl. Entomol. Zool.* **14:**159–163.

Matsumoto, S., Isogai, A., Suzuki, A., Ogura, N., and Sonobe, H., 1981, Purification and properties of the melanization and reddish colouration hormone (MRCH) in the armyworm, *Leucania separata* (Lepidoptera), *Insect Biochem.* **11:**725–733.

Matsumoto, S., Isogai, A., and Suzuki, A., 1985, N-terminal amino acid sequence of an insect neurohormone, melanization and reddish coloration hormone (MRCH): Heterogeneity and sequence homology with human insulin-like growth factor II, *FEBS. Lett.* **189:**115–118.

Matsumoto, S., Isogai, A., and Suzuki, A., 1986, Isolation and amino terminal sequence of melanization and reddish coloration hormone (MRCH) from the silkworm, *Bombyx mori, Insect Biochem.* **16:**775–779.

Matsuo, N., Aizono, Y., Funatsu, G., Funatsu, M., and Kobayashi, M., 1985, Purification and some properties of prothoracicotropic hormone in the silkworm, *Bombyx mori, Insect Biochem.* **15:**189–195.

Mattson, M. P., and Spaziani, E., 1985, 5-Hydroxytryptamine mediates release of molt-inhibiting hormone activity from isolated crab eyestalk ganglia, *Biol. Bull.* (Woods Hole, Mass.) **169:** 246–255.

Mattson, M. P., and Spaziani, E., 1986, Evidence for ecdysteroid feedback on release of molt-inhibiting hormone from crab eyestalk ganglia, *Biol. Bull.* (Woods Hole, Mass.) **171:**264–273.

Maxwell, G. D., Tait, J. F., and Hildebrand, J. G., 1978, Regional synthesis of neurotransmitter candidates in the CNS of the moth *Manduca sexta, Comp. Biochem. Physiol.* **61C:**109–119.

May, T. E., Brown, B. E., and Clements, A. N., 1979, Experimental studies upon a bundle of tonic fibres in the locust extensor tibialis muscle, *J. Insect Physiol.* **25:**169–181.

Mayer, R. J., and Candy, D. J., 1967, Changes in haemolymph lipoproteins during locust flight, *Nature* **215:**987.

Mayer, R. J., and Candy, D. J., 1969, Control of haemolymph lipid concentration during locust flight: An adipokinetic hormone from the corpora cardiaca, *J. Insect Physiol.* **15:**611–620.

Mayeri, A., Rothman, B. S., Brownell, P. H., Branton, W. D., and Padgett, L., 1985, Monosynaptic characteristics of neurotransmission mediated by egg laying hormone in the abdominal ganglion of *Aplysia, J. Neurosci.* **5:**2060–2077.

Mazzini, M., and Giorgi, F., 1985, The follicle cell-oocyte interaction in ovarian follicles of the stick insect *Bacillus rossius* (Rossi): (Insecta: Phasmatodea), *J. Morphol.* **185:**37–49.

McCaffery, A. R., 1976, Effects of electrocoagulation of cerebral neurosecretory cells and implantation of corpora allata on oöcyte development in *Locusta migratoria, J. Insect Physiol.* **22:**1081–1092.

McCaffery, A. R., and Highnam, K. C., 1975a, Effects of corpora allata on the activity of the cerebral neurosecretory system of *Locusta migratoria* R. and F., *Gen. Comp. Endocrinol.* **25:** 358–372.

McCaffery, A. R., and Highnam, K. C., 1975b, Effects of corpus allatum hormone and its mimics on the activity of the cerebral neurosecretory system of *Locusta migratoria migratorioides* R. and F., *Gen. Comp. Endocrinol.* **25:**373–386.

McCaffery, A. R., and McCaffery, V. A., 1983, Corpus allatum activity during overlapping cycles of oöcyte growth in adult female *Melanoplus sanguinipes, J. Insect Physiol.* **29:**259–266.

McCaleb, D. C., and Kumaran, A. K., 1980, Control of juvenile hormone esterase activity in *Galleria mellonella* larvae, *J. Insect Physiol.* **26:**171–177.

McCarthy, R., and Ralph, C. L., 1962, The effects of corpora allata and cardiaca extracts on hemolymph sugars of the cockroach, *Am. Zoologist* **2:**429–430.

McClure, J. B., and Steele, J. E., 1981, The role of extracellular calcium in hormonal activation of glycogen phosphorylase in cockroach fat body, *Insect Biochem.* **11:**605–613.

McDaniel, C. N., 1979, Haemolymph ecdysone concentrations in *Hyalophora cecropia* pupae, dauer pupae and adults, *J. Insect Physiol.* **25:**143–145.

McDaniel, C. N., and Berry, S. J., 1967, Activation of the prothoracic glands of *Antheraea polyphemus, Nature* **214:**1032–1034.

McDaniel, C. N., Johnson, E., Saum, T., and Berry, S. J., 1976, Ultrastructure of active an inhibited prothoracic glands, *J. Insect Physiol.* **22:**473–481.

McLeod, D. G. R., and Beck, S. D., 1963, The anatomy of the neuroendocrine complex of the European corn borer, *Ostrinia nubilalis,* and its relation to diapause, *Ann. Entomol. Soc. Am.* **36:**723–727.

Mechoulam, R., Brueggemeier, R. W., and Denlinger, D. L., 1984, Estrogens in insects, *Experientia* **40:**942–944.

Meidell, E. M., 1982, Effects of a synthetic juvenile hormone mimic on the reproduction of the tsetse fly *Glossina morsitans, Insect Sci. Appl.* **3:**263–266.

Meller, V. H., Aucoin, R. R., Tobe, S. S., and Feyereisen, R., 1985, Evidence for an inhibitory role of cyclic AMP in the control of juvenile hormone biosynthesis by cockroach corpora allata, *Mol. Cell. Endocrinol.* **43:**155–163.

Melnikova, E. J., 1985, The brain neurosecretory cells and retrocerebral complex of some Coccinellidae (Coleoptera), *Z. Mikrosk, Anat. Forsch.* **99:**503–509.

Melnikova, E. J., and Panov, A. A., 1981, Brain neurosecretory cells in larval and imaginal flour beetle *Tenebrio molitor* (Coleoptera, Tenebrionidae), *Dokl. Biol. Sci.* **260:**463–466.

Meneses, P., and Ortiz, M., 1975, A protein extract from *Drosophila melanogaster* with insulin-like activity, *Comp. Biochem. Physiol.* **51A:**483–485.

Meola, R. W., and Adkisson, P. L., 1977, Release of prothoracicotropic hormone and potentiation of developmental ability during diapause in the bollworm, *Heliothis zea, J. Insect Physiol.* **23:** 683–688.

Meola, R., and Lea, A. O., 1972, Humoral inhibition of egg development in mosquitoes, *J. Med. Entomol.* **9:**99–103.

Meola, S. M., 1988, Neurosecretory innervation of the accessory gland region of the male stable fly, *Stomoxys calcitrans, Int. J. Invert. Reprod. Develop.* **13:**205–223.

Meola, S., and Cook, B. J., 1986, Neuroendocrine plexus in the pericardial septum of the stable fly, *Stomoxys calcitrans* (L.) (Diptera: Muscidae), *Int. J. Insect Morphol. Embryol.* **15:**269–281.

Mercer, A. R., Mobbs, P. G., Davenport, A. P., and Evans, P. D., 1983, Biogenic amines in the brain of the honeybee, *Apis mellifera, Cell Tissue Res.* **234:**655–677.

Merle, J., 1968, Fonctionnement ovarien et réceptivité sexuelle de *Drosophila melanogaster* après implantation de fragments de l'appareil génital mâle, *J. Insect Physiol.* **14:**1159–1168.

Mesnier, M., 1972a, Recherches sur le déterminisme de la ponte chez *Galleria mellonella* (Lépidoptères), *C.R. Acad. Sci. Ser. D* **274:**708–711.

Mesnier, M., 1972b, Etude du contrôle neuroendocrine de l'oviposition chez *Clitumnus extradentatus* (Phasmides-Chéleutoptères), *C.R. Acad. Sci. Ser. D* **274:**564–567.

Mesnier, M., 1980, Study of oöcyte growth within the ovariole in a stick-insect, *Clitumnus extradentatus, J. Insect Physiol.* **26:**59–65.

Mesnier, M., 1981, Study of the ovulation process in the stick insect, *Clitumnus extradentatus, J. Insect Physiol.* **27:**425–433.

Mesnier, M., 1984, Patterns of laying behaviour and control of oviposition in insects: Further experiments on *Sphodromantis lineola* (Dictyoptera), *Int. J. Invert. Reprod. Develop.* **7:**23–32.

Mesnier, M., 1985, Origin and release sites of a hormone stimulating oviposition in the stick insect *Clitumnus extradentatus, J. Insect Physiol.* **31:**299–306.

Mesnier, M., and Provansal, A., 1975, Rôle des organes périsympathiques dans l'induction de la ponte chez *Galleria mellonella* (Lépidoptère), *C.R. Acad. Sci. Ser. D* **281:**905–907.

Mesnier, M., and Thomas, A., 1981, Influence de la β-ecdysone sur l'activité choriogène des cellules folliculaires de *Carausius morosus* (Insecte, Chéleutoptère), *C.R. Acad. Sci. Ser. D* **293:**155–158.

Meyer, A. S., Schneiderman, H. A., Hanzmann, E., and Ko, J. H., 1968, The two juvenile hormones from the *Cecropia* silk-moth, *Proc. Natl. Acad. Sci. USA* **60:**853–860.

Michel, R., 1973, Variations de la tendance au vol soutenu du criquet pélerin *Schistocerca gregaria* après implantation de corpora cardiaca, *J. Insect Physiol.* **19:**1317–1325.

Michel, R., and Bernard, A., 1973, Influence de la pars intercerebralis sur l'induction au vol soutenu chez le criquet pélerin *Schistocerca gregaria, Acrida* **2:**139–149.

Milburn, N. S., and Roeder, K. D., 1962, Control of efferent activity in the cockroach terminal abdominal ganglion by extracts of corpora cardiaca, *Gen. Comp. Endocrinol.* **2:**70–76.

Milburn, N., Weiant, E. A., and Roeder, K. D., 1960, The release of efferent nerve activity in the roach *Periplaneta americana,* by extracts of the corpus cardiacum, *Biol. Bull. (Woods Hole, Mass.)* **118:**111–119.

Miller, T. A., 1968a, Role of cardiac neurons in the cockroach heartbeat, *J. Insect Physiol.* **14:** 1265–1275.

Miller, T. A., 1968b, Response of cockroach cardiac neurons to cholinergic compounds, *J. Insect Physiol.* **14:**1713–1717.

Miller, T. A., 1985, Structure and physiology of the circulatory system, in *Comprehensive Insect Physiology, Biochemistry and Pharmacology* (G. A. Kerkut and L. I. Gilbert, eds.), Vol. 7, Pergamon Press, New York, pp. 289–353.

Miller, T. A., and James, J., 1976, Chemical sensitivity of the hyperneural nerve-muscle preparation of the American cockroach, *J. Insect Physiol.* **22:**981–988.

Miller, T. A., and Metcalf, R. L., 1968, Site of action of pharmacologically active compounds on the heart of *Periplaneta americana* L., *J. Insect Physiol.* **14:**383–394.

Miller, T. A., and S.-Rózsa, K., 1981, Control of the alary muscles of locust dorsal diaphragm, *Physiol. Entomol.* **6:**51–59.

Miller, T. A., and Thomson, W. W., 1968, Ultrastructure of cockroach cardiac innervation, *J. Insect Physiol.* **14:**1099–1104.

Miller, T. A., and Usherwood, P. N. R., 1971, Studies of cardio-regulation in the cockroach, *Periplaneta americana, J. Exp. Biol.* **54:**329–348.

Mills, R. R., 1965, Hormonal control of tanning in the American cockroach. II. Assay for the hormone and the effect of wound healing, *J. Insect Physiol.* **11:**1269–1275.

Mills, R. R., 1966, Hormonal control of tanning in the American cockroach, III. Hormone stability and post-ecdysial changes in hormone titre, *J. Insect Physiol.* **12:**275–280.

Mills, R. R., 1967a, Hormonal control of excretion in the American cockroach, I. Release of a diuretic hormone from the terminal abdominal ganglion, *J. Exp. Biol.* **46:**35–41.

Mills, R. R., 1967b, Control of cuticular tanning in the cockroach: Bursicon release by nervous stimulation, *J. Insect Physiol.* **13:**815–820.

Mills, R. R., and Lake, C. R., 1966, Hormonal control of tanning in the American cockroach. IV. Preliminary purification of the hormone, *J. Insect Physiol.* **12:**1395–1401.

Mills, R. R., and Nielsen, D. J., 1967, Hormonal control of tanning in the American cockroach. V. Some properties of the purified hormone, *J. Insect Physiol.* **13:**273–280.

Mills, R. R., and Whitehead, D. L., 1970, Hormonal control of tanning in the American cockroach: Changes in blood cell permeability during ecdysis, *J. Insect Physiol.* **16:**331–340.

Mills, R. R., Lake, C. R., and Alworth, W. L., 1967, Biosynthesis of N-acetyl dopamine by the American cockroach *Periplaneta americana, J. Insect Physiol* **13:**1539–1546.

Mitsuhashi, J., 1963, Histological studies on the neurosecretory cells of the brain and on the corpus allatum during diapause in some lepidopterous insects, *Bull. Natl. Inst. Agric. Sci. C, Tokyo* **16:** 67–118.

Mitsuhashi, J., and Fukaya, M., 1961, The hormonal control of larval diapause in the rice stem borer, *Chilo suppressalis:* Histological studies on the neurosecretory cells of the corpora allata during diapause and host diapause, *Jpn. J. Appl. Entomol. Zool.* **4:**127.

Miwa, H., and Yamashita, O., 1985, Controlled accumulation of vitellin in *Bombyx* eggs, *Int. J. Invert. Reprod. Develop.* **8:**193–196.

Miyazaki, S. I., 1980, The ionic mechanism of action potentials in neurosecretory cells and non-neurosecretory cells of the silkworm, *J. Comp. Physiol.* **140:**43–52.

Mizoguchi, A., Ishizaki, H., Nagasawa, H., Kataoka, H., Isogai, A., Tamura, S., Suzuki, A., Fujino, M., and Kitada, C., 1987, A monoclonal antibody against a synthetic fragment of bombyxin (4K-prothoracicotropic hormone) from the silk-moth, *Bombyx mori:* Characterization and immunohistochemistry, *Mol. Cell. Endocrinol.* **51:**227–235.

Mokia, G. G., 1941, Contribution to the study of hormones in insects, *Dokl. Akad. Nauk. SSSR* **30:** 368–370.

Moore, D., and Loher, W., 1988, Axonal projections within the brain–retrocerebral complex of the cricket, *Teleogryllus commodus, Cell Tissue Res.* **252:**501–514.

Mordue (Luntz), A. J., Evans, K. A., and Charlet, M., 1986, Azadirachtin, ecdysteroids and ecdysis in *Locusta migratoria, Comp. Biochem. Physiol.* **85C:**297–301.
Mordue, W., 1965, Studies on oöcyte production and associated histological changes in the neuroendocrine system in *Tenebrio molitor* L., *J. Insect Physiol.* **11:**493–503.
Mordue, W., 1967, The influence of feeding upon the activity of the neuroendocrine system during oöcyte growth in *Tenebrio molitor, Gen. Comp. Endocrinol.* **9:**406–415.
Mordue, W., 1969, Hormonal control of Malpighian tube and rectal function in the desert locust, *Schistocerca gregaria, J. Insect Physiol.* **15:**273–285.
Mordue, W., 1970, Evidence for the existence of diuretic and anti-diuretic hormones in locusts, *J. Endocrinol.* **46:**119–120.
Mordue, W., 1972, Hormones and excretion in locusts, *Gen. Comp. Endocrinol.* **3:**289–298.
Mordue, W., and De Kort, C. A. D., 1978, Energy substrates for flight in the Colorado beetle, *Leptinotarsa decemlineata* Say., *J. Insect Physiol.* **24:**221–224.
Mordue, W., and Goldsworthy, G. J., 1969, The physiological effects of corpus cardiacum extracts in locusts, *Gen. Comp. Endocrinol.* **12:**360–369.
Mordue, W., and Goldsworthy, G. J., 1973, Transaminase levels and uric acid production in adult locusts, *Insect Biochem.* **3:**419–427.
Mordue, W., and Morgan, P. J., 1985, Chemistry of peptide hormones, in *Comprehensive Insect Physiology, Biochemistry and Pharmacology* (G. A. Kerkut and L. I. Gilbert, eds.), Vol. 7, Pergamon Press, New York, pp. 153–183.
Mordue, W., and Stone, J. V., 1976, Comparison of the biological activities of an insect and a crustacean neurohormone that are structurally similar, *Nature* **264:**287–289.
Mordue, W., and Stone, J. V., 1977, Relative potencies of locust adipokinetic hormone and prawn red pigment-concentrating hormone in insect and crustacean systems, *Gen. Comp. Endocrinnol.* **33:**103–108.
Moreau, R., Raoelison, C., and Sutter, B. C. J., 1981, An intestinal insulin-like molecule in *Apis mellifica* L. (Hymenoptera), *Comp. Biochem. Physiol.* **69A:**79–83.
Moreau, R., Gourdoux, L., Lequellec, Y., and Dutrieu, J., 1982, Endocrine control of hemolymph carbohydrates in *Locusta migratoria:* Comparison between effects of two endogenous hormonal extracts and effects of insulin and glucagon, *Comp. Biochem. Physiol.* **73A:**669–673.
Moreau, R., Gourdoux, L., and Girardie, J., 1988, Neuroparsin: A new energetic neurohormone in the African locust, *Arch. Insect Biochem. Physiol.* **8:**135–145.
Moreteau-Lévita, B., 1972a, Rôle des lobes optiques dans la réalisation de l'homochromie d'*Oedipoda coerulescens* L. (Acridien, Orthoptère), *C.R. Acad. Sci. Ser. D* **274:**2690–2691.
Moreteau-Lévita, B., 1972b, Rôle des corpora cardiaca dans la pigmentation d'*Oedipoda coerulescens* L. (Acridien, Orthoptère), *C.R. Acad. Sci. Ser. D* **275:**2699–2701.
Moreteau-Lévita, B., 1972c, Rôle de la pars intercerebralis dans la réalisation de l'homochromie d'*Oedipoda coerulescens* L. (Acridien, Orthoptère), *C.R. Acad. Sci. Ser. D* **274:**3277–3279.
Morgan, P. J., and Mordue, W., 1981, Stimulated fluid secretion is sodium dependent in the Malpighian tubules of *Locusta migratoria, J. Insect Physiol.* **27:**271–279.
Morgan, P. J., and Mordue, W., 1983a, Electrochemical gradients across *Locusta* Malpighian tubules, *J. Comp. Physiol.* **151:**175–183.
Morgan, P. J., and Mordue, W., 1983b, Separation and characteristics of diuretic hormone from the corpus cardiacum of *Locusta, Comp. Biochem. Physiol.* **75B:**75–80.
Morgan, P. J., and Mordue, W., 1984, Diuretic hormone: Another peptide with widespread distribution within the insect CNS, *Physiol. Entomol.* **9:**197–206.
Morgan, P. J., and Mordue, W., 1985a, Cyclic AMP and locust diuretic hormone action, *Insect Biochem.* **15:**247–257.
Morgan, P. J., and Mordue, W., 1985b, The role of calcium in diuretic hormone action on locust Malpighian tubules, *Mol. Cell. Endocrinol.* **40:**221–231.

Morgan, P. J., Siegert, K. J., and Mordue, W., 1987, Preliminary characterization of locust diuretic peptide (DP-1) and another corpus cardiacum peptide (LCCP), *Insect Biochem.* **17:** 383–388.

Morita, M., Hatakoshi, M., and Tojo, S., 1988, Hormonal control of cuticular melanization in the common cutworm *Spodoptera litura, J. Insect Physiol.* **34:**751–758.

Morohoshi, S., and Fugo, H., 1977, The control of growth and development in *Bombyx mori,* XXXVII. Some aspects on a hormone controlling adult eclosion, *Proc. Jpn. Acad.* **53:**75–78.

Morohoshi, S., and Oshiki, T., 1969, Effect of the brain on the suboesophageal ganglion and determination of voltinism in *Bombyx mori, J. Insect Physiol.* **15:**167–175.

Morris, G. P., and Steel, C. G. H., 1975, Ultrastructure of neurosecretory cells in the pars intercerebralis of *Rhodnius prolixus* (Hemiptera), *Tissue Cell* **7:**73–90.

Morrison, P. E., Venkatesh, K., and Thompson, B., 1982, The role of male accessory-gland substance on female reproduction with some observations of spermatogenesis in the stable fly, *J. Insect Physiol.* **28:**607–614.

Morton, D. B., and Evans, P. D., 1983, Octopamine distribution in solitarious and gregarious forms of the locust, *Schistocerca americana gregaria, Insect Biochem.* **13:**177–183.

Morton, D. B., and Evans, P. D., 1984, Octopamine release from an identified neurone in the locust, *J. Exp. Biol.* **113:**269–287.

Morton, D. B., and Truman, J. W., 1985, Steroid regulation of the peptide-mediated increase in cyclic GMP in the nervous system of the hawkmoth, *Manduca sexta, J. Comp. Physiol. A* **157:** 423–432.

Moshitzky, P., Yamashiro, D., Stuve, L., Ramachandran, J., and Applebaum, S. W., 1987a, Determination of locust AKH-I by radioimmunoassay and the identification of an AKH-like factor in the locust brain, *Insect Biochem.* **17:**765–769.

Moshitzky, P., Henzel, W. J., Rafaeli, A., Ramachandran, J., and Applebaum, S. W., 1987b, Synthesis of adipokinetic hormone (AKH-I) in the locust brain, *Insect Biochem.* **17:**1133–1137.

Moss, J. I., and Miller, T. A., 1988, A pharmacological study of the hyperneural muscle of *Periplaneta americana, J. Insect Physiol.* **34:**997–1003.

Mothes, G., 1960, Weitere Untersuchungen über den physiologischen Farbwechsel von *Carausius morosus* (Br.), *Zool. Jahrb. Abt. Allg. Zool. Physiol. Tiere* **69:**133–162.

Mouton, J., 1968, Effet de la castration sur les cellules neurosécrétrices ventrales du ganglion infraoesophagien chez le phasme *Carausius morosus* (Phasmides-Orthoptères), *C.R. Acad. Sci. Ser. D* **266:**2120–2121.

Müller, H. P., 1965, Zur Frage der Steuerung des Paarungsverhaltens und der Eireifung bei der Feldheuschrecke *Euthystira brachyptera* Ocsk. unter besonderer benicksichtigung der Rolle der Corpora allata, *Z. Vgl. Physiol.* **50:**447–497.

Müller, H. P., and Engelmann, F., 1968, Studies on the endocrine control of metabolism in *Leucophaea maderae* (Blattaria) II. Effect of the corpora cardiaca on fat-body respiration, *Gen. Comp. Endocrinol.* **11:**43–50.

Mundall, E. C., Tobe, S. S., and Stay, B., 1979, Induction of vitellogenin and growth of implanted oocytes in male cockroaches, *Nature* **282:**97–98.

Muneoka, Y., and Saitoh, H., 1986, Pharmacology of FMRFamide in *Mytilus* catch muscle, *Comp. Biochem. Physiol.* **85C:**207–214.

Muraleedharan, D., and Prabhu, V. K. K., 1979, Role of the median neurosecretory cells in secretion of protease and invertase in the red cotton bug, *Dysdercus cingulatus, J. Insect Physiol.* **25:**237–240.

Muraleedharan, D., and Prabhu, V. K. K., 1981, Hormonal influence on feeding and digestion in a plant bug, *Dysdercus cingulatus,* and a caterpillar, *Hyblaea puera, Physiol. Entomol.* **6:**183–189.

Murphy, T. A., and Wyatt, G. R., 1965, The enzymes of glycogen and trehalose synthesis in silkmoth fat body, *J. Biol. Chem.* **240:**1500–1508.

Murtaugh, M. P., and Denlinger, D. L., 1987, Regulation of long-term oviposition in the house cricket, *Acheta domesticus:* Roles of prostaglandin and factors associated with sperm, *Arch. Insect Biochem. Physiol.* **6:**59–72.

Muszyńska-Pytel, M., 1987, Some aspects of activity regulation of *Galleria mellonella* PTTH cells, *Arch. Insect Biochem. Physiol.* **5:**211–224.

Mwangi, R. W., and Goldsworthy, G. J., 1977a, Age related variations in the response to adipokinetic hormone in *Locusta, Physiol. Entomol.* **2:**37–42.

Mwangi, R. W., and Goldsworthy, G. J., 1977b, Diglyceride-transporting lipoproteins in *Locusta, J. Comp. Physiol.* **114:**177–190.

Myers, C. M., and Evans, P. D., 1985, An FMRFamide antiserum differentiates between populations of antigens in the ventral nervous system of the locust, *Schistocerca gregaria, Cell Tissue Res.* **242:**109–114.

Myers, C. M., and Evans, P. D., 1987, An FMRFamide antiserum differentiates between populations of antigens in the brain and retrocerebral complex of the locust, *Schistocerca gregaria, Cell Tissue Res.* **250:**93–99.

Myers, C. M., and Evans, P. D., 1988, Peripheral neurosecretory cells on the thoracic median nerves of the locust, *Schistocerca gregaria, J. Morph.* **195:**45–58.

Nachman, R. J., Holman, G. M., Haddon, W. F., and Ling, N., 1986a, Leucosulfakinin, a sulfated insect neuropeptide with homology to gastrin and cholecystokinin, *Science* **234:**71–73.

Nachman, R. J., Holman, G. M., Cook, B. J., Haddon, W. F., and Ling, N., 1986b, Leucosulfakinin-II, a blocked sulfated insect neuropeptide with homology to cholecystokinin and gastrin, *Biochem. Biophys. Res. Commun.* **140:**357–364.

Nagasawa, H., Isogai, A., Suzuki, A., Tamura, S., and Ishizaki, H., 1979, Purification and properties of the prothoracicotripic hormone of the silkworm, *Bombyx mori, Dev. Growth Differ.* **21:**29–38.

Nagasawa, H., Fugo, H., Takahashi, S. Y., Kamito, T., Isogai, A., and Suzuki, A., 1983, Purification and properties of eclosion hormone of the silkworm, *Bombyx mori, Agric. Biol. Chem.* **47:** 1901–1906.

Nagasawa, H., Kataoka, H., Hori, Y., Isogai, A., Tamura, S., Suzuki, A., Guo, F., Zhong, X., Mizoguchi, A., Fujishita, M., Takahashi, S. Y., Ohnishi, E., and Ishizaki, H., 1984a, Isolation and some characterization of the prothoracicotropic hormone from *Bombyx mori, Gen. Comp. Endocrinol.* **53:**143–152.

Nagasawa, H., Kataoka, H., Isogai, A., Tamura, S., and Suzuki, A., 1984b, Amino-terminal amino acid sequence of the silkworm prothoracicotropic hormone: Homology with insulin, *Science* **226:**1344–1345.

Nagasawa, H., Kataoka, H., Isogai, A., Tamura, S., Mizoguchi, A., Fujiwara, Y., Suzuki, A., Takahashi, S. Y., Suzuki A., and Ishizaki, H., 1986, Amino acid sequence of a prothoracicotropic hormone of the silkworm *Bombyx mori, Proc. Natl. Acad. Sci. USA* **83:**5840–5843.

Nair, K. S. S., 1974, Studies on the diapause of *Trogoderma granarium:* Effects of juvenile hormone analogues on growth and metamorphosis, *J. Insect Physiol.* **20:**231–244.

Nair, V. S. K., 1981, Hormonal control on the mating behaviour in the females of *Spodoptera mauritia* (Lepidoptera: Noctuidae), *Curr. Sci.* **50:**690–691.

Naisse, J., 1966, Contrôle endocrinien de la différenciation sexuelle chez *Lampyris noctiluca* (Coléoptère Lampyride). III. Influence des hormones de la pars intercerebralis, *Gen. Comp. Endocrinol.* **7:**105–110.

Naisse, J., 1969, Rôle des neurohormones dans la différenciation sexuelle de *Lampyris noctiluca, J. Insect Physiol.* **15:**877–892.

Nässel, D. R., and Cantera, R., 1985, Mapping of serotonin-immunoreactive neurons in the larval nervous system of the flies *Calliphora erythrocephala* and *Sarcophaga bullata*. A comparison with ventral ganglia in adult animals, *Cell Tissue Res.* **239:**423–434.

Nässel, D. R., and Elekes, K., 1985, Serotonergic terminals in the neural sheath of the blowfly nervous system: Electron microscopical immunocytochemistry and 5,7-dihydroxytryptamine labelling, *Neuroscience* **15:**293–307.

Nässel, D. R., and Klemm, N., 1983, Serotonin-like immunoreactivity in the optic lobes of three insect species, *Cell Tissue Res.* **232:**129–140.

Nässel, D. R., and Laxmyr, L., 1983, Quantitative determination of biogenic amines and dopa in the CNS of adult and larval blowflies, *Calliphora erythrocephala, Comp. Biochem. Physiol.* **75C:** 259–265.

Nässel, D. R., and O'Shea, M., 1987, Proctolin-like immunoreactive neurons in the blowfly central nervous system, *J. Comp. Neurol.* **265:**437–454.

Nässel, D. R., Ohlsson, L. G., and Cantera, R., 1988, Metamorphosis of identified neurons innervating thoracic neurohemal organs in the blowfly: Transformation of cholecystokinin-like immunoreactive neurons, *J. Comp. Neurol.* **267:**343–356.

Natalizi, G. M., and Frontali, N., 1966, Purification of insect hyperglycaemic and heart accelerating hormones, *J. Insect Physiol.* **12:**1279–1287.

Natalizi, G. M., Pansa, M. C., D'Ajello, V., Casaglia, O., and Frontali, N., 1970, Physiologically active factors of *Periplaneta americana, J. Insect Physiol.* **16:**1827–1836.

Nathanson, J. A., and Greengard, P., 1973, Octopamine-sensitive adenylate cyclase: Evidence for a biological role of octopamine in nervous tissue, *Science* **180:**308–310.

Nayar, K. K., 1958, Probable endocrine mechanism controlling oviposition in the insect *Iphita limbata* Stal., in: *Zweites International Symposium über Neurosekretion,* Springer-Verlag, Berlin, pp. 102–104.

Newitt, R. A., and Hammock, B. D., 1986, Relationship between juvenile hormone and ecdysteroids in larval-pupal development of *Trichoplusia ni* (Lepidoptera: Noctuidae), *J. Insect Physiol.* **32:**835–844.

Nicolson, S. W., 1976, The hormonal control of diuresis in the cabbage white butterfly *Pieris brassicae, J. Exp. Biol.* **65:**565–575.

Nicolson, S. W., 1980, Water balance and osmoregulation in *Onymacris plana,* a tenebrionid beetle from the Namib desert, *J. Insect Physiol.* **26:**315–320.

Nicolson, S. W., and Hanrahan, S. A., 1986, Diuresis in a desert beetle? Hormonal control of the Malpighian tubules of *Onymacris plana* (Coleoptera: Tenebrionidae), *J. Comp. Physiol. B* **156:** 407–413.

Nicolson, S. W., and Millar, R. P., 1983, Effects of biogenic amines and hormones on butterfly Malpighian tubules: Dopamine stimulates fluid secretion, *J. Insect Physiol.* **29:**611–615.

Nijhout, H. F., 1975, Axonal pathways in the brain-retrocerebral neuroendocrine complex of *Manduca sexta* (L.) (Lepidoptera: *Sphingidae*), *Int. J. Insect Morphol. Embryol.* **4:**529–538.

Nijhout, H. F., and Carrow, G. M., 1978, Diuresis after a bloodmeal in female *Anopheles freeborni, J. Insect Physiol.* **24:**293–298.

Nijhout, H. F., and Williams, C. M., 1974a, Control of moulting and metamorphosis in the tobacco hornworm, *Manduca sexta* (L.): Growth of the last-instar larva and the decision to pupate, *J. Exp. Biol.* **61:**481–491.

Nijhout, H. F., and Williams, C. M., 1974b, Control of moulting and metamorphosis in the tobacco hornworm, *Manduca sexta* (L.): Cessation of juvenile hormone secretion as a trigger for pupation, *J. Exp. Biol.* **61:**493–501.

Nishiitsutsuji-Uwo, J., 1961, Further studies on the developmental features of the male germ cells in the diapausing and nondiapausing swallowtails, *Papilio xuthus, Annot. Zool. Jpn.* **34:** 30–37.

Nishiitsutsuji-Uwo, J., 1972, Purification and some properties of insect brain hormone extracted from silkworm heads, *Botyu-Kagaku* **37**:93–102.

Nishiitsutsuji-Uwo, J., and Pittendrigh, C. S., 1968, Central nervous system control of circadian rhythmicity in the cockroach. III The optic lobes, locus of the driving oscillation? *Z. Vgl. Physiol.* **58**:14–46.

Nishiitsutsuji-Uwo, J., Petropulos, S. F., and Pittendrigh, C. S., 1967, Central nervous system control of circadian rhythmicity in the cockroach. I. Role of the pars intercerebralis, *Biol. Bull. (Woods Hole, Mass.)* **133**:679–696.

Nkouka, E., 1977a, Recherches sur le contrôle endocrine cérébral de la reproduction chez un insecte vivipare, *Glossina fuscipes fuscipes* (Diptère, Muscidae), *C.R. Acad. Sci. Ser. D* **284**:2381–2383.

Nkouka, E., 1977b, Contrôle neuro-endocrine de l'ovulation et de la parturition chez un insecte vivipare, *Glossina fuscipes fuscipes* (Diptera, Muscidae), *C.R. Acad. Sci. Ser. D* **285**:77–80.

Normann, T. C., 1965, The neurosecretory system of the adult *Calliphora erythrocephala.* I. The fine structure of the corpus cardiacum with some observations on adjacent organs, *Z. Zellforsch.* **67**:461–501.

Normann, T. C., 1969, Experimentally induced exocytosis of neurosecretory granules, *Exp. Cell Res.* **55**:285–287.

Normann, T. C., 1970, The mechanism of hormone release from neurosecretory axon endings in the insect *Calliphora erythrocephala,* in *Aspects of Neuroendocrinology,* 5th International Symposium on Neurosecretion (W. Bargmann and B. Scharrer, eds.), Springer-Verlag. Berlin, pp. 30–42.

Normann, T. C., 1972, Heart activity and its control in the adult blowfly, *Calliphora erythrocephala, J. Insect Physiol.* **18**:1793–1810.

Normann, T. C., 1973, Membrane potential of the corpus cardiacum neurosecretory cells of the blowfly, *Calliphora erythrocephala, J. Insect Physiol.* **19**:303–318.

Normann, T. C., 1974, Calcium-dependence of neurosecretion by exocytosis, *J. Exp. Biol.* **61**:401–409.

Normann, T. C., 1975, Neurosecretory cells in insect brain and production of hypoglycaemic hormone, *Nature* **254**:259–261.

Normann, T. C., and Duve, H., 1969, Experimentally induced release of a neurohormone influencing hemolymph trehalose level in *Calliphora erythrocephala* (Diptera), *Gen. Comp. Endocrinol.* **12**:449–459.

Novak, F. J., and Pipa, R. L., 1986, Anatomy and fine structure of nerves associated with the corpus allatum and foregut musculature of *Periplaneta americana* (L.) (Dictyoptera: Blattidae), *Int. J. Insect Morphol. Embryol.* **15**:171–181.

Novak, F., De Clerk, J., Paesen, G., Swevers, L., and De Loof, A., 1987, Radioimmunological quantification of C_{21}, C_{19}, and C_{18} steroids in haemolymph of the insect *Locusta migratoria, Int. J. Invert. Reprod. Dev.* **11**:255–264.

Nowock, J., 1973, Growth and metamorphosis in the testes of *Ephestia kuhniella* in vitro, *J. Insect Physiol.* **19**:941–949.

Numata, H., and Hidaka, T., 1984, Termination of adult diapause by a juvenile hormone analogue in the bean bug, *Riptortus clavatus, Zool. Sci.* **1**:751–754.

Nunez, J. A., 1956, Untersuchungen über die Regelung das Wasserhaushaltes bei *Anisotarsus cupripennis, Z. Vgl. Physiol.* **36**:341–354.

Oberlander, H. S., Berry, J., Krishnakumarann, A., and Schneiderman, H. A., 1965, RNA and DNA synthesis during activation and secretion of the prothoracic glands of saturniid moths, *J. Exp. Zool.* **159**:15–32.

O'Brien, M. A., Granger, N. A., Agui, N., Gilbert, L. I., and Bollenbacher, W. E., 1986, Prothoracicotropic hormone in the developing brain of the tobacco hornworm, *Manduca sexta:* Relative amounts of two molecular forms, *J. Insect Physiol.* **32**:719–725.

O'Connor, K. J., and Baxter, D., 1985, The demonstration of insulin-like material in the honey bee, *Apis Mellifera, Comp. Biochem. Physiol.* **81B:**755–760.

Odhiambo, T. R., 1966, The metabolic effects of the corpus allatum hormone in the male desert locust. I. Lipid metabolism, *J. Exp. Biol.* **45:**45–50.

Ogiso, M., and Ohnishi, E., 1986, Does estradiol play a role in ovarian maturation or embryonic development of the silkworm? *Gen. Comp. Endocrinol.* **61:**82–86.

Ogura, N., 1975, Hormonal control of larval coloration in the armyworm, *Leucania separata, J. Insect Physiol.* **21:**559–576.

Ogura, N., and Mitsuhashi, J., 1979, Melanization of integuments cultured in vivo and in vitro, *Appl. Entomol. Zool.* **14:**118–120.

Ogura, N., and Saito, T., 1973, Induction of embryonic diapause in the silkworm, *Bombyx mori* L. (Lepidoptera: Bombycidae), by implantation of ganglia of the common armyworm larvae, *Leucania separata* Walker (Lepidoptera: Noctuidae), *Appl. Entomol. Zool.* **8:**46–48.

Ohashi, M., Tsusué, M., and Kiguchi, K., 1983, Juvenile hormone control of larval colouration in the silkworm, *Bombyx mori:* Characterization and determination of epidermal brown colour induced by the hormone, *Insect Biochem.* **13:**123–127.

Ohlsson, L. G., and Nässel, D. R., 1987, Postembryonic development of serotonin-immunoreactive neurons in the central nervous system of the blowfly, *Calliphora erythrocephala.* I. The optic lobes, *Cell Tissue Res.* **249:**669–679.

Ohnishi, E., and Hidaka, T., 1957, Effet des facteurs du milieu ambiant sur la détermination des types de chrysalides de certains papillons diurnes nommés "queue d'hirondelle," *Bull. Soc. Hist. Nat. Toulouse* **92:**177–180.

Ohnishi, E., Ogiso, M., Wakabayashi, K., Fujimoto, Y., and Ikekawa, N., 1985, Identification of estradiol in the ovaries of the silkworm, *Bombyx mori, Gen. Comp. Endocrinol.* **60:**35–38.

Ohtaki, T., 1960, Humoral control of pupal coloration in the cabbage white butterfly, *Pieris rapae crucivora, Annot. Zool. Jpn.* **33:**97–103.

Ohtaki, T., 1963, Further studies on the development of pupal coloration in the cabbage white butterfly, *Pieris rapae crucivora, Annot. Zool. Jpn.* **36:**78–82.

Ohtaki, T., and Takahashi, M., 1972, Induction and termination of pupal diapause in relation to the change of ecdysone titer in the fleshfly, *Sarcophaga peregrina, Jpn. J. Med. Sci. Biol.* **25:**369–376.

Okelo, O., 1971, Physiological control of oviposition in the female desert locust *Schistocerca gregaria* Forsk., *Can. J. Zool.* **49:**969–974.

Okuda, M., Sakurai, S., and Ohtaki, T., 1985, Activity of the prothoracic gland and its sensitivity to prothoracicotropic hormone in the penultimate and last-larval instar of *Bombyx mori, J. Insect Physiol.* **31:**455–461.

Oldfield, A. C., and Huddart, H., 1982, Spontaneous activity of foregut and hindgut visceral muscle of the locust, *Locusta migratoria.* I. Normal activity and the effect of KC1 depolarization and glutamate, *Comp. Biochem. Physiol.* **73C:**297–302.

Oltmer, A., 1968, Die Steuerung des Melanineibaus in das Farbmuster der Kohlweiblingspuppe *Pieris brassicae* L., *Wilhelm Roux Arch. Entw. Mech. Org.* **160:**401–427.

O'Meara, G. F., and Evans, D. G., 1977, Autogeny in saltmarsh mosquitoes induced by a substance from the male accessory gland, *Nature* **267:**342–343.

Ono, S., 1982, Effect of juvenile hormone on the caste determination in the ant, *Pheidole fervida* Smith (Hymenoptera: Formicidae), *Appl. Ent. Zool.* **17:**1–7.

Opoczynska-Sembratowa, Z., 1936, Recherches sur l'anatomie et l'innervation du coeur de *Carausius morosus, Bull. Int. Acad. Pol. Sci. Lett. Cl. Sci. Math. Nat. Ser. B* **1936:** 411–436.

Orchard, I., 1987, Adipokinetic hormones: An update, *J. Insect Physiol.* **33:**451–463.

Orchard, I., and Finlayson, L. H., 1976, The electrical activity of mechanoreceptive and neurosecretory neurons in the stick insect *Carausius morosus, J. Comp. Physiol.* **107:**327–338.

Orchard, I., and Finlayson, L. H., 1977, Studies on peripheral neurons and neurohaemal tissue in the thorax of the stick insect (*Carausius morosus*), *Experientia* **33:**1440–1442.

Orchard, I., and Lange, A. B., 1983a, Release of identified adipokinetic hormones during flight and following neural stimulation in *Locusta migratoria, J. Insect Physiol.* **29:**425–429.

Orchard, I., and Lange, A. B., 1983b, The hormonal control of haemolymph lipid during flight in *Locusta migratoria, J. Insect Physiol.* **29:**639–642.

Orchard, I., and Lange, A. B., 1985, Evidence for octopaminergic modulation of an insect visceral muscle, *J. Neurobiol.* **16:**171–181.

Orchard, I., and Lange, A. B., 1986, Pharmacological profile of octopamine receptors on the lateral oviducts of the locust, *Locusta migratoria, J. Insect Physiol.* **32:**741–745.

Orchard, I., and Lange, A. B., 1987, Cockroach oviducts: The presence and release of octopamine and proctolin, *J. Insect Physiol.* **33:**265–268.

Orchard, I., and Loughton, B. G., 1980, A hypolipaemic factor from the corpus cadiacum of locusts, *Nature* **286:**494–496.

Orchard, I., and Loughton, B. G., 1981, The neural control of release of hyperlipaemic hormone from the corpus cardiacum of *Locusta migratoria, Comp. Biochem. Physiol.* **68A:**25–30.

Orchard, I., and Loughton, B. G., 1985, Neurosecretion, in *Comprehensive Insect Physiology, Biochemistry and Pharmacology* (G. A. Kerkut and L. I. Gilbert, eds.), Vol. 7, Pergamon Press, New York, pp. 61–108.

Orchard, I., and Shivers, R., 1986, Electrical coupling and gap junctions between neurosecretory cells in an insect, *Brain Res.* **397:**359–364.

Orchard, I., and Steel, C. G. H., 1980, Electrical activity of neurosecretory axons from the brain of *Rhodnius prolixus:* Relation of changes in the pattern of activity to endocrine events during the moulting cycle, *Brain Res.* **191:**53–65.

Orchard, I., Friedel, T., and Loughton, B. G., 1981, Release of a neurosecretory protein from the corpora cardiaca of *Locusta migratoria* induced by high potassium saline and compound action potentials, *J. Insect Physiol.* **27:**297–304.

Orchard, I., Carlisle, J. A., Loughton, B. G., Gole, J. W. D., and Downer, R. G. H., 1982, *In vitro* studies on the effects of octopamine on locust fat body, *Gen. Comp. Endocrinol.* **48:**7–13.

Orchard, I., Loughton, B. G., Gole, J. W. D., and Downer, R. G. H., 1983a, Synaptic transmission elevates adenosine 3′,5′-monophosphate (cyclic AMP) in locust neurosecretory cells, *Brain Res.* **258:**152–155.

Orchard, I., Gole, J. W. D., and Downer, R. G. H., 1983b, Pharmacology of aminergic receptors mediating an elevation in cyclic AMP and release of hormone from locust neurosecretory cells, *Brain Res.* **288:**349–353.

Orchard, I., Ruegg, R. P., and Davey, K. G., 1983c, The role of central aminergic neurons in the action of 20-hydroxyecdysone on neurosecretory cells of *Rhodnius prolixus, J. Insect Physiol.* **29:**387–391.

Orchard, I., Martin, R. J., Sloley, B. D., and Downer, R. G. H., 1986, The association of 5-hydroxytryptamine, octopamine, and dopamine with the intrinsic (glandular) lobe of the corpus cardiacum of *Locusta migratoria, Can. J. Zool.* **64:**271–274.

Orchard, I., Lange, A. B., and Barrett, F. M., 1988, Serotonergic supply to the epidermis of *Rhodnius prolixus:* Evidence for serotonin as the plasticising factor, *J. Insect Physiol.* **34:**873–879.

Orr, G. L., Gole, J. W. D., Jahagirdar, A. P., Downer, R. G. H., and Steele, J. E., 1985, Cyclic AMP does not mediate the action of synthetic hypertrehalosemic peptides from the corpus cardiacum of *Periplaneta americana, Insect Biochem.* **15:**703–709.

Osanai, M., 1966, Über die Umfärbung der Raupen von *Hestina japonica* zu Beginn der Uberwinterung II, *Hoppe-Seyler's Z. Physiol. Chem.* **347:**145–155.

Osborne, D. J., Carlisle, D. B., and Ellis, P. E., 1968, Protein synthesis in the fat body of the female desert locust, *Schistocerca gregaria* Forsk., in relation to maturation, *Gen. Comp. Endocrinol.* **11:**347–354.

Osborne, M. P., Finlayson, L. H., and Rice, M. J., 1971, Neurosecretory endings associated with striated muscles in three insects (*Schistocerca, Carausius,* and *Phormia*) and a frog (*Rana*), *Z. Zellforsch,* **116:**391–404.

O'Shea, M., and Adams, M. E., 1981, Pentapeptide (proctolin) associated with an identified neuron, *Science* **213:**567–569.

O'Shea, M., and Bishop, C. A., 1982, Neuropeptide proctolin associated with an identified skeletal motoneuron, *J. Neurosci.* **2:**1242–1251.

O'Shea, M., and Evans, P. D., 1979, Potentiation of neuromuscular transmission by an octopaminergic neurone in the locust, *J. Exp. Biol.* **79:**169–190.

O'Shea, M., Witten, J., and Schaffer, M., 1984, Isolation and characterization of two myoactive neuropeptides: Further evidence of an invertebrate peptide family, *J. Neurosci.* **4:**521–529.

Ozbas, S., and Hodgson, E. S., 1958, Action of insect neurosecretion upon central nervous system in vitro and upon behavior, *Proc. Natl. Acad. Sci. USA* **44:**825–830.

Ozeki, K., 1968, Experimental studies on the regression of the ventral glands of the earwig, *Anisolabis maritima,* during metamorphosis, *Sci. Pap. Coll. Gen. Educ. Univ. Tokyo,* **18:**199–219.

Ozeki, K., 1980, Further studies on the mechanisms controlling the corpus allatum function of the earwig, *Anisolabis maritima, Sci. Pap. Coll. Gen. Educ. Univ. Tokyo (Biol. Part)* **30:**49–54.

Padgham, D. E., 1976a, Control of melanization in first instar larvae of *Schistocerca gregaria, J. Insect Physiol.* **22:**1409–1419.

Padgham, D. E., 1976b, Bursicon-mediated control of tanning in melanizing and non-melanizing first instar larvae of *Schistocerca gregaria, J. Insect Physiol.* **22:**1447–1452.

Page, T. L., 1982, Transplantation of the cockroach circadian pacemaker, *Science* **216:**73–75.

Pagés, M., Jiménez, F., Ferrus, A., Peralta, E., Ramirez, G., and Gelpi, E., 1983, Enkephalin-like immunoreactivity in *Drosophila melanogaster, Neuropeptides* **4:**87–98.

Palay, L., 1957, The fine structure of the neurohypophysis, in *Ultrastructure and Cellular Chemistry of Neural Tissue* (H. Waelsch, ed.), Hoeber, New York, pp. 31–49.

Pallen, C. J., and Steele, J. E., 1988, A putative role for calmodulin in corpus cardiacum stimulated trehalose synthesis in fat body of the American cockroach (*Periplaneta americana*), *Insect Biochem.* **18:**577–584.

Pan, M. L., and Wyatt, G. R., 1971, Juvenile hormone induces vitellogenin synthesis in the monarch butterfly, *Science* **174:**503–505.

Pannabecker, T., and Orchard, I., 1986a, Octopamine and cyclic AMP mediate release of adipokinetic hormone I and II from isolated locust neuroendocrine tissue, *Mol. Cell. Endocrinol.* **48:**153–159.

Pannabecker, T., and Orchard, I., 1986b, Pharmacological properties of octopamine-2 receptors in locust neuroendocrine tissue, *J. Insect Physiol.* **32:**909–915.

Pannabecker, T., Orchard, I., and Lange, A. B., 1987, Adipokinetic hormones in neuroendocrine tissue of the larval locust: Quantification and regulation of in vitro release, *J. Insect Physiol.* **33:** 575–580.

Panov, A. A., 1964, Neurosecretory cells in the abdominal ganglia of orthoptera (Insecta, Orthoptera), *Ent. Rev.* **43:**403–409.

Panov, A. A., 1976, The composition of the medial neurosecretory cells of the pars intercerebralis in *Calliphora* and *Lucilia* adults, *Zool. Anz.* **196:**23–27.

Panov, A. A., 1979a, Comparison between pars intercerebralis neurosecretory cells of *Glossina* and those of other higher diptera, *J. Zool. London* **187:**455–463.

Panov, A. A., 1979b, Neurosecretory cells in the pars intercerebralis of Geocorisae (Heteroptera, Insecta), *Zool. Anz.* **203:**289–315.

Panov, A. A., 1979c, Brain neurosecretory cells and their axon pathways in the larva of *Chironomus plumosus* L. (Diptera: Chironomidae), *Int. J. Insect Morphol. Embryol.* **8:**203–212.

Panov, A. A., 1983, The median neurosecretory cells in the brain of *Lepidoptera* (Insecta), *Zool. Anz.* **210:**180–203.

Panov, A. A., 1985a, The cerebral neurosecretory cells and the corpora cardiaca of the blister-beetles (Coleoptera: Meloidea), *Zool. Anz.* **214:**101–110.

Panov, A. A., 1985b, The cerebral neurosecretory cells and retrocerebral endocrine complex of Carabidae (Coleoptera, Insecta), *Zool. Jahrb. Abt. Allg. Anat. Ont.* **113:**1–35.

Panov, A. A., 1986, The cerebral neurosecretory cells and retrocerebral endocrine complex in several representatives of staphyliniformic beetles (Coleoptera, Staphyliniformia), *J. Hirnforsch.* **27:**409–421.

Panov, A. A., and Bassurmanova, O. K., 1970, Fine structure of the gland cells in inactive and active corpus allatum of the bug, *Eurygaster integriceps, J. Insect Physiol.* **16:**1265–1281.

Panov, A. A., and Chudakova, I. V., 1987, Dopaminergic control of gonadotropic activity of the endocrine system in the house cricket, *Acheta domesticus* L. (Insecta, Orthoptera), *Zool. Jahrb. Abt. Allg. Zool. Physiol. Tiere* **91:**39–50.

Panov, A. A., and Davydova, E. D., 1976, Medial neurosecretory cells in the brain of Mecoptera and Neuropteroidea (Insecta), *Zool. Anz.* **197:**187–206.

Panov, A. A., and Kind, T. B., 1963, Neurosecretory cell system of the brain of some *Lepidoptera, Dokl. Akad. Nauk SSSR* **153:**1186–1189.

Panov, A. A., and Krjutchkova, E. E., 1983, The cerebral neurosecretory cells and cephalic neurohaemal organs of earwigs (Dermaptera, Insecta), *Zool. Jahrb. Abt. Allg. Anat. Ont.* **109:**503–519.

Panov, A. A., and Melnikova, E. J., 1986, The cerebral neurosecretory cells and corpora cardiaca of the longicorn beetles (Coleoptera, Cerambycidae), *Zool. Jahrb. Abt. Allg. Anat. Ont.* **114:**205–219.

Panov, A. A., and Melnikova, E. J., 1987, Histological studies on the cerebral neurosecretory cells and retrocerebral endocrine complex of *Lucanidae* and *Scarabaeidae* (Coleoptera, Insecta), *Zool. Jahrb. Abt. Allg. Anat. Ont.* **116:**323–350.

Pappas, C., and Fraenkel, G., 1978, Hormonal aspects of oögenesis in the flies *Phormia regina* and *Sarcophaga bullata, J. Insect Physiol.* **24:**75–80.

Pass, G., Sperk, G., Agricola, H., Baumann, E., and Penzlin, H., 1988a, Octopamine in a neurohaemal area within the antennal heart of the American cockroach, *J. Exp. Biol.* **135:**495–498.

Pass, G., Agricola, H., Birkenbeil, H., and Penzlin, H., 1988b, Morphology of neurones associated with the antennal heart of *Periplaneta americana* (Blattodea, Insecta), *Cell Tissue Res.* **253:** 319–326.

Passera, L., and Suzzoni, J. P., 1979, Le rôle de la reine de *Pheidole pallidula* (Nyl). (Hymenoptera, Formicadae) dans la sexualisation du couvain après traitement par l'hormone juvénile, *Insectes Soc.* **26:**343–353.

Paul, H., 1937, Transplantation und Regeneration der Flügel zur Üntersuchen ihrer Formbildung bei einen Schmetterling mit Geschlechts Dimorphismus *Orgya antiqua, Arch. Entwickl. Organ.* **136:**74–111.

Paulson, C. R., and Stay, B., 1987, Inhibition of corpora allata by larval brain extract in the cockroach *Diploptera punctata, Mol. Cell. Endocrinol.* **51:**243–252.

Paulson, C. R., Stay, B., Kikukawa, S., and Tobe, S. S., 1987, In vitro inhibition of the corpora allata by a larval brain factor in the cockroach *Diploptera punctata, Insect Biochem.* **17:**961–964.

Pearse, A. G. E., 1966, Common cytochemical properties of cells producing polypeptide hormones, with particular reference to calcitonin and the thyroid C cells, *Vet. Record.* **79:**587–590.

Pearse, A. G. E., and Takor-Takor, T., 1979, Embryology of the diffuse neuroendocrine system and its relationships to the common peptides, *Fed. Proc.* **38:**2288–2294.

Pener, M. P., 1965, On the influence of corpora allata on maturation and sexual behaviour of *Schistocerca gregaria, J. Zool.* **147:**119–136.

Pener, M. P., 1967, Effects of allactectomy and sectioning of the nerves of the corpora allata on oocyte growth, male sexual behaviour, and colour change in adults of *Schistocerca gregaria, J. Insect Physiol.* **13:**665–684.

Pener, M. P., 1985, Hormonal effects on flight and migration, in *Comprehensive Insect Physiology, Biochemistry and Pharmacology* (G. A. Kerkut and L. I. Gilbert, eds.), Vol. 8, Pergamon Press, New York, pp. 491–550.

Pener, M. P., and Broza, M., 1971, The effect of implanted, active corpora allata on reproductive diapause in adult females of the grasshopper *Oedipoda miniata, Ent. Exp. Appl.* **14:**190–202.

Pener, M. P., Girardie, A., and Joly, P., 1972, Neurosecretory and corpus allatum controlled effects on mating behaviour and color change in adult *Locusta migratoria migratorioides* males, *Gen. Comp. Endocrinol.* **19:**494–508.

Petzel, D. H., Hagedorn, H. H., and Beyenbach, K. W., 1986, Peptide nature of two mosquito natriuretic factors, *Am. J. Physiol.* **19:**R328–R332.

Petzel, D. H., Berg, M. M., and Beyenbach, K. W., 1987, Hormone-controlled cAMP-mediated fluid secretion in yellow-fever mosquito, *Am. J. Physiol.* **22:**701–711.

Pfeiffer, I. W., 1939, Experimental study of the function of the corpora allata in the grasshopper, *Melanoplus differentialis, J. Exp. Zool.* **82:**439–461.

Pflüger, H. J., and Watson, A. H. D., 1988, Structure and distribution of dorsal unpaired median (DUM) neurones in the abdominal nerve cord of male and female locusts, *J. Comp. Neurol.* **268:**329–345.

Pflugfelder, O., 1937, Bau, Entwicklung und Funktion der Corpora allata und cardiaca von *Dixippus morosus* Br., *Z. Wiss. Zool.* **149:**477–512.

Pflugfelder, O., 1938, Farbveränderungen und Gewebsentartigen nach Nervendurchschneidung und Extirpation der Corpora allata von *Dixippus morosus, Verh. Dtsch. Zool. Ges.* **40:**127–137.

Phillips, J. E., 1983, Osmoregulation and excretion, *Verh. Dtsch. Zool. Ges.* **1983:**55–68.

Phillips, J. E., Mordue, W., Meredith, J., and Spring, J., 1980, Purification and characteristics of the chloride transport stimulating factor from locust corpora cardiaca: A new peptide, *Can. J. Zool.* **10:**1851–1860.

Pickering, B. T., 1980, Lessons from a peptidergic neurone, *Nature* **288:**117–118.

Piek, T., and Mantel, P., 1977, Myogenic contractions in locust muscle induced by proctolin and by wasp, *Philanthus triangulum,* venom, *J. Insect Physiol.* **23:**321–325.

Piek, T., Visser, B. J., and Mantel, P., 1979, Effect of proctolin and related peptides on rhythmic contractions in *Locusta migratoria, Comp. Biochem. Physiol.* **62C:**151–154.

Pilcher, D. E. M., 1970, Hormonal control of the Malpighian tubules of the stick insect, *Carausius morosus, J. Exp. Biol.* **52:**653–665.

Pilcher, D. E. M., 1971, Stimulation of movements of Malpighian tubules of *Carausius* by pharmacologically active substances and tissue extracts, *J. Insect Physiol.* **17:**463–470.

Pimley, R. W., 1983, Neuroendocrine stimulation of uterine gland protein synthesis in the tsetse fly, *Glossina morsitans, Physiol. Entomol.* **8:**429–437.

Pimley, R. W., 1985, Cyclic AMP and calcium mediate the regulation of fat cell activity by octopamine and peptide hormones in *Glossina morsitans, Insect Biochem.* **15:**293–298.

Pimley, R. W., and Langley, P. A., 1982, Hormone stimulated lipolysis and proline synthesis in the fat body of the adult tsetse fly, *Glossina morsitans, J. Insect Physiol.* **28:**781–789.

Pipa, R., 1986, Disinhibition of oocyte growth in adult, virgin *Periplaneta americana* by corpus allatum denervation: Age dependency and relatedness to mating, *Arch. Insect Biochem. Physiol.* **3:**471–483.

Pipa, R., 1988, Commissural ring nerve: A female-specific neurosecretory tract supplied by bifurcating median neurons in the cockroach *Periplaneta americana* (L.) and the cricket *Teleogryllus commodus* (Walker), *Cell Tissue Res.* **251:**333–338.

Pipa, R. L., and Novak, F. J., 1979, Pathways and fine structure of neurons forming the nervi corporis allati II of the cockroach *Periplaneta americana* (L.), *Cell Tissue Res.* **201:**227–237.

Pitman, R. M., 1985, Nervous system, in *Comprehensive Insect Physiology, Biochemistry and Pharmacology* (G. A. Kerkut and L. I. Gilbert, eds.), Vol. 8, Pergamon Press, New York, pp. 491–550.

Pitman, R. M., Tweedle, C. D., and Cohen, D. J., 1973, The form of nerve cells: Determination by cobalt impregnation, in: *Intracellular Staining in Neurobiology* (B. Kater and C. Nicholson, eds.), Springer-Verlag, Berlin, pp. 83–97.

Platt, N., and Reynolds, S. E., 1985, Cardioactive peptides from the CNS of a caterpillar, the tobacco hornworm, *Manduca sexta, J. Exp. Biol.* **114:**397–414.

Platt, N., and Reynolds, S. E., 1986, The pharmacology of the heart of a caterpillar, the tobacco hornworm, *Manduca sexta, J. Insect Physiol.* **32:**221–230.

Plotnikova, S. I., and Govyrin, V. A., 1966, Distribution of catecholamine-containing nerve elements in some representatives of *Protostomia* and *Coelenterata, Arch. Anat. Hist. Embryol., Leningrad* **50:**79–87.

Pluot, D., 1979, Evolution régressive des ovarioles chez les coléoptères *Scarabeinae, Ann. Soc. Entomol. Fr.* **15:**575–587.

Pohlhammer, K., 1968, Zur Hormonalen Steuerung der Ovarentwicklung bei den weiblichen Larven der heterogonen Gallmücke *Heteropeza pygmaea* Winnertz 1846 (Diptera, Cecidomyiidae), *Zool. Jahrb. Abt. Allg. Zool. Physiol. Tiere* **74:**1–30.

Poras, M., 1982, Le contrôle endocrinien de la diapause imaginale des femelles de *Tetrix undulata* (Sowerby, 1806) (Orthoptère, Tetrigidae), *Gen. Comp. Endocrinol.* **46:**200–210.

Poras, M., 1983, L'innervation protocérébrale des corps allates chez *Locusta migratoria* (Orthoptères), *C. R. Acad. Sci. Ser. III* **297:**483–488.

Poras, M., Baehr, J. C., and Cassier, P., 1983, Control of corpus allatum activity during the imaginal diapause in females of *Locusta migratoria* L., *Int. J. Invert. Reprod. Dev.* **6:**111–122.

Possompès, B., 1953, Recherches expérimentales sur le déterminisme de la métamorphose de *Calliphora erythrocephala* Meig., *Arch. Zool. Exp. Gen.* **89:**203–364.

Possompès, B., 1957, Capacité d'accession à l'état imaginal des formes juvéniles du phasme *Sipyloidea sipylus* W., après ablation des corpora allata, *C. R. Acad. Sci.* **245:**2404–2406.

Postlethwait, J. H., and Handler, A. M., 1979, The role of juvenile hormone and 20-hydroxyecdysone during vitellogenesis in isolated abdomens of *Drosophila melanogaster, J. Insect Physiol.* **25:**455–460.

Postlethwait, J. H., Lauge, G., and Handler, A. M., 1980, Yolk protein synthesis in ovariectomized and genetically agametic (X^{87}) *Drosophila melanogaster, Gen. Comp. Endocrinol.* **40:**385–390.

Prasad, O., 1981, Optic lobe neurosecretory cells in *Poekilocerus pictus* (Orthoptera), *Periplaneta americana* (Dictyoptera), *Belostoma indicum* (Hemiptera), *Polistes hebraeus* (Hymenoptera) and *Sandrocottus de jeani* (Coleoptera), *Z. Mikrosk. Anat. Forsch.* **95:**477–483.

Prasad, O., and Srivastava, V. K., 1984, On the intracerebral neurohaemal organ in adult *Poekilocerus pictus* (Orthoptera: Acrididae), *Curr. Sci.* **53:**553–554.

Pratt, G. E., and Davey, K. G., 1972, The corpus allatum and ocogenesis in *Rhodnius prolixus* (Stal.). I. The effects of allatectomy, *J. Exp. Biol.* **56:**201–214.

Pratt, G. E., and Tobe, S. S., 1974, Juvenile hormones radiobiosynthesized by corpora allata of adult female locusts in vitro, *Life Sci.* **14:**575–586.

Price, D. A., and Greenberg, M. J., 1977, Structure of a molluscan cardioexcitatory neuropeptide, *Science* **197:**670–671.

Proux, B., Proux, J., and Phillips, J. E., 1984, Antidiuretic action of a corpus cardiacum factor (CTSH) on long-term fluid absorption across locust recta in vitro, *J. Exp. Biol.* **113:**409–421.

Proux, J., 1978a, Influence de saignées répétées sur le comportement pondéral et hydrique du criquet migrateur privé de cellules neurosécrétrices sous–ocellaires médianes diurétiques, *C.R. Acad. Sci. Ser. D* **286:**981–984.

Proux, J., 1978b, Section de la chaîne nerveuse ventrale et comportement hydrique après une saignée chez le criquet migrateur, *C.R. Acad. Sci. Ser. D* **287:**305–307.

Proux, J., 1979, Influence de l'électrocoagulation de la pars intercerebralis et des cellules neurosécrétrices sous-ocellaires médianes sur le comportement pondéral, hydrique et alimentaire du criquet migrateur, *Bull. Soc. Zool. Fr.* **104:**89–103.

Proux, J., and Buscarlet, L. A., 1976, Etude à l'aide d'eau tritiée, du rôle des cellules neurosécrétrices sous-ocellaires médianes dans la régulation hydrique des males de *Locusta migratoria migratorioides* (F. et F.), *Acrida* **5:**311–320.

Proux, J., and Girardie, A., 1982, Neurosecretory regulation of the haemolymphatic titre of a vasopressin-like substance in the migratory locust, *Neurosci. Letters* **33:**73–77.

Proux, J., and Rougon-Rapuzzi, G., 1980, Evidence for vasopressin-like molecule in migratory locust. Radioimmunological measurements in different tissues: Correlation with various states of hydration, *Gen. Comp. Endocrinol.* **42:**378–383.

Proux, J., Rougon-Rapuzzi, G., and Remy, C., 1980, Influence de la pars intercerebralis et des cellules neurosécrétrices sous-ocellaires médianes sur la teneur en substance apparentée à la vasopressine dans la chaîne nerveuse ventrale et l'hémolymphe du criquet migrateur. Etude radio-immunologique et immunohistologique, *J. Physiol. (Paris)* **76:**277–282.

Proux, J., Rougon, G., and Cupo, A., 1982, Enhancement of excretion across locust Malpighian tubules by a diuretic vasopressin-like hormone, *Gen. Comp. Endocrinol.* **47:**449–457.

Proux, J. P., Delaage, M. P., and Chauveau, J., 1987, Immunoaffinity purification of a diuretic hormone from the nervous ventral cord of the migratory locust, *Comp. Biochem. Physiol.* **88B:** 807–811.

Provansal, A., 1968, Mise en évidence d'organes neurohémaux métamériques associés à la chaîne nerveuse ventrale chez *Vespa crabro* L. et *Vespula germanica* Fabr., *C.R. Acad. Sci. Ser. D* **267:**864–867.

Provansal, A., Baudry, N., and Raabe, M., 1970, Recherches sur l'ultrastructure des organes neurohémaux périsympathiques des *Vespidae* (Hyménoptères). Les organes médians sphériques, *C.R. Acad. Sci. Ser. D* **271:**1115–1118.

Provansal, A., Baudry-Partiaoglou, N., and Sláma, K., 1977, Haemolymph pressure pulses in the metamorphosis of *Tenebrio molitor, Acta. Ent. Bohemoslov.* **74:**362–374.

Provansal-Baudez, A., 1983, Physiological roles of perisympathetic organs in insects, in *Neurohemal organs of Arthropods* (A. P. Gupta, ed.), Charles C Thomas, pp. 581–601.

Provansal-Baudez, A., and Baudry-Partiaoglou, N., 1983, Ultrastructural comparison of the perisympathetic organs in three coleoptera: *Chrysocarabus auronitens* F., *Oryctes rhinoceros* L. and *Tenebrio molitor* L., *Gen. Comp. Endocrinol.* **49:**383–403.

Provansal-Baudez, A., and Sláma, K., 1985, Effect of perisympathetic organs on extracardiac pulsations in *Tenebrio molitor* (Coleoptera), *Acta. Ent. Bohemoslov.* **82:**161–169.

Quistad, G. B., Adams, M. E., Scarborough, R. M., Carney, R. L., and Schooley, D. A., 1984, Metabolism of proctolin, a pentapeptide neurotransmitter in insects, *Life Sci.* **34:**569–576.

Quo, F., 1959, Studies on the reproduction of the oriental migratory locust: The physiological effects of castration and copulation, *Acta Entomol. Sinica.* **9:**464–476.

Quo, F., 1965, Studies on the reproduction of the oriental migratory locust: The role of the corpora allata, *Acta Entomol. Sinica.* **14:**211–224.

Raabe, M., 1961, Recherches sur le déterminisme des genèses de pigments chez le phasme, *Carausius morosus, C.R. Acad. Sci.* **252:**3663–3665.

Raabe, M., 1963a, Mise en évidence, chez des insectes d'ordres variés, d'éléments neurosécréteurs tritocérébraux, *C.R. Acad. Sci.* **257:**1171–1173.

Raabe, M., 1963b, Existence chez divers insectes d'une innervation tritocérébrale des corpora cardiaca, *C.R. Acad. Sci.* **257:**1552–1555.

Raabe, M., 1963c, Recherches expérimentales sur la localisation intracérébrale du facteur chromactif des insectes, *C.R. Acad. Sci.* **257:**1804–1806.

Raabe, M., 1965, Etudes des phénomènes de neurosécrétion au niveau de la chaîne nerveuse ventrale des phasmides, *Bull. Soc. Zool. Fr.* **90:**631–654.

Raabe, M., 1966, Recherches sur la neurosécrétion dans la chaîne nerveuse ventrale du phasme, *Carausius morosus:* liaison entre l'activité des cellules B1 et la pigmentation, *C.R. Acad. Sci. Ser. D* **263:**408–411.

Raabe, M., 1972, Les organes perisympathiques des Dermaptères, *C.R. Acad. Sci. Ser. D* **275:** 2925–2929.

Raabe, M., 1982, *Insect Neurohormones,* Plenum Press, New York.

Raabe, M., 1986a, Comparative immunocytochemical study of release sites of insulin, glucagon and AKH-like products in *Locusta migratoria, Periplaneta americana,* and *Carausius morosus, Cell Tissue Res.* **245:**267–271.

Raabe, M., 1986b, Insect reproduction: Regulation of successive steps, *Adv. Insect Physiol.* **19:**29–154.

Raabe, M., and Demarti-Lachaise, F., 1973, Etude de l'action des organes périsympathiques sur la protéinémie due phasme, *Carausius morosus, C.R. Acad. Sci. Ser. D* **277:**2037–2040.

Raabe, M., and Provansal, A., 1972, Les organes périsympathiques des paléoptères, *C.R. Acad. Sci. Ser. D* **275:**925–928.

Raabe, M., and Ramade, F., 1967, Observations sur l'ultrastructure des organes périsympathiques des phasmides, *C.R. Acad. Sci. Ser. D* **264:**77–80.

Raabe, M., Cazal, M., Chalaye, D., and De Besse, N., 1966, Action cardioaccélératrice des organes neurohémaux périsympathiques ventraux de quelques insectes, *C.R. Acad. Sci. Ser. D* **263:** 2002–2005.

Raabe, M., Baudry, N., and Provansal, A., 1970, Recherches sur l'ultrastructure des organes neurohémaux périsympathiques des Vespidae (Hyménoptères). Les organes latéraux longitudinaux, *C.R. Acad. Sci. Ser. D* **271:**1210–1213.

Raabe, M., Baudry, N., Grillot, J. P., and Provansal, A., 1974, The perisympathetic organs of insects, in *Neurosecretion. The Final Neuroendocrine Pathway,* VIth International Symposium on Neurosecretion, London (F. Knowles and L. Vollrath, eds.), Springer-Verlag, Berlin, pp. 59–71.

Raabe, M., Panov, A. A., Davydova, E. D., and Chervin, D., 1979, Diversité des produits de neurosécrétion de la pars intercerebralis des insectes, *Experientia* **35:**404–405.

Raabe, M., Hiripi, L., and Vero, M., 1985, Estimation of serotonin and its action on oviducts and uteri of some oviparous and viviparous insects, *Experientia* **41:**118–120.

Racioppi, J. V., Gemmill, R. M., Kogan, P. H., Calvo, J. M., and Hagedorn, H. H., 1986, Expression and regulation of vitellogenin messenger RNA in the mosquito, *Aedes aegypti, Insect Biochem.* **16:**255–262.

Rademakers, L. H. P. M., 1977a, Effects of isolation and transplantation of the corpus cardiacum on hormone release from its glandular cells after flight in *Locusta migratoria.* A quantitative electron microscopical study, *Cell Tissue Res.* **184:**213–224.

Rademakers, L. H. P. M., 1977b, Identification of a secretomotor centre in the brain of *Locusta migratoria,* controlling the secretory activity of the adipokinetic hormone producing cells of the corpus cardiacum, *Cell Tissue Res.* **184:**381–395.

Rademakers, L. H. P. M., and Beenakkers, A. M. T., 1977, Changes in the secretory activity of the glandular lobe of the corpus cardiacum of *Locusta migratoria* induced by flight, *Cell Tissue Res.* **180:**155–171.

Rafaeli, A., and Mordue, W., 1982, The responses of the Malpighian tubules of *Locusta* to hormones and other stimulants, *Gen. Comp. Endocrinol.* **46:**130–135.

Rafaeli, A., Pines, M., Stern, P. S., and Applebaum, S. W., 1984, Locust diuretic hormone-stimulated synthesis and excretion of cyclic-AMP: A novel Malpighian tubule bioassay, *Gen. Comp. Endocrinol.* **54:**35–42.

Rafaeli, A., Moshitzky, P., and Applebaum, S. W., 1987, Diuretic action and immunological cross-reactivity of corticotropin and locust diuretic hormone. *Gen. Comp. Endocrinol.* **67:**1–6.

Raikhel, A. S., 1987, Monoclonal antibodies as probes for processing of the mosquito yolk protein; a high-resolution immunolocalization of secretory and accumulative pathways, *Tissue Cell* **19:** 515–529.

Raikhel, A. S., and Lea, A. O., 1982, Abnormal vitelline envelope induced by unphysiological doses of ecdysterone in *Aedes aegypti, Physiol. Entomol.* **7:**55–64.

Raikhel, A. S., and Lea, A. O., 1985, Hormone-mediated formation of the endocytic complex in mosquito oocytes, *Gen. Comp. Endocrinol.* **57:**422–433.

Raikhel, A. S., and Lea, A. O., 1986, Internalized proteins directed into accumulative compartments of mosquito oocytes by the specific ligand, vitellogenin, *Tissue Cell* **18:**559–574.

Raina, A. K., and Bell, R. A., 1978, Morphology of the neuroendocrine system of the pink bollworm and histological changes in the neurosecretory cells of the brain during induction, maintenance, and termination of diapause, *Ann. Ent. Soc. Am.* **71:**375–382.

Raina, A. K., and Klun, J. A., 1984, Brain factor control of sex pheromone production in the female corn earworm moth, *Science* **225:**531–533.

Ralph, C. L., 1962a, Action of extracts of cockroach nervous system on fat bodies in vitro, *Am. Zoologist* **2:**362.

Ralph, C. L., 1962b, Heart accelerators and decelerators in the nervous system of *Periplaneta americana* L., *J. Insect Physiol.* **8:**431–440.

Ralph, C. L., and Matta, R. J., 1965, Evidence for hormonal effects on metabolism of cockroaches from studies of tissue homogenates, *J. Insect Physiol.* **11:**983–991.

Ralph, C. L., and McCarthy, R., 1964, Effects of brain and corpus cardiacum extracts on haemolymph trehalose of the cockroach, *Periplaneta americana, Nature* **203:**1195–1196.

Ramade, F., and L'Hermite, P., 1971, Mise en évidence de neurones adrénergiques par la microscopie de fluorescence dans le système central de *Calliphora erythrocephala* Meig., et de *Musca domestica* L., *C.R. Acad. Sci. Ser. D* **272:**3314–3317.

Ramamurty, P. S., and Engels, W., 1977, Allatektomie- und Juvenilhormon-Wirkungen auf Synthese und Einlagerung von Vitellogenin bei der Bienenkönigin (*Apis mellifica*), *Zool. Jahrb. Abt. Allg. Zool. Physiol. Tiere* **81:**165–176.

Ramsay, J. A., 1954, Active transport of water by the Malpighian tubules of the stick insect, *Dixippus morosus* (Orthoptera, Phasmidae), *J. Exp. Biol.* **31:**104–113.

Rane, S. G., Gerlach, P. H., and Wyse, G. A., 1984, Neuromuscular modulation in *Limulus* by both octopamine and proctolin, *J. Neurobiol.* **15:**173–196.

Rankin, M. A., and Jackle, H., 1980, Hormonal control of vitellogenin synthesis in *Oncopeltus fasciatus, J. Insect Physiol.* **26:**671–684.

Rankin, M. A., and Riddiford, L. M., 1978, Significance of haemolymph juvenile hormone titer changes in timing of migration and reproduction in adult *Oncopeltus fasciatus, J. Insect Physiol.* **24:**31–38.

Rankin, S. M., and Stay, B., 1984, The changing effect of the ovary on rates of juvenile hormone synthesis in *Diploptera punctata, Gen. Comp. Endocrinol.* **54:**382–388.

Rankin, S. M., and Stay, B., 1985, Ovarian inhibition of juvenile hormone synthesis in the viviparous cockroach, *Diploptera punctata, Gen. Comp. Endocrinol.* **59:**230–237.

Rankin, S. M., and Stay, B., 1987, Distribution of allatostatin in the adult cockroach, *Diploptera punctata* and effects on corpora allata in vitro, *J. Insect Physiol.* **33:**551–558.

Rankin, S. M., Stay, B., Aucoin, R. R., and Tobe, S. S., 1986, In vitro inhibition of juvenile hormone synthesis by corpora allata of the viviparous cockroach, *Diploptera punctata, J. Insect Physiol.* **32:**151–156.

Rasenick, M. M., Neuberg, M., and Berry, S. J., 1976, Brain cyclic levels and the initiation of adult development in the Cecropia silkmoth, *J. Insect Physiol.* **22:**1453–1456.

Ravetto, C., 1964, Alcian blue-alcian yellow: A new method for the identification of different acidic groups, *J. Histochem. Cytochem.* **12:**44–45.

Ray, A., Kumar, D., and Ramamurty, P. S., 1980, Histological study on the retrocerebral-endocrine complex with special reference to neurohaemal involvement of aorta and pericardial cells in *Coccinella septempunctata* (L.) (Coccinellidae - Coleoptera), *Z. Mikrosk. Anat. Forsch.* **94:** 1141–1148.

Raziuddin, M., Khan, T. R., and Singh, S. B., 1979, Neuropilar neurosecretory fibres in the grasshopper, *Poekilocerus pictus* (Fabr.) and the cockroach, *Periplaneta americana, Zool. Anz.* **202:**209–218.

Readio, J., and Meola, R., 1985, Two stages of juvenile hormone-mediated growth of secondary follicles in *Culex pipiens, J. Insect Physiol.* **31:**559–562.

Readio, J., Peck, K., Meola, R., and Dahm, K. H., 1988, Corpus allatum activity (*in vitro*) in female *Culex pipiens* during adult life cycle, *J. Insect Physiol.* **34:**131–135.

Reddy, G., and Kumaran, A. K., 1985, The effect of juvenile hormone and its antagonists on JH esterase activity in *Tenebrio molitor, Entomol. Exp. Appl.* **37:**213–218.

Reddy, G., Hwang-Hsu, K., and Kumaran, A. K., 1979, Factors influencing juvenile hormone esterase activity in the wax moth, *Galleria mellonella, J. Insect Physiol.* **25:**65–71.

Redfern, C. P. F., 1982, 20-Hydroxyecdysone and ovarian development in *Anopheles stephensi, J. Insect Physiol.* **28:**97–109.

Regis, L., 1977, Functional compensatory hypertrophy resulting from spontaneous or induced atrophy disconnecting one of the ovaries of *Triatoma infestans* (Heteroptera, Reduviidae, Triatominae), *Ann. Biol. Anim. Biochem. Biophys.* **17:**961–969.

Regis, L., 1979, The role of the blood meal in egg-laying periodicity and fecundity in *Triatoma infestans, Int. J. Invert. Reprod. Dev.* **1:**187–195.

Rehder, V., Bicker, G., and Hammer, M., 1987, Serotonin-immunoreactive neurons in the antennal lobes and suboesophageal ganglion of the honey-bee, *Cell Tissue Res.* **247:**59–66.

Rehm, M., 1955, Morphologische und Histochemische Untersuchungen an neurosekretorischen Zellen von Schmetterlingen, *Z. Zellforsch.* **42:**19–58.

Reichhart, J. M., and Charlet, M., 1986, Ecdysiotropic activity in brains and corpora cardiaca of larvae and adults of *Locusta migratoria:* In vitro assays, *Int. J. Invert. Reprod. Dev.* **10:**17–25.

Reinecke, J. P., Gerst, J., and O'Gara, B., and Adams, T. S., 1978, Innervation of hindgut muscle of larval *Manduca sexta* (L.) (Lepidoptera: Sphingidae) by a peripheral multinucleate neurosecretory neuron, *Int. J. Insect Morphol. Embryol.* **7:**435–453.

Reissig, W. H., and Kamm, T. A., 1975, Effect of synthetic juvenile hormone on reproduction of adult male *Draeculacephala crassicornis, Ann. Ent. Soc. Am.* **68:**353–354.

Rembold, H., 1976, The role of determination in caste formation in the honey-bee, in *Phase and Caste Determination in Insects* (M. Lüscher, ed.), Pergamon Press, New York, pp. 21–33.

Rembold, H., 1987, Caste specific modulation of juvenile hormone titers in *Apis mellifera, Insect Biochem.* **17:**1003–1006.

Rembold, H., Czoppelt, C., and Rao, P. J., 1974, Effect of juvenile hormone treatment on caste differentiation in the honey-bee, *Apis mellifera, J. Insect Physiol.* **20:**1193–1202.

Rembold, H., Eder, J., and Ulrich, G. M., 1980, Inhibition of allatotropic activity and ovary development in *Locusta migratoria* by anti-brain-antibodies, *Z. Naturforsch. Sect C. Biol. Sci.* **35:**1117–1119.

Rembold, H., Schlagintweit, B., and Ulrich, G. M., 1986, Activation of juvenile hormone synthesis in vitro by a corpus cardiacum factor from *Locusta migratoria, J. Insect Physiol.* **32:**91–94.

Rémy, C., 1982, Parentés immunochimiques entre produits de neurosécrétion d'invertébrés et neuropeptides de vertébrés, *J. Physiol. (Paris)* **78:**514–522.

Rémy, C., and Dubois, M. P., 1981, Immunohistological evidence of methionine enkephalin-like material in the brain of the migratory locust, *Cell Tissue Res.* **218:**271–278.

Rémy, C., and Girardie, J., 1980, Anatomical organization of two vasopressin-neurophysin-like neurosecretory cells throughout the central nervous system of the migratory locust, *Gen. Comp. Endocrinol.* **40:**27–35.

Rémy, C., Girardie, J., and Dubois, M. P., 1977, Exploration immunocytologique des ganglions cérébroïdes et sous-oesophagien du phasme *Clitumnus extradentatus:* Existence d'une neurosécrétion apparentée à la vasopressine-neurophysine, *C.R. Acad. Sci. Ser. D* **285:**1495–1497.

Rémy, C., Girardie, J., and Dubois, M. P., 1978, Présence dans le ganglion sous-oesophagien de la chenille processionnaire du pin (*Thaumetopoea pityocampa* Schiff) de cellules révélées en immunofluorescence par un anti-corps anti-α-endorphine, *C.R. Acad. Sci. Ser. D* **286:**651–653.

Rémy, C., Girardie, J., and Dubois, M. P., 1979, Vertebrate neuropeptide-like in the suboesophageal ganglion of two insects: *Locusta migratoria* R. et F. (Orthoptera) and *Bombyx mori* L. (Lepidoptera) immunocytological investigation, *Gen. Comp. Endocrinol.* **37:**93–100.

Rémy, C., Guy, J., Pelletier, G., and Boer, H. H., 1988, Immunohistological demonstration of a substance related to neuropeptide Y and FMRFamide in the cephalic and thoracic nervous systems of the locust *Locusta migratoria, Cell Tissue Res.* **254:**189–195.

Renucci, M., Strambi, A., and Augier, R., 1985, Physiological approach to the onset of receptivity in females *Acheta domesticus.* I. Role of the corpora allata and ovaries, *Hormones and Behavior* **19:**441–453.

Retnakaran, A., and Joly, P., 1976, Neurosecretory control of juvenile hormone inactivation in *Locusta migratoria* L., in *Actualités sur les Hormones d'Invertébrés,* Coll. Int. C.N.R.S. n° 251, Villeneuve d'Ascq 1975, pp. 317–323 (C.N.R.S., ed.).

Retnakaran, A., Grisdale, D., and Percy, J., 1985, Regulation of juvenile hormone metabolism in the oleander hawkmoth, *Deilephila nerii* (L.) *Comp. Biochem. Physiol.* **82A:**117–124.

Reynolds, S. E., 1974, Pharmacological induction of plasticization in the abdominal cuticle of *Rhodnius, J. Exp. Biol.* **61:**705–718.

Reynolds, S.E., 1976, Hormonal regulation of cuticle extensibility in newly emerged adult blowflies, *J. Insect Physiol.* **22:**529–534.

Reynolds, S. E., 1977, Controle of cuticle extensibility in the wings of adult *Manduca* at the time of eclosion: Effects of eclosion hormone and bursicon, *J. Exp. Biol.* **70:**27–39.

Reynolds, S. E., Taghert, P. H., and Truman, J. W., 1979, Eclosion hormone and bursicon titres and the onset of hormonal responsiveness during the last day of adult development in *Manduca sexta* (L.), *J. Exp. Biol.* **78:**77–86.

Richard, D. S., and Saunders, D. S., 1987, Prothoracic gland function in diapause and non-diapause *Sarcophaga argyrostoma* and *Calliphora vicina, J. Insect Physiol.* **33:**385–392.

Richard, D. S., Warren, J. T., Saunders, D. S., and Gilbert, L. I., 1987, Haemolymph ecdysteroid titres in diapause- and non-diapause-destined larvae and pupae of *Sarcophaga argyrostoma, J. Insect Physiol.* **33:**115–122.

Richard-Mercier, N., 1982, Le contrôle de la différenciation sexuelle des insectes: Etude chez un coléoptère Chrysomélidé, le Doryphore, *Année Biol.* **21:**359–394.

Richter, K., 1973, Struktur und Funktion der Herzen wirbelloser Tiere, *Zool. Jahrb. Abt. Allg. Zool. Physiol. Tiere* **77:**477–668.

Richter, K., 1983, Experimentelle Untersuchungen der nervösen Steuerung der Prothorakaldrüse bei *Periplaneta americana, Zool. Jahrb. Abt. Allg. Zool. Physiol. Tiere* **87:**65–78.

Richter, K., 1985, Physiological investigation on the role of the innervation in the regulation of the prothoracic gland in *Periplaneta americana, Arch. Insect Biochem. Physiol.* **2:**319–329.

Richter, K., and Gersch, M., 1981, Die elektrophysiologische Aktivität der Nervi corporis cardiaci I und II von *Periplaneta americana* in Beziehung zum Häutungszyklus, *Zool. Jahrb. Abt. Allg. Zool. Physiol. Tiere* **85:**412–425.

Richter, K., and Gersch, M., 1983, Electrophysiological evidence of nervous involvement in the control of the prothoracic gland in *Periplaneta americana, Experientia* **39:**917–918.

Richter, K., and Stürzebecher, J., 1969, Elektrophysiologische Untersuchungen zum Wirkungsmechanismus von Neurohormon D am lateralen Herznerven von *Periplaneta americana* (L.), *Z. Wiss. Zool. Abt. A* **180:**148.

Riddiford, L. M., 1980, Interaction of ecdysteroids and juvenile hormone in the regulation of larval growth and metamorphosis of the tobacco hornworm, in *Progress in Ecdysone Research* (J. A. Hoffman, ed.), Elsevier North-Holland, Amsterdam, pp. 409–430.

Riddiford, L. M., 1985, Hormone action at the cellular level, in *Comprehensive Insect Physiology, Biochemistry and Pharmacology* (G. A. Kerkut and L. I. Gilbert, eds.), Pergamon Press, New York, pp. 37–84.

Riddiford, L. M., and Ashenhurst, J., 1973, The switchover from virgin to mated behavior in female Cecropia moths; the role of the bursa copulatrix, *Biol. Bull. (Woods Hole, Mass.)* **144:**162–171.

Riddiford, L. M., and Williams, C. M., 1971, Role of the corpora cardiaca in the behavior of saturniid moths, I. Release of sex pheromone, *Biol. Bull. (Woods Hole, Mass.)* **140:**1–7.

Riemann, J. G., Moen, D. J., and Thorson, B. J., 1967, Female monogamy and its control in houseflies, *J. Insect Physiol.* **13:**407–418.

Robert, A., 1979, Les premiers stades de l'ovogenèse et les variations de la neurosécrétion cérébrale chez deux espèces sympatriques *Roscius elongatus* Stal. et *R. brazzavilliensis* Robert (Heteroptera: Pyrrhocoridae), *Int. J. Insect Morphol. Embryol.* **8:**11–31.

Robert, A., Grillot, J. P., Guilleminot, J., and Raabe, M., 1984, Experimental and ultrastructural study of the control of ovulation and parturition in the tsetse fly *Glossina fuscipes* (Diptera), *J. Insect Physiol.* **30:**671–684.

Robert, A., Strambi, A., and Strambi, C., 1986a, Haemolymph ecdysteroid levels in female tsetse fly *Glossina fuscipes* (Diptera) during the first reproductive cycle: A comparison between virgin and mated females, *J. Insect Physiol.* **32:**665–671.

Robert, A., Strambi, A., Strambi, C., and Gonella, J., 1986b, Opposite effects of ecdysone and 20-hydroxyecdysone on in vitro uterus motility of a tsetse fly, *Life Sci.* **39:**2617–2622.

Roberts, B., and Gilbert., L. I., 1986, Ring gland and prothoracic gland sensitivity to interspecific prothoracicotropic hormone extracts, *J. Comp. Physiol. B.* **156:**767–771.

Roberts, B., Baker, M., Kotzman, M., and Wentworth, S. L., 1982, A possible role of ecdysteroids in pre-ecdysial tanning in larvae of *Sarcophaga bullata* (Diptera: Sarcophagidae), *J. Insect Physiol.* **28:**123–127.

Roberts, B., Gilbert, L. I., and Bollenbacher, W. E., 1984, In vitro activity of dipteran ring glands and activation by the prothoracicotropic hormone, *Gen. Comp. Endocrinol.* **54:**469–477.

Robertson, H. A., and Juorio, A. V., 1976, Octopamine and some related noncatecholic amines in invertebrate nervous system, *Int. Rev. Neurobiol.* **19:**173–224.

Robinson, N. L., and Goldsworthy, G. J., 1976, Adipokinetic hormone and flight metabolism in the locust, *J. Insect Physiol.* **22:**1559–1564.

Röhrkasten, A., and Ferenz, H. J., 1985, In vitro study of selective endocytosis of vitellogenin by locust oocytes, *Roux's Arch. Dev. Biol.* **194:**411–416.

Röhrkasten, A., and Ferenz, H. J., 1986, Properties of the vitellogenin receptor of isolated locust oocyte membranes, *Int. J. Invert. Reprod. Dev.* **10:**133–142.

Röller, H., 1962, Über den Einfluss der Corpora allata auf den Stoffwechsel der Wachsmotte, *Naturwissenschaften* **49:**524–525.

Röller, H., Piepho, H., and Holz, I., 1963, Zum Problem der Hormonabhängigkeit des Paasrungsverhaltens bei Insekten. Untersuchungen an *Galleria mellonella* L., *J. Insect Physiol.* **9:**187–194.

Röller, H., Dahm, K. H., Sweely, C. C., and Trost, B. M., 1967, The structure of the juvenile hormone, *Angewandte Chemie* **6:**179–180.

Romer, F., Emmerich, H., and Nowock, J., 1974, Biosynthesis of ecdysones in isolated prothoracic glands and oenocytes of *Tenebrio molitor* in vitro, *J. Insect Physiol.* **10:**1975–1987.

Romeuf, M., and Rémy, C., 1984, Early immunohistochemical detection of somatostatin-like and methionine-enkephalin-like neuropeptides in the brain of the migratory locust embryo, *Cell Tissue Res.* **236:**289–292.

Romeuf, M., Proux, J., and Rémy, C., 1986, The neurophysin-vasopressin-like system in migratory locust embryos. Immunohistological and radioimmunological studies, *Neurochem. Int.* **8:**299–302.

Röseler, P. F., 1977, Juvenile hormone control of oögenesis in bumblebee workers, *Bombus terrestris, J. Insect Physiol.* **23:**985–992.

Röseler, P. F., Röseler, I., and Strambi, A., 1980, The activity of corpora allata in dominant and subordinated females of the wasp *Polistes gallicus, Insect Soc.* **27:**97–107.

Röseler, P. F., Röseler, I., and Van Honk, C. G. J., 1981, Evidence for inhibition of corpora allata activity in workers of *Bombus terrestris* by a pheromone from the queen's mandibular glands, *Experientia* **37:**348–351.

Rosinski, G., and Gäde, G., 1988, Hyperglycaemic and myoactive factors in the corpora cardiaca of the mealworm, *Tenebrio molitor, J. Insect Physiol.* **34:**1035–1042.

Rossignol, P. A., Feinsod, F. M., and Spielman, A., 1981, Inhibitory regulation of corpus allatum activity in mosquitoes, *J. Insect Physiol.* **27:**651–654.

Roth, L. M., and Barth, R. H., 1964, The control of sexual receptivity in female cockroaches, *J. Insect Physiol.* **10:**965–975.

Rounds, H. D., 1963, A functional division in the nervous system of the cockroach, *Blaberus giganteus* (L.), *Acta Zool.* **44:**43–55.

Rounds, H. D., and Gardner, F., 1968, A quantitative comparison of the activity of cardioacceleratory extracts from various portions of the cockroach nerve cord, *J. Insect. Physiol.* **14:**495–497.

Rountree, D. B., and Bollenbacher, W. E., 1986, The release of the prothoracicotropic hormone in the tobacco hornworm, *Manduca sexta,* is controlled intrinsically by juvenile hormone, *J. Exp. Biol.* **120:**41–58.

Roussel, J. P., 1966, Rôle des corpora allata sur la pigmentation de *Gryllus bimaculatus* Deb., *J. Insect Physiol.* **12:**1085–1092.

Roussel, J. P., 1967, Fonctions des corpora allata et contrôle de la pigmentation chez *Gryllus bimaculatus* de Geer, *J. Insect Physiol.* **13:**113–130.

Roussel, J. P., 1972, Innervation du coeur et régulation nerveuse du rythme cardiaque chez *Locusta migratoria* L., *Arch. Zool. Exp. Gen.* **113:**265–294.

Rowell, C. H. F., 1967, Corpus allatum implantation and green/brown polymorphism in three African grasshoppers, *J. Insect Physiol.* **13:**1401–1412.

Rowell, C. H. F., 1976, The cells of the insect neurosecretory system: Constancy, variability, and the concept of the unique identifiable neuron, in *Advances in Insect Physiology,* Vol. 12, Academic Press, London, pp. 63–123.

Rubenstein, E. C., Kelly, T. J., Schwartz, M. B., and Woods, C. W., 1982, In vitro synthesis and secretion of ecdysteroids by *Drosophila melanogaster* ovaries, *J. Exp. Zool.* **223:**305–308.

Rubenstein, E. C., Kelly, T. J., Culbert, L., Woods, C. W., and Weeks, B., 1986, Ecdysteroids in developing ovarian follicles of *Hyalophora cecropia, Arch. Insect Biochem. Physiol.* **3:**253–263.

Ruegg, R. P., and Davey, K. G., 1979, The effect of C18 juvenile hormone and altosid on the efficiency of egg production in *Rhodnius prolixus, Int. J. Invert. Reprod.* **1:**3–8.

Ruegg, R. P., Kriger, F. L., Davey, K. G., and Steel, C. G. H., 1981, Ovarian ecdysone elicits release of a myotropic ovulation hormone in *Rhodnius* (Insecta: Hemiptera), *Int. J. Invert. Reprod.* **3:**357–361.

Ruegg, R. P., Orchard, I., and Davey, K. G., 1982, 20-hydroxy-ecdysone as a modulator of electrical activity in neurosecretory cells of *Rhodnius prolixus, J. Insect Physiol.* **28:**243–248.

Ruegg, R. P., Lococo, D. J., and Tobe, S. S., 1983, Control of the corpus allatum activity in *Diploptera punctata:* Roles of the pars intercerebralis and pars lateralis, *Experientia* **39:**1329–1334.

Rutz, W., Gerig, L., Wille, H., and Lüscher, M., 1976, The function of JH in adult worker honeybees, *Apis mellifera, J. Insect Physiol.* **22:**1485–1491.

Ryerse, J. S., 1978, Ecdysterone switches off fluid secretion at pupation in insect Malpighian tubules, *Nature (London)* **271:**745–746.

Sabaratnam, M., 1974, Hormonal control of resilin deposition in the blowfly, *Calliphora erythrocephala, J. Insect Physiol.* **20:**935–946.

Safranek, L., and Williams, C. M., 1980, Studies of the prothoracicotropic hormone in the tobacco hornworm, *Manduca sexta, Biol. Bull. (Woods Hole, Mass.).* **158:**141–153.

Safranek, L., Cymborowski, B., and Williams, C. M., 1980, Effects of juvenile hormone on ecdysone-dependent development in the tobacco hornworm, *Manduca sexta, Biol. Bull. (Woods Hole, Mass.)* **158:**248–256.

Safranek, L., Squire, C. R., and Williams, C. M., 1986, Precocious termination of diapause in neck- and abdomen- ligated pupal preparations of the tobacco hornworm, *Manduca sexta, Biol. Bull. (Lancaster)* **171:**126–134.

Saha, L. M., Mandal, S., and Choudhuri, D. K., 1986, The effect of juvenile hormone analogue and ecdysterone on the fat body of female *Chrysocoris stolli* W., *Zool. Jahrb. Abt. Allg. Zool. Physiol. Tiere* **90:**85–100.

Sahota, T. S., 1969, Hormonal control of ovarian development and metamorphosis in *Malacosoma pluviale, Can. J. Zool.* **47:**917–920.

Sakurai, H., 1977, Endocrine control of oögenesis in the housefly, *Musca domestica vicina, J. Insect Physiol.* **23:**1295–1302.

Sakurai, S., 1983, Temporal organization of endocrine events underlying larval-larval ecdysis in the silkworm, *Bombyx mori, J. Insect Physiol.* **29:**919–932.

Sakurai, S., 1984, Temporal organization of endocrine events underlying larval-pupal metamorphosis in the silkworm, *Bombyx mori, J. Insect Physiol.* **30:**657–664.

Samaranayaka-Ramasamy, M., and Denlinger, D. L., 1985, Blockage of ovulation and insemination in the tsetse *Glossina morsitans morsitans* with reserpine, *Insect Sci. Applic.* **6:**159–161.

Sasaki, M., and Riddiford, L. M., 1984, Regulation of reproductive behaviour and egg maturation in the tobacco hawk moth, *Manduca sexta, Physiol. Entomol.* **9:**315–327.

Sasaki, M., Riddiford, L. M., Truman, J. W., and Moore, J. K., 1983, Re-evaluation of the role of corpora cardiaca in calling and oviposition behaviour of giant silk moths, *J. Insect Physiol.* **29:**695–705.

Scarborough, R. M., Jamieson, G. C., Kalish, F., Kramer, S. J., McEnroe, G. A., Miller, C. A., and Schooley, D. A., 1984, Isolation and primary structure of two peptides with cardioacceleratory and hyperglycemic activity from the corpora cardiaca of *Periplaneta americana, Proc. Natl. Acad. Sci. USA* **81:**5575–5579.

Schaffer, M. H., 1985, Isolation and characterization of neuropeptides, in *Neurochemical Techniques in Insect Research* (H. Breer and T. A. Miller, eds.), Springer-Verlag, Berlin, pp. 47–78.

Schaffer, M. H., and Noyes, B. E., 1987, The peptides, mRNA and genes of the AKH family, in *Molecular Entomology* (J. Law, ed.), Alan R. Liss, New York, pp. 67–71.

Schaller, F., and Meunier, J., 1968, Etude du système neurosécréteur céphalique des insectes Odonates, *Bull. Soc. Zool. Fr.* **93:**233–249.

Schaller, F., Andries, J. C., Mouze, M., and Defossez, A., 1974, Nouveaux aspects du contrôle hormonal du cycle biologique des Odonates: recherches sur la larve d'*Aeshna cyanea* (Müller) (Anisoptera: Aeshnidae), *Odonatologica* **3:**49–62.

Schally, A. V., Arimura, A., and Kastin, A. J., 1973, Hypothalamic regulatory peptides, *Science* **179**:341–350.

Scharrer, B., 1946, The relationship between corpora allata and reproductive organs in adult *Leucophaea maderae* (Orthoptera), *Endocrinology* **38**:46–55.

Scharrer, B., 1952, Neurosecretion. XI. The effects of nerve section on the intercerebralis-cardiacum-allatum system of the insect *Leucophaea maderae, Biol. Bull. (Woods Hole, Mass.)* **102**:261–272.

Scharrer, B., 1955, "Castration cells" in the central nervous system of an insect (*Leucophaea maderae* Blattaria), *Trans. N.-Y. Acad. Sci. II* **17**:520–525.

Scharrer, B., 1963, Neurosecretion. XIII. The ultrastructure of the corpus cardiacum of the insect *Leucophaea maderae, Z. Zellforsch. Mikrosk. Anat.* **60**:761–796.

Scharrer, B., 1964a, Histophysiological studies on the corpus allatum of *Leucophaea maderae*. IV. Ultrastructure during normal activity cycle, *Z. Zellforsch. Mikrosk. Anat.* **62**:125–148.

Scharrer, B., 1964b, The fine structure of Blattarian prothoracic glands, *Z. Zellforsch.* **64**:301–326.

Scharrer, B., 1965a, Recent progress in the study of neuroendocrine mechanisms in insects, *Arch. Anat. Micr. Morphol. Exp.* **54**:331–342.

Scharrer, B., 1965b, Ultrastructural study of the regressing prothoracic glands of Blattarian insects, *Z. Zellforsch.* **69**:1–21.

Scharrer, B., 1968, Neurosécrétion. XIV. Ultrastructural study of sites of release of neurosecretory material in Blattarian insects, *Z. Zellforsch.* **89**:1–16.

Scharrer, B., 1978, Histophysiological studies on the corpus allatum of *Leucophaea maderae, Cell Tissue Res.* **194**:533–545.

Scharrer, B., and Kater, S. B., 1969, Neurosecretion. XV. An electron microscopic study of the corpora cardiaca of *Periplaneta americana* after experimentally induced hormone release, *Z. Zellforsch.* **95**:177–186.

Scharrer, B., and Von Harnack, M., 1958, Histophysiological studies on the corpus allatum of *Leucophaea maderae*. I. Normal life cycle in male and female adults, *Biol. Bull. (Woods Hole, Mass.)* **115**:508–520.

Scharrer, B., and Von Harnack, M., 1961, Histophysiological studies on the corpus allatum of *Leucophaea maderae*. III. The effect of castration, *Biol. Bull. (Woods Hole, Mass.)* **121**:193–208.

Scharrer, B., and Scharrer, E., 1944, Neurosecretion. VI. A comparison between the intercerebralis-cardiacum-allatum system of the insects and the hypothalamo-hypophyseal system of the vertebrates, *Biol. Bull. (Woods Hole, Mass.)* **87**:242–251.

Scharrer, E., 1928, Die Lichtempfindlichkeit blinder Elritzen (Untersuchungen über das Zwischenhirn der Fische) I, *Z. Vgl. Physiol.* **7**:1–38.

Scharrer, E., 1965, The final common path in neuroendocrine integration, *Arch. Anat. Micr. Morphol. Exp.* **54**:359–370.

Scharrer, E., and Scharrer, B., 1937, Über Drüsen-Nervenzellen und Neurosekretorische Organe bei Wirbellosen und Wirbeltieren, *Biol. Rev.* **12**:185–216.

Scharrer, E., and Scharrer, B., 1954, Hormones produced by neurosecretory cells, in *Recent Progress in Hormone Research,* Proc. Laurentian Hormone Conf. X, pp. 183–232.

Scheller, K., and Bodenstein, D., 1981, Effects of ecdysterone and the juvenile hormone analogue methoprene on protein, RNA and DNA synthesis in brains of the blowfly, *Calliphora vicina, Zool. Jahrb. Abt. Allg. Zool. Physiol. Tiere* **85**:1–19.

Scheller, R. H. J. F., Jackson, L. B., McAllister, J. H., Schwartz, E. R., Kandel, and Asd, R., 1982, A family of genes codes for ELH, a neuropeptide eliciting a stereotyped pattern of behavior in *Aplysia, Cell* **28**:707–719.

Scheurer, R., 1969a, Endocrine control of protein systhesis during oöcyte maturation in the cockroach *Leucophaea maderae, J. Insect Physiol.* **15**:1411–1420.

Scheurer, R., 1969b, Hemolymph proteins and yolk formation in the cockroach, *Leucophaea maderae, J. Insect Physiol.* **15:**1673–1682.

Schmidt, E. L., and Williams, C. H., 1953, Physiology of insect diapause. V. Assay of the growth and differentiation hormone of Lepidoptera by the method of tissue culture, *Biol. Bull. (Woods Hole, Mass.)* **105:**174–187.

Schnee, M. E., Ma, M., and Kelly, T. J., 1984, Hormonal basis of adult eclosion in the gypsy moth (Lepidoptera: Lymantriidae), *J. Insect Physiol.* **30:**351–356.

Schneiderman, H. A., and Williams, C. M., 1954, The physiology of insect diapause. IX. The cytochrome oxidase system in relation to the diapause and development of the cecropia silkworm, *Biol. Bull. (Woods Hole, Mass.)* **106:**238–252.

Schols, D., Verhaert, P., Huybrechts, R., Vaudry, H., Jegou, S., and De Loof, A., 1987, Immunocytochemical demonstration of proopiomelanocortin- and other opioid-related substances and a CFR-like peptide in the gut of the American cockroach, *Periplaneta americana L., Histochemistry* **86:**345–351.

Schoofs, L., Jegou, S., Vaudry, H., Verhaert, P., and De Loof, A., 1987, Localization of melanotropin-like peptides in the central nervous system of two insect species, the migratory locust, *Locusta migratoria,* and the fleshfly, *Sarcophaga bullata, Cell Tissue Res.* **248:**25–31.

Schoofs, L., Schrooten, S., Huybrechts, R., and De Loof, A., 1988a, Methionine-enkephalin immunoreactivity in the gonads and nervous system of two insect species: *Locusta migratoria* and *Sarcophaga bullata, Gen. Comp. Endocrinol.* **69:**1–12.

Schoofs, L., Jegou, S., Andersen, A. C., Tonon, M. C., Eberle, A. N., Huybrechts, R., De Loof, A., and Vaudry, H., 1988b, Coexistence of melanin-concentrating hormone and alpha MSH immunoreactivities in the central nervous system of the locust *Locusta-migratoria, Brain Res.* **450:**202–208.

Schooley, D. A., Miller, C. A., and Proux, J. P., 1987, Isolation of two arginine vasopressin-like factors from ganglia of *Locusta migratoria, Arch. Insect Biochem. Physiol.* **5:**157–166.

Schooneveld, H., 1972, Effects of juvenile hormone on the endocrine system during break of diapause in the Colorado beetle, *J. Endocrinol.* **57:**Iv–Ivi.

Schooneveld. H., 1974a, Ultrastructure of the neurosecretory system of the Colorado potato beetle, *Leptinotarsa decemlineata* (Say). I. Characterization of the protocerebral neurosecretory cells, *Cell Tissue Res.* **154:**275–288.

Schooneveld, H., 1974b, Ultrastructure of the neurosecretory system of the Colorado potato beetle *Leptinotarsa decemlineata* (Say) II. Pathways of axonal secretion transport and innervation of neurosecretory cells, *Cell Tissue Res.* **154:**289–301.

Schooneveld, H., Sanchez, A. O., and De Wilde, J., 1977, Juvenile hormone-induced break and termination of diapause in the Colorado potato beetle, *J. Insect Physiol.* **23:**689–696.

Schooneveld, H., Kramer, S. J., Privee, H., and Van Huis, A., 1979, Evidence of controlled corpus allatum activity in the adult Colorado potato beetle, *J. Insect Physiol.* **25:**449–453.

Schooneveld, H., Tesser, G. I., Veenstra, J. A., and Romberg-Privee, H. M., 1983, Adipokinetic hormone and AKH-like peptide demonstrated in the corpora cardiaca and nervous system of *Locusta migratoria* by immunocytochemistry, *Cell Tissue Res.* **230:**67–76.

Schooneveld, H., Romberg-Privee, H. M., and Veenstra, J. A., 1985, Adipokinetic hormone-immunoreactive peptide in the endocrine and central nervous system of several insect species: A comparative immunocytochemical approach, *Gen. Comp. Endocrinology* **57:**184–194.

Schooneveld, H., Romberg-Privee, H. M., and Veenstra, J. A., 1986, Immunocytochemical differentiation between adipokinetic hormone (AKH)-like peptides in neurons and glandular cells in the corpus cardiacum of *Locusta migratoria* and *Periplaneta americana* with C-terminal and N-terminal specific antisera to AKH, *Cell Tissue Res.* **243:**9–14.

Schooneveld, H., Romberg-Privee, H. M., and Veenstra, J. A., 1987a, Phylogenetic differentiation of glandular cells in corpora cardiaca as studied immunocytochemically with region-specific antisera to adipokinetic hormone, *J. Insect Physiol.* **33:**167–176.

Schooneveld, H., Van Herp, F., and Van Minnen, J., 1987b, Demonstration of substances immunologically related to the identified arthropod neuropeptides AKH/RPCH in the CNS of several invertebrate species, *Brain Res.* **406:**224–232.

Schreiner, B., 1977, The effect of the hormone(s) from the corpus allatum complex on the ovarian tissue of *Oncopeltus fasciatus*. A light and electron microscopic investigation, *J. Morphol.* **151:** 81–110.

Schwartz, L. M., and Reynolds, S. E., 1979, Fluid transport in *Calliphora* Malpighian tubules: A diuretic hormone from the thoracic ganglion and abdominal nerves, *J. Insect Physiol.* **25:**847–854.

Schwartz, L. M., and Truman, J. W., 1982, Peptide and steroid regulation of muscle degeneration in an insect, *Science* **215:**1420–1421.

Schwartz, L. M., and Truman, J. W., 1984a, Hormonal control of muscle atrophy and degeneration in the moth *Antheraea polyphemus, J. Exp. Biol.* **111:**13–30.

Schwartz, L. M., and Truman, J. W., 1984b, Cyclic GMP may serve as a second messenger in peptide-induced muscle degeneration in an insect, *Proc. Natl. Acad. Sci. USA* **81:**6718–6722.

Schwartz, M. B., Kelly, T. J., Imberski, R. B., and Rubenstein, E. C., 1985, The effects of nutrition and methoprene treatment on ovarian ecdysteroid synthesis in *Drosophila melanogaster, J. Insect Physiol.* **31:**947–957.

Schwartz, T. L., Lee, G. M. H., Siwicki, K. K., Standaert, D. G., and Kravitz, E. A., 1984, Proctolin in the lobster: The distribution, release, and chemical characterization of a likely neurohormone, *J. Neurosci.* **4:**1300–1311.

Sedlak, B. J., 1981, An ultrastructural study of neurosecretory fibers within the corpora allata of *Manduca sexta, Gen. Comp. Endocrinol.* **44:**207–218.

Sedlak, B. J., Marchione, L., Devokin, B., and Davino, R., 1983, Correlations between endocrine gland ultrastructure and hormone titers in the fifth larval instar of *Manduca sexta, Gen. Comp. Endocrinol.* **52:**291–310.

Seecof, R. L., and Dewhurst, S., 1974, Insulin is a *Drosophila* hormone and acts to enhance the differentiation of embryonic *Drosophila* cells, *Cell. Differ. Netherl.* **3:**63–70.

Sefiani, M., 1986, Regulation of egg laying and oviduct contractions *in vitro,* in *Gryllus bimaculatus, J. Insect Physiol.* **33:**215–222.

Sehnal, F., 1983, Juvenile hormone analogues, in *Endocrinology of Insects* (R. G. H. Downer and H. Laufer, eds.), Alan R. Liss, New York, pp. 657–672.

Sehnal, F., and Edwards, J. S., 1969, Body constraint and developmental arrest in *Galleria mellonella* L.: Further studies, *Biol. Bull. (Woods Hole, Mass.).* **137:**352–358.

Sehnal, F., and Granger, N. A., 1975, Control of corpora allata function of *Galleria mellonella* L. (Lepidoptera), *Biol. Bull. (Woods Hole, Mass.)* **148:**106–116.

Sehnal, F., and Rembold, H., 1985, Brain stimulation of juvenile hormone production in insect larvae, *Experientia* **41:**684–685.

Sehnal, F., Maróy, P., and Malá, J., 1981, Regulation and significance of ecdysteroid titre fluctuations in lepidopterous larvae and pupae, *J. Insect Physiol.* **27:**535–544.

Sehnal, F., Delbecque, J. P., Maróy, P., and Malá, J., 1986, Ecdysteroid titres during larval life and metamorphosis of *Galleria mellonella, Insect Biochem.* **16:**157–162.

Sekeris, C., and Karlson, P., 1962, Zum Tyrosinstoffwechsel der Insekten, VII. Der Katabolische Abbau des Tyrosins and die Biogenese der Sklerotisierungssubstanz, N-Acetyl-dopamin, *Biochim. Biophys. Acta.* **62:**103–113.

Seligman, M., and Doy, F. A., 1973, Hormonal regulation of disaggregation of cellular fragments in the haemolymph of *Lucilia cuprina, J. Insect Physiol.* **19:**125–135.

Seligman, M., Friedman, S., and Fraenkel, G., 1969, Bursicon mediation of tyrosine hydroxylation during tanning of the adult cuticle of the fly, *Sarcophaga bullata, J. Insect Physiol.* **15:**553–562.

Seligman, M., Blechl, A., Blechl, J., Herman, P., and Fraenkel, G., 1977, Role of ecdysone, pupariation factors, and cyclic AMP in formation and tanning of the puparium of the fleshfly *Sarcophaga bullata, Proc. Natl. Acad. Sci. USA* **74:**4697–4701.

Seshan, K. R., 1968, Les phénomènes de neurosécrétion dans la chaîne nerveuse ventrale d'*Iphita limbata, C.R. Acad. Sci. Ser. D* **266:**619–622.

Shaaya, E., and Karlson, P., 1965, Der Ecdysontiter während der Insektenentwicklung. IV. Die Entwicklung der Lepidopteren *Bombyx mori* L. und *Cerura vinula* L., *Dev. Biol.* **11:**424–432.

Shalom, U., and Pener, M. P., 1986, Endocrine manipulations and ontogenesis of male sexual behaviour in *Locusta:* Studies on precocene-induced early adultiforms, *Physiol. Entomol.* **11:** 441–452.

Shalom, U., and Pener, M. P., 1987, Endocrine manipulations and ontogenesis of male sexual behaviour in *Locusta:* Studies on azadirachtin-induced over-aged nymphs, *Physiol. Entomol.* **12:**197–208.

Shapiro, J. P., 1983, Ovarian cyclic AMP and response to a brain hormone from the mosquito *Aedes aegypti, Insect Biochem.* **13:**273–279.

Shapiro, J. P., and Hagedorn, H. H., 1982, Juvenile hormone and the development of ovarian responsiveness to a brain hormone in the mosquito, *Aedes aegypti, Gen. Comp. Endocrinol.* **46:** 176–183.

Shibuya, I., and Yagi, S., 1972, Effect of ecdysterone on cultivated ovaries of the greater wax moth larvae (Lepidoptera: Pyralidae), *Appl. Entomol. Zool.* **7:**97–98.

Shimada, S., and Yamashita, O., 1979, Trehalose absorption related with trehalase in developing ovaries of the silkworm, *Bombyx mori, J. Comp. Physiol.* **131:**333.

Shimizu, T., 1986, Culture of the peritoneal sheath from the diapausing cabbage armyworm (*Mamestra brassicae*) pupal testis, *Zool. Sci.* **3:**309–313.

Shimizu, T., and Yagi, S., 1978, Hormonal effect on cultivated insect tissues. IV. The role of testis in spermiogenesis of the cabbage armyworm, *Mamestra brassicae,* in vitro, *Appl. Entomol. Zool.* **13:**278–282.

Shimizu, T., and Yagi, S., 1982, Hormonal effect on cultivated insect tissues. V. The role of testis proteins in spermiogenesis of the cabbage armyworm, *Mamestra brassicae,* in vitro (Lepidoptera: Noctuidae), *Appl. Entomol. Zool.* **17:**385–392.

Shimizu, T., Moribayashi, A., and Agui, N., 1985, *In vitro* analyses of spermiogenesis and testicular ecdysteroids in cabbage armyworm, *Mamestra brassicae* L. (Lepidoptera: Noctuidae), *Appl. Entomol. Zool.* **20:**56–61.

Shinyaeva, L. I., 1981, Effect of the juvenile hormone analogues on spermatogenesis in *Eurygaster integriceps* during the period of prediapause and diapause formation, *Zool. Zh.* **60:**1198–1204.

Shukle, R. H., and Judson, C. L., 1984, Action of two corpus cardiacal peptides on muscles of the hingut of *Periplaneta americana* L. (Orthoptera: Blattidae), *J. Insect Physiol.* **30:**927–937.

Sieber, K. P., and Rembold, H., 1983, The effects of azadirachtin on the endocrine control of moulting in *Locusta migratoria, J. Insect Physiol.* **29:**523–527.

Sieber, R., and Benz, G., 1980a, The hormonal regulation of the larval diapause in the codling moth, *Laspeyresia pomonella* (Lep. *Tortricidae*), *J. Insect Physiol.* **26:**213–218.

Sieber, R., and Benz, G., 1980b, Hormonal regulation of pupation in the codling moth, *Laspeyresia pomonella, Physiol. Entomol.* **5:**283–290.

Siegert, K. J., and Mordue, W., 1986a, Quantification of adipokinetic hormones I and II in the corpora cardiaca of *Schistocerca gregaria* and *Locusta migratoria, Comp. Biochem. Physiol.* **84A:**279–284.

Siegert, K. J., and Mordue, W., 1986b, Elucidation of the primary structures of the cockroach hyperglycaemic hormones I and II using enzymatic techniques and gas-phase sequencing, *Physiol. Entomol.* **11:**205–211.

Siegert, K. J., and Mordue, W., 1987, Breakdown of locust adipokinetic hormone I by Malpighian tubules of *Schistocerca gregaria, Insect Biochem.* **17:**705–710.

Siegert, K. J., Morgan, P. J., and Mordue, W., 1985, Primary structure of locust adipokinetic hormones II, *Hoppe Seyler's Z. Physiol. Chem.* **366:**723–727.

Siew, Y. C., 1965, The endocrine control of adult reproductive diapause in the chrysomelid beetle *Galeruca tanaceti* (L.). II, *J. Insect Physiol.* **11:**463–479.

Siew, Y. C., and Gilbert, L. I., 1971, Effects of moulting hormone and juvenile hormone on insect endocrine gland activity, *J. Insect Physiol.* **17:**2095–2104.

Singh, H. H., and Sehnal, F., 1979, Lack of specific neurons in the ventral nerve cord for the control of prothoracic glands, *Experientia* **35:**1117–1119.

Singh, U. V., and Awasthi, V. B., 1981, The neurosecretory cells and retrocerebral endocrine glands of *Amsacta collaris* Hampson (Lepidoptera: Arctiidae) during post-embryonic development, *J. Hirnforsch.* **22:**265–278.

Singh, Y. N., and Arif, M., 1981, Morphology and secretory activities of the brain and ventral nerve cord neurosecretory cells during the post-embryonic development of *Poekilocerus pictus,* Fabr. (Acridoidea: Pyrgomorphidae), *Arch. Biol.* **92:**35–52.

Singh, Y. N., and Narain, R., 1980a, Neuroendocrine system and storage of neurosecretory material in the dorsal aorta in a bug *Chrysocoris stollii* Wolf (Heteroptera: Pentatomidae), *J. Hirnforsch.* **21:**443–448.

Singh, Y. N., and Narain, R., 1980b, Neurosecretory system and storage of paraldehyde fuchsin positive neurosecretory material in the dorsal aorta in an earwig, *Anisolabis annulipes* Lucas *(Dermaptera: Labiduridae), Experientia* **36:**834–835.

Sivasubramanian, P., Friedman, S., and Fraenkel, G., 1974, Nature and role of proteinaceous hormonal factors acting during puparium formation in flies, *Biol. Bull. (Woods Hole, Mass.)* **147:**163–185.

Siwicki, K. K., and Bishop, C. A., 1985, Mapping of proctolin-like immunoreactivity in the nervous systems of lobster and crayfish, *J. Comp. Neurol.* **243:**435–453.

Skinner, W. S., Quistad, G. B., Adams, M. E., and Schooley, D. A., 1987, Metabolic degradation of the peptide periplanetin CC-2 in the cockroach, *Periplaneta americana, Insect Biochem.* **17:** 433–437.

Sláma, K., 1964a, Hormonal control of respiratory metabolism during growth, reproduction, and diapause in female adults of *Pyrrhocoris apterus* L. (Hemiptera), *J. Insect Physiol.* **10:**283–303.

Sláma, K., 1964b, Hormonal control of respiratory metabolism during growth reproduction, and diapause in male adults of *Pyrrhocoris apterus* L. (Hemiptera), *Biol. Bull. (Woods Hole, Mass.)* **127:**499–510.

Sláma, K., 1976, Insect haemolymph pressure and its determination, *Acta Entomol. Bohemoslov.* **73:**65–75.

Sláma, K., 1980, Homeostatic function of ecdysteroids in ecdysis and oviposition, *Acta Entomol. Bohemoslov.* **77:**145–168.

Sláma, K., 1983, Illusive functions of the prothoracic gland in *Galleria, Acta Entomol. Bohemoslov.* **80:**161–176.

Sláma, K., 1984, Recording of haemolymph pressure pulsations from the insect body surface, *J. Comp. Physiol. B* **154:**635–643.

Sláma, K., Romañuk, M., and Sorm, F., 1974, *Insect Hormones and Bioanalogues,* Vienna: Springer.

Slinger, A. J., and Isaac, R. E., 1988, Synthesis of apolar ecdysone esters by ovaries of the cockroach *Periplaneta americana, Gen. Comp. Endocrinol.* **70:**74–82.

Smalley, K. N., 1970, Median nerve neurosecretory cells in the abdominal ganglia of the cockroach, *Periplaneta americana, J. Insect Physiol.* **16:**241–250.

Smith, N. A., 1969, Observations on the neural rhythmicity in the cockroach cardiac ganglion, *Experientia* suppl. **15:**200–205.

Smith, U., and Smith, D. S., 1966, Observations on the secretory processes in the corpus cardiacum of the stick insect, *Carausius morosus, J. Cell Sci.* **1:**59–66.

Smith, W. A., and Gilbert, L. I., 1986, Cellular regulation of ecdysone synthesis by the prothoracic glands of *Manduca sexta, Insect Biochem.* **16:**143–147.

Smith, W. A., and Nijhout, H. F., 1982, Ultrastructural changes accompanying secretion and cell death in the molting glands of an insect (Oncopeltus), *Tissue Cell* **14:**243–252.

Smith, W. A., and Nijhout, H. F., 1983, In vitro stimulation of cell death in the moulting glands of *Oncopeltus fasciatus* by 20-hydroxyecdysone, *J. Insect Physiol.* **29:**169–176.

Smith, W. A., Gilbert, L. I., and Bollenbacher, W. E., 1984, The role of cyclic AMP in the regulation of ecdysone synthesis, *Mol. Cell. Endocrinol.* **37:**285–294.

Smith, W. A., Gilbert, L. I., and Bollenbacher, W. E., 1985, Calcium-cyclic AMP interactions in prothoracicotropic hormone stimulation of ecdysone synthesis, *Mol. Cell. Endocrinol.* **39:**71–78.

Smith, W. A., Watson, R. D., Gilbert, L. I., and Bollenbacher, W. E., 1986, The steroidogenic action of haemolymph stimulatoryfactor in *Manduca sexta.* Comparisons with prothoracicotropic hormone, *Insect Biochem.* **16:**781–787.

Sokolove, P. G., and Loher, W., 1975, Role of eyes, optic lobes, and pars intercerebralis in locomotory and stridulatory circadian rhythms of *Teleogryllus commodus, J. Insect Physiol.* **21:** 785–799.

Sonobe, H., and Keino, H., 1975, Diapause factor in the brains, subesophageal ganglia and prothoracic ganglia of the silkworm, *Naturwissenschaften* **62:**348.

Sonobe, H., and Ohnishi, E., 1971, Silkworm *Bombyx mori* L.: Nature of diapause factor, *Science* **174:**835–838.

Sparks, T. C., Hammock, B. D., and Riddiford, L. M., 1983, The haemolymph juvenile hormone esterase of *Manduca sexta* (L.)—inhibition and regulation, *Insect Biochem.* **13:**529–541.

Speidel, C. C., 1919, Gland cells of internal secretion in the spinal cord of the skates, *Carnegie Inst. Washington Publ.* **13:**1–31.

Spencer, I. M., and Candy, D. J., 1976, Hormonal control of diacyl glycerol mobilization from fat body of the desert locust, *Schistocerca gregaria, Insect Biochem.* **6:**289–296.

Spielman, A., Gwadz, R. W., and Anderson, W. A., 1971, Ecdysone-initiated ovarian development in mosquitoes, *J. Insect Physiol.* **17:**1807–1814.

Spring, J. H., 1986, Hormonally-stimulated chloride transport in the Eastern lubber grasshopper, *Romalea microptera, J. Exp. Zool.* **237:**3–9.

Spring, J. H., and Gäde, G., 1987, Factors regulating carbohydrate and lipid metabolism isolated from the corpus cardiacum of the Eastern lubber grasshopper, *Romalea microptera, J. Exp. Zool.* **241:**41–50.

Spring, J. H., and Hazelton, S. R., 1987, Excretion in the house cricket (*Acheta domesticus*): Stimulation of diuresis by tissue homogenates, *J. Exp. Biol.* **129:**63–81.

Spring, J. H., and Phillips, J. E., 1980, Studies on locust rectum. II. Identification of specific ion transport processes regulated by corpora cardiaca and cyclic-AMP, *J. Exp. Biol.* **86:**225–236.

Spring, J., Hanrahan, J., and Phillips, J., 1978, Hormonal control of chloride transport across locust rectum, *Can. J. Zool.* **56:**1879–1882.

Sridevi, R., Ray, A., and Ramamurty, P. S., 1988, 20-Hydroxyecdysone stimulated DNA synthesis in early larval testes of *Spodoptera litura* (Noctuidae-Lepidoptera), *Int. J. Invert. Reprod. Develop.* **13:**199–201.

Srivastava, B. B. L., and Hopkins, T. L., 1975, Bursicon release and activity in haemolymph during metamorphosis of the cockroach, *Leucophaea maderae, J. Insect Physiol.* **21:**1985–1993.

Srivastava, K. P., Singh, H. H., and Tiwari, R. K., 1975, Neuroendocrine organs of the lemon-butterfly, *Papilio demoleus* L. I. The corpora cardiaca-allata complex of the larva, *Z. Mikrosk. Anat. Forsch.* **89:**415–422.

Srivastava, K. P., Tiwari, R. K., and Kumar, P., 1977, Effect of sectioning the prothoracic gland nerves in the larva of the lemon butterfly, *Papilio demoleus* L., *Experientia* **33:**98–99.

Srivastava, V. K., and Prasad, O., 1985, Histology, secretory activity and metamorphic changes in the ventral nerve cord neurosecretory cells in the larvae of *Samia cynthia ricini* (Lepidoptera: Saturnidae), *Z. Mikrosk. Anat. Forsch.* **99:**95–106.

Staal, G. B., 1961, Studies on the physiology of phase induction in *Locusta migratoria migratorioides* R. et F., *Publicatie 40 LEB Fonds Wageningen* **72.**

Staal, G. B., 1986, Anti juvenile hormone agents, *Annu. Rev. Entomol.* **31.**

Staal, G. B., Henrick, C. A., Bergot, B. J., Cerf, D. C., Edwards, J. P., and Kramer, S. J., 1981, Relationships and interactions between JH and anti-JH analogues in Lepidoptera, in *Regulation of Insect Development and Behaviour* (F. Sehnal, A. Zabza, J. J. Menn, and B. Cymborowski, eds.), Part I, Wroclaw Technical University Press, Wroclaw, pp. 323–430.

Stangier, J., Dircksen, H., and Keller, R., 1986, Identification and immunocytochemical localization of proctolin in pericardial organs of the shore crab, *Carcinus maenas, Peptides* **7:**67–72.

Stanley-Samuelson, D. W., Peloquin, J. J., and Loher, W., 1986, Egglaying in response to prostaglandin injections in the Australian field cricket, *Teleogryllus commodus, Physiol. Entomol.* **11:**213–219.

Stanley-Samuelson, D. W., Jurenka, R. A., Blomquist, G. J., and Loher, W., 1987, Sexual transfer of prostaglandin precursor in the field cricket *Teleogryllus commodus, Physiol. Entomol.* **12:** 347–354.

Starratt, A. N., and Brown, B. E., 1975, Structure of the pentapeptide proctolin, a proposed neurotransmitter in insects, *Life Sci.* **17:**1253–1256.

Starratt, A. N., and Brown, B. E., 1977, Synthesis of proctolin, a pharmacologically active pentapeptide in insects, *Can. J. Chem.* **55:**4238–4242.

Starratt, A. N., and Brown, B. E., 1979, Analogs of the insect myotropic peptide proctolin: Synthesis and structure-activity studies, *Biochem. Biophys. Res. Commun.* **90:**1125–1130.

Starratt, A. N., and Steele, R. W., 1984, In vivo inactivation of the insect neuropeptide proctolin in *Periplaneta americana, Insect Biochem.* **14:**97–102.

Stay, B., and Tobe, S. S., 1977, Control of juvenile hormone biosynthesis during the reproductive cycle of a viviparous cockroach. I. Activation and inhibition of corpora allata, *Gen. Comp. Endocrinol.* **33:**531–540.

Stay, B., and Tobe, S. S., 1978, Control of juvenile hormone biosynthesis during the reproductive cycle of a viviparous cockroach. II. Effects of unilateral allatectomy, implantation of supernumerary corpora allata and ovariectomy, *Gen. Comp. Endocrinol.* **34:**276–286.

Stay, B., Friedel, T., Tobe, S. S., and Mundall, E. C., 1980, Feedback control of juvenile hormone synthesis in cockroaches: Possible role for ecdysterone, *Science* **207:**898–900.

Stay, B., Tobe, S. S., Mundall, E. C., and Rankin, S., 1983, Ovarian stimulation of juvenile hormone biosynthesis in the viviparous cockroach, *Diploptera punctata, Gen. Comp. Endocrinol.* **52:**341–349.

Stay, B., Ostedgaard, L. S., Tobe, S. S., Strambi, A., and Spaziani, E., 1984, Ovarian and haemolymph titres of ecdysteroid during the gonadotrophic cycle in *Diploptera punctata, J. Insect Physiol.* **30:**643–651.

Steel, C. G. H., 1977, The neurosecretory system in the aphid *Megoura viciae,* with reference to unusual features associated with long distance transport of neurosecretion, *Gen. Comp. Endocrinol.* **31:**307–322.

Steel, C. G. H., 1978, Some functions of identified neurosecretory cells in the brain of the aphid, *Megoura viciae, Gen. Comp. Endocrinol.* **34:**219–228.

Steel, C. G. H., and Lees, A. D., 1977, The role of neurosecretion in the photoperiodic control of polymorphism in the aphid *Megoura viciae, J. Exp. Biol.* **67:**117–135.

Steel, C. G. H., Bollenbacher, W. E., Smith, S. L., and Gilbert, L. I., 1982, Haemolymph ecdysteroid titres during larval-adult development in *Rhodnius prolixus:* Correlations with moulting hormone action and brain neurosecretory cell activity, *J. Insect Physiol.* **28:** 519–525.

Steele, J. E., 1961, Occurrence of a hyperglycaemic factor in the corpus cardiacum of an insect, *Nature* **92:**680–681.

Steele, J. E., 1963, The site of action of insect hyperglycemic hormone, *Gen. Comp. Endocrinol.* **3:** 46–52.

Steele, J. E., 1964, The activation of phosphorylase in an insect by adenosine 3′,5′-phosphate and other agents, *Am. Zool.* **4:**328.

Steele, J. E., 1985, Control of metabolic processes, in *Comprehensive Insect Physiology, Biochemistry and Pharmacology* (G. A. Kerkut and L. I. Gilbert, eds.), Vol. 4, Pergamon Press, New York, pp. 99–145.

Steele, J. E., and Tolman, J. H., 1980, Regulation of water transport in the cockroach rectum by the corpora cardiaca-corpora allata system: The requirement for Na^+, *J. Comp. Physiol.* **138:**357–365.

Steele, J. E., McDougall, G. E., and Shadwick, R., 1988, Trehalose efflux from cockroach fat body *in vitro:* Paradoxical effects of the corpus cardiacum and methylxanthines, *Insect Biochem.* **18:** 585–590.

Steele, R. W., and Starratt, A. N., 1985, In vitro inactivation of the neuropeptide proctolin in haemolymph from *Periplaneta americana, Insect Biochem.* **15:**511–519.

Stefano, G. B., and Scharrer, B., 1981, High affinity binding of an enkephalin analog in the cerebral ganglion of the insect *Leucophaea maderae* (Blattaria), *Brain Res.* **225:**107–114.

Steinbusch, H. W., Verhofstad, A. A. J., and Joosten, H. W. J., 1978, Localization of serotonin in the central nervous system by immunohistochemistry: Description of a specific and sensitive technique and some applications, *Neuroscience* **3:**811–819.

Stephen, W. F., and Gilbert, L. I., 1970, Alterations in fatty acid composition during the metamorphosis of *Hyalophora cecropia:* Correlations with juvenile hormone titre, *J. Insect Physiol.* **16:**851–864.

Sterba, G., 1961, über eine sehr spezifische neue Methode zum Nachweis von Neurosekret, *Acta Biol. Med. Germ.* **7:**228–231.

Sternburg, J., Chang, S. C., and Kearns, C. W., 1959, The release of a neuroactive agent by the American cockroach after exposure to DDT or electrical stimulation, *J. Econ. Entomol.* **52:** 1070–1076.

Stewart, W. W., 1981, Lucifer dyes—highly fluorescent dyes for biological tracing. *Nature* **292:**17–21.

Stone, J. V., and Mordue, W., 1980, Adipokinetic hormone, in *Neurohormonal Techniques in Insects* (T. A. Miller, ed.), Springer-Verlag, New York, pp. 31–80.

Stone, J. V., Mordue, W., Batley, K. E., and Morris, H. R., 1976, Structure of locust adipokinetic hormone, a neurohormone that regulates lipid utilisation during flight, *Nature* **263:**207–211.

Stone, J. V., Mordue, W., Broomfield, C. E., and Hardy, P. M., 1978, Structure-activity relationships for the lipid mobilising action of locust adipokinetic hormone, *Eur. J. Biochem.* **89:** 195–202.

Stout, J. F., Gerard, G., and Hasso, S., 1976, Sexual responsiveness mediated by the corpora allata and its relationship to phonotaxis in the female cricket, *Acheta domestica* L., *J. Comp. Physiol.* **108:**1–9.

Strambi, C., Strambi, A., Cupo, A., Rougon-Rapuzzi, G., and Martin, N., 1978, Etude des taux d'une substance apparentée à la vasopressine dans le système nerveux de grillons soumis à différentes conditions hygrométriques, *C.R. Acad. Sci. Ser. D* **287:**1227–1230.

Strambi, C., Rougon-Rapuzzi, G., Cupo, A., Martin, N., and Strambi, A., 1979, Mise en évidence immunocytologique d'un composé apparenté à la vasopressine dans le système nerveux du grillon *Acheta domesticus, C.R. Acad. Sci. Ser. D* **288:**131–133.

Strambi, C., Strambi, A., de Reggi, M. L., Hirn, M. H., and Delaage, M. A., 1981, Radioimmunoassay of insect juvenile hormones and of their diol derivatives, *Eur. J. Biochem.* **118:**401–406.

Strambi, A., Strambi, C. Röseler, P. F., and Röseler, I., 1984, Simultaneous determination of juvenile hormone and ecdysteroid titers in the hemolymph of bumblebee prepupae (*Bombus hypnorum* and *B. terrestris*), *Gen. Comp. Endocrinol.* **55:**83–88.

Strejvkova, A., Servit, Z., and Novak, V. J. A., 1964, Effect of neurohormones C1 and D1 on spontaneous electrical activity of the central nervous system of the cockroach, *J. Insect Physiol.* **11:**889–896.

Stretton, A. O. W., and Kravitz, E. A., 1968, Neuronal geometry: Determination with a technique of intracellular dye injection, *Science* **162:**132–134.

Strong, L., 1965a, The relationships between the brain, corpora allata, and oocyte growth in the central American locust, *Schistocerca sp.* I. The cerebral neurosecretory system, the corpora allata, and oocyte growth, *J. Insect Physiol.* **11:**135–146.

Strong, L., 1965b, The relationships between the brain, corpora allata, and oocyte growth in the central American locust, *Schistocerca sp.* II. The innervation of the corpora allata the lateral neurosecretory complex and oocyte growth, *J. Insect Physiol.* **11:**271–280.

Strong, L., 1966, On the occurrence of neuroglandular axons within the sympathetic nervous system of a locust, *Locusta migratoria migratorioides, J. Roy. Microsc. Soc.* **86:**141–149.

Strong, L., 1968, Locomotor activity, sexual behaviour, and the corpus allatum hormone in males of *Locusta, J. Insect Physiol.* **14:**1685–1692.

Strong, L., and Amerasinghe, F. P., 1977, Allatectomy and sexual receptivity in females of *Schistocerca gregaria, J. Insect Physiol.* **23:**131–135.

Sugawara, T., 1986, Oviposition behaviour of the cricket *Teleogryllus commodus:* The site of action of an oviposition-stimulating factor and the role of the nervous system, *J. Insect Physiol.* **32:** 485–492.

Sullivan, R. E., 1979, A proctolin-like peptide in crab pericardial organs, *J. Exp. Zool.* **210:**543–552.

Sullivan, R. E., and Miller, M. W., 1984, Dual effects of proctolin on the rhythmic burst activity of the cardiac ganglion, *J. Neurobiol.* **15:**173–196.

Sullivan, R. E., and Newcomb, R. W., 1982, Structure function analysis of an arthropod peptide hormone: Proctolin and synthetic analogues compared on the cockroach hindgut receptor, *Peptides* **3:**337–344.

Suzuki, A., 1986, Prothoracicotropic hormones and neurohormones, in *Insect Neurochemistry and Neurophysiology,* (A. B. Borkovec and D. B. Gelman, eds.), The Humana Press, pp. 29–51.

Suzuki, A., Isogai, A., Horii, T., Ishizaki, H., and Tamura, S., 1975, A simple procedure for partial purification of silkworm brain hormone, *Agric. Biol. Chem.* **39:**2157–2162.

Suzuki, A., Matsumoto, S., Ogura, N., Isogai, A., and Tamura, S., 1976, Extraction and partial purification of the hormone in during cuticular melanization in armyworm larvae, *Agric. Biol. Chem.* **40:**2307–2309.

Suzuki, A., Nagasawa, H., Kataoka, H., Hori, Y., Isogai, A., Tamura, S., Guo, F., Zhong, X., Ishizaki, H., Fujishita, M., and Mizoguchi, A., 1982, Isolation and characterization of prothoracicotropic hormone from silkworm, *Bombyx mori, Agric. Biol. Chem.* **46:**1107–1109.

Suzuki, C., and Ishizaki, H., 1986, Prothoracicotropic hormone bioassay: *Bombyx* larva assay, *Int. J. Insect Reprod. Dev.* **10:**259–274.

Suzzoni, J. P., Passera, L., and Strambi, A., 1983, Ecdysteroid production during caste differentiation in larvae of the ant, *Plagiolepis pygmaea, Physiol. Entomol.* **8:**93–96.

Szibbo, C. M., and Tobe, S. S., 1983, Nervous and humoral inhibition of C_{16} juvenile hormone synthesis in last instar females of the viviparous cockroach, *Diploptera punctata, Gen. Comp. Endocrinol.* **49:**437–445.

Szöllösi, A., 1975, Imaginal differentiation of the spermiduct in acridids: Effects of juvenile hormone, *Acrida* **4:**205–216.

Szöllösi, A., and Landureau, J. C., 1977, Imaginal cell differentiation in spermiduct of *Samia cynthia* (Lepidoptera). Responses in vitro to ecdysone and ecdysterone, *Biol. Cell.* **28:**23–36.

Szopa, T. M., 1982, Cytodifferentiation of the accessory glands of *Tenebrio molitor.* IX. Differentiation of the spermathecal accessory gland in vitro, *Cell Tissue Res.* **222:**269–281.

Szopa, T. M., Lenoir-Rousseaux, J. J., Yuncker, C., and Happ, G. M., 1985, Ecdysteroids accelerate mitoses in accessory glands of beetle pupae, *Dev. Biol.* **107:**325–336.

Tager, H. S., and Kramer, K. J., 1980, Insect glucagon-like peptides: Evidence for a high molecular weight form in midgut from *Manduca sexta* (L.), *Insect Biochem.* **10:**617–619.

Tager, H. S., Markese, J., Kramer, K. J., Speirs, R. D., and Childs, C. N., 1976, Glucagon-like and insulin-like hormones of the insect neurosecretory system, *Biochem. J.* **156:**515–520.

Taghert, P. H., and Truman, J. W., 1982a, The distribution and molecular characteristics of the tanning hormone, bursicon, in the tobacco hornworm *Manduca sexta, J. Exp. Biol.* **98:**373–383.

Taghert, P. H., and Truman, J. W., 1982b, Identification of the bursicon-containing neurones in abdominal ganglia of the tobacco hornworm, *Manduca sexta, J. Exp. Biol.* **98:**385–401.

Taghert, P. H., Truman, J. W., and Reynolds, S. E., 1980, Physiology of pupal ecdysis in the tobacco hornworm, *Manduca sexta.* II. Chemistry, distribution, and release of eclosion hormone at pupal ecdysis, *J. Exp. Biol.* **88:**339–349.

Taghert, P. H., Carr, J. N., Wall, J. B., and Copenhaver, P. F., 1986, Embryonic formation of a simple neurosecretory nerve in the moth, *Manduca sexta,* in *Insect Neurochemistry and Neurophysiology* (A. B. Borkovec and D. B. Gelman, eds.), Plenum Publishing, New York.

Takeda, M., Endo, Y., Saito, H., Nishimura, M., and Nishiitsutsuji-Uwo, J., 1985, Neuropeptide and monoamine immunoreactivity of the circadian pacemaker in *Periplaneta, Biomed. Res.* **6:** 395–406.

Takeda, N., 1972a, Activation of neurosecretory cells in *Monema flavescens* (Lepidoptera) during diapause break, *Gen. Comp. Endocrinol.* **18:**417–427.

Takeda, N., 1972b, Effect of ecdysterone on spermatogenesis in the diapausing slug moth pharate pupa, *Monema flavescens, J. Insect Physiol.* **18:**571–580.

Takeda, N., 1978, Hormonal control of prepupal diapause in *Monema flavescens* (Lepidoptera), *Gen. Comp. Endocrinol.* **34:**123–131.

Takeda, S., 1977, Induction of egg diapause in *Bombyx mori* by some cephalo-thoracic organs of the cockroach, *Periplaneta americana, J. Insect Physiol.* **23:**813–816.

Takeda, S., and Girardie, A., 1985, An active principle from the corpora cardiaca, corpora allata and suboesophageal ganglion of the locust, inducing egg diapause in *Bombyx mori, J. Insect Physiol.* **31:**761–766.

Takeda, S., and Ogura, N., 1976, Induction of egg diapause in *Bombyx mori* by implantation of corpora cardiaca and corpora allata, *J. Insect Physiol.* **22:**941–944.

Takeda, S., Vieillemaringe, J., Geffard, M., and Remy, C., 1986, Immunohistological evidence of dopamine cells in the cephalic nervous system of the silkworm *Bombyx mori.* Coexistence of dopamine and α-endorphin-like substance in neurosecretory cells of the suboesophageal ganglion, *Cell Tissue Res.* **243:**125–128.

Takeuchi, S., 1969, Endocrinological studies of spermatogenesis in the silkworm *Bombyx mori* (L.), *Dev. Growth Different* **11:**8–28.

Tamarelle, M., Romeuf, M., and Vanderhaeghen, J. J., 1988, Immunohistochemical localization of gastrin-cholecystokinin-like material in the central nervous system of the migratory locust, *Histochemistry* **89:**201–207.

Tang, J. D., Charlton, R. E., Cardé, R. T., and Yin, C. M., 1987, Effect of allatectomy and ventral nerve cord transection on calling, pheromone emission and pheromone production in *Lymantria dispar, J. Insect Physiol.* **33:**469–476.

Taylor, D. P., Dyer, K. A., and Niewburgh, R. W., 1976, Cyclic nucleotides in neuronal and glial-enriched fractions from the nerve cord of *Manduca sexta, J. Insect Physiol.* **22:**1303–1304.

Tedesco, J. L., Courtright, J. B., and Kumaran, A. K., 1981, Ultrastructural changes induced by juvenile hormone analogue in oöcyte membranes of *Apterous Drosophila melanogaster, J. Insect Physiol.* **27:**895–902.

Teissier, G., 1947, Fonctionnement des chromatophores de la larve de Corèthre, *C.R. Acad. Sci. Paris* **225:**204–205.

Telfer, W. H., 1961, The route of entry and localization of blood proteins in the oocytes of saturniid moths, *J. Biophys. Biochem. Cytol.* **9:**747–759.

Telfer, W. H., Huebner, E., and Smith, D. S., 1982, The cell biology of vitellogenic follicles in *Hyalophora* and *Rhodnius,* in *Insect Ultrastructure* (R. C. King and H. Akai, eds.), Vol. 1, Plenum Press, New York, pp. 118–149.

Teller, J. K., and Pilc, L., 1985, Insulin in insects: Analysis of immunoreactivity in tissue extracts, *Comp. Biochem. Physiol.* **81B:**493–497.

Teller, J. K., and Pilc, L., 1986, Heterogeneity of insulin immunoreactivity in acid-ethanol extracts from mealworm tissues, *Bull. Soc. Amis. Sci. Lettr. Poznan. Ser. D. Sci. Biol.* **25:**47–50.

Teller, J. K., Rosinski, G., Pilc, L., Kasprzyk, A., and Lesicki, A., 1983, The presence of insulin-like hormone in heads and midguts of *Tenebrio molitor* L. (Coleoptera) larvae, *Comp. Biochem. Physiol.* **74A:**463–465.

Ter Wee, T. J., and Stay, B., 1987, Termination of pregnancy and milk synthesis in the viviparous cockroach *Diploptera punctata:* A role of juvenile hormone, *Int. J. Invert. Reprod.* **11:**59–75.

Thibout, E., 1982, Le comportement sexuel du doryphore, *Leptinotarsa decemlineata Say* et son possible contrôle par l'hormone juvénile et les corps allates, *Behaviour* **80:**199–217.

Thomas, A., 1964, Etude expérimentale relative au contrôle endocrine de l'ovogénèse chez *Gryllus domesticus* L., *Bull. Soc. Zool. Fr.* **89:**835–854.

Thomas, A., 1979a, Nervous control of egg progression into the common oviduct and genital chamber of the stick-insect *Carausius morosus, J. Insect Physiol.* **25:**811–823.

Thomas, A., 1979b, Inhibition of the onset of vitellogenesis by abnormal retention of mature eggs in the stick insect, *Carausius morosus, Int. J. Invert. Reprod.* **1:**347–354.

Thomas, A., and Mesnier, M., 1973, Le rôle du système nerveux central sur les mécanismes de l'oviposition chez *Carausius morosus* et *Clitumnus extradentatus, J. Insect Physiol.* **19:**383–396.

Thomas, A., and Raabe, M., 1974, Les organes périsympathiques des Orthoptères, *Bull. Soc. Zool. Fr.* **99:**187–206.

Thompson, M. J., Svoboda, J. A., Lozano, R., and Wilzer, K. R., 1988, Profile of free and conjugated ecdysteroids and ecdysteroid acids during embryonic development of *Manduca sexta* (L.) following maternal incorporation of [$_{14}$C]cholesterol, *Arch. Insect Biochem. Physiol.* **7:** 157–172.

Thomsen, E., 1942, An experimental and anatomical study of the corpus allatum in the blow-fly *Calliphora erythrocephala* Meig., *Vidensk. Medd. Naturh. Foren. Kbh.* **106:**320–405.

Thomsen, E., 1952, Functional significance of the neurosecretory brain cells and the corpus cardiacum in the female blow-fly, *Calliphora erythrocephala, J. Exp. Biol.* **29:**137–172.

Thomsen, E., 1964, Hormonal control of egg development in *Calliphora, Science* **143:**973.

Thomsen, E., and Lea, A. O., 1969, Control of the medial neurosecretory cells by the corpus allatum in *Calliphora erythrocephala, Gen. Comp. Endocrinol.* **12:**51–57.

Thomsen, E., and Møller, I., 1959, Neurosecretion and intestinal proteinase activity in an insect, *Calliphora, Nature* **183:**1401–1402.

Thomsen, E., and Møller, I., 1963, Influence of neurosecretory cells and of corpus allatum on intestinal protease activity in the adult *Calliphora erythrocephala* Meig., *J. Exp. Biol.* **40:**301–321.

Thomsen, E., and Thomsen, M., 1969, Fine structure of the corpus allatum of the female *Calliphora,* with special regard to hormone formation, *Gen. Comp. Endocrinol.* **13:**534.

Thomsen, E., Hansen, B. L., Hansen, G. N., and Jensen, P. V., 1980, Ultrastructural immunocytochemical localization of vitellogenin in the fat body of the blowfly, *Calliphora vicina* Rob.-Desv. (*erythrocephala* Meig.) by use of the unlabeled antibody-enzyme method, *Cell Tissue Res.* **208:**445–455.

Thomsen, M., 1943, Effect of corpus cardiacum and other insect organs on the colour-change of the shrimp, *Leander adspersus, Biol. Medd. K. Dan. Vidensk. Selsk.* **19:**1–38.

Thomsen, M., 1965, The neurosecretory system of the adult *Calliphora erythrocephala.* II. Histology of the neurosecretory cells of the brain and some related structures, *Z. Zellforsch.* **67:**693–717.

Thomsen, M., 1969, The neurosecretory system of the adult *Calliphora erythrocephala.* IV. A histological study of the corpus cardiacum and its connections with the nervous system, *Z. Zellforsch.* **94:**205–219.

Thorson, B. J., and Riemann, J. G., 1982, Effects of 20-hydroxyecdysone on sperm release from the testes of the mediterranean flour moth, *Anagasta Kuehniella* (Zeller), *J. Insect Physiol.* **28:** 1013–1019.

Tobe, S. S., and Stay, B., 1979, Modulation of juvenile hormone synthesis by an analogue in the cockroach, *Nature* **281:**481–482.

Tobe, S. S., and Stay, B., 1980, Control of juvenile hormone biosynthesis during the reproductive cycle of a viviparous cockroach. III. Effects of denervation and age on compensation with unilateral allatectomy and supernumerary corpora allata, *Gen. Comp. Endocrinol.* **40:**89–98.

Tobe, S. S., and Stay, B., 1986, Structure and regulation of the corpus allatum, *Adv. Insect Physiol.* **18:**305–432.

Tobe, S. S., Clarke, N., Stay, B., and Ruegg, R. P., 1984, Changes in cell number and activity of the corpora allata of the cockroach *Diploptera punctata:* A role for mating and the ovary, *Can. J. Zool.* **62:**2178–2182.

Tobe, S. S., Ruegg, R. P., Stay, B., Baker, F. C., Miller, C. A., and Schooley, D. A., 1985, Juvenile hormone titre and regulation in the cockroach *Diploptera punctata, Experientia* **41:** 1028–1034.

Tolman, J. H., and Steele, J. E., 1980, The effect of the corpora cardiaca-corpora allata system on oxygen consumption in the cockroach rectum: The role of Na^+ and K^+, *J. Comp. Physiol.* **138:**347–355.

Tombes, A. S., and Smith, D. S., 1970, Ultrastructural studies on the corpora cardiaca-allata complex of the adult alfalfa weevil, *Hypera postica, J. Morphol.* **132:**137–148.

Traina, M. E., Bellino, M., and Frontali, N., 1974, A comparison between neurohormone D and peak I extracted from cockroach corpora cardiaca, *Zool. Jahrb. Abt. Allg. Zool. Physiol. Tiere* **78:**424–428.

Traina, M. E., Bellino, M., Serpietri, L., Massa, A., and Frontali, N., 1976, Heart-accelerating peptides from cockroach corpora cardiaca, *J. Insect Physiol.* **22:**323–329.

Triepel, J., and Grimmelikhuijzen, C. J. P., 1984, A critical examination of the occurrence of FMRFamide immunoreactivity in the brain of guinea pig and rat, *Histochemistry* **80:**63–71.

Truman, J. W., 1971, Physiology of insect ecdysis. I. The eclosion behaviour of saturniid moths and its hormonal release, *J. Exp. Biol.* **54:**805–814.

Truman, J. W., 1973a, How moths "turn on": A study of the action of hormones on the nervous system, *Am. Sci.* **61:**700–706.

Truman, J. W., 1973b, Physiology of insect ecdysis. III. Relationship between the hormonal control of eclosion and of tanning in the tobacco hornworm, *Manduca sexta, J. Exp. Biol.* **58:**821–829.

Truman, J. W., 1973c, Physiology of insect ecdysis. II. The assay and occurrence of the eclosion hormone in the Chinese oak silkmoth, *Antheraea pernyi, Biol. Bull. (Woods Hole, Mass.)* **144:** 200–211.

Truman, J. W., 1978, Hormonal release of stereotyped motor programmes from the isolated nervous system of the *Cecropia* silkmoth, *J. Exp. Biol.* **74:**151–173.

Truman, J. W., 1979, Interaction between abdominal ganglia during the performance of hormonally triggered behavioural programmes in moths, *J. Exp. Biol.* **83:**239–253.

Truman, J. W., 1981, Interaction between ecdysteroid, eclosion hormone and bursicon titers in *Manduca sexta, Am. Zool.* **21:**655–661.

Truman, J. W., 1983, Programmed cell death in the nervous system of an adult insect, *J. Comp. Neurol.* **216:**445–452.

Truman, J. W., 1985, Hormonal control of ecdysis, in *Comprehensive Insect Physiology, Biochemistry and Pharmacology* (G. A. Kerkut and L. I. Gilbert, eds.), Vol. 8, Pergamon Press, New York, pp. 289–353.

Truman, J. W., and Riddiford, L. M., 1970, Neuroendocrine control of ecdysis in silkmoths, *Science* **167:**1624–1626.

Truman, J. W., and Riddiford, L. M., 1971, Role of the corpora cardiaca in the behavior of saturniid moths. II. Oviposition, *Biol. Bull. (Woods Hole, Mass.)* **140:**8–14.

Truman, J. W., and Riddiford, L. M., 1974, Physiology of insect rhythms. III. The temporal organization of the endocrine events underlying pupation of the tobacco hornworm, *J. Exp. Biol.* **60:**371–382.

Truman, J. W., and Sokolove, P. G., 1972, Silk moth eclosion: Hormonal triggering of a centrally programmed pattern of behavior, *Science* **175:**1491–1493.

Truman, J. W., Fallon, A. M., and Wyatt, G. R., 1976, Hormonal release of programmed behavior in silk moths: Probable mediation by cyclic AMP, *Science* **194:**1432–1434.

Truman, J. W., Mumby, S. M., and Welch, S. K., 1979, Involvement of cyclic GMP in the release of stereotyped behaviour patterns in moths by a peptide hormone, *J. Exp. Biol.* **84:**201–212.

Truman, J. W., Taghert, P. H., and Reynolds, S. E., 1980, Physiology of pupal ecdysis in the tobacco hornworm, *Manduca sexta*. I. Evidence for control by eclosion hormone, *J. Exp. Biol.* **88:**327–337.

Truman, J. W., Taghert, P. H., Copenhaver, P. F., Tublitz, N. J., and Schwartz, L. M., 1981, Eclosion hormone may control all ecdyses in insects, *Nature* **291:**70–71.

Truman, J. W., Rountree, D. B., Reiss, S. E., and Schwartz, L. M., 1983, Ecdysteroids regulate the release and action of eclosion hormone in the tobacco hornworm, *Manduca sexta* (L.), *J. Insect Physiol.* **29:**895–900.

Tsuchida, K., Nagata, M., and Suzuki, A., 1987, Hormonal control of ovarian development in the silkworm, *Bombyx mori, Arch. Insect Biochem. Physiol.* **5:**167–177.

Tsumuki, H., and Kanehisa, K., 1981, Effect of JH and ecdysone on glycerol and carbohydrate contents in diapausing larvae of the rice stem borer, *Chilo suppressalis* Walker (Lepidoptera: Pyralidae), *Appl. Entomol. Zool.* **16:**7–15.

Tublitz, N. J., and Evans, P. D., 1986, Insect Cardioactive peptides: Cardioacceleratory peptide (CAP) activity is blocked *in vivo* and *in vitro* with a monoclonal antibody, *J. Neurosci.* **6:**2451–2456.

Tublitz, N. J., and Truman, J. W., 1985a, Identification of neurons containing cardioacceleratory peptides (CAPs) in the ventral nerve cord of the tobacco hawkmoth, *Manduca sexta, J. Exp. Biol.* **116**:395–410.

Tublitz, N. J., and Truman, J. W., 1985b, Insect cardioactive peptides. I. Distribution and molecular characteristics of two cardioacceleratory peptides in the tobacco hawkmoth, *Manduca sexta, J. Exp. Biol.* **114:**365–379.

Twarog, B. M., and Roeder, K. D., 1957, Pharmacological observations on the desheathed last abdominal ganglion of the cockroach, *Ann. Entomol. Soc. Am.* **50:**231–237.

Ude, J., Eckert, M., and Penzlin, H., 1978, The frontal ganglion of *Periplaneta americana* L. (Insecta). An electron microscopic and immunohistochemical study, *Cell Tissue Res.* **191:**171–182.

Ulrich, G. M., Schlagintweit, B., Eder, J., and Rembold, H., 1985, Elimination of the allatotropic activity in locusts by microsurgical and immunological methods: Evidence for humoral control of the corpora allata, hemolymph proteins, and ovary development, *Gen. Comp. Endocrinol.* **59:**120–129.

Unger, H., 1956, Neurohormonale Steuerung der Herztätigkeit bei Insekten, *Naturwissenschaften* **43:**66–67.

Unger, H., 1967, Der Einfluss der Neurohormone C und D auf den Wasserhaushalt der Stabheuschrecke (*Carausius morosus* Br.), *Biol. Rundsch.* **5:**31–33.

Vallés, A. M., and White, K., 1988, Serotonin-containing neurons in *Drosophila melanogaster:* Development and distribution, *J. Comp. Neurol.* **268:**414–428.

Vancassel, M., Foraste, M., Strambi, A., and Strambi, C., 1984, Normal and experimentally induced changes in hormonal hemolymph titers during parental behavior of the earwig *Labidura riparia, Gen. Comp. Endocrinol.* **56:**444–456.

Vandenberg, R. D., and Mills, R. R., 1974, Hormonal control of tanning by cockroach: Cyclic AMP as a probable intermediate, *J. Insect Physiol.* **20:**623–627.

Van Handel, E., and Lea, A. O., 1965, Medial neurosecretory cells as regulators of glycogen and triglyceride synthesis, *Science* **149:**298–300.

Van Heusden, M. C., Van Der Horst, D. J., and Beenakkers, A. M. T., 1984, *In vitro* studies on hormone-stimulated lipid mobilization from fat body and interconversion of haemolymph lipoproteins of *Locusta migratoria, J. Insect Physiol.* **30:**685–693.

Van Marrewijk, W. J. A., Van Den Broek, A. T. M., and Beenakkers, A. M. T., 1980, Regulation of glycogenolysis in the locust fat body during flight, *Insect Biochem.* **10:**675–679.

Van Marrewijk, W. J. A., Van Den Broek, A. T. M., and Beenakkers, A. M. T., 1983, Regulation of glycogen phosphorylase activity in fat body of *Locusta migratoria* and *Periplaneta americana, Gen. Comp. Endocrinol.* **50:**226–234.

Van Marrewijk, W. J. A., Van Den Broek, A. T. M., Van Der Horst, D. J., and Beenakkers, A. M. T., 1984, Hypertrehalosaemic and hyperlipaemic responses to adipokinetic hormone in fifth larval instar locusts, *Locusta migratoria, Insect Biochem.* **14:**151–157.

Vedeckis, W. V., and Gilbert, L. I., 1973, Production of cyclic AMP and adenosine by the brain and prothoracic glands of *Manduca sexta, J. Insect Physiol.* **19:**2445–2457.

Vedeckis, W. V., Bollenbacher, W. E., and Gilbert, L. I., 1974, Cyclic AMP as a possible mediator of prothoracic gland activation, *Zool. Jahrb. Abt. Allg. Zool. Physiol. Tiere* **78:**440–448.

Vedeckis, W. V., Bollenbacher, W. E., and Gilbert, L. I., 1976, Insect prothoracic glands: A role for cyclic AMP in the stimulation of α-ecdysone secretion, *Mol. Cell. Endocrinol.* **5:** 81–88.

Veenstra, J. A., 1984, Immunocytochemical demonstration of a homology in peptidergic neurosecretory cells in the suboesophageal ganglion of a beetle and a locust with antisera to bovine pancreatic polypeptide. FMRFamide, vasopressin and α-MSH, *Neurosci. Lett.* **48:**185–190.

Veenstra, J. A., 1988, Effects of 5-hydroxytryptamine on the Malpighian tubules of *Aedes aegypti, J. Insect Physiol.* **34:**299–304.

Veenstra, J. A., and Schooneveld, H., 1984, Immunocytochemical localization of neurons in the nervous system of the Colorado potato beetle with antisera against FMRFamide and bovine pancreatic polypeptide, *Cell Tissue Res.* **235:**303–308.

Veenstra, J. A., and Yanaihara, N., 1984, Immunocytochemical localization of gastrin-releasing peptide/bombesin-like immunoreactive neurons in insects, *Histochemistry* **81:**133–138.

Veenstra, J. A., Romberg-Privee, H. M., and Schooneveld, H., 1984, Immunocytochemical localization of peptidergic cells in the neuro-endocrine system of the Colorado potato beetle, *Leptinotarsa decemlineata,* with antisera against vasopressin, vasotocin and oxytocin, *Histochemistry* **81:**29–34.

Veenstra, J. A., Romberg-Privee, H. M., Schooneveld, H., and Polak, J. M., 1985a, Immunocytochemical localization of peptidergic neurons and neurosecretory cells in the neuro-endocrine system of the Colorado potato beetle with antisera to vertebrate regulatory peptides, *Histochemistry* **82:**9–18.

Veenstra, J. A., Romberg-Privee, H. M., and Schooneveld, H., 1985b, A proctolin-like peptide and its immunocytochemical localization in the Colorado potato beetle, *Leptinotarsa decemlineata, Cell Tissue Res.* **240:**535–540.

Vejbjerg, K., and Normann, T. C., 1974, Secretion of hyperglycaemic hormone from the corpus cardiacum of flying blowflies, *Calliphora erythrocephala, J. Insect Physiol.* **20:**1189–1192.

Venkatesh, K., and Roe, R. M., 1988, The role of juvenile hormone and brain factor(s) in the regulation of plasma juvenile hormone esterase activity during the last larval stadium of the tobacco hornworm, *Manduca sexta, J. Insect Physiol.* **34:**415–425.

Verhaert, P., and De Loof, A., 1985a, Immunocytochemical localization of a methionine-enkephalin-resembling neuropeptide in the central nervous system of the American cockroach *Periplaneta americana,* L., *J. Comp. Neurol.* **239:**54–61.

Verhaert, P., and De Loof, A., 1985b, Substance P-like immunoreactivity in the central nervous system of the blattarian insect *Periplaneta americana* L. revealed by a monoclonal antibody, *Histochemistry* **83:**501–507.

Verhaert, P., and De Loof, A., 1986, Substances resembling peptides of the vertebrate gonadotropin system occur in the central nervous system of *Periplaneta americana* L. Immunocytological and some biological evidence, *Insect Biochem.* **16:**191–197.

Verhaert, P., Marivoet, S., Vandesande, F., and De Loof, A., 1984a, Localization of CRF immunoreactivity in the central nervous system of three vertebrate and one insect species, *Cell Tissue Res.* **238:**49–53.

Verhaert, P., Geysen, J., De Loof, A., and Vandesande, F., 1984b, Immunoreactive material resembling vertebrate neuropeptides and neurophysins in the bain, suboesophageal ganglion, corpus cardiacum and corpus allatum of the dictyopteran *Periplaneta americana L., Cell Tissue Res.* **238:**55–59.

Verhaert, P., Grimmelilhuijzen, C. J. P., and De Loof, A., 1985, Distinct localization of FMRFamide- and bovine pancreatic polypeptide- like material in the brain, retrocerebral complex and suboesophageal ganglion of the cockroach *Periplaneta americana L., Brain Res.* **348:** 331–338.

Veron, J. E. N., 1973, Physiological control of the chromatophores of *Austrolestes annulosus* (Odonata), *J. Insect Physiol.* **19:**1689–1703.

Vieillemaringe, J., Duris, P., Bensch, C., and Girardie, J., 1982, Co-localization of amines and peptides in the same median neurosecretory cells of locusts, *Neurosci. Lett.* **31:**237–240.

Vieillemaringe, J., Duris, P., Geffard, M., Le Moal, M., Delaage, M., Bensch, C., and Girardie, J., 1984, Immunohistochemical localization of dopamine in the bain of the insect *Locusta migratoria migratorioides* in comparison with the catecholamine distribution determined by the histofluorescence technique, *Cell Tissue Res.* **237:**391–394.

Vietinghoff, U., 1966a, Untersuchungen über die Funktion der Rektaldrüsen der Stabheuschrecke *Carausius morosus* Br, *Zool. Anz.* **29:**157–162.

Vietinghoff, U., 1966b, Einfluss der Neurohormone C1 und D1 auf die Absorptionsleistung der Rektaldrüsen der Stabheuschrecke (*Carausius morosus*), *Naturwissenschaften* **52:**162–163.
Vietinghoff, U., 1967, Neurohormonal control of renal function in *Carausius morosus* Br., *Gen. Comp. Endocrinol.* **9:**503.
Vincent, J. F. V., 1971, Effects of bursicon on cuticular properties in *Locusta migratoria migratorioides, J. Insect Physiol.* **17:**625–636.
Vincent, J. F. V., 1972, The dynamics of release and the possible identity of bursicon in *Locusta migratoria migratorioides, J. Insect Physiol.* **18:**757–780.
Vincent, J. F. V., and Ablett, S., 1987, Hydration and tanning in insect cuticle, *J. Insect Physiol.* **33:**973–979.
Von Knorre, D., Gersch, M., and Kusch, T., 1972, Zur Frage der Beeinflussung des "tanning" Phänomens durch zyklisches 3′, 5′-AMP, *Zool. Jahrb. Abt. Allg. Zool. Physiol. Tiere* **76:**434–440.
Vroman, H. E., Kaplanis, J. N., and Robbins, W. E., 1965, Effect of allatectomy on lipid biosynthesis and turnover in the female American cockroach, *Periplaneta americana* L., *J. Insect Physiol.* **11:**897–904.
Vuillaume, M., Seugé, S., and Bergerard, J., 1971, Photopériode et pigment tégumentaire vert des chenilles de *Pieris brassicae:* Conditionnement de la diapause, *C.R. Acad. Sci. Ser. D* **273:** 1608–1610.
Wakayama, E. J., Dillwith, J. W., and Blomquist, G. J., 1986a, Occurrence and metabolism of prostaglandins in the housefly, *Musca domestica* (L.), *Insect Biochem.* **16:**895–902.
Wakayama, E. J., Dillwith, J. W., and Blomquist, G. J., 1986b, Characterization of prostaglandin synthesis in the housefly, *Musca domestica* (L.), *Insect Biochem.* **16:**903–909.
Waku, Y., 1960, Studies on the hibernation and diapause in insects. 4. Histological observations of the endocrine organs in the diapause and non-diapause larvae of the Indian mealmoth, *Plodia interpunctella* Hübner, *Sci. Rep. Tôhoku Univ. Ser. 4 (Biol.).* **26:**327–340.
Walker, G. P., and Denlinger, D. L., 1980, Juvenile hormone and moulting hormone titres in diapause and non-diapause destined flesh flies, *J. Insect Physiol.* **26:**661–664.
Walker, R. J., 1984, 5-Hydroxytryptamine in invertebrates, *Comp. Biochem. Physiol.* **79C:**231–235.
Wall, B. J., 1965, Regulation of water metabolism by the Malpighian tubules and rectum in the cockroach *Periplaneta americana* L., *Zool. Jahbr. Abt. Zool. Physiol. Tiere* **71:**702–709.
Wall, B. J., 1966, Diuresis in the cockroach, *Am. Zoologist* **6:**340.
Wall, B. J., and Ralph, C. L., 1962, Factors from the nervous system of the cockroach that influence excretion rates of Malpighian tubules (15 mn), *Am. Zoologist* **2:**103.
Walther, C., and Schäfer, S., 1988, FMRFamide-like immunoreactivity in the metathoracic ganglion of the locust (*Schistocerca gregaria*), *Cell Tissue Res.* **253:**489–491.
Walther, C., Schiebe, M., and Voigt, K. H., 1984, Synaptic and non-synaptic effects of molluscan cardioexcitatory neuropeptides on locust skeletal muscle, *Neurosci. Letters* **45:**99–104.
Warren, J. T., Sakurai, S., Rountree, D. B., Gilbert, L. I., Lee, S. S., and Nakanishi, K., 1988, Regulation of the ecdysteroid titer of *Manduca sexta:* Reappraisal of the role of the prothoracic glands, *Proc. Natl. Acad. Sci. USA* **85:**958–962.
Warton, S., 1981, Effect of disulfiram on the ultrastructure of the peptidergic and aminergic cells in the pars intercerebralis of *Galleria mellonella* (Lepidoptera), *Cell Tissue Res.* **215:**417–424.
Wasserman, A. J., 1985, Central and peripheral neurosecretory pathways to an insect flight motor nerve, *J. Neurobiol.* **16:**329–346.
Wasserthal, W., and Wasserthal, L. T., 1980, Multinucleate neurons with neurohaemal and synapsing axons at the heart and alary muscles of the butterfly *Caligo beltrao* Illiger (Lepidoptera), *Cell. Tissue Res.* **212:**351–362.
Watanabe, K., and Ohnishi, E., 1984, The mode of ecdysteroid accumulation in ovaries of *Bombyx mori* during the pupal and pharate adult period, *Zool. Sci.* **1:**114–119.

Watson, R. D., and Bollenbacher, W. E., 1988, Juvenile hormone regulates the steroidogenic competence of *Manduca sexta* prothoracic glands, *Mol. Cell. Endocr.* **57**:251–259.

Watson, R. D., Ciancio, M. J., Gunnar, W.P., Gilbert, L. I., and Bollenbacher, W. E., 1985, Regulation of insect prothoracic glands: Existence of a haemolymph stimulatory factor in *Manduca sexta, J. Insect Physiol.* **31**:487–494.

Watson, R. D., Whisenton, L. R., Bollenbacher, W. E., and Granger, N. A., 1986, Interendocrine regulation of the corpora allata and prothoracic glands of *Manduca sexta, Insect Biochem.* **16**: 149–155.

Watson, R. D., Agui, N., Haire, M. E., and Bollenbacher, W. E., 1987a, Juvenile hormone coordinates the regulation of the hemolymph ecdysteroid titer during pupal commitment in *Manduca sexta, Insect Biochem.* **17**:955–959.

Watson, R. D., Williams, T. K., and Bollenbacher, W. E., 1987b, Regulation of ecdysone biosynthesis in the tobacco hornworm, *Manduca sexta:* Titre of the haemolymph stimulatory factor during the last larval instar, *J. Exp. Biol.* **128**:159–173.

Watson, R. D., Haire, M. E., and Bollenbacher, W. E., 1988, Juvenile hormone regulates the titer of a hemolymph factor that enhances ecdysone synthesis by insect (*Manduca sexta*) prothoracic glands *in vitro, Arch. Insect Biochem. Physiol.* **9**:157–165.

Watson, W. H., Augustine, G. J., Benson, J. A., and Sullivan, R. E., 1983, Proctolin and an endogenous proctolin-like peptide enhance the contractility of the *Limulus* heart, *J. Exp. Biol.* **103**:55–73.

Weaver, R. J., 1981, Radiochemical assays of corpus allatum activity in adult female cockroaches following ovariectomy in the last nymphal instar, *Experientia* **37**:435–436.

Weaver, R. J., 1984, Effects of food and water availability and of NCA-1 section, upon juvenile hormone biosynthesis and oocyte development in adult female *Periplaneta americana, J. Insect Physiol.* **10**:831–838.

Weaver, R. J., Pratt, G. E., and Finney, J. R., 1975, Cyclic activity of the corpus allatum related to gonotrophic cycles in adult female *Periplaneta americana, Experientia* **31**:597–598.

Weaver, R. J., Strambi, A., and Strambi, C., 1984, The significance of free ecdysteroids in the haemolymph of adult cockroaches, *J. Insect Physiol.* **30**:705–711.

Weber, V., and Emmerich, H., 1976, RNA biosynthesis in isolated prothoracic glands of *Tenebrio molitor* in vitro, *Experientia* **32**:1609–1610.

Webster, R. P., and Cardé, R. T., 1984, The effects of mating, exogenous juvenile hormone and a juvenile hormone analogue on pheromone titre, calling and oviposition in the omnivorous leafroller moth (*Platynota stultana*), *J. Insect Physiol.* **30**:113–118.

Weeda, E., 1981, Hormonal regulation of proline synthesis and glucose release in the fat body of the Colorado potato beetle, *Leptinotarsa decemlineata, J. Insect Physiol.* **27**:411–417.

Weeda, E., De Kort, C. A. D., and Beenakkers, A. M. T., 1979, Fuels for energy metabolism in the Colorado beetle *Leptinotarsa decemlineata* Say, *J. Insect Physiol.* **25**:951–955.

Weeks, J. C., and Truman, J. W., 1986, Hormonally mediated reprogramming of muscles and motoneurones during the larval-pupal transformation of the tobacco hornworm, *Manduca sexta, J. Exp. Biol.* **125**:1–13.

Weinbörmair, G., Pohlhammer, K., and Durnberger, H., 1975, Das Axonsystem des ventro-medianen neurosekretorischen Zellpaares im Suboesophageal Ganglion bei *Teleogryllus commodus* Walker. Darstellung im Totalpräparat mit Hilfe von Resorcinfuchsin, *Mikroskopie.* **31**:147–154.

Weiss, A., Beichert, K. H., and Penzlin, H., 1987, Einfluss von Corpora Cardiaca-Rohextrakt auf die Acquisition von zwei verschiedenen assoziativen Lernprozessen bei der Schabe *Periplaneta americana* (L.), *Zool. Jahrb. Abt. Allg. Zool. Physiol. Tiere* **91**:229–241.

Went, D. F., 1978, Ecdysone stimulates and juvenile hormone inhibits follicle formation in a gall midge ovary *in vitro, J. Insect Physiol.* **24**:53–59.

Wentworth, S. L., and Roberts, B., 1984, Ecdysteroid levels during adult reproductive and embryonic developmental stages of *Sarcophaga bullata* (Sarcophagidae: Diptera), *J. Insect Physiol.* **30:**157–163.

Weyer, F., 1935, Uber drüsenartige Nervenzellen im Gehirn der Honigbiene, Apis mellifica L. *Zool. Anz.* **112:**137–141.

Wheeler, C. H., and Goldsworthy, G. J., 1983a, Qualitative and quantitative changes in *Locusta* haemolymph proteins and lipoproteins during ageing and adipokinetic hormone action, *J. Insect Physiol.* **29:**339–347.

Wheeler, C. H., and Goldsworthy, G. J., 1983b, Protein-lipoprotein interactions in the haemolymph of *Locusta* during the action of adipokinetic hormone: The role of C_L-proteins, *J. Insect Phyisol.* **29:**349–354.

Wheeler, D. E., and Nijhout, H. F., 1983, Soldier determination in *Pheidole bicarinata:* Effect of methoprene on caste and size within castes, *J. Insect Physiol.* **29:**847–854.

Wheelock, G. D., and Hagedorn, H. H., 1985, Egg mutaration and ecdysiotropic activity in extracts of mosquito (*Aedes aegypti*) heads, *Gen. Comp. Endocrinol.* **60:**196–203.

Whisenton, L. R., Bowen, M. F., Granger, N. A., Gilbert, L. I., and Bollenbacher, W. E., 1985, Brain-mediated 20-hydroxyecdysone regulation of juvenile hormone synthesis by the corpora allata of the tobacco hornworm, *Manduca sexta, Gen. Comp. Endocrinol.* **58:**311–318.

Whisenton, L. R., Kelly, T. J., and Bollenbacher, W. E., 1987a, Multiple forms of cerebral peptides with steroidogenic functions in pupal and adult brains of the yellow fever mosquito, *Aedes aegypti, Mol. Cell. Endocrinol.* **50:**3–14.

Whisenton, L. R., Granger, N. A., and Bollenbacher, W. E., 1987b, A kinetics analysis of brain-mediated 20-hydroxyecdysone stimulation of the corpora allata of *Manduca sexta, Mol. Cell. Endocrinol.* **54:**171–178.

Whitmore, D., Whitmore, E., and Gilbert, L. I., 1972, Juvenile hormone induction of esterases: A mechanism for the regulation of juvenile hormone titer, *Proc. Natl. Acad. Sci. USA* **69:**1592–1595.

Wicker, C., 1987, Effects of brain, suboesophageal ganglion and corpus cardiacum factors on juvenile hormone biosynthesis *in vitro* in *Acheta domesticus* (Orthoptera), *Comp. Biochem. Physiol.* **87C:**161–165.

Wicker, C., and Wicker, C., 1987, Isolation and structure of a peptide isolated from the suboesophageal ganglion of *Acheta domesticus* (Orthoptera), *Comp. Biochem. Physiol.* **88C:**185–187.

Wielgus, J. J., 1983, Stimulation of intermoult cuticle deposition by a haemolymph trophic factor in the tobacco hornworm, *Manduca sexta, Insect Biochem.* **13:**313–322.

Wiens, A. W., and Gilbert, L. I., 1965, Regulation of cockroach fat body metabolism by the corpus cardiacum in vitro, *Science* **150:**614–616.

Wiens, A. W., and Gilbert, L. I., 1967, The phosphorylase system of the silkmoth, *Hyalophora cecropia, Comp. Biochem. Physiol.* **21:**145–159.

Wigglesworth, V. B., 1934, The physiology of ecdysis in *Rhodnius prolixus* (Hemiptera). II. Factors controlling moulting and metamorphosis, *Q. J. Microsc. Sci.* **77:**191–222.

Wigglesworth, V. B., 1936, The function of the corpus allatum in the growth and reproduction of *Rhodnius prolixus, Q. J. Microsc. Sci.* **79:**91–121.

Wigglesworth, V. B., 1940, The determination of characters at metamorphosis in *Rhodnius prolixus* (Hemiptera), *J. Exp. Biol.* **17:**201–222.

Wigglesworth, V. B., 1952, The thoracic gland in *Rhodnius prolixus* (Hemiptera) and its role in moulting, *J. Exp. Biol.* **29:**561–570.

Wigglesworth, V. B., 1954, The breakdown of the thoracic gland in the adult insect, *Rhodnius prolixus, J. Exp. Biol.* **32:**485–491.

Wigglesworth, V. B., 1955, The role of the haemocytes in the growth and moulting of an insect, *Rhodnius prolixus* (Hemiptera), *J. Exp. Biol.* **32:**649–663.

Wigglesworth, V. B., 1957, The action of growth hormones in insects, *The Biological Action of Growth Substances, Symp. Soc. Exp. Biol. XI.* pp. 204–227.

Wigglesworth, V. B., 1985, Historical perspectives, in *Comprehensive Insect Physiology, Biochemistry and Pharmacology* (G. A. Kerkut and L. I. Gilbert, eds.), Vol. 7, Pergamon Press, New York, pp. 1–24.

Wilhelm, R., Riechsteiner, R., and Lanzrein, B., 1987, On the competence of corpora allata to synthesise juvenile hormone and the dependence of their activity on haemolymph factors in the cockroach, *Nauphoeta cinerea, Insect Biochem.* **17:**971–975.

Wilkens, J. L., 1968, The endocrine and nutritional control of egg maturation in the fleshfly *Sarcophaga bullata, J. Insect Physiol.* **14:**927–943.

Wilkens, J. L., 1969, The endocrine control of protein metabolism as related to reproduction in the fleshfly *Sarcophaga bullata, J. Insect Physiol.* **15:**1015–1024.

Williams, C. M., 1946, Physiology of insect diapause: The role of the brain in the production and termination of pupal dormancy in the giant silkworm, *Platysamia cecropia, Biol. Bull. (Woods Hole, Mass.)* **90:**234–243.

Williams, C. M., 1947, Physiology of insect diapause. II. Interaction between the pupal brain and prothoracic glands in the metamorphosis of the giant silkworm, *Platysamia cecropia, Biol. Bull. (Woods Hole, Mass.)* **93:**89–98.

Williams, C. M., 1948, Physiology of insect diapause. III. The prothoracic glands in the Cecropia silkworm, with special reference to their significance in embryonic and post-embryonic development, *Biol. Bull. (Woods Hole, Mass.)* **94:**60–65.

Williams, C. M., 1951, Morphogenesis and the metamorphosis of insects, *Harvey Lectures* **47:**126–156.

Williams, C. M., 1952, Physiology of insect diapause. IV. The brain and prothoracic glands as an endocrine system in the Cecropia silkworm, *Biol. Bull. (Woods Hole, Mass.)* **103:**120–138.

Williams, C. M., 1959, The juvenile hormone. I. Endocrine activity of the corpora allata of the adult Cecropia silkworm, *Biol. Bull. (Woods Hole, Mass.)* **116:**323–328.

Williams, C. M., 1961, The juvenile hormone. II. Its role in the endocrine control of molting, pupation and adult development in the Cecropia silkworm, *Biol. Bull. (Woods Hole, Mass.)* **121:**572–585.

Williams, C. M., 1967, The present status of the brain hormone, in *Insects and Physiology* (J. W. L. Beament and J. E. Treherne, eds.), Olivier and Boyd, Edinburgh, pp. 133–139.

Williams, C. M., and Adkisson, P. L., 1964, Physiology of insect diapause in the oak silkworm, *Antheraea pernyi, Biol. Bull. (Woods Hole, Mass.)* **127:**511–525.

Williams, J. C., and Beyenbach, K. W., 1983, Differential effects of secretagogues on Na and K secretion in the Malpighian tubules of *Aedes aegypti, J. Comp. Physiol.* **149:**511–517.

Williams, J. C., and Beyenbach, K. W., 1984, Differential effects of secretagogues on the electrophysiology of the Malpighian tubules of the yellow fever mosquito, *J. Comp. Physiol.* **154:** 301–309.

Willis, J. H., and Brunet, P. C. J., 1966, The hormonal control of colleterial gland secretion, *J. Exp. Biol.* **44:**363–378.

Wilson, T. G., 1982, A correlation between juvenile hormone deficiency and vitellogenic oocyte degeneration in *Drosophila melanogaster, Wilhelm Roux Arch. Entw. Mech. Org.* **191:**257–263.

Wirtz, P., and Beetsma, J., 1972, Induction of caste differentiation in the honey-bee (*Apis mellifera L.*) by juvenile hormone, *Entomol. Exp. Appl.* **15:**517–520.

Witten, J. L., Schaffer, M. H., O'Shea, M., Cook, J. C., Hemling, M. E., and Rinehart, K. L. J., 1984, Structures of two cockroach neuropeptides assigned by fast atom bombardment mass spectrometry, *Biochem. Biophys. Res. Commun.* **124:**350–358.

Wright, J. E., Chamberlain, W. F., and Barrett, C. C., 1971, Ovarian maturation in stable flies: Inhibition by 20-hydroxyecdysone, *Science* **172:**1247–1248.

Wu, S. J., Zhang, J. Z., and Ma, M., 1987, Monitoring the effects of juvenile hormones and 20-

hyroxyecdysone on yolk polypeptide production of *Drosophila melanogaster* with enzyme immunoassay, *Physiol. Entymol.* **12:**355–361.

Wyatt, G. R., 1961, Effects of experimental injury on carbohydrate metabolism in silkmoth pupae, *Fed. Proc.* **20:**81.

Wyss-Huber, M., and Lüscher, M., 1966, Über die hormonale Beeinflussbarkeit der Proteinsynthese *in vitro* im Fettkörper von *Leucophaea maderae* (Insecta), *Revue Suisse Zool.* **73:**517–521.

Yagi, K., and Bern, H. A., 1963, Electrophysiological indications of the osmoregulatory role of the teleost urophysis, *Science* **142:**491–493.

Yagi, S., 1980, Hormonal control of larval colouration associated with phase variation in the armyworm *Spodoptera exempta, ICIPE Annu. Rep.* **7:**14.

Yagi, S., and Akaike, N., 1976, Regulation of larval diapause by juvenile hormone in the European corn borer, *Ostrinia nubilalis, J. Insect Physiol.* **22:**389–392.

Yagi, S., and Fukaya, M., 1974, Juvenile hormone as a key factor regulating larval diapause of the rice stem borer, *Chilosuppressalis* (Lepidoptera: Pyralidae), *Appl. Entomol. Zool.* **9:** 247–255.

Yagi, S., Kondo, E., and Fukaya, M., 1969, Hormonal effect of cultivated insect tissue. I. Effect of ecdysterone on cultivated testes of diapausing rice stem borer larvae, *Appl. Entomol. Zool.* **4:** 70–78.

Yamamoto, D., and Fukami, J. I., 1980, Neuromuscular mechanisms in the insect foregut, *J. Exp. Zool.* **211:**71–85.

Yamaoka, K., and Hirao, T., 1971, Role of nerves from the last abdominal ganglion in oviposition behaviour of *Bombyx mori, J. Insect Physiol.* **17:**2327–2336.

Yamashita, O., and Hasegawa, K., 1966, The response of the pupal ovaries of the silkworm to the diapause hormone with special reference to their physiological age, *J. Insect Physiol.* **12:**325–330.

Yamashita, O., and Hasegawa, K., 1967, The effect of the diapause hormone on the trehalase activity in pupal ovaries of the silkworm (Preliminary note), *Proc. Jpn. Acad.* **43:**547–551.

Yamazaki, M., and Kobayashi, M., 1969, Purification of the proteinic brain hormone of the silkworm, *Bombyx mori, J. Insect Physiol.* **15:**1981–1990.

Yin, C. M., and Chippendale, G. M., 1973, Juvenile hormone regulation of the larval diapause of the southwestern corn borer, *Diatraea grandiosella, J. Insect Physiol.* **19:**2403–2420.

Yin, C. M., and Chippendale, G. M., 1976, Hormonal control of larval diapause and metamorphosis of the southwestern corn borer *Diatraea grandiosella, J. Exp. Biol.* **64:**303–310.

Yin, C. M., and Chippendale, G. M., 1979, Diapause of the southwestern corn borer, *Diatraea grandiosella:* Further evidence showing juvenile hormone to be the regulator, *J. Insect Physiol.* **25:**513–523.

Yin, C. M., Wang, Z. S., and Chaw, W. D., 1985, Brain neurosecretory cell and ecdysiotropin activity of the non-diapausing, pre-diapausing and diapausing southwestern corn borer, *Diatraea grandiosella* Dyar, *J. Insect Physiol.* **31:**659–667.

Yokum, L. B., Denlinger, D. L., Katlic, N. E., Brueggemeier, R. W., and Mechoulam, R., 1987, A developmental profile of estrogen and androgen radioimmunoactive substances in the flesh fly, *Sarcophaga crassipalpis, Insect. Biochem.* **17:**1149–1153.

Yui, R., Fujita, T., and Ito, S., 1980, Insulin-, gastrin-, pancreatic polypeptide-like immunoreactive neurons in the brain of the silkworm, *Bombyx mori, Biomed. Res.* **1:**42–46.

Zachary, D., Goltzené, F., Holder, F. C., Berchtold, J. P., Nagasawa, H., Suzuki, C., Misoguchi, H., Ishizaki, H., and Hoffmann, J. A., 1988, Presence of bombyxin-(4K-PTTH)-like molecules in neurosecretory granules of brain-corpora cardiaca complexes of *Locusta migratoria* developmental aspects, *Int. J. Invert. Reprod. Dev.* **14:**1–10.

Zaretsky, M., and Loher, W., 1983, Anatomy and electrophysiology of individual neurosecretory cells of an insect brain, *J. Comp. Neurol.* **216:**253–263.

Ždárek, J., 1968, Le comportement d'accouplement à la fin de la diapause imaginale et son contrôle hormonal dans le cas de la punaise *Pyrrhocoris apterus* L., *Ann. Endocrinol.* **29:**703–707.

Ždárek, J., 1985, Regulation of pupariation in flies, in *Comprehensive Insect Physiology, Biochemistry and Pharmacology* (G. A. Kerkut and L. I. Gilbert, eds.), Vol. 8, Pergamon Press, New York, pp. 301–333.

Ždárek, J., and Denlinger, D. L., 1975, Action of ecdysteroids, juvenoids, and non-hormonal agents on termination of pupal diapause in the flesh fly, *J. Insect Physiol.* **21:**1193–1202.

Ždárek, J., and Denlinger, D., 1987, Pupal ecdysis in flies: The role of ecdysteroids in its regulation, *J. Insect Physiol.* **33:**123–128.

Ždárek, J., and Fraenkel, G., 1969, The effect of neurosecretion and ecdysone in the puparium formation of flies, *Am. Zool.* **9:**1080.

Ždárek, J., and Fraenkel, G., 1972, The mechanism of puparium formation in flies, *J. Exp. Zool.* **179:**315–324.

Ždárek, J., and Sláma, K., 1968, Mating activity in adultoids or supernumerary larvae induced by agents with high juvenile hormone activity, *J. Insect Physiol.* **14:**563–567.

Ždárek, J., Sláma, K., and Fraenkel, G., 1979, Changes in internal pressure during puparium formation in flies, *J. Exp. Zool.* **207:**187–196.

Ždárek, J., Rohlf, R., Blechl, J., and Fraenkel, G., 1981, A hormone effecting immobilization in pupariating fly larvae, *J. Exp. Biol.* **93:**51–63.

Zhu, X. X., Gfeller, H., and Lanzrein, B., 1983, Ecdysteroids during oögenesis in the ovoviviparous cockroach *Nauphoeta cinerea, J. Insect Physiol.* **29:**225–235.

Ziegler, R., 1979, Hyperglycaemic factor from the corpora cardiaca of *Manduca sexta* (L.) (Lepidoptera: Sphingidae), *Gen. Comp. Endocrinol.* **39:**350–357.

Ziegler, R., 1984, Developmental changes in the response of the fat body of *Manduca sexta* to injections of corpora cardiaca extracts, *Gen. Comp. Endocrinol.* **54:**51–58.

Ziegler, R., and Gäde, G., 1984, Preliminary characterization of glycogen phosphorylase activating hormone and adipokinetic hormone from *Manduca sexta* corpora cardiaca, *Physiol. Entomol.* **9:** 229–236.

Ziegler, R., Eckert, K., Schwarz, H., and Keller, R., 1985, Amino acid sequence of *Manduca sexta* adipokinetic hormone elucidated by combined fast atom bombardment (FAB)/tandem mass spectrometry, *Biochem. Biophys. Res. Commun.* **133:**337–342.

Zimowska, G., Handler, A. M., and Cymborowski, B., 1985, Cellular events in the prothoracic glands and ecdysteroid titres during the last-larval instar of *Spodoptera littoralis, J. Insect Physiol.* **31:**331–340.

Species Index

Subject Index